# Eco-Environmental Informatics

TianXiang Yue · Na Zhao · WenJiao Shi ·
ZeMeng Fan

# Eco-Environmental Informatics

Springer

TianXiang Yue
Institute of Geographic Sciences
and Natural Resources Research
Chinese Academy of Sciences (CAS)
University of Chinese Academy of Sciences
Beijing, China

Na Zhao
Institute of Geographic Sciences
and Natural Resources Research
Chinese Academy of Sciences (CAS)
University of Chinese Academy of Sciences
Beijing, China

WenJiao Shi
Institute of Geographic Sciences
and Natural Resources Research
Chinese Academy of Sciences (CAS)
University of Chinese Academy of Sciences
Beijing, China

ZeMeng Fan
Institute of Geographic Sciences
and Natural Resources Research
Chinese Academy of Sciences (CAS)
University of Chinese Academy of Sciences
Beijing, China

ISBN 978-981-96-1976-4 ISBN 978-981-96-1977-1 (eBook)
https://doi.org/10.1007/978-981-96-1977-1



This Springer imprint is published by the registered company Springer Nature Singapore Pte Ltd.
The registered company address is: 152 Beach Road, #21-01/04 Gateway East, Singapore 189721, Singapore

# Preface

*Eco-Environmental Informatics* (EEI) refers to the methodological and knowledge systems used to support information acquisition, data pre-processing, model construction, spatial and temporal interpolations, scale transformations, data fusion, model-data assimilation, model coupling and system simulations in studies connecting ecology, environmental science, geographical science and economics. The theoretical cores are the methods for high accuracy surface modelling (HASM) and the fundamental theorem for eco-environmental surface modelling (FTEEM). A surface is a raster expression of a region or one of its eco-environmental properties in the region. Surface modelling is a process of constructing a surface model for dynamically describing an Earth's surface system or a specific component of the Earth's surface environment. An eco-environmental surface is a unified term, used to represent a surface of nature, a surface of nature's contribution to people, or a surface of a driving force of natural changes.

In Chap. 1, we review the groundworks of knowledge, data and models for the EEI emergence since the early 1850s. In Chap. 2, the core principles of EEI are described, such as the methods for high accuracy surface modelling and the fundamental theorem for eco-environmental surface modelling. From FTEEM, seven corollaries have been deduced, corresponding to interpolation, upscaling, downscaling, data fusion and data assimilation, respectively. Spatial interpolation is used to predict the values of a primary variable at points within the same region of sampled locations in terms of spatial data in the form of discrete points or some form of a surface with data voids. Upscaling refers to the transfer of knowledge from a finer to a coarser resolution in order to reduce computational costs. Downscaling approaches are developed to obtain information at a finer resolution from coarser resolution models and data. Data fusion is the process of integrating multiple data and knowledge streams representing the same real-world object into a consistent, accurate, and useful representation. The aim of data fusion is to improve the quality of the information so that it is more accurate than one using the data sources individually. Model-data assimilation seeks to improve the performance of a model by either optimizing parameters or initial state variables according to a given data set.

Then, we describe in details HASM algorithms for spatial interpolation in Chap. 3, HASM algorithms for upscaling and HASM approaches to downscaling in Chap. 4, HASM methods for data fusion in Chap. 5, HASM algorithms for model-data assimilation in Chap. 6. We exhibit three typical case studies of HASM and FTEEM finally: surface modelling of climatic change in Chap. 7, surface modelling of ecosystems and ecological diversity in Chap. 8, and surface modelling of soil properties in Chaps. 9 and 10.

It can be used as a textbook for undergraduate and graduate students in the field of EEI. It can also be used as reference book for people involved in geographical information science, ecological informatics, environmental informatics, computer-aided design, and planetary surface modeling.

This work was supported by the Textbook Publishing Center of University of Chinese Academy of Science.

Beijing, China

TianXiang Yue
Na Zhao
WenJiao Shi
ZeMeng Fan

**Acknowledgements** This work was supported by the Textbook Publishing Center of University of Chinese Academy of Science, the National Natural Science Foundation of China (42330707 and 41930647), National Key Research & Development Program of China (2024YFD1700904), and the Science Fund for Creative Research Groups of the National Natural Science Foundation of China (72221002).

Textbook Series of University of Chinese Academy of Sciences (Graduate Level) (YJC0705011).

**Competing Interests** The authors have no competing interests to declare that are relevant to the content of this manuscript.

# Brief Introduction

This book describes the basic contents of eco-environmental informatics (EEI) systematically. It consists of 10 chapters: development history of the groundworks of knowledge, data and models for the EEI emergence since the early 1850s, the core principles of EEI including methods for high accuracy surface modelling, the fundamental theorem for eco-environmental surface modelling, and approaches to spatial interpolations, scale-transformations, data fusion and model-data assimilation, as well as their applications connecting ecology, environmental science, and geographical science.

It can be used as a textbook for undergraduate and graduate students in the field of EEI. It can also be used as a reference book for people involved in geographical information science, ecological informatics, environmental informatics, computer-aided design, and planetary surface modeling.

# Contents

# Chapter 1
# Conception of Eco-Environmental Informatics and Its Historical Roots

**Learning Objectives**

- To appreciate a basic concept of Eco-Environmental Informatics and its historical roots.
- To know the historical development of the Intergovernmental Science-Policy Platform on Biodiversity and Ecosystem Services (IPBES) and the Intergovernmental Panel on Climate Change (IPCC) as well as related international programs.
- To get knowledges about progresses in digital terrain models, the general circulation model, ecological modelling, integrated models and modelling platforms.
- To be aware of challenges faced by Eco-Environmental Informatics.

Eco-Environmental Informatics refers to the methodological and knowledge systems used to support information acquisition, data pre-processing, model construction, spatial and temporal interpolations, scale-transformations, data fusion, model-data assimilation, model coupling and system simulations in studies connecting ecology, environmental science, geographical science and economics. Since the early 1850s, launches and growth of a number of international organizations and international programs, especially the Intergovernmental Panel on Climate Change (IPCC) and Intergovernmental Science-Policy Platform on Biodiversity and Ecosystem Services (IPBES), have laid the groundworks of knowledges, data and models for the emergence of Eco-Environmental Informatics.

 
T. Yue et al., *Eco-Environmental Informatics*,
https://doi.org/10.1007/978-981-96-1977-1_1

## 1.1 International Organizations

### *1.1.1 IPCC*

In 1851, Matthew Fontaine Maury, an American naval officer, proposed the first plan for coordinating worldwide weather observations at sea and ashore (Sullivan 1961). This plan was discussed in an international conference held in Brussels in 1853. Representatives from Belgium, Britain, Denmark, France, the Netherlands, Norway, Portugal, Russia, Sweden, and the United States agreed that their warships would use standard procedures to record weather and oceanic phenomena wherever they went. A second conference was then proposed to map a program of international weather observations on land, which was not realized because of the Crimean War (1853–1856).

Different organizations were formed thereafter: (1) in 1864, the International Association of Geodesy was established to test the theorem that fixed the equatorial bulge of the earth in terms of competition between the inward pull of gravity and the outward tug of centrifugal force, (2) in 1875, the International Bureau of Weights and Measures was established to measure the earth's shape, and (3) in 1878, the International Meteorological Organization was established, a forerunner of today's World Meteorological Organization (WMO).

In the late 1880s, weather charts were drawn for each day covering North America and Europe and extending deep into the Arctic. By the late 1930s, weather maps of the entire Northern Hemisphere were compiled, covering every day from August 1932 to August 1933.

In the first half of the twentieth century, two key international organizations were founded: (1) in 1931, the International Council of Scientific Union (ICSU) was established, bringing together scientists from various international scientific unions with a common area of interest, and (2) in 1950, the WMO was established, which in 1951 became a specialized agency of meteorology, operational hydrology, and related geophysical sciences for the United Nations.

In 1952, the ICSU General Assembly approved the proposal for the International Geophysical Year (IGY). The IGY conference held in Uccle, Belgium, in April 1955 decided to establish three world data centers that housed a complete set of IGY data. World data centers A, B, and C was planned for the United States, the Soviet Union, Western Europe, Australia, and Japan, respectively. The IGY, jointly sponsored by WMO and ICSU, began on July 1, 1957 and ended on December 31, 1958 (Odishaw 1958a, b). This period coincided with an expected peak of sunspot activity as well as several eclipses. The IGY investigations, for which about 60,000 scientists and technicians in 67 countries made observations and measurements at more than 4000 stations scattered across the world, were classified into three categories (Sullivan 1961). The first category, the physics of the upper atmosphere, included solar activity, geomagnetism, aurora and airglow, ionosphere physics and cosmic rays, as well as rockets and satellites. The second category, the Earth's heat and water regime, included meteorology, glaciology, and oceanography. The third category, the Earth's structure and interior, included seismology, latitudes, and longitudes. IGY stimulated an interdisciplinary approach in which men trained in different fields worked together to resolve problems beyond the reach of any single specialist group.

In 1980 United Nations Environment Programme (UNEP), ICSU, and WMO jointly initiated the first international assessment of available knowledge on possible future human-induced climatic changes. A joint UNEP/WMO/ICSU international conference was organized in Villach of Austria in October 1985 by the Austrian government after the completion of the assessment report, "The Assessment of the Role of Carbon Dioxide and of Other Greenhouse Gases in Climate Variations and Associated Impacts". The UNEP, WMO, and ICSU agreed to form the advisory group on green-house gases (AGGG). The AGGG held its first meeting in July 1986 in Geneva.

The scientific assessments in the late 1970s and 1980s, especially the report, "Our Common Future" (World Commission on Environment and Development 1987), brought the climate change issue to the attention of the UN General Assembly in 1987. At this point, both the WMO congress and the UNEP governing council agreed that the executive heads of the two organizations should take steps to jointly organize an IPCC. Three working groups were formed at the first meeting of IPCC held in Geneva in November 1988, namely: (1) Assessment of Available Scientific Information on Climate Change, (2) Assessment of Environmental and Socio-Economic Impacts of Climate Change, and (3) Formulation of Response Strategies. The WMO took responsibility for the secretariat. The first assessment reports of the three working groups were titled: (1) Scientific Assessment (IPCC 1990a), (2) Impact Assessment (IPCC 1990b), and (3) Response Strategies (IPCC 1990c). These assessments were completed in June 1990, and played a decisive role in the establishment of the United Nations Framework Convention on Climate Change (UNFCCC). UNFCCC opened for signature at the Rio de Janeiro Summit of 1992 and came into force in 1994.

The second assessment reports (IPCC 1996a, b, c) were completed in 1995 after the reorganization of the IPCC, which provided key input for the negotiation of the Kyoto Protocol in 1997. The working groups I, II, and III, and the assessment reports, (a) Science of Climate Change, (b) Impacts, Adaptation and Mitigation, and (c) Economic and Social Dimensions of Climate Change, were approved in November, October, and July 1995, respectively. When the IPCC's second assessment reports were published, they faced severe criticism, but nonetheless gave impetus to subsequent work on the climate convention.

In early 2001, the third assessment reports were completed (IPCC 2001a, b). The contributions of working groups I, II, and III were adopted in January, February, and March 2001, respectively. In 2007, the fourth assessment was completed under the umbrella title "Climate Change 2007." The most comprehensive and up-to-date scientific assessment of past, present, and future climate change was obtained from working group I in February 2007. It was titled "The Physical Science Basis (IPCC 2007a)". In April 2007, working group II gave an assessment of the impacts of climate change, the vulnerability of natural and human environments, and the potential for response through its report, "Impacts, Adaptation and Vulnerability". The report of working group III, "Mitigation of Climate Change," which was launched in May

2007 in Bangkok, provided an up-to-date overview of the characteristics of various sectors, mitigation measures that could be employed, costs, and specific barriers, and policy implementation issues.

In 2015, IPCC published the Synthesis Report of the Fifth Assessment Report (AR5). It integrates the findings of the three Working Group contributions to the Fifth Assessment Report (AR5) (IPCC 2013, 2014a, b, c), and of two Special Reports (IPCC 2012a, b). The IPCC is currently in its Sixth Assessment cycle, during which the IPCC will produce the Assessment reports of its three Working Groups, three Special Reports, a refinement to the methodology report and the Synthesis Report. The Synthesis Report will be the last of the AR6 products, due for release in 2022 (https://www.ipcc.ch/report).

The IPCC continues to be a major source of information for negotiations under the United Nations Framework Convention on Climate Change (UNFCCC). It has 197 Parties and is the parent treaty of the 2015 Paris Agreement. The main aim of the Paris Agreement is to hold the increase in the global average temperature to below 2 °C above pre-industrial levels and to pursue efforts to limit the temperature increase to 1.5 °C above pre-industrial levels (https://unfccc.int/).

### *1.1.2 IPBES*

The Millennium Ecosystem Assessment (MA) was called for by the United Nations Secretary-General Kofi Annan in 2000. Initiated in 2001, the objective of the MA was to assess the consequences of ecosystem change for human well-being and the scientific basis for action needed to enhance the conservation and sustainable use of those systems and their contribution to human well-being.

As a result of the response to the Millennium Ecosystem Assessment consultation process, a resolution establishing "an intergovernmental science-policy platform for biodiversity and ecosystem services (IPBES)" was passed by the 65th United Nations General Assembly in 2010. The resolution was discussed at the 26th Council meeting of the United Nations Environment Programme in February 2011. Subsequently, IPBES was established in Panama City, on 21 April 2012 by 94 Governments, with the seat of the secretariat located in Bonn, Germany. IPBES is an independent intergovernmental body established by States to strengthen the science-policy interface for biodiversity and ecosystem services for the conservation and sustainable use of biodiversity, long-term human well-being and sustainable development (https://ipbes.net/).

The IPBES conceptual framework includes five interlinked elements constituting a social-ecological system that operates at various scales in time and space: nature, nature's contributions to people (NCP), anthropogenic assets, drivers of change, and good quality of life.

Nature refers to the natural world with an emphasis on its living components. It includes categories such as biodiversity, ecosystems (both structure and functioning), evolution, the biosphere, humankind's shared evolutionary heritage, and biocultural diversity. Within the context of other knowledge systems, it includes categories such as Mother Earth and systems of life, and it is often viewed as inextricably linked to humans, not as a separate entity.

Nature's contributions to people (NCP) are all the contributions, both positive and negative, of living nature (i.e. diversity of organisms, ecosystems, and their associated ecological and evolutionary processes) to the quality of life for people. Beneficial contributions from nature include such things as food provision, water purification, flood control, and artistic inspiration, whereas detrimental contributions include disease transmission and predation that damages people or their assets. Many NCP may be perceived as benefits or detriments depending on the cultural, temporal or spatial context.

Anthropogenic assets include built-up infrastructure, health facilities, or knowledge—including indigenous and local knowledge systems and technical or scientific knowledge—as well as formal and non-formal education, work, technology (both physical objects and procedures), and financial assets. Anthropogenic assets have been highlighted to emphasize that a good quality of life is achieved by a co-production of benefits between nature and people.

Drivers of change are all the factors that, directly or indirectly, cause changes in nature, anthropogenic assets, nature's contributions to people and a good quality of life. Direct drivers of change can be both natural and anthropogenic. Direct drivers have direct physical (mechanical, chemical, noise, light etc.) and behaviour-affecting impacts on nature. They include, inter alia, climate change, pollution, different types of land use change, invasive alien species and zoonoses, and exploitation. Indirect drivers are drivers that operate diffusely by altering and influencing direct drivers, as well as other indirect drivers. They do not impact nature directly. Rather, they do it by affecting the level, direction or rate of direct drivers. Interactions between indirect and direct drivers create different chains of relationship, attribution, and impacts, which may vary according to type, intensity, duration, and distance. These relationships can also lead to different types of spill-over effects. Global indirect drivers include economic, demographic, governance, technological and cultural ones. Special attention is given, among indirect drivers, to the role of institutions (both formal and informal) and impacts of the patterns of production, supply and consumption on nature, nature's contributions to people and good quality of life.

Good quality of life is the achievement of a fulfilled human life, a notion which may varies strongly across different societies and groups within societies. It is a context-dependent state of individuals and human groups, comprising aspects such as access to food, water, energy and livelihood security, and also health, good social relationships and equity, security, cultural identity, and freedom of choice and action. "Living in harmony with nature", "living-well in balance and harmony with Mother Earth" and "human well-being" are examples of different perspectives on a "Good quality of life".

IPBES performs regular and timely assessments of knowledge on biodiversity and ecosystem services and their interlinkages, which include comprehensive thematic, global and regional assessments. To date, eight IPBES assessments have been completed: the methodological assessment report on scenarios and models of biodiversity and ecosystem services (IPBES 2016a), the assessment report of the Intergovernmental Science-Policy Platform on Biodiversity and Ecosystem Services on pollinators, pollination and food production (IPBES 2016b), the IPBES regional assessment report on biodiversity and ecosystem services for Africa (IPBES 2018a), the IPBES regional assessment report on biodiversity and ecosystem services for the Americas (IPBES 2018b), the IPBES regional assessment report on biodiversity and ecosystem services for Asia and the Pacific (IPBES 2018c), the IPBES regional assessment report on biodiversity and ecosystem services for Europe and Central Asia (IPBES 2018d), the IPBES assessment report on land degradation and restoration (IPBES 2018e), and Global assessment report on biodiversity and ecosystem services of the Intergovernmental Science-Policy Platform on Biodiversity and Ecosystem Services (IPBES 2019).

The Kunming-Montreal Global Biodiversity Framework seeks to respond to the Global Assessment Report of Biodiversity and Ecosystem Services issued by the Intergovernmental Science-Policy Platform on Biodiversity and Ecosystem Services (IPBES), the fifth edition of the Global Biodiversity Outlook, and many other scientific documents that provide ample evidence that, despite ongoing efforts, biodiversity is deteriorating worldwide at rates unprecedented in human history. The Framework supports the achievement of the Sustainable Development Goals and has four long-term goals for 2050 and 23 targets to be achieved by 2030. In other words, the actions set out in each target need to be initiated immediately and completed by 2030. Together, the results will enable achievement towards the outcome-oriented goals for 2050. Actions to reach these targets would be implemented consistently and in harmony with the Convention on Biological Diversity and its Protocols and other relevant international obligations, taking into account national circumstances, priorities and socioeconomic condition (https://www.cbd.int/gbf).

In addition, the eighth session of the Intergovernmental Science-Policy Platform on Biodiversity and Ecosystem Services in June 2021 approved the undertaking of the nexus assessment as outlined in the "Scoping report for a thematic assessment of the interlinkages among biodiversity, water, food, and health", and of the transformative change assessment as outlined in the "Scoping report for a thematic assessment of the underlying causes of biodiversity loss, and the determinants of transformative change and options for achieving the 2050 Vision for Biodiversity".

## 1.2 International Programs

Since the early 1950s, several international programs have been promoted, which has greatly driven the worldwide collaboration. The international programs include International Geophysical Year (IGY), International Program on Biology (IBP), the Man and Biosphere Program (MAB), International Geosphere–Biosphere Program (IGBP), International Human Dimensions Program on Global Environmental Change

(IHDP), World Climate Research Program (WCRP), International Program of Biodiversity Science (DIVERSITAS), Millennium Ecosystem Assessment (MA) and Earth System Science Partnership (ESSP).

### 1.2.1 *International Geophysical Year*

In 1952, the ICSU General Assembly approved the proposal for the IGY. The IGY conference held in Uccle, Belgium, in April 1955 decided to establish three world data centers that housed a complete set of IGY data. World data centers A, B, and C was planned for the United States, the Soviet Union, Western Europe, Australia, and Japan, respectively. The IGY, jointly sponsored by WMO and ICSU, began on July 1, 1957 and ended on December 31, 1958 (Odishaw 1958a, b). This period coincided with an expected peak of sunspot activity as well as several eclipses. The IGY investigations, for which about 60,000 scientists and technicians in 67 countries made observations and measurements at more than 4000 stations scattered across the world, were classified into three categories (Sullivan 1961). The first category was the physics of the upper atmosphere, which included solar activity, geomagnetism, aurora and airglow, ionosphere physics and cosmic rays, as well as rockets and satellites. The second category was the Earth's heat and water regime, including meteorology, glaciology, and oceanography. The third category was the Earth's structure and interior, including seismology, latitudes, and longitudes. IGY stimulated an interdisciplinary approach in which men trained in different fields worked together to resolve problems beyond the reach of any single specialist group.

### 1.2.2 *International Program on Biology and the Man and Biosphere Program*

The IGY highlighted the advantages of international collaboration in geophysical science, which inspired biologists to think up similar international efforts (Worthington 1975). The possibility of an IBP to sponsor the International Biological Year (IBY) was first considered by the Executive Board of ICSU in October 1959. The International Union of Biological Sciences (IUBS) presented a formal scheme for an International Program on Biology (IPB) to the executive committee of ICSU at its meeting in Lisbon in 1960. After a series of meetings, ICSU launched IBP at the first General Assembly of IUBS in Paris in July 1964. This 10-year IBP was limited to basic studies related to biological productivity and human welfare.

In September 1968, a conference on the "scientific basis for rational use and conservation of resources of the biosphere," held as a part of the intergovernmental conference convened by the United Nations Educational, Scientific and Cultural Organization (UNESCO) in Paris, led to the IBP process being transferred to the

Man and Biosphere Program (MAB). In contrast to the nongovernmental IBP, MAB was officially launched in 1970 as an intergovernmental program, which underwent an intensive phase of planning before it began after the first session held by its international coordinating council in 1971. It was likely to involve many scientists who had not been a part of IBP, although many of MAB's projects had their origins in IBP.

The general objective of MAB was to develop a basis for the rational use of the natural resources of the biosphere, and to improve relationships between man and his environment on a global scale. It focused on the general study of the structure and functioning of the biosphere and its ecological divisions, systematic observation, changes brought about by man, the overall effects of these changes upon the human species, and providing education and information on the above issues (UNESCO 1972).

MAB was intended to develop the following specific projects: (1) identifying and assessing the changes in biosphere resulting from human activities, and assessing the effect of these changes on man, (2) studying and comparing the structure, functioning, and dynamics of natural, modified, and managed ecosystems, (3) studying and comparing the dynamic interrelationships between natural ecosystems and socio-economic processes, especially the impact of changes in human population, settlement patterns, and technology on the future viability of these systems, (4) developing ways and means to measure quantitative and qualitative changes in the environment in order to establish scientific criteria to serve as a basis for the rational management of natural resources, including the protection of nature, and the establishment of standards of environmental quality, (5) helping to bring about greater global coherence in environmental research, (6) promoting the development and application of simulation and other techniques for prediction as tools for environmental management, and (7) promoting environmental education in its broadest sense.

### *1.2.3 World Climate Research Program*

On April 1, 1960, the United States launched its first meteorological satellite, TIROS 1, which allowed scientists to view the earth and its atmosphere from space. A very effective new tool for observing the weather had become available (Bolin 2007). A WMO advisory committee on research was constituted to use this new tool operationally. In 1961, the United Nations agreed on a resolution to use satellites for observing the weather from space. An additional resolution of the United Nations in 1962 asked the WMO to develop its plan for an expanded program to strengthen methodological services and research. Responsibility for developing the expanded program of research was given to ICSU, specifically its International Union of Geodesy and Geophysics (IUGG). At the general assembly of IUGG held in the summer of 1963 in San Francisco, it was decided that IUGG should launch a truly international effort to prepare for the use of satellite technology in studies of the general circulation of the atmosphere to develop new methods for weather forecasting. In November 1963, an interunion committee on atmospheric science (CAS) was formed by IUGG that worked as the parent organization of the CAS. The creation of CAS could be seen as the beginning of a series of global research programs in the field of environmental sciences. Its first meeting took place in 1965 at the WMO Headquarters in Geneva. Defined as an entirely research-oriented cooperative international meteorological and analytical program, its goal was to vastly improve our understanding of the general circulation of the global atmosphere.

In 1967, an agreement between ICSU and WMO on the formation of Global Atmospheric Research Program (GARP) was reached in Rome and a Joint Organizing Committee (JOC) of GARP was established. At the eighth session of JOC in 1973, a set of guiding principles was formulated, which concentrated on mathematical models, the mechanisms of climate fluctuations and the nature of climate change.

In 1980, ICSU and WMO agreed that GARP would be transformed into a committee for international cooperation in climate research, and the WCRP was born. Since 1993, WCRP has also been sponsored by the Intergovernmental Oceanographic Commission (IOC) of UNESCO. WCRP, sponsored by the ICSU, WMO, and IOC of UNESCO, is uniquely positioned to draw on the totality of climate-related systems, facilities and intellectual capabilities of more than 185 countries.

The two overarching objectives of WCRP are, (1) to determine the predictability of climate, and (2) to determine the effect of human activities on climate. These two objectives underpin and directly address the requirements of the United Nations Framework Convention on Climate Change (UNFCCC), as well as contributing to many other international policy instruments. The WCRP is currently implementing a strategic framework titled Coordinated Observation and Prediction of the Earth System. Launched in 2005, this initiative aims to facilitate the analysis and prediction of earth system variability for use in an increasing range of practical applications of direct relevance, benefit, and value to society. In moving to provide a broader suite of products and services to a larger group of users, the WCRP is re-prioritizing its activities to optimize societal benefits. One of the primary WCRP pathways to application and end-user benefits will continue to be the integration of observations and models to generate new understanding, leading to enhanced benefits from climate predictions (http://wcrp.wmo.int).

### *1.2.4 International Geosphere–Biosphere Program*

In September 1986, the ICSU General Assembly began planning a new major transdisciplinary research programme, entitled the IGBP. The overall objectives of the IGBP were to describe and understand the interactive physical, chemical, and biological processes that regulate the total earth system, to understand the unique environment that it provides for life, to comprehend the changes that are currently occurring in the earth system, and to assess the manner in which these changes are influenced by human actions.

A primary goal of the IGBP was to advance our capacity to predict change in the global environment. The development of this capacity would require the cooperation and complementary efforts of climate modelers, incorporating an appropriate level of understanding of relevant global biological, geological, and chemical processes into physical models of the earth system.

The major scientific IGBP components and their interdependencies are terrestrial biosphere–atmospheric chemistry interactions, marine biosphere–atmosphere interactions, biospheric aspects of the hydrological cycle, and the effects of climate change on terrestrial ecosystems (Special Committee for the IGBP 1988).

### 1.2.5 International Human Dimensions Program on Global Environmental Change

In 1990, the IHDP was initially launched by the International Social Science Council (ISSC) as the human dimensions program. In 1996, ICSU joined ISSC as cosponsors of the IHDP, and the secretariat was moved to Bonn, Germany. The IHDP is an international, interdisciplinary, nongovernmental science program dedicated to promoting and coordinating research. Its aims are to describe, analyze, and understand the human dimensions of global environmental change (Jaeger 2003).

The IHDP is designed around three main objectives of research, capacity building, and networking. IHDP has four core research projects: (1) Land-Use and Land-Cover Change (LUCC), (2) global environmental change and human security (GECHS), (3) institutional dimensions of global environmental change (IDGEC), and (4) industrial transformation (IT). The LUCC's objectives are to obtain a better understanding of land-use and land-cover changes, and the physical and human driving forces behind them.

Regarding the GECHS, its main goal is to advance interdisciplinary, international research and policy efforts in the area of human security and environmental change. The IDGEC analyzes the roles that social institutions play as determinants of the course of human/environment interactions. The IT research project focuses on the relationship between changes in industrial systems and changes in the environment. It also aims to discover ways of decreasing the environmental impact of industrial activities. In the meantime, three themes, carbon, water, and food systems have been selected for collaborative projects among the Global Environmental Change Research Programs of IGBP, WCRP, and the DIVERSITAS.

### 1.2.6 International Program of Biodiversity Science

The International Program of Biodiversity Science, DIVERSITAS, was established in 1991, aimed at developing an international, nongovernmental umbrella program that would address the complex scientific questions posed by changes in global biodiversity (http://www.diversitas-international.org). Its founding sponsors, the UNESCO, the Scientific Committee on Problems of the Environment (SCOPE) and the IUBS, identified three projects: (1) the effects of biodiversity on ecosystem functioning (bioDISCOVERY), (2) origins, maintenance, and loss of biodiversity (ecoSERVICES), and (3) inventory and classification of biodiversity (bioSUSTAINABILITY).

In 1996, ICSU and the International Union of Microbiological Societies (IUMS) joined DIVERSITAS as new sponsors, and seven core projects were added to this initial list. The seven core projects included, (1) the monitoring of biodiversity, (2) the conservation, restoration, and sustainable use of biodiversity, (3) soil and sediment biodiversity, (4) marine biodiversity, (5) microbial biodiversity, (6) inland water biodiversity, and (7) human dimensions of biodiversity.

A partnership with the Convention on Biological Diversity was initiated in 1996. During the period 1991 to 1998, the DIVERSITAS Secretariat was hosted by MAB. In 2001, the five sponsors of DIVERSITAS, ICSU, IUBS, IUMS, SCOPE, and UNESCO, decided to launch a second phase of the program. They opened a new Secretariat, hosted by ICSU in Paris, and called upon a task force of scientists to develop an international framework for biodiversity research.

The resulting Science Plan, published at the end of 2002, clearly identified three interrelated areas for further development: (1) discovering biodiversity and predicting its changes, (2) assessing the impacts of biodiversity change on ecosystem functioning and services, and (3) developing the science of the conservation and sustainable use of biodiversity. In addition to the three thematic core projects, two integrated cross-cutting networks, the global invasive species program (GISP) and the global mountain biodiversity assessment (GMBA), were developed, which embraced issues addressed in all three core projects. In 2004, in addition to its three core projects and its two cross-cutting networks, DIVERSITAS established two new cross-cutting networks on agro-biodiversity and freshwater biodiversity. Since 2005, the convention for biological diversity (CBD) has been an ex-officio member of the DIVERSITAS scientific committee.

In 2006, the DIVERSITAS Scientific Committee recognized the need to create a new core project aiming at providing an evolutionary framework for biodiversity science, as the scope of bioDISCOVERY was too large to efficiently manage the major issues related to evolution and systematics. This new core project, titled bioGENESIS, and bioDISCOVERY will focus on issues related to the monitoring of biodiversity, understanding the drivers of biodiversity changes and predicting those changes.

### 1.2.7 Millennium Ecosystem Assessment

The history of the MA can be traced to demands from both scientists and policy-makers. By the mid-1990s, many individuals, who involved in the work of international conventions such as the Convention on Biological Diversity (CBD) and the Convention to Combat Desertification (CCD), realized that the extensive needs for scientific assessments within the conventions were not being met through the mechanisms then in place. In contrast, effective assessment process like the Intergovernmental Panel on Climate Change (IPCC) did exist for such treaties as the Framework Convention on Climate Change and the Montreal Protocol on Substances that Deplete the Ozone Layer.

Scientists had also identified a need for an international ecosystem assessment. Although major advances had been made in ecological sciences, resource economics and other fields during the 1980s and 1990s, these new findings appeared to be poorly reflected in policy discussions concerning ecosystems. Recognizing these shortcomings, a panel of 40 leading scientists prepared a draft international assessment—"Protecting our Planet, Securing our Future: Linkages Among Global Environmental Issues and Human Needs". The study, published in November 1998 by UNEP, NASA, and the World Bank, called for "a more integrative assessment process for selected scientific issues, a process that can highlight the linkages between questions relevant to climate, biodiversity, desertification, and forest issues".

The specific proposal for the Millennium Ecosystem Assessment (MA) arose during a brainstorming meeting held at World Resources Institute (WRI) on May 17, 1998 to discuss plans for the biennial World Resources Report, published by WRI, UNEP, World Bank, and the United Nations Development Programme (UNDP). Between October 1999 and July 2000, the Conference of Parties of the CBD and CCD formally endorsed the MA as a mechanism to meet some of their assessment needs and a similar endorsement was provided by the Standing Committee of the Convention on Wetlands (Ramsar). United Nations Secretary General Kofi Annan features the MA as one of five major initiatives for "Sustaining our Future" in his Millennium Report to the United Nations General Assembly in April 2000.

In 2001, the millennium ecosystem assessment (MA) was initiated to respond to the needs described in the Millennium Report to the UN General Assembly in April 2000. In this Millennium Report, Kofi Annan stated that effective environmental policy must be based on sound scientific information, but there had never been a comprehensive global assessment of the world's major ecosystems (MA 2003). The MA was designed to meet some of the assessment needs of the convention on biological diversity, the convention to combat desertification, and the wetlands convention, as well as the needs of other users in the private sector and civil society.

The MA was carried out through four working groups on conditions and trends, scenarios, responses, and subglobal assessments. The millennium assessment focused on how humans had altered ecosystems, how changes in ecosystem services had affected human well-being, how ecosystem changes might affect people in future decades, and what types of responses could be adopted at local, national, or global levels to improve ecosystem management and, thereby, contribute to human well-being and poverty alleviation. Ecosystem services were defined as the benefits people obtain from ecosystems. These include the provisioning of services such as food and water; the regulation of services such as flood prevention, drought prevention, the prevention of land degradation, and the prevention of disease; the provision of supporting services such as soil formation and nutrient cycling; and the provision of cultural services such as recreational, spiritual, religious, and other nonmaterial benefits.

The MA has involved the work of more than 1360 experts worldwide. Their findings, contained in five technical volumes and six synthesis reports, provide a state-of-the-art scientific appraisal of the condition and trends in the world's ecosystems and the services they provide (such as clean water, food, forest products, flood control, and natural resources) and the options to restore, conserve or enhance the sustainable use of ecosystems (Millennium Ecosystem Assessment 2005a, b, c, d, e, f). On March 30, 2005, MA findings publicly launched at press conferences and seminars in Beijing, Brasilia, Cairo, Delhi, Lisbon, London, Nairobi, Paris, Rome, Sao Paulo, Stockholm, Tokyo, and Washington DC (http://www.millenniumassessment.org/).

### *1.2.8 Earth System Science Partnership*

At the first Global Environmental Change Open Science Conference in Amsterdam in 2001, the 1400 participants signed the Amsterdam Declaration on Global Environmental Change. The declaration called for the strengthening of cooperation amongst the global environmental change research programs. In response to the declaration, the four international global environmental change research programs, DIVERSITAS, IGBP, IHDP, and WCRP joined together to form the ESSP. The ESSP brings together researchers from diverse fields from across the globe, to undertake an integrated study of the earth system (www.essp.org). The declaration defines the earth system as a single, self-regulating system comprised of physical, chemical, biological, and human components (Loevbrand et al. 2009).

Earth system science views the earth as a synergistic physical system of interrelated phenomena, processes, and cycles (Johnson et al. 2000). The goal of earth system science is to obtain a scientific understanding of the entire earth system on a global scale by describing how its component parts and their interactions have evolved, how they function and how they may be expected to continue to evolve on all timescales. Its challenge is to develop our capability to predict changes that will occur anytime between the next decade and the next century, both naturally and in response to human activities. Its fundamental purpose, which builds upon the traditional disciplines, is to emphasize relevant interactions of physical and dynamical properties that could extend over a spatial scale of anything from millimeters to the circumference of the earth, and over a timescale ranging from seconds to billions of years.

## 1.3 Eco-Environmental Modelling

Models are qualitative or quantitative representations of key components of a system and of relationships between these components (IPBES 2016). Benchmarking of models is the process of systematically comparing sets of model predictions against measured data in order to evaluate model performance. Validation of models typically refers to checking model outputs for consistency with observations. However, since models cannot be validated in the formal sense of the term (i.e., proven to be true), some scientists prefer to use the words "benchmarking" or "evaluation".

Eco-environmental models can be classified into statistical models, process-based models, and hybrid models. A statistical model is a model, in which available data and statistical approaches are used to estimate values for parameters linking eco-environmental variables; the parameters do not have predefined ecological meaning and the modelling processes are implicit rather than explicit. A process-based model (also known as "mechanistic model") is a model in which relationships among eco-environmental variables are described in terms of explicitly stated processes or mechanisms based on established scientific understanding, and model parameters therefore have clear ecological interpretation, defined beforehand. Hybrid models are models that combine correlative and process-based modelling approaches. Eco-environmental modelling, which usually have specific application purposes, includes ecological modelling, surface modelling, and model coupling.

### *1.3.1 Ecological Modelling*

Ecological modeling has been through various phases of development, beginning in the middle of the nineteenth century. Little progress in applying mathematics to ecological issues was made until the latter half of the nineteenth century (Israel and Gasca 2002). Beginning in the 1880s, several key achievements were made thereafter. In 1884, statistical analysis and elementary quantitative techniques started to be applied to the handling of biological information (Galton 1884). In 1916, the probability theory was applied to study a priori pathometry (Ross 1916). In 1920, the logistic curve was introduced to theoretical biology (Pearl and Reed 1920). In 1925, statistical methods were developed for dealing with problems in population genetics (Fisher 1925). In 1926, ordinary differential equations and integro-differential equations were applied to build the rational mechanics of biological association (Volterra 1926). Ecological modeling was schematized as follows: (1) characteristic properties of the natural phenomenon under study were identified; (2) a general description was set up in mathematical terms to obtain a certain number of differential equations, (3) qualitative methods were used to determine parameters, and (4) results were compared with empirical facts. During the late 1920s and the early 1930s, mathematical models were proposed for natural selection and evolution (Haldane 1927;

Wright 1936). However, these specific studies focused on small-scaled systems and relatively few processes were modeled.

After a brief hiatus during World War II, studies on ecological models were eventually resumed in the late 1940s. However, it was not only the classical biometrical methods and differential equations that were in use; new mathematical tools such as game theory, system theory, and information theory were also employed. For instance, Leslie (1945, 1948) developed a population growth model, while Chapman (1950) proposed the idea of constructing a model of the earth's atmosphere and the space surrounding it on the basis of detecting the pull of the moon on the atmosphere by studying 6457 hourly records of air pressure made at Greenwich, England (Chapman 1957). Sullivan (1961) and Skellam (1951) developed a model for spatial population distribution, while Beverton and Holt (1957) developed a dynamic relation model between fish population, which accounted not only for the reactions of particular fish population but also for the interactions between them. The success of these early studies encouraged the development of a series of quantitative models. For instance, a model using the electrical analogue circuit was developed to analyze photosynthetic production, community metabolism, biomass, and species variety (Odum 1960). Methods to increase the realism of Lotka–Volterra equations were also explored (Garfinkel 1962, 1967a, b; Garfinkel and Sack 1964), although Lotka–Volterra equations had previously been criticized (Smith 1952). An animal population process was analyzed by developing a simple mathematical model (Holling 1964). A one-dimensional model of radiative–convective equilibrium was developed (Manabe and Strickler 1964). Plankton population dynamics were simulated by a linear differential equation set (Davidson and Clymer 1966) and digital simulation models were developed for a salmon resource system, which were coded in DYNAMO (Paulik and Greenough 1966). In the 1970s, numerous computer-oriented mathematical models were developed for the analysis of entire ecosystems. For instance, grassland ecosystem models were developed, which involved about 40 state variables and several hundred parameters (Bledsoe et al. 1971; Anway et al. 1972; Patten 1972).

The first semi-mechanistic computer model of forest growth was developed to reproduce the population dynamics of trees in mixed-species forest type (Botkin et al. 1972). A generalized model for simulating lake ecosystems was developed in response to the growing need for models to enable humans to manage their environment (Park et al. 1974). A world model was developed for simulating the relationship between gross national product and human population growth (Jørgensen 1975a). This period also saw the development of various model types to describe specific ecosystem properties; for instance, a linear differential model was developed to analyze productivity–stability relationships (Rosenzweig 1971). Models which analyzed the diversity–stability relationship also became popular (Gardner and Ashby 1970; May 1972). In 1975, the first peer-reviewed journal on ecosystem modeling, Ecological Modelling, was published. The journal attempted to combine three fields of science, namely mathematical modeling, system analysis, and computer techniques, along with ecology and environmental management (Jørgensen 1975b). However, this generation of models suffered various problems, such as a lack of data and methods needed for obtaining measurements, inadequate modeling theory for macrosystem applications, and an inability to halt the propagation of errors, as well as uncertainties in model predictions (Patten 1972; Shugart and O'Neill 1979). In addition, these models assumed that processes occurring within the borders of a system were sufficiently uniform that one could adequately describe their dynamics by considering the total or mean properties of the system alone.

One of the most common procedures of ecological modeling includes some or all of the following nine steps: (1) problem definition, (2) system identification, (3) decision on model type, (4) mathematical formulation, (5) decision on computing methods, (6) programming, (7) parameter estimation, (8) parameter validation, and (9) experimentation (Kitching 1983). Four fundamental phases, that is, conceptual-model formulation, quantitative-model specification, model evaluation, and model use, were identified in the process of developing usable system models (Grant et al. 1997). Phase 1 steps include stating the model objectives, bounding the system-of-interest, categorizing the components within the system-of-interest, identifying the relationships among the components of interest, representing the conceptual model, and describing the expected patterns of model behavior. Phase 2 steps include selecting the general quantitative structure for the model, choosing the basic time unit for the simulations, identifying the functional forms of the model equations, estimating the parameters of the model equations, coding the model equations for computer, executing the baseline simulation, and presenting the model equations. Phase 3 steps include assessing the reasonableness of the model structure and the interpretability of functional relationships within the model, evaluating the correspondence between model predictions and data from the real system and determining the sensitivity of model predictions to changes in the value of important parameters. Phase 4 steps include developing and executing the experimental design for the simulations, analyzing and interpreting the simulation results, examining additional types of management policy or environmental situations, and communicating the simulation results.

The models emphasized structure and function in point-type systems that incorporated time but not space. Ecosystem models devoid of spatial consideration were ultimately found to be too abstract for use (Neuhold 1975). Since the middle of the 1980s, the translation of model results into geographical patterns has been rapidly developing (Jørgensen 2002). For instance, a dynamic spatial simulation model composed of interacting cells was designed to simulate habitat changes as a function of marsh type, hydrology, subsidence, and sediment transport for a generalized coastal wetland area (Sklar et al. 1985; Costanza et al. 1990). Geographical information system (GIS) was used to spatially verify the close statistical relationship between the dependent variable and each of the independent variables selected by logistic regression modeling to better understand the dimensions of anthropogenic deforestation occurring throughout the tropical world (Ludeke et al. 1990). It was found that the storage and manipulation of large sets of data in GIS, which was facilitated by rapid advances in computing technology, presented opportunities for modeling spatial distribution and changing distribution on a regional scale (Elston and Buckland 1993). The likely impact of climate change on the distribution and abundance of wildlife species was analyzed by means of spatial interpolation techniques (Aspinall and Matthews 1994). Four predictive mathematical models were combined with GIS for population estimation in urban areas (Al-Garni 1995). A spatially explicit forest community model was developed to generate estimates of potential natural vegetation for the entire potential forest area of Switzerland under today's climate as well as under altered climate regimes (Kienast et al. 1996). GIS and remote sensing were applied to aquatic botany in conjunction with simulation models (Caloz and Collet 1997). The scale, pattern, and process relationships in fluvial and alpine environments were studied by means of optical and microwave remote sensing systems and GIS (Walsh et al. 1998). GIS was employed to improve landscape classifications which could be useful for climate, socio-economic, or biodiversity modelers (Lioubimtseva and Defourny 1999). A linkage between the GIS and water resources management model was developed to simulate river basin water allocation problem as a collection of spatial and thematic objects (McKinney and Cai 2002). Forest ecosystems were spatially simulated and simulation uncertainty was analyzed (Larocque et al. 2006a, 2008).

The revival of interest in differential equations in the early 2000s had a great impact in the field of ecological modeling (Israel and Gasca 2002), such as the method for high accuracy surface modelling (HASM) (Yue et al. 2007, 2010; Yue and Song 2008; Yue and Wang 2010). A model, developed in combination with GIS, dealing with soil, water quality, superficial material, and aquifer properties was constructed to identify all areas of groundwater, vulnerable to nitrate pollution (Lake et al. 2003). Methods for surface modeling were used to simulate the change trends and scenarios of major terrestrial ecosystems (Yue et al. 2005, 2006). It was suggested that the development of spatially explicit disturbance historic information from historic accounts might enable future research to develop more precise quantitative disturbance history inputs at appropriate resolutions for ecosystem process models (McNeil et al. 2006). A data model, a statistical model and an ecological model were integrated into a GIS environment to support fine-scale vegetation community modeling (Accad and Neil 2006). Spatially explicit models were used to estimate ecosystem services (Troy and Wilson 2006; Yue et al. 2008). A simulation program that runs on GIS was developed to predict the multispecies size-structure dynamics of forest stands (Umeki et al. 2008). Several existing process-based models linked to economic valuation methods were integrated into a GIS platform to value ecosystem services (Gret-Regamey et al. 2008). Four different GIS-based habitat models were comparatively applied to analyze impacts of land-use changes, urbanization, and infrastructure developments on the fragmentation of natural habitats and biodiversity loss (Gontier et al. 2010). A spatial decision support system was developed to support land-use planning and local forestry by providing spatially explicit information on carbon sequestration and economic benefits under various scenarios of the carbon credit market (Wang et al. 2010). GIS has now become a popular tool for spatially ecological modelling.

### 1.3.2 Surface Modeling

A raster expression of a region or one of its eco-environ mental properties can be abstracted to a mathematical surface (Yue et al. 2007). Surface modeling refers to the process of simulating a surface through a scattered point-form data set, a grid-form data set and/or an area-form data set. Surface modeling is aimed at formulating an object in a grid system, in which each grid cell contains an estimate of the object that is representative for that particular location. Representing data in grid form has at least four advantages (Martin and Bracken 1991; Deichmann 1996; Yue et al. 2008, 2009, 2010): (1) a regular grid can be easily reaggregated to any areal arrangement required; (2) producing ecological data in grid form is one way of ensuring compatibility between heterogeneous data sets; (3) data in grid form makes the fusion of multiresolution and multisource information easier, and (4) converting data into grid form can provide a way to avoid some of the problems imposed by artificial boundaries.

Krige (1951), for example, proposed the basic idea of what today is the Kriging suite of geostatistical methods, which considers position relations between given rasters and ones to be estimated as well as spatial relations among variables. The Kriging method is a linear least squares regression. Birkhoff and Garabedian (1960) expanded the spline approximation, which had been developed by Schoenberg in 1946, to a spline function in two dimensions. DeBoor (1962) successfully constructed a bicubic spline function using a two-dimension mesh. Bengtsson and Nordbeck (1964) established an irregular triangular network (TIN) for creating a surface by using irregularly distributed point data. Zienkiewlcz (1967) proposed a finite element method. Akima (1978a, b) completed the computer programming of the finite element method (FEM) for fitting a surface. Shepard (1968) published the method for inverse distance weighting (IDW), for which the results are closely related to the selected weight (Franke 1982). Hardy (1971) formulated a multiquadric method (MQM) under the assumption that any smooth mathematical surface and smooth irregular surface might be approximated to any desired degree of exactness by the summation of mathematically defined surfaces. Harder and Desmarais (1972) proposed an interpolation method using splines, which is based on the infinite plate deflection equation. Maude (1973) proposed a Rectangle Based Blending method that generalized the idea of quintic splines to several variables. Triangle-based blending methods were developed for surface estimation by improving IDW (Gold et al. 1977; McLain 1976). Talmi and Gilat (1977) constructed a method for smooth approximation of data based on the Spline idea, which had many advantages compared with polynomial interpolation. Foley and Nielson (1980) proposed an iteration method for producing a surface approximation, which is based on Bernstein's polynomial and bicubic spline.

Surface modeling began to be used in the 1960s, with the general availability of computers (Lo and Yeung 2002), but because it requires powerful software and a large amount of spatially explicit data, its development was limited before the 1990s. The major advances that enabled the use of surface modeling included trend surface analysis (TSA) (Ahlberg et al. 1967; Schroeder and Sjoquist 1976; Legendre and Legendre 1983), the digital terrain model (Stott 1977), surface approximation (Long 1980), spatial simulation of wetland habitats (Sklar et al. 1985), spatial pattern matching (Costanza 1989), spatial prediction (Turner et al. 1989), and modeling coastal landscape dynamics (Costanza et al. 1990). Surface modeling has greatly progressed since the early 1990s, with the rapid development of remote sensing (RS) and geographical information systems, as well as the accumulation of spatially explicit data.

The accuracy of the classic methods for surface modeling has been comparatively analyzed by many scholars since 1980s. For instance, Yakowitz and Szidarovsky (1985) claimed that nonparametric regression was more robust and Kriging was better when the data contained various types of trends. Having compared IDW, Kriging, and spline by using a carefully and specially designed survey of soil pH, Laslett and McBratney (1990) concluded that Kriging was in fact the most effective method. However, Weber and Englund (1992) contested this, believing IDW to be far superior. Wood and Fisher (1993) assessed the simulation accuracy of IDW, contour flood filling, simultaneous over-relaxation and spline, and concluded that each of the techniques emphasized a particular aspect of the elevation model and no single technique was sufficient. Hutchinson and Gessler (1994) showed that spline was more accurate than kriging, while Myers (1994) concluded that interpolation from spatial data could be performed in various ways by invoking different assumptions and models. Mitas and Mitasova (1999) found that different methods could produce quite different spatial representations and in-depth knowledge of the phenomenon was needed to evaluate which one was the closest to reality. Laslett and McBratney (1990) calculated accuracies of Kriging and spline in terms of several real data sets, and reported that Kriging never performed worse than spline. Brus et al. (1996) used Kriging, IDW, and spline to estimate soil properties and found that Kriging was more reliable in the sense that it estimated all soil properties well. Desmet (1997) applied IDW, Kriging, and spline and constructed a digital elevation model from irregularly spaced sample points and found that spline yielded the best results with regard to accuracy. Borga and Vizzaccaro (1997) compared Kriging with MQM and concluded that Kriging performed better at lower gauge density, while the accuracy of both estimators is similar at higher gauge density. Carrara et al. (1997) found that each method exhibited advantages and pitfalls; their respective applicability is largely dependent on the way in which source data are collected and stored. Caruso and Uarta (1998) compared IDW, Kriging, MQM, and the tension finite difference method and found that each method had its own advantages and drawbacks, which strongly depended on the characteristics of the set of point data. Zimmerman et al. (1999) demonstrated that Kriging was substantially superior to IDW over all levels of surface type, sampling pattern, noise, and correlation. Kravchenko (2003) found that Kriging with known variogram parameters, performed significantly better than IDW for most of the case studies; however, it was very less accurate than IDW when a reliable sample variogram could not be obtained because of either an insufficient number of data points or too large a distance between the data points. Aguilar et al. (2005) comparatively analyzed IDW and radial basis functions (RBF) groups, including MQM, the inverse multiquadric method (IMQM), the multilog method (MLM), natural cubic splines (NCS), and thin plate splines (TPS); the results showed that MQM in the RBF group was the best method in terms of accuracy and numerical stability. Chaplot et al. (2006) evaluated the performance of IDW, ordinary Kriging (OK), universal Kriging (UK), MQM, and regularized spline with tension (RST). Their results indicated that a few differences existed between the methods when the sampling density was high; Kriging had a higher accuracy if the spatial structure of altitude was strong, IDW and RST performed better when the spatial structure of height was weak, and MQM performed well in the mountainous areas.

Several scholars have also examined the role of computational speed in their comparative studies. Franke (1982) evaluated IDW, spline, TIN, FEM, and MQM and concluded that IDW performed well because it required moderate storage and computation time, while Pringle et al. (2009) argued that the primary disadvantage of Kriging was its relatively slow computing speed. Katzil and Doytsher (2000) compared the computational speed of different methods, and concluded that improved cubic spline was 85 times faster than Kriging when applied to a large area grid. Mitas and Mitasova (1999) pointed out that accuracy and high computational speed are seen as high priorities, and related comparative studies demonstrate that the high computational cost remains a significant problem of surface modeling (Lee et al. 1997; Aguilar et al. 2005). Cornford (2005) reviewed the comparison studies on different methods for surface modeling, and concluded that a good prior knowledge of the desired effect is crucial, and that the method of surface modeling chosen should be based on computational speed, cost of implementation, scaling with data size, and the ability to make probabilistic predictions.

The dominating classical approaches include TSA, IDW, TIN, Kriging, and spline (Table 1). TSA is the simplest approach to generating a surface, which uses sparse data to fit a trend surface by least-squares regression on the spatial coordinates. The trend surface could be approximately formulated as a polynomial surface. The shortcoming of this simple method is that it loses detail because powerful smoothing and variation in one part of the region affects the fit of the surface everywhere. TSA assumes that the residuals from the fitted surface are independent of one another, which is almost always violated and as a result no true estimation variance can be calculated (Oliver and Webster 1990).

IDW uses an inverse distance weighting function to determine the interpolation value for any given point within the calculated area (Lee and Angelier 1994). IDW allows an individual spatial analysis of each independent variable (Julià et al. 2004) and is a good method when all the variables have similar weight (Sinowski et al. 1997). However, IDW fails to incorporate the spatial structure and ignore information beyond the neighborhood (Zhao et al. 2005; Magnussen et al. 2007).

TIN is a popular model for representing surface models (Bengtsson and Nordbeck 1964), because it has a simple data structure and can easily be rendered using common graphics hardware (Yang et al. 2000). TIN is composed of nodes, edges, triangles, topology, and a hull. It is one of the basic models for representing digital terrain, a piece of land on the Earth's surface. TIN models of the terrain surface are well known in GIS, but they ignore nonlinear information and are unable to represent cliffs, caves,

**Table 1** Basic ideas and shortcomings of the classical methods

| Method | Basic ideas | Shortcomings |
|---|---|---|
| TSA | Sparse data are used to fit a trend surface by least-squares regression on the spatial coordinates | The assumption of the residuals from the fitted surface being independent of one another is almost always violated |
| IDW | An inverse distance weighting function is used to determine the interpolation value for any given point within the calculated area | It fails to incorporate the spatial structure and ignores information beyond the neighborhood |
| TIN | The value of each point within a triangle is calculated by a linear function based on its location | It ignores nonlinear information and is unable to represent cliffs, caves, or holes |
| Kriging | It tries to have the mean residual or error equal to zero, and aims at minimizing the variance of the errors | The goals are practically unattainable, since the mean error and the variance of the errors are always unknown |
| Spline | The specific types of spline include uniform rational basis-spline, uniform nonrational basis-spline, nonuniform nonrational basis-spline, and nonuniform rational basis-spline. Rational curves enjoy all the properties of nonrational curves | The control points of a nonrational curve can only be transformed under an affine transformation. The perspective transformation used in computer graphics is not affine and the nonrational curves whose control points are transformed into image space cannot be correctly generated in this space |

Modified after Yue (2011)

or holes (Tse and Gold 2004). TIN models cause well-known artifacts that are easily recognized as linear breaks to occur in a regular rectangular or triangular pattern (Schneider 2001).

Kriging is the term used for a family of generalized least-squares regression algorithms (Krige 1951) and is named after the South African miningengineer, D. G. Krige (Kleijnen and van Beers 2005). It is a fundamental tool in geostatistics. It includes ordinary Kriging (OK), co-Kriging, and disjunctive Kriging (Kleijnen 2009). OK is expected to be the best linear unbiased estimator. It is linear, because its estimates are weighted linear combinations of the available data; it is unbiased, because it tries to have the mean residual or error equal to zero. It is a good linear unbiased estimator, because it aims at minimizing the variance of the errors. However, the goals of OK are ambitious and practically unattainable, since the mean error and the variance of errors are always unknown (Isaaks and Srivastava 1989). Therefore, it could not be guaranteed that the mean error is exactly zero and that the variance of errors is minimized. The best we can do is to build a model of the available data, in which both the error and the error variance can be calculated and then choose weights for the nearby samples that ensure the average error of the model to be zero and the error variance is minimized.

The specific types of spline include uniform rational basis-spline, uniform nonrational basis-spline, nonuniform nonrational basis-spline, and nonuniform rational basis-spline. A basis-spline, is a complete piecewise cubic polynomial consisting of any number of curve segments (Watt 2000). It is a cubic segment over a certain interval, which goes from one interval to the next. Each curve segment is defined by four control points and each control point influences four and only four curve segments, which is the local control property of the basis-spline curve. If knots, that is the join points on the values of x between segments, are spaced at equal intervals of the parameter x, the basis-spline is termed a uniform one. A nonuniform basis-spline (NURBS) is a curve where parametric intervals between successive knot values are not necessarily equal. A rational curve is a curve defined in four-dimensional space (known as projective space), which is then projected into three-dimensional space. The rational form of the NURBS is one of the commonest forms used in practice. Rational curves enjoy all the properties of nonrational curves. The control points of a nonrational curve can only be transformed under an affine transformation. However, the perspective transformation used in computer graphics is not affine and the nonrational curves whose control points are transformed into image space cannot be correctly generated in this space.

The method for high accuracy surface modelling (HASM) in the light of the fundamental theorem of surfaces has given solutions to error and multi-scale problems that plagued surface modelling prior to the development of HASM. Almost all the classical methods for surface modelling rely on local weighting, while HASM provides a finite difference method (Yue et al. 2007; Yue 2011).

### *1.3.3 Coupled Models*

According to the degree of coupling, loose coupling and integration of models locate at the two ends of the coupling spectrum. The loose coupling means that model output can be computed separately. The loose coupling of specialized models has the advantage that the specific strengths of each model are retained. However, the loose coupling approach lacks representation of feedback between the modelled components, and has a risk of propagation of error and uncertainty between the coupled models, which is difficult to track and quantify (Verburg et al. 2013).

Integrated models are interdisciplinary models that aim to describe the complex relationships between environmental, social, and economic drivers that determine current and future state of the ecosystem and the effects of global change, in order to derive policy-relevant insights. One of their essential characteristics is the simultaneous consideration of the multiple dimensions of environmental problems. The choice of integrated modelling versus a loose coupling of models depends on the specific requirements but also on the system being studied. An integrated modelling approach is required when feedback between the system components or spatial scales studied is important to system outcomes (IPBES 2016).

The mostly used integrated models in MA, IPCC, and IPBES include General Circulation Model, World Model for the Limits to Growth, Ecopath with Ecosim, Integrated Model to Assess the Global Environment, CENTURY Model, Dynamic

Integrated Climate–Economy Model, Terrestrial Ecosystem Model, Asian Pacific Integrated Model, Model for Water–Global Assessment and Prognosis, Global Unified Metamodel of the Biosphere, and Grid-Enabled Integrated Earth System Modeling.

#### 1.3.3.1 General Circulation Models

General Circulation Models (GCMs; also known as Global Climate Models). The development of GCMs was divided into three phases (Arakawa 2000): the early numerical weather prediction model (NWP), the early general circulation model (EGCM), and the recent coupled atmosphere–ocean GCM.

In 1904, Bjerknes first proposed the idea of weather prediction. In 1922, Richardson presented a weather prediction model with a basic structure not dissimilar to the current model; however, its forecast of the surface pressure at two points over Europe failed. In 1939, Rossby recognized the relevance of absolute vorticity advection to large-scale wave motions. In the late 1940s, the theories of baroclinic and barotropic instability (Charney 1947), and the concept of equivalent-barotropy (Charney and Eliassen 1949) represented major developments.

In 1950, Charney et al. successfully predicted 24-h weather by using the adiabatic quasi-geostrophic equivalent-barotropic model, which indicated the appearance of NWP. In 1953, an adiabatic three-level quasi-geotrophic model was successfully used to predict the rapid development of a storm observed over the United States in November 1950 (Charney and Phillips 1953). Although natural forecasts were not always successful, NWP became operational, first in the United States and then in many other countries. In 1956, Philips' numerical general circulation experiment recognized the close relationship between the dynamics of cyclones and that of general circulation. In 1959, Phillips demonstrated that nonlinear computational instability may occur in solutions of the nondivergent barotropic vorticity equation, which was the simplest nonlinear dynamical equation applicable to the real atmosphere.

Development of EGCM was distinguished into five generations. In 1958, Mintz described generation I of EGCM, which later became known as the Mintz–Arakawa model. This model included seasonal changes of solar radiation and long-wave cooling for each layer given as a function of the temperature at the lower level of the EGCM. Generation II of EGCM had the same two-level vertical structure as that of generation I; the horizontal domain covered the entire globe with uniform grid intervals in both longitude and latitude (Arakawa 1969). Generation III of EGCM was the first multilevel model, which included a change of horizontal grid structure, the implementation of a bulk model for the planetary boundary layer (PBL), the inclusion of the Arakawa–Schubert cumulus parameterization, and the prediction of ozone mixing ratio with interactive photochemistry (Schlesinger and Mintz 1979). In generation IV of EGCM, the vertical discretization followed Arakawa and Suarez (1983) for the troposphere above the PBL. The horizontal differencing of the momentum equation was based on the energy and potential enstrophy conserving scheme for the shallow water equations (Arakawa and Lamb 1981). Generation V of EGCM was characterized by a change in the radiation scheme (Harshvardan et al. 1987), inclusion of an orographic gravity wave drag parameterization (Kim and Arakawa 1995), inclusion of convective downdraft effects in the cumulus parameterization (Cheng and Arakawa 1997), the revision of PBL moist processes (Li et al. 1999) and the inclusion of explicit predictions on ice and liquid clouds (Koehler et al. 1997). For most applications of EGCMs, the geographical distribution of sea surface temperature (SST) was prescribed as an external condition, meaning that calculations of the surface heat flux and the vertical distribution of radiative heating did not have to be very accurate.

NWPs and EGCMs instead concentrated on synthesizing hydrological processes, cloud processes, boundary-layer processes, radiation processes, and land processes. GCMs focused on interactions among ocean and land processes, radiation, and chemical processes, boundary-layer processes, cloud processes, and hydrological processes (Arakawa 2000).

#### 1.3.3.2 World Model for the Limits to Growth

In the early 1970s, a world model of limits to growth was developed and used as the prototype model for investigating five major trends of global concern: accelerating industrialization, rapid population growth, widespread malnutrition, the depletion of nonrenewable resources, and a deteriorating environment (Forrester 1971; Meadow et al. 1972). The first version of the Limits to Growth called World3, was written in the computer simulation language, DYNAMO. In 1992, on the basis of studying global developments between 1970 and 1990, the world model was updated and the results published in a book, Beyond the Limits (Meadows et al. 1992). World3 was updated into World3-91 for Beyond the Limits and the computer language was converted from DYNAMO to STELLA. In 2004, the 30-year update of limits to growth indicated that humanity was already in unsustainable territory but the human ecological footprint was still increasing (Meadows et al. 2004). The ecological footprint measures the amount of biologically productive land and water area required to support the demands of a population or productive activity, which is then compared to the carrying capacity. Productive land and sea areas support human demands for food, fiber, timber, and energy as well as providing space for infrastructure. These areas also absorb waste products from the human economy. The ecological footprint measures the sum of these areas that are often weighted according to their relative productivity and expressed in global hectares (Kitzes and Wackernagel 2009). World3-01 was slightly updated and turned into World3-03, which showed the behavior of the human welfare index and the human ecological footprint for the period between 1900 and 2100. It concluded that it would take a long time to obtain political support for reversing current trends, bringing the ecological footprint back below the long-term carrying capacity.

#### 1.3.3.3 Ecopath with Ecosim

A mathematical ecosystem model, ECOPATH, was constructed for an entire coral reef ecosystem and was used to estimate its standing stock and production budget (Polovina 1984). ECOPATH was based on assuming mass balance in trophic flows and the idea that if some biomasses and/or flows are known, then others can be calculated under the mass-balance assumption. ECOPATH partitions the ecosystem into species groups and produces estimates of mean annual biomass, annual biomass production, and annual biomass consumption for each of the species groups, because each species group is an aggregation of species having common physical habitats, similar diets, and similar life history characteristics.

However, ECOPATH only provides a static picture of ecosystem trophic structure by means of a linear equation system. Thus, dynamic ecosystem models, termed ECOSIMs, were derived from ECOPATH results by transferring the linear equation system to a system of differential equations. ECOSIM is a modeling tool for representing the spatially aggregated dynamics of whole ecosystems by combining relatively simple differential equations for biomass dynamics of some ecosystem components with delay-difference, age-size-structured equations for some key populations that have complex trophic ontogenies and use selective harvesting for older animals. The incorporation of ECOSIM into ECOPATH enabled a wide range of potential users to conduct fisheries policy analyzes that explicitly accounted for ecosystem trophic interactions. But ECOSIM had many limitations, such as overestimating the potential productivity of low-fecundity species, producing overly optimistic assessments of maximum fishing mortality rates for intermediate trophic levels, producing indeterminate outcomes in complex food webs, and producing misleading parameter estimates due to ECOPATH's equilibrium assumptions (Walters et al. 1997).

ECOSPACE represents biomass dynamics over two-dimensional space as well as time. The Lagrangian approach and Eulerian approach are two possible ways to discretize dynamic relationships for the practical simulation of space–time patterns. The Lagrangian approach is to divide biomasses into large numbers of parcels assumed homogeneous, and to move these parcels about in space, while the Eulerian approach treats movement as flows of organisms among fixed spatial reference points or cells, without retaining information about the history of the organisms present at any point in any given moment. ECOSPACE takes the Eulerian approach. ECOSPACE is a spatially explicit model for policy evaluation that allows consideration of the impact of marine protected areas in an ecosystem context, relying on the ECOPATH mass-balance approach for most of its parameterization.

The modeling approach, Ecopath with Ecosim (EwE), combines models for ecosystem trophic mass balance analysis (ECOPATH) with a dynamic modeling capability (ECOSIM), as a means of exploring the past and future impacts of fishing and environmental disturbances as well as optimal fishing policies. Ecosim models can be replicated over a spatial map grid (ECOSPACE) to allow the exploration of policies such as marine protected areas, while accounting for spatial dispersal effects (Christensen and Walters 2004).

#### 1.3.3.4 Integrated Model to Assess the Global Environment

According to Kram and Stehfest (2006), development of the integrated model to assess the global environment (IMAGE) can be distinguished into IMAGE 1.0, IMAGE 2.0, IMAGE 2.1, IMAGE 2.2, IMAGE 2.3, IMAGE 2.4, and IMAGE 3.0. In 1984, the National Institute for Public Health and the Environment (RIVM) of the Netherlands initiated a project, Reference Function for Global Air Pollution/$CO_2$, within which global modeling was defined as a quintessential activity (Rotmans 1990).

In 1986, IMAGE 1.0 was developed on the basis of prototypes constructed in the previous two years. IMAGE 1.0 was the first version of IMAGE. The primary objective of IMAGE 1.0 was to create a comprehensive picture of global climate change by integrating separate components relating to the world economy, atmospheric chemistry, marine and terrestrial biogeochemistry, ecology, climatology, hydrology, and glaciology into a synthetic framework. IMAGE 1.0 was based on the hypothesis of the greenhouse effect, and so its underlying principles would be greatly tarnished if the greenhouse hypothesis was to be rejected in future. IMAGE 1.0 is a policy-oriented model that tries to capture as much as possible of the cause–effect relationship with respect to climate change. It was developed for the calculation of historical and future emissions of greenhouse gases on global temperatures and sea level rise, taking into account ecological and socio-economic interests in specific regions. IMAGE 1.0 consisted of interlinked modules that included emission modules, concentration modules, a climate module, a sea level rise module, and a socio-economic impact module. Each module described a specific element of climate change and the output of one module served as the input to the next. In the emission modules, the trace gases included $CO_2$, $CH_4$, CO, $N_2O$, CFC11, and CFC-12. Annual estimates of historical emissions were incorporated for the period 1900 to 1985. Four sets of scenarios were chosen for the period 1985 to 2100. The underlying scenario assumptions were based on different sources of trace gas emissions grouped as nature, energy, agriculture and industry. The world was divided into nine regions, the United States, OECD west, OECD Asia, Centrally Planned Europe, Centrally Planned Asia, the Middle East, Africa, Latin America, and South and East Asia. The main parameters included population, economic growth, end use and conversion efficiency, cost development for nonfossil fuels, environmental costs and carbon taxes. The input of the concentration module was taken from the output of the emission modules. The concentration module was linked with an ocean module and a deforestation module. The calculated trace gas concentrations provided input for the climate module. The resulting global mean equilibrium surface temperature rise was calculated by a climate feedback factor, in which the water vapor factor was explicitly taken into account. The resulting temperature changes provided input for the sea level rise module. The effect of global warming on potential sea levels was determined by five processes: the thermal expansion of ocean water, the melting of alpine glaciers, the ablation of the Greenland ice caps, the accumulation of the Antarctic, and natural trends. The socio-economic impact module was developed for describing the consequences of an accelerated sea level rise for the coastal defense systems of dikes, dunes, and intertidal areas. IMAGE 1.0 was performed with four sets of scenarios. Scenario A assumed that economic growth would continue without being limited by environmental constraints. Scenario B assumed that environmental measurespresently being considered to control environmental problems would be implemented. Scenario C assumed the enforcement of stricter environmental controls. Scenario D calculated the possibility of maximum efforts toward global sustainable development. In all scenarios, the world population was assumed to approach 10.8 billion in 2100, and it was assumed not to be influenced by greenhouse policies.

IMAGE's focus was then shifted to a regional base, and became known as version 2.0 (Alcamo 1994). In this version, land-cover and land-use modeling were done on a resolution of $0.5 \times 0.5°$, drawing on experience with geographically explicit global models. IMAGE 2.0 was the first published global integrated model to have a geographic resolution. IMAGE 2.0 consisted of three linked clusters of modules: the Energy-Industry System (EIS), the Terrestrial Environment System (TES), and the Atmosphere–Ocean System (AOS).

Guided by the recommendations of international review meetings, further refinements and extensions were implemented in IMAGE 2.1 (Alcamo et al. 1998). Major steps included an improved computation of future regional energy use in EIS. Since the development of IMAGE 2.1, future fuel prices have influenced the selection of fuels in the model, depending on resource depletion on the supply side and price-dependent energy conservation on the demand side.

The third session of the IMAGE Advisory Board in 1999 resulted in a list of recommendations and suggestions for further development work on IMAGE. One of the major changes of IMAGE 2.2 concerned the recommended two-track strategy for climate modeling. The earlier zonal-meanclimate-ocean model in IMAGE was replaced by a combination of the simple Model for the Assessment of Greenhouse-gas Induced Climate Change (MAGICC) and the Bern ocean model. For the new approach, the resulting global average temperature and precipitation changes were scaled using temperature and precipitation patterns generated by complex, coupled Global Circulation Models (GCMs). The widely accepted method for scaling patterns of aerosol-induced climate change was also adopted. A specific advantage of this method is that patterns from different GCMs can be used to explore uncertainties in the behavior of the global climate system. This IMAGE 2.2 version was used to contribute to work on the SRES of the IPCC.

Participation in the IPCC-SRES process has greatly enhanced the capacity of both the model and IMAGE teams to explore scenarios and obtain results geared to the requirements of various international assessment processes. Experiences gained from the SRES process have, however, reinforced the desire to seriously reconsider the future of the IMAGE model. A set of model enhancements was identified and later initiated; these enhancements taken jointly will constitute the next generation model, IMAGE 3. However, parallel to this, a small set of model changes, internally referred to as version 2.3, and mainly concerned with the integration of energy crops and carbon plantations, was implemented for the analysis of mitigation options.

In IMAGE 2.4, grid-based population dynamics have been improved by introducing a new downscaling algorithm. The population within a grid cell is calculated by a proportional method using available country-specific data combined with world regional trends. Aggregated economic indicators like GDP, household consumption and value added in industry, services, and agriculture are used to estimate the demand for energy services. Energy supply chains with substantial technological detail are then selected on the basis of relative costs to meet the resulting final energy demand after autonomous and price-induced energy savings. Demand for and production of agricultural products on the basis of population changes and economic developments are simulated through a linkage to the global trade analysis project model. One of the most striking elements of IMAGE 2.4 is its geographically explicit land-use modeling, which considers both cropping and livestock systems on the basis of agricultural demand and demand for energy crops. However, important nonlinear interactions between the land, atmosphere, and ocean cannot be studied with IMAGE 2.4 due to limitations of the current climate model and the natural vegetation module.

In July 2014, Netherlands Environmental Assessment Agency released IMAGE 3.0. In IMAGE 3.0, the descriptions of several critical areas of global environmental change have been improved, including the dynamics of land-use change, water use, and energy demand and production (https://www.pbl.nl/).

#### 1.3.3.5 CENTURY Model

In 1987, a rangeland and cropland version of the CENTURY model was developed to simulate the effects of macro-environmental gradients as a first step toward simulating the effects of climatic change. The model incorporated multiple compartments of soil organic matter (SOM), simulated decomposition rates that vary as a function of monthly soil temperature and precipitation, and included both C and N flows (Parton et al. 1987). The model simulated primary productivity, soil nutrient dynamics, and soil water but focused mainly on changes in SOM. The first objective of the modeling exercise was to simulate the effects of climatic gradients on productivity and SOM over large areas; the second objective was to identify key soil properties that would allow the difference between soils within a single climatic zone to be simulated.

The CENTURY model consisted of a soil and decomposition submodel, a plant submodel, and a nitrogen submodel. The soil and decomposition submodel contained three fractions of SOM: (1) an active fraction of soil C and N consisting of live microbes and microbial products; (2) a pool of C and N that is physically protected and/ or in chemical forms with more biological resistance to decomposition; (3) a fraction that is chemically recalcitrant and can be physically protected. The plant submodel simulated the monthly dynamics of C and N in live and dead aboveground plant material, live roots, structural and metabolic surfaces, and soil residual pools. The nitrogen submodel assumed that most N was bonded to C and that the C/N ratio of structural, active, slow, and passive fractions remained fixed.

In 1993, a grassland version of the CENTURY model was tested using observed data from 11 temperate and tropical grasslands around the world (Parton et al. 1993). In this version, biophysical submodels were added to calculate monthly evaporation and transpiration water loss, water content of soil layers, snow water content, and the saturated flow of water between soil layers. This improved version of the CENTURY model could be used to simulate the dynamics of carbon (C), nitrogen (N), phosphorus (P), and sulfur (S) for different plant–soil systems. Different plant production submodels for grassland, forests, and crops were linked to a common SOM submodel. The SOM submodel simulated the flow of selected elements through different inorganic and organic pools in the soil.

In 1994, a cropland version of the CENTURY model was developed to simulate long-term (10–1000 year) patterns in the dynamics of SOM (0–30 cm depth), plant production, and nutrient cycling (N, P, and S) using data collected on a monthly basis. The driving variables included monthly average maximum air temperature (°C at 2 m), monthly precipitation (cm), soil texture (sand, silt, and clay content), dead plant material nutrient and lignin content, and atmospheric and soil inputs of N (Parton and Rasmussen 1994).

In 2002, the SOM submodel was modified to provide a better description of the litter and soil C and N dynamics of forests. The modifications included: (1) incorporation of additional woody litter pools, (2) allowing the N content of the SOM pool to vary, (3) constraining N mineralization to the active SOM pool, (4) incorporation of a small flux of mineral N to the resistant SOM pool, (5) allowance of mycorrhizal uptake of N, and (6) reformulation of temperature and moist effects on decomposition. This modified version of the CENTURY model was able to simulate a range of complex observations related to the C and N dynamics reported in a number of published studies (Kirschbaum and Paul 2002).

In 2003, the point-based CENTURY model was integrated with spatially distributed data. Spatially explicit estimates of historical, current, and future soil C were produced for a semi-arid area in the Sudan using readily available data for temperature, precipitation, soil texture, and land cover (Ardoe and Olsson 2003). The integration of the CENTURY model with spatially explicit data provided a flexible and powerful way to assess how different scenarios for land use, management, and climatic change can affect carbon dynamics on a regional scale.

#### 1.3.3.6 Terrestrial Ecosystem Model

The terrestrial ecosystem model (TEM) is a process-based biogeochemical model that uses spatially referenced information on climate, elevation, soils, and vegetation to make monthly estimates of important carbon and nitrogen fluxes and pool sizes for the terrestrial biosphere. The first version of TEM (Raich et al. 1991) was used to describe and analyze the spatial and temporal patterns of terrestrial NPP in South America. The second TEM version (McGuire et al. 1992) was used to estimate NPP for potential vegetation in North America. The third version examined the response of NPP to temperature and carbon dioxide for temperate forests (McGuire et al. 1993), and its response to climate change predicted by the GCM for global terrestrial biosphere (Melillo et al. 1993). Version 4.0 of TEM was improved so that it was able to evaluate global and regional responses of NPP and carbon storage to elevated carbon dioxide (McGuire et al. 1997). TEM 4.1 was applied to investigating the dynamics of terrestrial carbon fluxes and storage in potential vegetation of the conterminous United States during the period 1900 to 1994 (Tian et al. 1999). TEM 4.2, in which the GPP equation of TEM 4.1 had been improved, was used to simulate the combined effects of climate variability, increased atmospheric $CO_2$ concentration, cropland establishment, and abandonment of $CO_2$ between the atmosphere and monsoon in Asian ecosystems (Tian et al. 2003).

It simulated changes in carbon storage during three stages of disturbance: conversion from natural vegetation to cultivation, production, and harvest on cultivated land, and abandonment of cultivated land. The net carbon exchange between the terrestrial biosphere and the atmosphere in TEM is represented by net ecosystem production (NEP), which is calculated as the difference between NPP and heterotrophic respiration (RH). NPP is calculated as the difference between gross primary production (GPP) and plant respiration (RA). Gross primary production represents the uptake of atmospheric $CO_2$ during photosynthesis and is influenced by light availability, atmospheric $CO_2$ concentration, temperature, and the availability of water and nitrogen (Tian et al. 2000). Plant respiration includes both maintenance and construction respiration, and is calculated as a function of temperature and vegetation carbon. The flux RH represents the microbially mediated decomposition of organic matter in an ecosystem and is influenced by the amount of reactive soil organic carbon, temperature, and soil moisture. The annual NEP of an ecosystem is equivalent to its net carbon storage for the year. For regional or global extrapolation with TEM, spatially explicit input datasets include vegetation, elevation, soil texture, mean monthly temperature, monthly precipitation and mean monthly solar radiation. The input data sets are gridded at a resolution of 0.5° latitude and 0.5° longitude. In addition to the input datasets, TEM requires soil- and vegetation-specific parameters appropriate to a grid cell. Although many of the parameters in the model are defined from published information, some of the vegetation-specific parameters are determined by calibrating the model to the fluxes and pool sizes of an intensively studied field site.

#### 1.3.3.7 Dynamic Integrated Climate–Economy Model

Nordhaus (1993) developed a dynamic integrated climate–economy (DICE) model to simulate the long-term interactions between global warming damages and world economic growth. Because the DICE model does not involve energy flow systems,

Mori (1995) developed a DICE + e model, which imposed energy flows upon the DICE model to estimate the value of exhaustive resources, nuclear technology, and other options. The DICE + e was then extended to the multiregional approach for resource and industry allocation (MARIA) model to generate the international market prices of resources as well as the tradable carbon emission permits under certain constraints. MARIA has been applied to assess the potential contribution of fossil, biomass, nuclear, and other energy technologies and land-use changes to future greenhouse gas (GHG) emissions, global and regional GHG mitigation policies, and future GHG reduction options (Mori and Saito 2004).

Owing to the DICE model omitting induced innovation and overstating the costs of complying with carbon change policies, an ENTICE model (for Endogenous Technological change) was developed. It was used to compare the overall global welfare costs of carbon policy to the results of models in which technological change is exogenous (Popp 2004). ENTICE makes use of empirical results on technological change in the energy industry to incorporate endogenous technological change into the DICE model of climate change, and simulates technological change that enhances energy efficiency. The ENTICE framework provides a detailed sensitivity analysis of different assumptions on features such as imperfect research markets and the potential crowding out effects of energy research and development (R&D). ENTICE model limitations include the fact that it dramatically simplifies policy by modeling the world as a single region, that it only designs innovations to improve energy efficiency, and that does not consider the uncertainty factor.

A DICE94 + THC model was developed by amending the DICE Model to include an indicator of an emissions-dependent collapse of North Atlantic Thermohaline Circulation (THC) (Keller et al. 2004; Lempert et al. 2006). DICE94 + THC were run by repeatedly invoking the Stochastic Ranking and Evolution Strategy (SRES) algorithm developed by Runarsson and Yao (2000).

#### 1.3.3.8 Model for Evaluating Regional and Global Effects of GHG Reduction Policies

Manne et al. (1995) proposed a model for evaluating the regional and global effects of GHG reduction policies (MERGE). The model was designed to be sufficiently flexible when used to explore alternative views on a wide range of contentious issues such as costs, damages, valuation, and discounting. It consisted of a series of linked modules representing the major processes of interest, which included the costs of reducing emissions of radiatively important gases, natural system disposition, reactions to the emission of these gases, and the reaction of human and natural systems to changes in the atmospheric/climatic system.

MERGE had three major submodels: Global 2200, the climate submodel, and the damage assessment submodel. Global 2200 divides the world into five major geopolitical regions, the USA, other OECD nations (Western Europe, Japan, Canada, Australia, and New Zealand), FSU (the former USSR), China and the ROW (rest of world); Global 2200 is used to assess the economy-wide costs of alternative emission constraints at a regional and global level. The climate submodel describes the relationship between manmade emissions and atmospheric concentrations and their resulting impact on temperature. The damage assessment submodel divides damages into two categories, market and nonmarket, to quantify the impacts of climate change.

Kypreos (2007) introduced two stylized backstop systems with endogenous technological learning (ETL) in the MERGE, one for electric markets and the other for nonelectric markets, to analyze the impacts of ETL on carbon-mitigation policy. He contrasted the results of the ETL backdrop system with a situation in which ETL was not used.

#### 1.3.3.9 Asian Pacific Integrated Model

Matsuoka et al. (1995) developed the Asian Pacific Integrated Model (AIM). AIM was developed mainly to examine global warming response measures in the Asia–Pacific region, but was later linked to a world model enabling global estimates to be made. AIM comprises three main models, the GHG emission model (AIM/emission), the global climate change model (AIM/climate), and the climate change impact model (AIM/impact). The AIM/emission model consists of country-level, bottom-up type energy models and global-level, top-down type energy and land-use models. Emissions of $SO_2$, $NO_x$, and SPM are calculated within the AIM/emission model, and then used as input for the AIM/climate model. The absorption of $CO_2$ and heat by the oceans is calculated using an upwelling-diffusion model, which is a part of the AIM/climate model, with the oceans divided into a surface mixed layer and an intermediate layer that extends down to about 1000 m. Global mean temperature changes are calculated with an energy balance/upwelling diffusion ocean model, and used as input for the regional models. Data from the GCM experiments are used in order to estimate the regional distribution of climate parameters. They are coupled with the global mean temperature change calculated in the AIM/climate model. The interpolated climate distributions are used in the AIM/impact model, which calculates global and regional climatic impacts. The AIM/impact model mainly treats the impact on primary production industries, such as water supply, agriculture, forest products, and human health.

It is necessary to consider many uncertainties in human activities such as population growth, economic development, and technological innovation as well as uncertainties in natural processes to estimate future $CO_2$ emissions and to make plans for climate stabilization. A variety of global and regional assumptions about population growth, economic growth, technological improvement, and government policies are entered into the models to provide estimates of energy consumption and land-use changes, which ultimately provide predictions of GHG emissions. The main goal of AIM is to assess policy options for stabilizing the global climate, particularly in the Asia–Pacific region; from the two perspectives of reducing GHG emissions and avoiding the impacts of climate change (Kainuma et al. 2004).

#### 1.3.3.10 Model for Water–Global Assessment and Prognosis

The comprehensive assessment of the freshwater resources of the world was the first global-scale assessment of water resources (Kjellén and Mcgranahan 1997). However, the assessment had many limitations. For instance, it lacked a global modeling approach for water availability and use; the smallest spatial units for which water stress indicators could be computed were whole countries; the important impact of climate variability on water stress could not be taken into account, and the impact of climate change on future water availability and irrigation requirements could not be assessed.

To overcome these limitations and to achieve an improved assessment of the present and future water resources situation, a model for Water–Global Assessment and Prognosis (WaterGAP) was developed (Doell et al. 1999). It simulated the impact of demographic, socio-economic, and technological change on water use as well as the impact of climate change and variability on water availability and irrigation water requirements on a spatial resolution of $0.5° \times 0.5°$. It consists of the Global Water Use Model and the Global Hydrology Model. The Global Water Use Model comprises submodels for each of the water use sectors such as irrigation, livestock, households, and industry. Irrigation water requirements are modeled as a function of the cell-specific irrigated area, crop and climate, and livestock water use is calculated by multiplying livestock numbers by livestock-specific water use. Household and industrial water use in grid cells are computed by downscaling published country values based on population density, urban population, and access to safe drinking water (Doell et al. 2003). The Global Hydrology Model is used to compute time series of monthly runoff and river discharge on a spatial resolution of $0.5° \times 0.5°$. Runoff is determined by calculating the daily water balances of soil and canopies as well as lakes, wetlands, and large reservoirs. The storage capacity of groundwater, lakes, wetlands, rivers, and routes rivers discharge through basins are taken into account according to a global drainage direction map.

WaterGAP can be used in global assessments of water security, food security, and freshwater ecosystems. Time-variable gravity data of the Gravity Recovery and Climate Experiment (GRACE) satellite mission were incorporated directly into the tuning process of the Global Hydrological Model to improve simulations of the continental water cycle. By mapping time variations of the earth's gravitational field with GRACE since its launch in 2002, a global data set of mass variations close to the earth's surface has become available. After the removal of mass variations due to tidal and nontidal atmospheric and oceanic transport processes, the time-variable gravity data mainly represents water mass variations in continental hydrology, that is, the total water storage change on the continents (Werth et al. 2009).

#### 1.3.3.11 Global Unified Metamodel of the Biosphere

A global model is not meant to predict every possible aspect of the world, and does not support either a purely optimistic or pessimistic view of the future. It represents a mathematical expression of the interrelationships between global concerns such as population, industrial output, natural resources, and pollution. Global modelers investigate what might happen if policies continue along present lines, or if specific changes are instituted. With this assumption in mind, Boumans et al. (2002) described the framework of a Global Unified Metamodel of the Biosphere (GUMBO). A metamodel is interpreted as a regression model of an actual simulation model (Kleijnen 1987).

GUMBO consists of five distinct modules: atmosphere, lithosphere, hydrosphere, biosphere, and anthroposphere. These are further divided into 11 biomes: Open Ocean, coastal ocean, forest, grassland, wetland, lakes/rivers, desert, tundra, ice/rock, cropland, and urban areas. The module of atmosphere facilitates exchanges of carbon, water, and nutrients across biomes and accounts for global energy balances; atmospheric dynamics are calibrated against atmospheric carbon and global temperatures. The module of lithosphere represents the solid uppermost shell of the earth, which includes soils and deposited sediments; lithosphere stocks are represented by silicate rocks, carbon reserves, and ore and fossil fuel deposits in rock and soil. The module of hydrosphere accounts for biome-specific stocks of water, carbon, and nutrients in surface and subsurface water bodies; surface water exchanges between continental and oceanic biomes are calculated to compensate for uneven distributions among biome-specific evapotranspiration. The module of biosphere defines biosphere as a self-regulating system sustained by large-scale cycles of energy and materials such as carbon, oxygen, nitrogen, certain minerals, and water; many of the ecosystem services provided by the biosphere are associated with the rate of photosynthesis; important factors for achieving optimum productivity are the nutrient availability in the lithosphere, temperature, light levels, carbon pressure in the atmosphere, soil moisture in the hydrosphere, and waste levels generated within the anthroposphere. The module of anthroposphere represents human social and economic systems; the anthroposphere harvests large amounts of material and energy from the larger system and discards waste at each phase along a production chain.

The enormous number of variables in the GUMBO model makes it impractical to use analytical techniques to solve an optimum. In addition, with so many variables, it is virtually certain that there are many local optima for the model, and seeking a global optimum would probably be both futile and pointless. Therefore, GUMBO calculates the shadow prices for each of these optima and presents prices as a range.

#### 1.3.3.12 Grid-Enabled Integrated Earth System Modeling

The long-term response of the carbon cycle has previously been assessed with simple box models, but these had lacked spatial dimensions, a hydrological cycle, and other climate variables. To address this problem, a Grid Enabled Integrated Earth system model, termed GENIE, was developed (Lenton et al. 2006, 2007).

GENIE was composed of a 3-D frictional geostrophic ocean model, a model for energy moisture balance atmosphere, a dynamic and thermodynamic sea-ice model, a new land surface physics and terrestrial carbon model, and an ocean biogeochemical model. GENIE was designed to enable paleoclimate simulations, long-term future projections, large ensemble sensitivity studies, and coupling in an integrated assessment system. The frictional geostrophic ocean model was 3-D, had eight depth levels, and shared a latitude–longitude horizontal grid of $36 \times 36$ equal area grid cells (about $5^\circ \times 10^\circ$) with other components of the coupled model. It included realistic bathymetry, an isoneutral and eddy-induced mixing scheme and spatially varying drag. The atmospheric energy moisture balance model was based on an atmospheric model incorporating energy and moisture balance equations (Fanning and Weaver 1996).

Given the SST and specified surface wind field, the atmospheric model calculated the surface fields of air temperature, specific humidity, and heat and freshwater fluxes. The main changes of the model incorporated were the inclusion of seasonality, the separation of surface and atmosphere/cloud albedo, the inclusion of a relaxation time for precipitation and revision of the wind fields. The new land surface physics and terrestrial carbon model was designed as a minimal representation of land surface physics, hydrology, and carbon cycling for use in a spatially resolved context. It had single pools of vegetation carbon, soil carbon, and soil water at each land grid point. The ocean biogeochemical model was based on a single nutrient ($PO_4$) representation of ocean biogeochemical cycling. Phosphate concentrations in each surface ocean cell were restored with a characteristic time constant toward a prescribed target concentration. The excess $PO_4$ removed was converted into organic matter, while any deficit of model $PO_4$ compared to the target was left unchanged. A double exponential formulation was used to partition down through the water column the solute release arising from the remineralization of particulate matter. A double exponential formulation was also used to remineralize $CaCO_3$ in the water column. A uniform air–sea $CO_2$ transfer coefficient was employed to exchange $CO_2$ with a well-mixed atmosphere (Cameron et al. 2005).

## 1.4 Modelling Platforms

Many modelling platforms have been developed, which take into account hardware, software and data sets as a whole. For example, Earth simulator, Digital Earth System, Planet Simulator, Global Earth Observation System of Systems, Earth System Numerical Simulation Facility, and the FTEEM-based modelling platform.

### *1.4.1 Earth Simulator*

In 1997, an Earth simulator was developed to simulate global warming. The primary function was to create a huge-scale supercomputer, which would achieve at least 5 teraflop/s for an atmospheric GCM. It first conducted the simulations of global atmospheric circulation and oceanic circulation on a horizontal resolution of 10 km. According to Descartes' paradigm, nature loves equilibrium and stability; natural system conditions seldom move away from the equilibrium point or the point of stability. However, since nature is an open system through which information, energy, material, and flow are always passing in and out, it would appear that nature actually loves nonequilibrium, instability and therefore, nonlinearity. The earth simulator provides an important tool necessary for creating a new paradigm of nonequilibrium/ nonlinearity /openness (Sato 2004).

## 1.4.2 Digital Earth System

Digital Earth was defined as a system providing access to what is known about the planet and its inhabitants' activities—currently and for any period in history—via responses to queries and exploratory tools (Gore 1998). During the period 1998 to 2001, the Digital Earth Initiative, coordinated by the NASA-led Interagency Digital Earth Working Group (IDEW), tried to realize the integration of existing data from multiple sources in order to improve the application of geospatial data for visualization, decision support, and analysis (IDEW 2001). During this period, the results of the 3-year IDEW effort included the Web Mapping Service (WMS) standard, a Digital Earth Alpha Version prototype, and a Digital Earth Reference Model (DERM). The Geospatial Interoperability Reference Model was produced as an extension of the Digital Earth Initiative. Since 1998, many advances have been made. For instance, Earthviewer and GeoPlayer appeared in 2001, while World Wind and Google Earth were released in 2003 and 2005, respectively.

Grossner et al. (2008) proposed the concept of a digital earth system and defined it as a comprehensive, mass distributed geographic information and knowledge organization system. The digital earth system was distributed because: (1) there were necessarily multiple, geographically dispersed data stores providing content; and (2) the processing load of server-based query and analytical processes needed to be shared for performance reasons. Geographic information refers to very broad information about well-defined locations on the Earth's surface.

## 1.4.3 Planet Simulator

The Planet Simulator was a Model of Intermediate Complexity (MIC) that could be used to run climate and paleoclimate simulations for timescales up to ten thousand years or more in an acceptable real time (Fraedrich et al. 2005a, b). The Planet Simulator was built to support numerical experiments for understanding the dynamics of the climate of the earth and earth-like planets. The model was composed of a dynamical core, parameterizations and subsystems, and a graphical user interface. The dynamical core of the Planet Simulator was based on the moist primitive equations representing the conservation of momentum, mass, and energy. The dimensionless set of equations consisted of the prognostic equations for the vertical component of vorticity and horizontal divergence, the first law of thermodynamics, the equation of state with hydrostatic approximation, the continuity equation, and the prognostic equation for water vapor. The parameterizations and subsystems included boundary layers and diffusions, radiation, moist processes, clouds and dry convection, land surface and soil, and ocean and sea ice.

The Planet Simulator was configured and set up by the module graphical user interface and Model Starter. Model Starter was the fastest way to get the model running, allowing access to the most important parameters of the model preset to the most frequently used values. The graphical user interface for running the Planet Simulator had two main purposes. The first was to display model arrays in suitable representations; the second was to facilitate the interaction part of the graphical user interface, allowing the user to change selected model variables during the model run.

### 1.4.4 Global Earth Observation System of Systems

The GEOSS is currently being built by the Group on Earth Observations (GEO) on the basis of a 10-year implementation plan running from 2005 to 2015. The main purpose of GEOSS is to improve the monitoring of the state of the earth, increase the understanding of the earth's processes, and enhance the prediction of the behavior of earth systems. The focus of GEOSS is to produce societal benefits through more coordinated observations and better data sharing. GEOSS will provide an important scientific basis for sound policy and decision making in energy, public health, agriculture, transportation, and numerous other areas (Lautenbacher 2006). It will contribute to reducing loss of life and property from natural and human-induced disasters, increasing our understanding of environmental factors affecting human health and well-being, improving the management of energy resources, assessing and adapting to climate variability and climate change, improving water resource management through a better understanding of the water cycle, improving weather forecasting and weather warnings, improving the management and protection of ecosystems, supporting sustainable agriculture, combating desertification, and helping to monitor and conserve biodiversity (Christian 2005; MacPhail 2009).

### 1.4.5 Earth System Numerical Simulation Facility

Since November 2018, the Institute of Atmospheric Physics under the Chinese Academy of Sciences and Tsinghua University have started to jointly construct the Earth-system numerical simulation facility (ESNSF). The China's first ESNSF was launched on June 23, 2021, for modelling the physical, chemical and biological processes and their interactions in the atmosphere, hydrosphere, the cryosphere and biosphere as a whole with a high-performance scientific computing platform. It is expected to become fully operational and open to universities and researchers across the world in 2022.

### *1.4.6 The FTEEM-Based Modelling Platform*

A raster expression of a region or one of its eco-environmental properties can be abstracted to a mathematical surface. An eco-environmental surface is, as a unified term, used to represent a surface of nature, a surface of nature's contribution and a surface of a driving-force of nature changes. The FTEEM-based modelling platform (FBMP) is developed on the basis of FTEEM and Nature Future Framework (NFF). The FBMP consists of four modules, i.e. a FTEEM module including five submodules respectively for interpolation, upscaling, downscaling, data fusion and model-data assimilation, and three NFF modules for nature, for nature's contribution to people and for driving forces of nature changes. The submodule for interpolation is used to predict the values of a primary variable at points within the same region of sampled locations in terms of data in the form of discrete points or some form of a surface with data voids. The one for upscaling is designed to transfer data from a finer to a coarser resolution in order to mainly reduce computational costs. The one for downscaling is developed to obtain information at finer resolution from the coarser resolution models and data. The one for data fusion is to integrate multiple data and knowledge streams representing the same real-world object into a consistent, accurate, and useful representation. The one for model-data assimilation seeks to improve the performance of a model by either optimizing parameters or initial state variables according to a given data set. The module for nature focuses on species diversity and ecotope diversity as well as digital terrain model. The one for nature's contribution to people pays attention to food, fresh water and carbon stocks. The one for driving forces emphasizes carbon dioxide concentrations, climate change, land-use change and policies.

## 1.5 Summary

Issues, which need to be further studied, include model coupling, scale-transformation, real-time visualization in three dimensions, quantum computing, and intelligent computing.

### *1.5.1 Model Coupling*

No single model can capture all dynamics at a high level of detail, and the coupling of models has become a popular tool to combining the different dimensions (Pichs-Madruga et al. 2016). Loose coupling and integration of models are located at two ends in terms of the coupling degree. The choice of integration versus a loose coupling of models depends on the specific requirements of the system under consideration. The loose coupling of specialized models has the advantage that the specific strengths of each model are retained. The disadvantage of loose coupling models is the lack of representation of feedback between the modelled components (Verburg et al. 2013). An integrated modelling approach is required when feedback between the system components is important to system outcomes. However, current integrated models contain insufficient sub-models of ecological processes for accurately representing environmental conditions (Harfoot et al. 2014).

### *1.5.2 Multi-Scale Issues*

In the early 1980s, the multiscale issue was recognized as a fundamental problem in geographical information systems. In 1983, a small group of leading scientists was gathered by National Aeronautics and Space Administration (NASA) to define research priorities in GIS and multiscale representation was identified as one of them. Since the early 1990s, multiscale representation has become a popular research topic in geographical information science. In 1996, the University Consortium for Geographic Information Science (UCGIS) selected multiscale representation as one of the 10 research priorities. By the late 1990s, multiscale representation gained a boost with the funding of the project of Automated Generalization New Technology (AGENT) by the European Union. In 2000, the International Society for Photogrammetry and Remote Sensing established a working group on multiscale representation. By 2003 this was adopted as one of the 14-longterm research priorities of UCGIS (Usery and McMaster 2005). Since then, special sessions on multiscale representation have been organized at conferences, and special issues on this topic have been published in various journals.

Characterizing relationships between ecological processes and spatial patterns of ecosystems plays a foundational role in eco-environmental modelling, which manifest across a myriad of spatial scales (Stuber and Gruber 2020). Relation between spatial scale and temporal scale is another significant issue that needs to be further investigated.

### *1.5.3 Real-Time Visualization in Three Dimensions*

Time can be characterized as the fourth dimension of the physical space–time continuum. Real-time means momentary, that is, the same moment as any given event happens (Asproth et al. 1995). From a human point of view, a concrete system can move in any direction on the spatial dimensions, but only forward on the temporal dimension (Miller 1978). In real-time systems, this implies momentary updates. However, it is impossible to achieve momentary updates, as there is always some delay. The acceptable delay length for a real-time system depends on how dynamic the processes are and how time-critical the decisions are (Valsecchi et al. 1999).

The rapid development of computing technology in recent years has made real-time spatial analysis and real-time data visualization realizable.

### 1.5.4 *Quantum Computing*

It is urgent to speed up computation of eco-environmental surface modelling for realizing finer-resolution simulation on global level and real-time visualization. For instances, the general circulation models on global level are currently operated on very coarser resolutions because of slow computational speed of classical computers and the requirement of the simulation for huge memory. The spatial resolutions of the simulated results are too coarse to be applied into evaluating impacts of climatic change on ecosystems on regional and local levels (Washington and Parkinson 2005).

It was demonstrated that a 54-qubit quantum computer can solve in minutes a problem that would take a classical machine 10,000 years (Matthews 2021). The quantum computers now are something like classical computing at the stage in the late 1980s. A 1000-qubit quantum machine would have appeared by 2025. In other words, the slow-computational-speed problem would be solved at the root by further development of quantum computing.

### 1.5.5 *Intelligent Computing*

Intelligent computing, machine learning, and deep learning are three concepts that have been used frequently together (Abigail and Downs 2021) (Fig. 1). Intelligent computing is a discipline generating tools able to perform tasks that normally require human intelligence. Machine learning refers to the process of teaching a computer to perform a particular task by improving its measure of performance with experience. Deep learning is a specific type of machine learning methods and aims to learn the relationships among variables and the knowledge governing the relationships. A challenge faced by machine learning is lack of a large model for spatially explicit learning.

## 1.6 Further Reading

(1) Yue TX. 2011. Surface Modelling: High Accuracy and High Speed Methods. New York: CRC Press.

(2) IPBES. 2016. The methodological assessment report on scenarios and models of biodiversity and ecosystem services. Bonn, Germany: Secretariat of the Intergovernmental Science-Policy Platform on Biodiversity and Ecosystem Services.

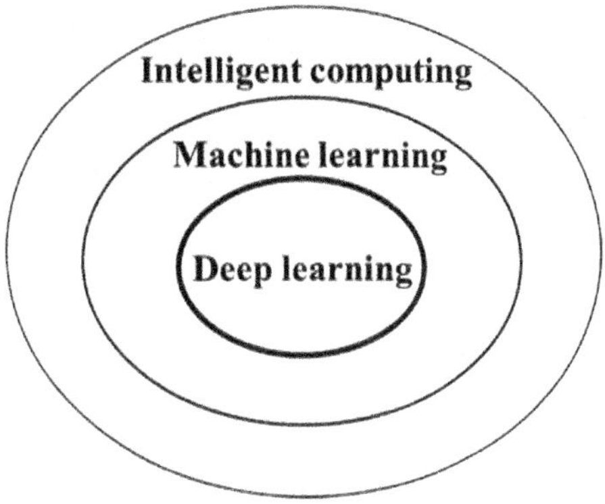


**Fig. 1** The relationship among intelligent computing, machine learning and deep learning (a modification based on Abigail and Downs 2021)

(3) Yue TX, Liu Y, Zhao MW, Du ZP, Zhao N. 2016. A fundamental theorem of Earth's surface modelling. Environmental Earth Sciences 75(9): article 751 (pages 1–12).
(4) Yue TX, Zhao N, Liu Y, Wang YF, Zhang B, Du ZP, Fan ZM, Shi WJ, Chen CF, Zhao MW, Song DJ, Song YJ, Yan CQ, Li QQ, Sun XF, Zhang LL, Tian YZ, Wang W, Wang YA, Ma SN, Huang HS, Lu YM, Wang Q, Wang CL, Wang YZ, Lu M, Zhou W, Liu Y, Yin XZ, Wang Z, Bao ZY, Zhao MM, Zhao YP, Jiao YM, Naseer U, Fan B, Li SB, Yang Y, Wilson J P. 2020. A fundamental theorem for eco-environmental surface modelling and its applications. Science China-Earth Sciences 63(8): 1092–1112.

### Exercises

(1) What is eco-environmental informatics?
(2) What is the difference between the model and the large model?
(3) What are the challenges faced by Eco-Environmental Informatics?
(4) Why currently available scenarios and models can not fit the requirement of decision-supporting for sustainable development?

## References

Accad, A., and D.T. Neil. 2006. Modelling pre-clearing vegetation distribution using GIS-integrated statistical, ecological and data models: A case study from the wet tropics of Northeastern Australia. *Ecological Modelling* 198: 85–100.

Aguilar, F.J., F. Agüera, M.A. Aguilar, and F. Carvajal. 2005. Effects of terrain morphology, sampling density, and interpolation methods on grid DEM accuracy. *Photogrammetric Engineering and Remote Sensing* 71: 805–816.

Ahlberg, J.H., E.N. Nilson, and J.L. Walsh. 1967. *The theory of splines and their application.* New York: Academic Press.

Akima, H. 1978a. A method of bivariate interpolation and smooth surface fitting for irregularly distributed data points. *ACM Transaction on Mathematical Software* 4: 148–159.

Akima, H. 1978b. Algorithm 526: Bivariate interpolation and smooth surface fitting for irregularly distributed data points. *ACM Transaction on Mathematical Software* 4: 160–164.

Alcamo, J. 1994. IMAGE 2.0: Integrated modeling of global change. Dordrecht: Kluwer Academic Publishers.

Alcamo, J., R. Leemans, and E. Kreileman. 1998. Global change scenarios of the 21st century—Results from the image 2.1 model. Oxford: Elsevier Science.

Al-Garni, A.M. 1995. Mathematical predictive models for population estimation in urban areas using space products and GIS technology. *Mathematical and Computer Modelling* 22 (1): 95–107.

Alkemade, R., M. Van Oorschot, L. Miles, et al. 2009. GLOBIO3: A framework to investigate options for reducing global terrestrial biodiversity loss. Ecosystems 12 (3): 374–390.

Alt, E. 1929. Der Stand des meteorologischen Strahlungsproblems. *Meteorologische Zeitschrift* 46: 50–54.

Anway, J.C., E.G. Brittain, H.W. Hunt, G.S. Innis, W.J. Parton, C.F. Rodell, and R.H. Sauer. 1972. ELM: Version 1.0. U.S. IBP Grassland Biome Tech. Rep. No. 156. Fort Collins: Colorado State University.

Arakawa, A. 1969. Parameterization of cumulus clouds. Proceedings of the WMO/IUGG Symposium on Numerical Weather Prediction, Tokyo, 1968. IV-8–1–IV-8–6

Arakawa, A. 2000. A personal perspective on the early years of general circulation modeling at UCLA. In *General circulation model development*, ed. D.A. Randall, 1–65. New York, NY: Academic Press.

# Chapter 2
# The Fundamental Theorem for Eco-Environmental Surface Modelling

**Learning Objectives**

- To appreciate the process of innovatively constructing a new model or method.
- To get knowledges about concepts and algorithms for spatial interpolation, up-scaling, down-scaling, data fusion and model-data assimilation.
- To know the historical development of the Fundamental Theorem for Eco-Environmental Surface Modelling.
- To be able to apply the Fundamental Theorem for Eco-Environmental Surface Modelling into practices of spatially explicit simulation of eco-environmental elements.
- To get knowledges about HASM quantum machine learning.

## 2.1 Introduction

Ground observation can obtain high accuracy data at observation points, but observations at fixed positions are confined within some limited dispersal points and not able to directly calculate relative parameters at regional scale. Satellite remote sensing can frequently supply surface information of geographical processes and ecological processes, but remote sensing description is not able to directly obtain process parameters. Remote-sensing data can generate information about earth surface that is impossible from ground-based studies. The timing and extent of land-cover change highlight unique information that is available only from satellite and airborne sensors (Chambers et al. 2007). However, maps derived from satellites observations are patchy and can not be used reliably as an independent source of information for earth surface monitoring because of the well-known limitations of satellite retrievals, such as missing data for cloud-covered pixels (Emili et al. 2011).

In fact, earth surface systems are controlled by a combination of global factors and local factors, which cannot be understood without accounting for both the local

 
T. Yue et al., *Eco-Environmental Informatics*,
https://doi.org/10.1007/978-981-96-1977-1_2

and global components. The system dynamic cannot be recovered from the global or local controls alone (Phillips 2002). In terms of fundamental theorem of surfaces, a surface is uniquely defined by the first fundamental coefficients and the second fundamental coefficients (Somasundaram 2005). The first fundamental coefficients express the information observed when we stay on the surface, about the details of the surface. The second fundamental coefficients express the change of the surface observed from outside the surface (Yue et al. 2015a, b, c).

Many studies are based on either satellite observations (e.g., Gupta et al. 2014) or ground observations (e.g., Alkhasawneh et al. 2014; Liu et al. 2013). It has been proven that simulation results would have bigger errors when only satellite measurements or ground measurements are considered. For instances, no matter what kind of support vector machine (SVR) models and what parameter optimization method were used, the relative errors of many points exceeded 20% when predicting mining subsidence in the water area using original observation data of the subsidence area without water directly in Eastern China (Li et al. 2014). A comprehensive survey in Guyana indicated that estimates only based on remotely sensed data may be inaccurate (Butt et al. 2015).

The need to combine extrinsic information with intrinsic information is a frequently discussed topic in eco-environmental surface modelling (Haber 2021). For instance, Brill et al. (1991) believed that a global model, to be as accurate as possible, must supplement information from the currently available ground observations. Dobermann and Ping (2004) found that utilizing a satellite image as secondary information decreased errors associated with yield monitor data and also allowed better prediction in areás where no reliable yield measurements were available. Gross primary production (GPP) and net ecosystem exchange (NEE) were simulated by assimilating meteorological data derived from the measurements from existing weather stations, forest volume data derived from a previous investigation, satellite data, flux tower data, and other ancillary data, which rendered the simulation more stable and accurate (Chiesi et al. 2011).

During the last decade, three global land cover datasets have been derived from satellite remotely sensed images, i.e., the GLC-2000, MODIS and GlobCover, but comparison studies have shown that there are large spatial discrepancies between these three products. One of the reasons for these discrepancies is the lack of sufficient in-situ data for the development of these products (Fritz et al. 2012). The integration of local observations and remote sensing products can provide a more complete view of the responses of biodiversity to environmental change and can improve the modeling of ecosystem processes across multiple scales (Pereira et al. 2013).

Errors of water vapor estimation were reduced up to certain extent using both satellite observations and ground observations over the Walnut Creek region in USA (Srivastava et al. 2014). The phytoplankton community structure is potentially influenced by both extrinsic effects originating from the surrounding environment and intrinsic effects based on interactions between two phytoplankton species. Yang et al. (2018) used variance partitioning analysis to show that the extrinsic factors explained 31% of the variation in the phytoplankton community and how looking beyond these extrinsic effects, there were intrinsic feedback mechanisms and interspecific interactions that determined community response to changes in extrinsic factors, thereby shaping phytoplankton community structure.

Ponsar et al. (2016) stated that global models and ground observations provide abundant information, but neither provides the complete picture. In relation to nature conservation, IPBES have highlighted the particular importance of including indigenous and local knowledge (ILK), which has long been marginalized or invisible in global scenarios and models (Pereira et al. 2020). Climate simulation results, when combined with observational data from climate stations, and remotely sensed satellite data, contribute significantly toward our most complete understanding of the inter-related complexities associated with the climate system (Bush et al. 2020).

All of the findings above described the essential significance of both extrinsic and intrinsic information, but the challenge is how to combine these two kinds of information. In other words, extrinsic information, from satellite observations and the outputs of spatial models, and intrinsic information, from ground observations and spatial sampling, provide two different but complementary streams of information about the Earth's surface. These extrinsic and intrinsic information are indispensable for the eco-environmental surface modeling along with an appropriate method for integrating these two kinds of information. Building on this idea, the method for high accuracy surface modelling (HASM) was constructed innovatively and then the fundamental theorem for eco-environmental surface modeling (FTEEM) were proposed on the basis of theoretical development and wide applications of HASM. Several corollaries have been deduced from FTEEM, corresponding to spatial interpolation, spatial upscaling, spatial downscaling, data fusion and model-data assimilation, respectively (Yue et al. 2020). FTEEM provide a solution to this challenge. FTEEM and their corollaries for interpolation, upscaling, downscaling, data fusion and model-data assimilation together form the theoretical basis of eco-environmental informatics (Haber 2021).

## 2.2 The Method for High Accuracy Surface Modelling

HASM is a method based on partial differential equation (PDE). It is the outcome of a multidisciplinary crossover of surface theory, system theory, and optimal control theory.

### *2.2.1 PDF-Based Approaches and Surface Theory*

The approach to representing surfaces as solutions for PDEs was initially developed in the early 1970s (Thompson et al. 1974). It demonstrated that a wide variety of surface shapes can be made accessible by varying the boundary conditions and parameter in the PDE (Bloor and Wilson 1990, 1989; Protopopescu et al. 1989). PDE surfaces have demonstrated many modeling advantages in many fields of design such as surface blending, free-form surface modeling, and surface functional specifications (Pasadas and Rodríguez 2009). The PDE approach has been used in a number of research fields. For instance, sculptured surfaces, ship hulls and propeller blades were modeled using PDE (Bloor and Wilson 1996), and PDE application to image analysis has become a research topic of increasing interest in the past few years. A direct map between two cortical surfaces was iteratively computed by solving a PDE defined on the source cortical surface, starting from a properly designed initial map (Shi et al. 2007). Elliptic PDEs have been used to model various kinds of real physical situations such as the flow of air pollutants, temperature deflection, electrostatic potential, velocity potential and stream function (Modani et al. 2008).

PDE based approaches, such as implicit surface modeling (Walder et al. 2006) and fast surface modeling, are effective for both surface boundary constraints and the generation of families of free-form surfaces. Compared with classical approaches, the PDE approach can model complex objects. According to Zhang and You (2004), an appropriate form of PDE needs to be properly specified and a suitable solution function constructed before the PDE based approach can be effectively applied to complex surface modeling. The higher the order of a PDE, the more boundary conditions it can meet, and therefore the more requirements the modeled surface can satisfy. These different PDEs are treated separately. To handle different classes of geometric PDEs such as biharmonic equations, second order geometric equations and higher order geometric equations, Xu and Zhang (2008) proposed a general framework for surface modeling.

However, solving a PDE subject to arbitrary boundary conditions is a classic complex mathematical problem. In many cases, numerical measures have to be taken resulting in low and sometimes unstable processes because many forms of PDEs are analytically unsolvable. From the literatures, the fundamental theorem of surfaces was not be considered for the surface modelling and little achieved to obtain accurate PDE solutions before HASM was developed.

In terms of the fundamental theorem of surfaces, a surface is uniquely defined by the first fundamental coefficients and the second fundamental coefficients (Henderson 1998). The first fundamental coefficients are used to express how the surface inherits the natural inner product of $R^3$, in which $R^3$ is the set of triples $(x, y, z)$ of real numbers (Carmo 2006). The first fundamental coefficients of a surface yield information about some geometric properties of the surface, by which we can calculate lengths of curves, angles of tangent vectors, areas of regions and geodesics on the surface. Those geometric properties and objects that can be determined only in terms of the first fundamental coefficients of a surface are called the intrinsic geometric properties. The collection of these geometric properties and objects forms the subject of intrinsic geometry of a surface; its properties do not depend on the shape of the surface, but only on measurements that we can carry out while staying on a surface itself (Toponogov 2006). The second fundamental coefficients reflect the local warping of the surface, namely its deviation from tangent plane at the point under consideration, which can be observed from outside the surface (Liseikin 2004; Yue et al. 2015a).

### 2.2.2 Optimal Control Theory

The mathematical theory of optimal control began in 1940s as a special topic within the discipline of differential equations (Lee and Markus 1986). Optimal control theory is a branch of mathematics developed to find optimal ways to control a dynamic system. The objective of optimal control theory is to determine the control signals that will cause a process to satisfy the physical constraints and at the same time minimize (or maximize) some performance criterion (Kirk 2004). In other words, optimal control theory deals with the problem of optimizing dynamic systems. The problem must be well posed before any solution can be attempted. This requires a clear mathematical description of the system to be optimized, the constraints imposed on the system, and the objective function to be minimized.

A significant part of any control problem is the process of modeling the dynamic system under consideration to obtain a mathematical description that adequately predicts the response to all anticipated inputs. After the mathematical model is selected, the next step is to define the physical constraints on the state. A performance measure of a system can be selected in order to evaluate the performance quantitatively. In selecting the performance measure, a mathematical expression is defined, which indicates that the system is performing in the most desirable manner when the mathematical expression is minimized. In other words, choosing a performance measure is a translation of the system's physical requirement into mathematical terms. Once the performance measure for a system has been chosen, the further task is to determine a control function for minimizing the criterion. An optimal control is defined as one that minimizes (or maximizes) the performance measure.

The maximum principle as a necessary condition must be satisfied by any optimal control for the basic problem having constraints only on control variables. In many applied models, it is necessary to impose constraints involving both control and state variables, which are called mixed inequality constraints. Some optimal control problems have constraints only on state variables, which are known as pure state constraints. Pure state constraints are more difficult to deal with than the mixed constraints because only control variables are under the direct influence of the decision maker. This enables the decision maker, when a mixed constraint becomes tight, to choose from the controls that would keep it tight for as long as required for optimality. However, the situation is different and more complicated for pure state constraints. When a constraint becomes tight, it does not provide any direct information to the decision maker on how to choose values for the control variable so as not to violate the constraint (Sethi 2019).

To control a dynamical system with optimality is still a challenging task in practices because more and more control elements have been involved due to the rapid progress in related technologies (Shen 2017).

### 2.2.3 System Theory

A system refers to a set of interacting elements (Seidl and Schoeneborn 2016). Systems on each lower level jointly form systems on the next higher level and systems on each level produce their own differentiated subsystems (Laszlo 1975). System theory was conceived as a science of 'wholeness' to overcome mechanistic and reductionist approaches. That the whole is more than the sum of parts is simply that constitutive characteristics are not explainable from the characteristics of isolated parts (Marin 1997).

The relationship between systemic openness and closure has been a central concern in the work of diverse system theorists (Assche et al. 2019). Closed systems are considered to be isolated from their environment. Thermodynamics expressly declares that its laws apply only to closed systems. In particular, the second principle of thermodynamics states that entropy must increase to a maximum and eventually the process comes to a stop at a state of equilibrium in a closed system. However, every living organism is essentially an open system. It maintains itself in a continuous inflow and outflow so long as it is alive. The living systems, maintaining themselves in a stead state, can avoid the increase of entropy.

System theory is a general theory of systems, in which the world is regarded as systems in systems and systems besides systems. In 1920s, a few of scientists became aware of the potentials of developing a general system theory of complex phenomena (Kohler 1920; Koehler 1927; Lotka 1925). System theory has made a great progress since the 1930s when the scientists faced a barrage of criticism for advocating theoretical biology. In 1949, a concept of general system theory was introduced (von Bertalanffy 1949).

System theory provides a framework for creating synergies between different disciplines. In fact, similar concepts, models and laws have often appeared in widely different fields, independently and based upon totally different facts. A general theory of systems would be a useful tool providing, on the one hand, models that can be used in and transferred to, different fields, and safeguarding, on the other hand from vague analogies which often have marred the progress in these fields (von Bertalanffy 1969). Factors favoring the development of system theory operate both internally and externally to science. There is an intrinsic trend within science itself to maximize the scope of theories consistently with their precision. There are also extrinsic pressures on science to overcome traditional boundaries in producing multidisciplinary theories applicable to societal problems. Incorporating various scientific dimensions may provide a good starting point for adherents of system theories to bring together approaches from different fields. However, a current problem within the field of system theories is that adherents subscribe to different notions of the philosophy behind system approaches (Drack and Schwarz 2010).

Development of system theory can be divided into three generations historically (Seidl and Schoeneborn 2016). The first generation can be described as the theory of closed systems; it focused on how the parts of a system relate to each other and to the system as a whole. The second generation, developed in the 1950s, is described as the theory of open systems; it focused on the relation between the system and its environment. The third generation, emerged in the 1970s, is described as the theory of autopoietic systems; it mainly concerned with how the system reproduces itself through its own operation, which placed emphasis on the internal interactions between the system's components and treating environmental influences as perturbations in the internal processes. It should be noticed that the open systems approach regards external influences as crucial and internal influences merely as disturbance, while and the autopoietic systems approach regards internal influences as crucial and external influences as disturbance.

### 2.2.4 HASM Formulation

In terms of the fundamental theorem of surfaces, a surface in $R^3$ is uniquely (up to the isometric transformation of $R^3$) defined by the first fundamental form and the second fundamental form (Henderson 1998). The first fundamental form is used to express how the surface inherits the natural inner product of $R^3$ (Carmo 2006), and gives information about some geometric properties of the surface, by which we can calculate lengths of curves, angles of tangent vectors, areas of regions and geodesics on the surface. Those geometric properties and objects that can be determined only in terms of the first fundamental form are called the intrinsic geometric properties. The collection of these geometric properties and objects forms the subject of intrinsic geometry of surfaces; these properties do not depend on the shape of the surface, but only on the measurements that we can carry out while staying on the surface (Toponogov 2006). The second fundamental form reflects the local warping of the surface, namely the deviation from tangent plane at the point under consideration (Liseikin 2004), which should be observed from the outside of the surface.

If a surface is the graph of a function $z = f(x, y)$ or $\mathbf{r} = (x, y, f(x, y))$, then the basis of tangent space can be represented by $\mathbf{r}_x = (1, 0, f_x)$ and $\mathbf{r}_y = (0, 1, f_y)$. The coefficients of the first fundamental form, $E$, $F$ and $G$, which are the intrinsic properties of the surface (Fig. 1), can be formulated as,

$$\begin{cases} E = \mathbf{r}_x \cdot \mathbf{r}_x = 1 + f_x^2 \\ G = \mathbf{r}_y \cdot \mathbf{r}_y = 1 + f_y^2 \\ F = \mathbf{r}_x \cdot \mathbf{r}_y = f_x \cdot f_y \end{cases} \tag{1}$$

If the normal vector of the surface at the point $(x, y)$ is expressed as $\mathbf{n}(x, y) = \left(1 + f_x^2 + f_y^2\right)^{-\frac{1}{2}} (-f_x, -f_y, 1)$, the coefficients of the second fundamental form $L$, $M$ and $N$, , which are the extrinsic properties of the surface (Fig. 2), can be formulated as,

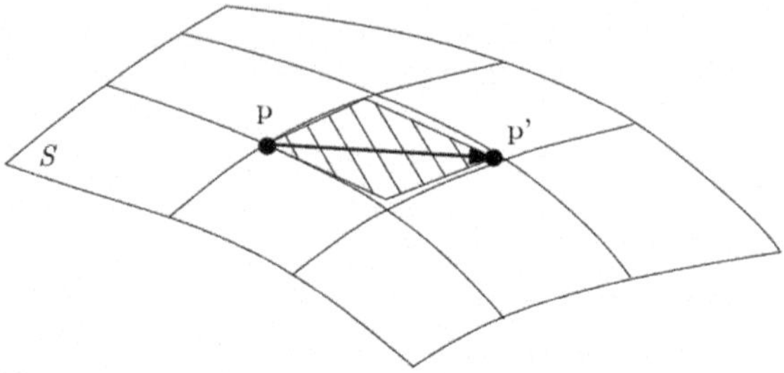


**Fig. 1** A diagram of the intrinsic geometric properties

$$\begin{cases} L = \mathbf{n}_x \cdot (-\mathbf{r}_x) = \frac{f_{xx}}{\sqrt{1+f_x^2+f_y^2}} = \frac{f_{xx}}{\sqrt{E+G-1}} \\ N = \mathbf{n}_y \cdot \left(-\mathbf{r}_y\right) = \frac{f_{yy}}{\sqrt{1+f_x^2+f_y^2}} = \frac{f_{yy}}{\sqrt{E+G-1}} \\ M = \mathbf{n}_x \cdot \left(-\mathbf{r}_y\right) = \frac{f_{xy}}{\sqrt{1+f_x^2+f_y^2}} = \frac{f_{xy}}{\sqrt{E+G-1}} \end{cases} \quad (2)$$

The structure equations, namely the differential equations of the vector basis, $\left(\mathbf{r}_x, \mathbf{r}_y, \mathbf{n}\right)$, are reduced to the following equations, which is collected as the master equation set of HASM,

$$\begin{cases} f_{xx} = \Gamma_{11}^1 \cdot f_x + \Gamma_{11}^2 \cdot f_y + L \cdot (E + G - 1)^{-\frac{1}{2}} \\ f_{yy} = \Gamma_{22}^1 \cdot f_x + \Gamma_{22}^2 \cdot f_y + N \cdot (E + G - 1)^{-\frac{1}{2}} \\ f_{xy} = \Gamma_{12}^1 \cdot f_x + \Gamma_{12}^2 \cdot f_y + M \cdot (E + G - 1)^{-\frac{1}{2}} \end{cases} \quad (3)$$

where the coefficients, $\Gamma_{i,j}^k$ $(i, j, k = 1 \text{ } or \text{ } 2)$, are the Christoffel symbols, which can be defined by the following equations,

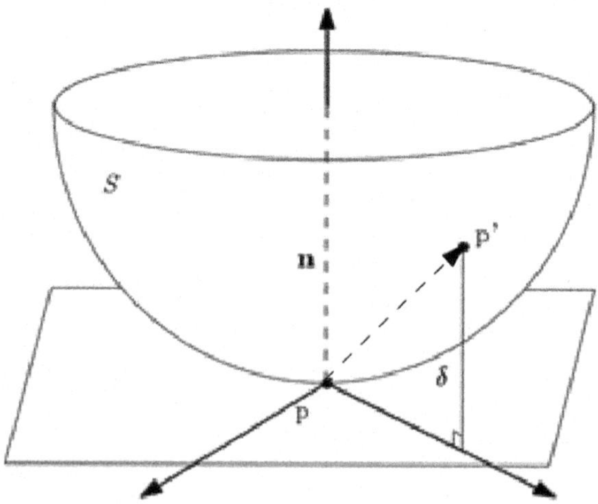


**Fig. 2** A diagram of the extrinsic geometric properties

$$\begin{cases} \Gamma_{11}^{1} = \frac{L}{\sqrt{E+G-1}} \cdot f_x = \frac{1}{2}\left(G \cdot E_x - 2F \cdot F_x + F \cdot E_y\right) \cdot \left(E \cdot G - F^2\right)^{-1} \\ \Gamma_{11}^{2} = \frac{L}{\sqrt{E+G-1}} \cdot f_y = \frac{1}{2}\left(2E \cdot F_x - E \cdot E_y - F \cdot E_x\right) \cdot \left(E \cdot G - F^2\right)^{-1} \end{cases} \tag{4}$$

$$\begin{cases} \Gamma_{12}^{1} = \frac{M}{\sqrt{E+G-1}} \cdot f_x = \frac{1}{2}\left(G \cdot E_y - F \cdot G_x\right) \cdot \left(E \cdot G - F^2\right)^{-1} \\ \Gamma_{12}^{2} = \frac{M}{\sqrt{E+G-1}} \cdot f_y = \frac{1}{2}\left(E \cdot G_x - F \cdot E_y\right) \cdot \left(E \cdot G - F^2\right)^{-1} \end{cases} \tag{5}$$

$$\begin{cases} \Gamma_{22}^{1} = \frac{N}{\sqrt{E+G-1}} \cdot f_x = \frac{1}{2}\left(2G \cdot F_y - G \cdot G_x - F \cdot G_y\right) \cdot \left(E \cdot G - F^2\right)^{-1} \\ \Gamma_{22}^{2} = \frac{N}{\sqrt{E+G-1}} \cdot f_y = \frac{1}{2}\left(E \cdot G_y - 2F \cdot F_y + F \cdot G_x\right) \cdot \left(E \cdot G - F^2\right)^{-1} \end{cases} \tag{6}$$

The finite difference of the HASM master equation set can be formulated as,

$$\begin{cases} f_{i+1,j} - 2f_{i,j} + f_{i-1,j} = (\Gamma_{11}^{1})_{i,j} \cdot \frac{f_{i+1,j} - f_{i-1,j}}{2} \cdot h + \left(\Gamma_{11}^{2}\right)_{i,j} \cdot \frac{f_{i,j+1} - f_{i,j-1}}{2} \cdot h \\ \qquad + \frac{L_{i,j}}{\sqrt{E_{i,j} + G_{i,j} - 1}} \cdot h^2 \\ f_{i,j+1} - 2f_{i,j} + f_{i,j-1} = (\Gamma_{22}^{1})_{i,j} \cdot \frac{f_{i+1,j} - f_{i-1,j}}{2} \cdot h + \left(\Gamma_{22}^{2}\right)_{i,j} \cdot \frac{f_{i,j+1} - f_{i,j-1}}{2} \cdot h \\ \qquad + \frac{N_{i,j}}{\sqrt{E_{i,j} + G_{i,j} - 1}} \cdot h^2 \\ f_{i+1,j+1} - f_{i-1,j+1} + f_{i-1,j-1} - f_{i+1,j-1} = 2 \cdot h \cdot (\Gamma_{12}^{1})_{i,j} \cdot \left(f_{i+1,j} - f_{i-1,j}\right) \\ \qquad + 2 \cdot h \cdot \left(\Gamma_{12}^{2}\right)_{i,j} \cdot \left(f_{i,j+1} - f_{i,j-1}\right) + \frac{4 \cdot M_{i,j} \cdot h^2}{\sqrt{E_{i,j} + G_{i,j} - 1}} \end{cases} \tag{7}$$

In terms of numerical mathematics (Quarteroni et al.2000), the iterative formulation of equation set (4) could be expressed as

$$\begin{cases} f_{\mathrm{i+1,j}}^{(n+1)} - 2f_{\mathrm{i,j}}^{(n+1)} + f_{\mathrm{i-1,j}}^{(n+1)} = (\Gamma_{11}^{1})_{\mathrm{i,j}}^{(n)} \cdot \dfrac{f_{\mathrm{i+1,j}}^{(n)} - f_{\mathrm{i-1,j}}^{(n)}}{2} \cdot h \\ +(\Gamma_{11}^{2})_{\mathrm{i,j}}^{(n)} \cdot \dfrac{f_{\mathrm{i,j+1}}^{(n)} - f_{\mathrm{i,j-1}}^{(n)}}{2} \cdot h \\ \qquad + \dfrac{L_{\mathrm{i,j}}^{(n)}}{\sqrt{E_{\mathrm{i,j}}^{(n)} + G_{\mathrm{i,j}}^{(n)} - 1}} \cdot h^2 \\ f_{\mathrm{i,j+1}}^{(n+1)} - 2f_{\mathrm{i,j}}^{(n+1)} + f_{\mathrm{i,j-1}}^{(n+1)} = (\Gamma_{22}^{1})_{\mathrm{i,j}}^{(n)} \cdot \dfrac{f_{\mathrm{i+1,j}}^{(n)} - f_{\mathrm{i-1,j}}^{(n)}}{2} \cdot h \\ +(\Gamma_{22}^{2})_{\mathrm{i,j}}^{(n)} \cdot \dfrac{f_{\mathrm{i,j+1}}^{(n)} - f_{\mathrm{i,j-1}}^{(n)}}{2} \cdot h \\ \qquad + \dfrac{N_{\mathrm{i,j}}^{(n)}}{\sqrt{E_{\mathrm{i,j}}^{(n)} + G_{\mathrm{i,j}}^{(n)} - 1}} \cdot h^2 \\ f_{\mathrm{i+1,j+1}}^{(n+1)} - f_{\mathrm{i-1,j+1}}^{(n+1)} + f_{\mathrm{i-1,j-1}}^{(n+1)} - f_{\mathrm{i+1,j-1}}^{(n+1)} = 2 \cdot h \cdot (\Gamma_{12}^{1})_{\mathrm{i,j}}^{(n)} \cdot \left(f_{\mathrm{i+1,j}}^{(n)} - f_{\mathrm{i-1,j}}^{(n)}\right) \\ \qquad + 2 \cdot h \cdot (\Gamma_{12}^{2})_{\mathrm{i,j}}^{(n)} \cdot \left(f_{\mathrm{i,j+1}}^{(n)} - f_{\mathrm{i,j-1}}^{(n)}\right) + \dfrac{4 \cdot h^2 \cdot M_{\mathrm{i,j}}^{(n)}}{\sqrt{E_{\mathrm{i,j}}^{(n)} + G_{\mathrm{i,j}}^{(n)} - 1}} \end{cases} \tag{8}$$

Suppose that $\mathbf{z}^{(n+1)} = (f_{1,1}^{(n+1)}, \cdots, f_{1,J}^{(n+1)}, \cdots, f_{I-1,1}^{(n+1)}, \cdots, f_{I-1,J}^{(n+1)}, f_{I,1}^{(n+1)}, \cdots, f_{I,J}^{(n+1)})^T$, $n \geq 0$; $\mathbf{z}^{(0)} = (\tilde{f}_{1,1}, \cdots, \tilde{f}_{1,J}, \cdots, \tilde{f}_{I-1,1}, \cdots, \tilde{f}_{I-1,J}, \tilde{f}_{I,1}, \cdots, \tilde{f}_{I,J})^T$ indicates an initial field. Then, HASM matrix formulation with equality constraints can be expressed as a constraint least squares approximation,

$$\begin{cases} \min \left\| \begin{pmatrix} \mathbf{A} \\ \mathbf{B} \\ \mathbf{C} \end{pmatrix} \cdot \mathbf{z}^{(n+1)} - \begin{pmatrix} \mathbf{d}^{(n)} \\ \mathbf{q}^{(n)} \\ \mathbf{p}^{(n)} \end{pmatrix} \right\|_2 \\ s.t. \\ \mathbf{S}_1 \cdot \mathbf{z}^{(n+1)} = \mathbf{k}_1 \end{cases} \tag{9}$$

where $\mathbf{S}_1$ and $\mathbf{k}_1$ are respectively ground observation positions and the values of $z = f(x, y)$ at the corresponding positions; the "min" means "minimize"; the "s.t." is the abbreviation of "subject to".

For sufficient large $\lambda$, HASM with equality constraints can be formulated as an unconstrained least squares approximation

$$\mathbf{z}^{(n+1)} = \begin{pmatrix} \mathbf{A}^T \cdot \mathbf{A} + \mathbf{B}^T \cdot \mathbf{B} \\ +\mathbf{C}^T \cdot \mathbf{C} + \lambda^2 \cdot \mathbf{S}_1^T \cdot \mathbf{S}_1 \end{pmatrix}^{-1} \cdot \begin{pmatrix} \mathbf{A}^T \cdot \mathbf{d}^{(n)} + \mathbf{B}^T \cdot \mathbf{q}^{(n)} \\ +\mathbf{C}^T \cdot \mathbf{p}^{(n)} + \lambda^2 \cdot \mathbf{S}_1^T \cdot \mathbf{k}_1 \end{pmatrix} \tag{10}$$

HASM matrix formulation with mixed constraints of equality and inequality can be formulated as,

$$\begin{cases} \min \left\| \begin{pmatrix} \mathbf{A} \\ \mathbf{B} \\ \mathbf{C} \end{pmatrix} \cdot \mathbf{z}^{(n+1)} - \begin{pmatrix} \mathbf{d}^{(n)} \\ \mathbf{q}^{(n)} \\ \mathbf{p}^{(n)} \end{pmatrix} \right\|_2 \\ s.t. \\ \mathbf{S}_1 \cdot \mathbf{z}^{(n+1)} = \mathbf{k}_1 \\ \mathbf{k}_2 \le \mathbf{z}^{(n+1)} \le \mathbf{k}_3 \end{cases} \tag{11}$$

where $\mathbf{k}_2$ and $\mathbf{k}_3$ are respectively lower and upper bounds of inequality constraints.

The formulation (11) can first be solved by the same way as formulation (9) or (10). If the solution satisfies $\mathbf{k}_2 \le \mathbf{z}^{(n+1)} \le \mathbf{k}_3$, , then the solution of (11) equals to the one of (10). If the solution of (10) does not satisfy $\mathbf{k}_2 \le \mathbf{z}^{(n+1)} \le \mathbf{k}_3$, then the solution of (11) is $\mathbf{z}^{(n+1)} = \mathbf{k}_2$ or $\mathbf{z}^{(n+1)} = \mathbf{k}_3$.

Owing to the solution of (8) requiring huge memory and computational cost, we simplify it as follows,

$$\begin{cases} f_{i+1,j}^{(n+1)} - 2f_{i,j}^{(n+1)} + f_{i-1,j}^{(n+1)} = (\Gamma_{11}^1)_{i,j}^{(n)} \cdot \dfrac{f_{i+1,j}^{(n)} - f_{i-1,j}^{(n)}}{2} \cdot h \\ +(\Gamma_{11}^2)_{i,j}^{(n)} \cdot \dfrac{f_{i,j+1}^{(n)} - f_{i,j-1}^{(n)}}{2} \cdot h \\ \qquad + \dfrac{L_{i,j}^{(n)}}{\sqrt{E_{i,j}^{(n)} + G_{i,j}^{(n)} - 1}} \cdot h^2 \\ f_{i,j+1}^{(n+1)} - 2f_{i,j}^{(n+1)} + f_{i,j-1}^{(n+1)} = (\Gamma_{22}^1)_{i,j}^{(n)} \cdot \dfrac{f_{i+1,j}^{(n)} - f_{i-1,j}^{(n)}}{2} \cdot h \\ +(\Gamma_{22}^2)_{i,j}^{(n)} \cdot \dfrac{f_{i,j+1}^{(n)} - f_{i,j-1}^{(n)}}{2} \cdot h \\ \qquad + \dfrac{N_{i,j}^{(n)}}{\sqrt{E_{i,j}^{(n)} + G_{i,j}^{(n)} - 1}} \cdot h^2 \end{cases} \tag{12}$$

The corresponding matrix formulation of HASM with equality constraints can be simplified as,

$$\mathbf{z}^{(n+1)} = \left(\mathbf{A}^T \cdot \mathbf{A} + \mathbf{B}^T \cdot \mathbf{B} + \lambda^2 \cdot \mathbf{S}_1^T \cdot \mathbf{S}_1\right)^{-1}\left(\mathbf{A}^T \cdot \mathbf{d}^{(n)} + \mathbf{B}^T \cdot \mathbf{q}^{(n)} + \lambda^2 \cdot \mathbf{S}_1^T \cdot \mathbf{k}_1\right) \tag{13}$$

The one with mixed constraints of equality and inequality can be simplified as,

$$
\begin{cases}
\min \left\| \begin{pmatrix} \mathbf{A} \\ \mathbf{B} \end{pmatrix} \cdot \mathbf{z}^{(n+1)} - \begin{pmatrix} \mathbf{d}^{(n)} \\ \mathbf{q}^{(n)} \end{pmatrix} \right\|_2 \\
s.t. \\
\mathbf{S}_1 \cdot \mathbf{z}^{(n+1)} = \mathbf{k}_1 \\
\mathbf{k}_2 \leq \mathbf{z}^{(n+1)} \leq \mathbf{k}_3
\end{cases}
\quad (14)
$$

## 2.3 HASM Quantum Machine Learning

### *2.3.1 Quantum Computing*

Both classical personal computers and supercomputers are based on binary bits in a state of either 0 or 1. But a qubit could be a 0, a 1, or a superposition of both states. The quantum phenomena used for computation is known as quantum computing. The qubits serve as the basic elements of quantum computing and the goal is to drive the qubits from one state to another in a controlled manner and to combine qubits in ways that lead to entanglement. The entanglement occurs when two or more quantum systemscannot be described independently, even when they are separated by distance (Mooij 2005).

Quantum mechanics provides the basis for understanding quantum computation, which is a mathematical framework for the construction of physical theories. It was developed in the first two decades of the twentieth century and assumed its current form in the late 1920s (Nielsen and Chuang 2010). In 1928, Hilbert et al. (1928) introduced a simple quantum mechanical model, and Turing (1937) proposed a computing machine a decade later. The Turing machine consists of a machine and an infinite tape that was divided into squares. Each square of the tape may be blank or it may contain one of a finite number of symbols. The finite symbol string is called the tape expression. As the machine scans the tape squares one by one, it can change the tape symbol, print a symbol on an empty square, shift a square left or right, or do nothing (Davis 1958). In 1980, a quantum mechanical model was constructed for standard Turing machines (Benioff 1980). These findings make it possible to put quantum computing into practice. Quantum computing is based on quantum phenomena such as superposition and entanglement beyond classical computing (Nimbe et al. 2021).

In 1982, Feynman found that the ability of quantum computing would be far stronger than classical computing to process information, which might generate new high-performance computers. Since the first quantum algorithm was presented using quantum parallelism (Deutsch 1985), many quantum algorithms have been proposed. For instance, an algorithm for factoring large integers demonstrated the computing power of a quantum computer (Shor 1994). Quantum random walks were developed as the quantum mechanical analogues of the well-known classical random walks, for which roles have been established in quantum information processing (Farhi and Gutmann 1998; Kempe 2003). A quantum algorithm was developed for solving linear systems of equations, which provided an exponential speedup compared with classical alternatives (Harrow et al. 2009). A quantum algorithm was proposed for efficiently determining the quality of a least-squares fit over an exponentially large data set (Wiebe et al. 2012). A quantum support vector machine algorithm was proposed for big data classification and provided an exponential speedup compared to classical algorithms (Rebentrost et al. 2014). Quantum algorithms were employed for the Hopfield network, which showed that an exponentially large network can be stored in a polynomial number of quantum bits (Rebentrost et al. 2018). A quantum approximate optimization algorithm was developed for solving combinatorial optimizations with prospects of quantum speedup on nearterm devices (Medvidović and Carleo 2021).

Moore's law states that the number of transistors on a microprocessor chip will double every two years or so. It was worldwide acknowledged that Moore's law was nearing its end (Waldrop 2016). The "more than Moore strategy" has further motivated the exploration of quantum algorithms, and quantum machine learning algorithms in particular. For instance, an artificial neural network was trained by using qubits as artificial neurons to develop a quantum algorithm, which was carried out on a classical computer (Chalumuri et al. 2020). A novel quantum algorithm based on the entanglement principle was proposed to solve these new types of problems by integrating entanglement measurement and concurrence with quantum parallelism and interference (Zidan et al. 2021). A classical-quantum hybrid algorithm was presented to compute nuclear structure functions in the highenergy Regge limit of Quantum Chromodynamics (QCD) (Mueller et al. 2021). Several quantum algorithms and machine learning methods have been constructed and used to forecast Gross Domestic Product (GDP) growth and to determine the model with higher accuracy. The results demonstrated that intelligent computing could improve the accuracy of quantum algorithms (Alaminos et al. 2022).

In the present era of quantum computing, one of the major goals is the demonstration of quantum computational advantage (QCA) (Deshpande et al. 2022). Since Feynman (1982) proposed that it is exponentially costly to simulate large quantum systems with classical computers, there have appeared many demonstrations of QCA (Deshpande et al. 2022). For instance, an experiment simulating quantum mechanics on a classical computer, by imagining the machine in some superposition of states at every step of the computation, illuminated that a quantum algorithm for factoring of integers is exponentially faster than the classical algorithm (Shor 1994). The Grover search algorithm (Grover 1997) exhibited a provable quadratic speedup over the best possible classical algorithm (Bennett et al. 1997). How to define and measure quantum speedup was illustrated with data from tests run on a D-Wave Two device (Rønnow et al. 2014). Biamonte et al. (2017) showed that small quantum information processors can produce statistical patterns in data mining that are computationally difficult for a classical computer, if efficient quantum algorithms can be found for machine learning. The roots of QCA are genuine quantum superposition, the entanglement of observables, and constructive interference (Khrennikov 2021).

Gaussian boson sampling (GBS) is a feasible approach to demonstrate QCA. For instance, an experiment using a programmable superconducting processor achieved QCA (or quantum supremacy) in the performance of random quantum circuit sampling (Arute et al. 2019). The experiment demonstrated that a 53-qubit processor generated a million noisy samples in 200 s with about 0.2% fidelity, for which a supercomputer would take 10,000 years. Gaussian boson sampling (GBS) in 200 s, with a Jiuzhang photonic quantum computer, would take 2.5 billion years on the Sunway TaihuLight supercomputer (Zhong et al. 2020). Evidence was provided for the hardness of classically simulating GBS and a new programmable architecture was proposed for demonstrating QCA by means of GBS, which bolstered the evidence for QCA in the experiment of Zhong et al. (2020), Deshpande et al. (2022). A superconducting quantum computing system, Zuchongzhi 2.1, achieved larger-scale random quantum circuit sampling with 60 qubits and 24 cycles, yielding a Hilbert space dimension of up to $2^{60}$ (Zhu et al. 2022). In other words, the development of quantum computing clearly demonstrates its advantage over classical computers for artificial problems.

The practical quantum advantage (PQA) has been discussed in recent years as well (Daley et al. 2022), which means quantum devices can be used to solve problems of practical interest that are not tractable for traditional supercomputers. To demonstrate a PQA, it is required to show a quantum simulator can generate a reliable solution for a relevant problem beyond the bounds of classical simulation with controlled errors, and to implement techniques to quantify the accuracy of a quantum simulation when it is operating beyond the capabilities of current classical simulations.

### 2.3.2 HASM Quantum Machine Learning

Intelligent computing has progressed dramatically over the past three decades and is one of today's most rapidly growing technical fields. Intelligent computing, machine learning, and deep learning are three concepts that have been used frequently together (Abigail and Downs 2021). Intelligent computing is the theory by which computers can perform tasks that normally require human intelligence. Machine learning is a sub-field within Intelligent computing and refers to the process of teaching a computer to perform a particular task by improving its performance with experience. Deep learning is a specific type of machine learning and aims to learn the relationships among variables and the knowledge governing the relationships. Machine learning methods can be generally classified into supervised, unsupervised, and reinforcement learning as well as mixtures of them. HASM is a method for reinforcement machine learning (Yue et al. 2022, 2023). It is interpretable. The Eq. (14) can be divided into two parts. The first part is an optimization procedure to achieve a target surface by an iterative process, which is a learning algorithm of the reinforcement machine learning and formulated as,

$$\min \left\| \begin{bmatrix} \mathbf{A} \\ \mathbf{B} \\ \mathbf{C} \end{bmatrix} \cdot \mathbf{z}^{(n+1)} - \begin{bmatrix} \mathbf{d}^{(n)} \\ \mathbf{q}^{(n)} \\ \mathbf{p}^{(n)} \end{bmatrix} \right\| \tag{15}$$

The second part is a rule for controlling the learning process, i.e. the iterative process, through ground observations and prior knowledge, which is expressed as,

$$\begin{cases} \mathbf{S}_1 \cdot \mathbf{z}^{(n+1)} = \mathbf{k}_1 \\ \mathbf{k}_2 \leq \mathbf{S}_2 \cdot \mathbf{z}^{(n+1)} \leq \mathbf{k}_3 \end{cases} \tag{16}$$

The vectors in the Eq. (14) can be replaced by quantum states. We term it a HASM method for quantum machine learning. The Eq. (14) can be rewritten as,

$$\begin{cases} \min \left\| \begin{bmatrix} \mathbf{A} \\ \mathbf{B} \\ \mathbf{C} \end{bmatrix} \cdot \left|z^{(n+1)}\right\rangle - \begin{bmatrix} \left|d^{(n)}\right\rangle \\ \left|q^{(n)}\right\rangle \\ \left|p^{(n)}\right\rangle \end{bmatrix} \right\| \\ s.t. \\ \mathbf{S}_1 \cdot \left|z^{(n+1)}\right\rangle = |k_1\rangle \\ |k_2\rangle \leq \mathbf{S}_2 \cdot \left|z^{(n+1)}\right\rangle \leq |k_3\rangle \end{cases} \tag{17}$$

HASM can be transformed into a large linear system by means of the Lagrangian factor method,

$$\mathbf{W}\left|z^{(n+1)}\right\rangle = \left|r^{(n)}\right\rangle \tag{18}$$

where $\mathbf{W} = \left(\mathbf{A}^T\mathbf{A} + \mathbf{B}^T\mathbf{B} + \beta\mathbf{S}_1^T\mathbf{S}_1\right)$ and $\left|r^{(n)}\right\rangle = \mathbf{A}^T\left|d^{(n)}\right\rangle + \mathbf{B}^T\left|q^{(n)}\right\rangle + \beta\mathbf{S}_1^T|k_1\rangle$.

In eco-environmental surface modelling, the sparse matrix $\mathbf{W}$ is usually a huge matrix which is the primary cause of the long computational time when a classical computer is used to run HASM. For a classical computer, among all classical algorithms, the best-performed one takes time $\mathcal{O}(Ns\kappa/\log(\epsilon))$ to solve $\mathbf{A}\cdot\mathbf{x} = \mathbf{b}$ with error $\epsilon$ for an $s$-sparse $N \times N$ matrix $\mathbf{A}$ with $\kappa$ condition numbers. But using the Harrow-Hassidim-Lloyd (HHL) quantum algorithm to solve the same matrix equation, the operation speed can be improved exponentially, which means the runtime is reduced to $\mathcal{O}(\log(N)s^2\kappa^2/\epsilon)$ (Harrow et al. 2009). Comparing the runtime expressions of the two algorithms, it is obvious that the benefit of the HHL quantum algorithm in shortening the runtime tends to be more significant when the matrix size $N$ becomes larger. Accordingly, we combined HASM machine learning with the HHL quantum algorithm to establish the HASM-HHL algorithm for quantum machine learning. Ideally, HASM-HHL can maintain the high accuracy of classical algorithms and simultaneously achieve exponential speedup compared to the classical algorithms (Yue et al. 2022).

When the HHL quantum algorithm is combined with HASM, it should be noted that the HHL quantum algorithm has certain limitations due to the properties accompanying the quantum phase estimation (QPE). Using QPE to get the eigenvalue $\lambda$

of the matrix **A** is one of the key steps of the HHL quantum algorithm, and in this process, it is necessary to find a suitable parameter $t$ such that $\lambda t \in [0,2\pi]$ (Morrell and Wong 2021). The selection of the parameter $t$ is directly related to the size of the final calculation error, and a large parameter $t$ will always lead to an obvious increase in calculation time. But in the process of quantum computing, it is more like a black box where we often do not know how this parameter $t$ is selected, so it is impossible to control the resulting errors and calculation time (Shao 2018). In addition, how to reduce the depth of quantum circuits used in operations is also a key issue that must be faced when using HHL quantum algorithms to solve large-scale matrix equations on real machines. The hybrid HHL algorithm (h-HHL) formed by combining the HHL quantum algorithm with the classical matrix operation has proved to be a relatively effective method, which can solve the $32 \times 32$ matrix equation on the Rigetti computer (Bravo-Prieto et al. 2019). This algorithm has also been run on IBM quantum computers and is able to solve 217-dimensional linear equations (Perelshtein et al. 2022). The limitations of the HHL quantum algorithm discussed above will become the problems that need to be considered and solved in the process of optimizing HASM quantum machine learning. At the same time, the h-HHL quantum algorithm also provides good ideas for the subsequent optimization of HASM quantum machine learning.

HASM-HHL would have practical quantum advantages for the future when large-scale quantum computers exist with enough qubits for quantum error correction. However, the capabilities of quantum computers in the near future will continue to be constrained by limitations such as the number of qubits and the effects of noise (Bennewitz et al. 2022). The effect of noise on quantum systems was thought to pose severe obstacles towards the experimental realization of the exponential speed-ups promised by quantum computers over classical computers in the early stage of quantum computation. In other words, whilst it is clear that quantum computers are much faster than the fastest conventional supercomputers at certain tasks, a fault-tolerant architecture is required to identify and correct errors in noisy quantum hardware.

To find a solution to error correction, Bacon (2006) presented two subsystems of operator quantum error correction (OQEC) to remove the need for a quantum microarchitecture performing quantum error correction and to offer a self-correcting quantum memory. The OQEC is a unified approach to error correction for noise represented by a completely positive trace-preserving linear map that is generally a function of time. Mandayam and Ng (2012) tried to develop a unifying framework for approximate quantum error correction. Cohen et al. (2022) coupled an ancillary system to a quantum error-correcting code in a fault-tolerant way. Bravo-Prieto et al. (2019) presented a hybrid quantum–classical algorithm for solving quantum linear systems, called the variational quantum linear solver (VQLS), by which quantum circuit depth was reduced and noisy intermediate-scale quantum computers could be operated. Bennewitz et al. (2022) introduced a method for neural error mitigation and found that the method provided accurate estimates of more-complex observables without requiring additional quantum resources.

It has been proven that machine learning provides an alternative route to quantum error mitigation and quantum computing speed-up (Caro et al. 2022; Yue et al. 2022). Fawzi et al. (2022) reported a framework for deep reinforcement learning, AlphaTensor, to discover efficient algorithms for the multiplication of arbitrary matrices, which was built on the AlphaZero algorithm that can achieve superhuman performance in many challenging games (Silver et al. 2018). AlphaTensor was operated on a graphics processing unit (GPU) and tensor processing unit (TPU) and made multiplying large matrices 23.9% faster than human-designed algorithms on the same hardware. This idea is helpful for optimizing HASM quantum reinforcement learning.

HASM quantum machine learning might be improved further by: (1) considerably reducing the number of gates required for operating HASM-HHL through the establishment of a hybrid classical-quantum platform for HASM reinforcement learning; (2) proposing a suitable scheme for solving the large sparse linear system based on training experiments on classical supercomputers to optimize quantum circuits and algorithms; (3) finding solutions to the four fundamental problems of quantum machine learning algorithms, i.e., input, output, costing and benchmark problems; and (4) introducing variational quantum eigen-solvers, quantum walks, quantum search, and quantum approximate optimization algorithm into the development of a new HASM machine learning system.

In the meanwhile, two libraries of classical and quantum algorithms will be built for running interpolation, upscaling, downscaling, data fusion, model-data assimilation, and model coupling on classical computers and quantum computers respectively. The performance of the quantum and classical algorithms will then be compared to clarify their advantages and disadvantages in terms of accuracy and computational speed, and the algorithms would be added to a cloud platform to support applications and gather active feedback from users of the algorithms.

## 2.4 The Fundamental Theorem for Eco-Environmental Surface Modelling

HASM was successfully applied for constructing digital elevation models (Chen and Yue 2010; Chen et al. 2013a, b), filling voids in the Shuttle Radar Topography Mission (SRTM) dataset (Yue et al. 2012), simulating climate change (Yue et al. 2013b, 2016b; Zhao and Yue 2014a, b; Zhao et al. 2018; Yue et al., 2019), estimating carbon stocks (Yue et al. 2016c), fusing satellite observations and the Total Carbon Column Observing Network (TCCON) measurements of the column-averaged dry air mole fraction of $CO_2$ ($XCO_2$) (Zhang et al. 2017; Zhao et al. 2017), filling voids on remotely sensed $XCO_2$ surfaces (Yue et al. 2015b), modeling surface soil properties (Shi et al. 2011) and soil pollution (Shi et al. 2009), and analyzing ecosystem responses to climatic change (Yue et al. 2015c). In all these applications, HASM produced more accurate results than the classical methods.

The fundamental theorem for earth surface system modeling (FTESM) was proposed on the basis of developing HASM methods and their successful applications. HASM and FTESM is based on a combination of surface theory, system theory, and optimal control theory (Somasundaram 2005; Djaferis and Schick 2000; Hull 2003). The FTESM can be summarized as follows: an Earth's surface system or a component of the Earth's surface environment, is uniquely defined by both extrinsic and intrinsic properties of the surface, which can be simulated with an appropriate method for integrating the extrinsic and intrinsic properties, such as high accuracy surface modeling (HASM), when the spatial resolution of the surface is sufficient to capture the attribute(s) of interest (Yue et al. 2016a). In other words, accurate modeling of the Earth's surface systems requires both extrinsic and intrinsic information as well as appropriate methods.

If an eco-environmental surface is, as a unified term, used to represent a surface of nature, a surface of nature's contribution to people, or a surface of a driving-force of natural changes, the fundamental theorem for eco-environmental surface modelling (FTEEM) can be created after the FTESM (Yue et al. 2016a; IPBES 2016, 2019). The FTEEM can be stated as follows: an eco-environmental surface is uniquely defined by both extrinsic information and intrinsic information of the surface, which can be simulated with an appropriate method for integrating the global and local information, such as HASM, when the spatial resolution of the surface is fine enough to capture the attribute(s) of interest.

In terms of the FTESM and FTEEM, the following seven corollaries can be derived:

- **Corollary 1 (Interpolation)**: If an intrinsic information is available only, HASM can be employed to create an eco-environment surface with higher accuracy after the necessary extrinsic information have been extracted from the intrinsic information by geo-statistics.
- **Corollary 2 (Upscaling)**: If a surface on a finer spatial resolution is transferred to a one on a coarser spatial resolution, ground-based observations are necessary to operate HASM for obtaining higher accuracy.
- **Corollary 3 (Downscaling)**: If a surface at a coarse spatial resolution is available, it is necessary to supplement ground-based observations in order to obtain a corresponding surface at finer spatial resolution with higher accuracy by means of HASM.
- **Corollary 4 (Data fusion)**: When remotely sensed data from satellites are available, ground measurements have to be obtained and incorporated before HASM can be used to generate a more accurate surface. This corollary emphasizes the necessity of information completeness for data fusion.
- **Corollary 5 (Data fusion)**: When both remotely sensed data from satellites and ground measurements are available, HASM can be used to generate a surface that is more accurate than the one from either the satellite observations or the ground measurements. This corollary states the necessity of an appropriate method for data fusion with higher accuracy.

- **Corollary 6 (Model-data assimilation)**: When a global model is available, a more accurate surface can be produced when ground observations are incorporated into the system model. This corollary emphasizes the importance of intrinsic information for the model-data assimilation because extrinsic information can be obtained from output of the global model.
- **Corollary 7 (Model-data assimilation)**: When both a system model and ground observations are available, a surface can be produced by using HASM to incorporate the ground observations into the system model, which is more accurate than the one either from system model or from the ground observations. This corollary pays attention to the significance of an appropriate method for the model-data assimilation.

## 2.5 Summary

The development of HASM can be divided into 6 stages (Fig. 3). In the first stage from 1986 to 2001, a surface was regarded as a combination of its profiles that were fitted in terms of the fundamental theorem for curves (FTC). It was learnt that slope and curvature are significant variables of surface analysis (Evans 1980). In fact, a profile of the Earth's surface is uniquely determined by its elevation, slope and curvature in terms of the curve-theorem in the plane (Spivak 1979). A model was constructed for a cirque surface in terms of the FTC (Yue and Ai 1990) and was then used for environmental change detection (Yue et al. 2002). In this first stage, the idea was limited to conventionally geographical thinking.

However, the modeling of the Earth's surface should be based on fundamental theorem for surfaces (FTS) instead of FTC. It has been shown that the equation of the Earth's surface can be formulated as $z = f(x, y)$, where z is an attribute value of the Earth's surface at location $(x, y)$ (Kerimov 2009). For the surface, a master equation

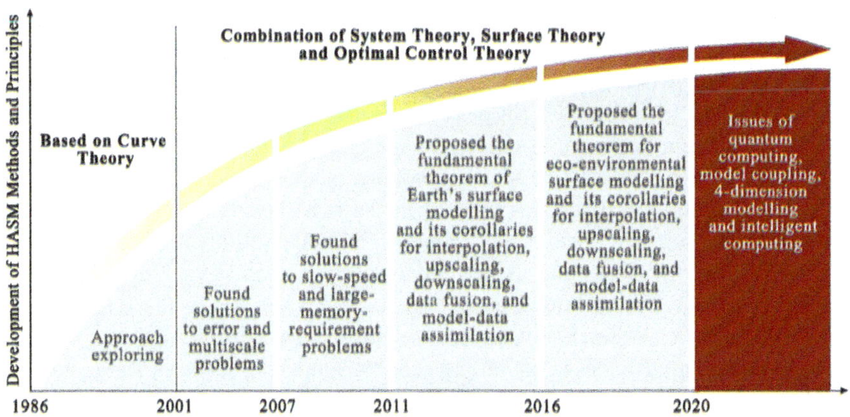


**Fig. 3** Historical trajectory of HASM and FTEEM development

set of HASM was constructed in terms of FTS (Yue et al. 2004, 2007, 2010a), which was transformed into a symmetric positive-definite and large sparse linear system. In this second stage from 2001 to 2007, solutions were found for error and multi-scale problems that plagued surface modelling prior to the development of HASM.

Although the error and multi-scale problems were solved during the second stage, HASM could only be used with small areas because it must use the master equation set for simulating each lattice of a surface, which incurs a huge computation cost. To speed up the computation of HASM, several fast algorithms were developed, such as a multi-grid algorithm of HASM (HASM-MG) (Yue and Song 2008; Yue et al. 2013a), an adaptive method of HASM (HASM-AM) (Yue et al. 2010a), an adjustment computation of HASM (HASM-AC) (Yue and Wang 2010), and a preconditioned conjugate gradient algorithm of HASM (HASM-PCG) (Yue et al. 2010b). Multigrid method is the fastest numerical method for solving partial differential equations, which is based on two principles that are error smoothing and coarse grid correction. The principle of the adaptive method is that grid cells where the error is large will be marked for refinement, while grid cells with a satisfied accuracy are left unchanged. The adjustment computation permits all observations, regardless of their number or type, to be entered into the adjustment and used simultaneously in the computations by means of least squares. A conjugate gradient algorithm was originally viewed as an acceleration technique for the effective solution of large linear systems by a succession of well-convergent approximations; the preconditioned conjugate gradient algorithm can be developed by introducing a preconditioner to ensure faster convergence of the conjugate gradient method. In this third stage from 2008 to 2011, the low computational speed and large memory problems was solved for simulation issues on national and global levels, but HASM is not fast enough yet. Quantum computing might be an alternative way for the solution.

During the fourth stage from 2011 to 2016, the fundamental theorem for earth surface system modeling (FTESM) was generated as an inevitable outcome of development of HASM methods and their successful applications. HASM has been involved in the Methodological Assessment Report on Scenarios and Models of Biodiversity and Ecosystem Services (IPBES 2016). The role of this methodological assessment is defined by IPBES as "guiding the use of scenario analysis and modelling in all work under IPBES to ensure the policy relevance of its deliverables" [Annex I, IPBES-2/5] (https://ipbes.net/). FTESM was referenced by the Global Assessment Report on Biodiversity and Ecosystem Services (IPBES 2019).

In the fifth stage from 2016 to 2020, a fundamental theorem for eco-environmental surface modelling (FTEEM) was created in the context of IPBES after FTESM. FTEEM fits in with the conceptual system of IPBES and has involved in the Nature Futures Framework (NFF) and its methodological guidance (https://ipbes.net/). In addition, FTEEM is contributing to the Thematic Assessment of the Interlinkages among Biodiversity, Water, Food and Health (https://ipbes.net/). HASM and FTEEM are transformative advances in modelling eco-environmental surfaces. For instances, all classical methods for spatial interpolation are local weighting approaches, but HASM provides a finite difference method.

Since 2020, we have started to work at four significant issues to further improve the HASM and FTEEM system: (1) how to capture and model the features, such as cliffs and fracture zones, which change quickly and the organic transitions between these and smooth areas; (2) how to incorporate advanced computational methods such as intelligent computing and quantum computing in HASM to support work with large datasets; (3) how to find one or more solutions for modeling in four dimensions and supporting simulations in three dimensions; (4) how to integrate software and hardware to generate a FTEEM-based modelling platform.

## 2.6 Further Reading

(1) Yue TX. 2011. Surface Modelling: High Accuracy and High Speed Methods. New York: CRC Press.
(2) IPBES. 2016. The methodological assessment report on scenarios and models of biodiversity and ecosystem services. Bonn, Germany: Secretariat of the Intergovernmental Science-Policy Platform on Biodiversity and Ecosystem Services.
(3) Yue TX, Liu Y, Zhao MW, Du ZP, Zhao N. 2016. A fundamental theorem of Earth's surface modelling. Environmental Earth Sciences 75(9): article 751 (pages 1–12).
(4) Yue TX, Zhao N, Liu Y, Wang YF, Zhang B, Du ZP, Fan ZM, Shi WJ, Chen CF, Zhao MW, Song DJ, Song YJ, Yan CQ, Li QQ, Sun XF, Zhang LL, Tian YZ, Wang W, Wang YA, Ma SN, Huang HS, Lu YM, Wang Q, Wang CL, Wang YZ, Lu M, Zhou W, Liu Y, Yin XZ, Wang Z, Bao ZY, Zhao MM, Zhao YP, Jiao YM, Naseer U, Fan B, Li SB, Yang Y, Wilson JP. 2020. A fundamental theorem for eco-environmental surface modelling and its applications. Science China-Earth Sciences 63(8): 1092–1112.
(5) Yue TX, Wu CC, Liu Y, Du ZP, Zhao N, Jiao YM, Xu Z, Shi WJ. 2023. HASM quantum machine learning. Science China-Earth Sciences 66(9): 1937–1945.

**Exercise**

(1) Define the terms "Surface", "Surface modelling", "Intrinsic properties" and "Extrinsic properties".
(2) What is the advantage and disadvantage of satellite observations?
(3) What is the advantage and disadvantage of ground observations?
(4) What is the high accuracy surface modeling (HASM)?
(5) What is the fundamental theorem for eco-environmental surface modelling (FTEEM)?
(6) Explain the Historical trajectory of HASM and FTEEM development.
(7) How are intelligent computing, machine learning and deep learning related?
(8) What is HASM-HHL quantum machine learning?

## References

Abigail, L., and J.A. Downs. 2021. Machine learning in geography—Past, present, and future. *Geography Compass* 15. https://doi.org/10.1111/gec3.12563

Alaminos, D., M.B. Salas, and M.A. Fernández-Gámez. 2022. Quantum machine learning algorithms: Read the fine print. *Computational Economics* 59: 803–829.

Alkhasawneh, M.S., U.K. Ngah, L.T. Tay, and N.A.M. Isa. 2014. Determination of importance for comprehensive topographic factors on land-slide hazard mapping using artificial neural network. *Environment and Earth Science* 72 (3): 787–799.

Arute, F., K. Arya, R. Babbush, D. Bacon, J.C. Bardin, R. Barends, R. Biswas, S. Boixo, F.G.S.L. Brandao, D.A. Buell, B. Burkett, Y. Chen, Z. Chen, B. Chiaro, R. Collins, W. Courtney, A. Dunsworth, E. Farhi, B. Foxen, A. Fowler, C. Gidney, M. Giustina, R. Graff, K. Guerin, S. Habegger, M.P. Harrigan, M.J. Hartmann, A. Ho, M. Hoffmann, T. Huang, T.S. Humble, S.V. Isakov, E. Jeffrey, Z. Jiang, D. Kafri, K. Kechedzhi, J. Kelly, P.V. Klimov, S. Knysh, A. Korotkov, F. Kostritsa, D. Landhuis, M. Lindmark, E. Lucero, D. Lyakh, S. Mandrà, J.R. McClean, M. McEwen, A. Megrant, X. Mi, K. Michielsen, M. Mohseni, J. Mutus, O. Naaman, M. Neeley, C. Neill, M.Y. Niu, E. Ostby, A. Petukhov, J.C. Platt, C. Quintana, E.G. Rieffel, P. Roushan, N.C. Rubin, D. Sank, K.J. Satzinger, V. Smelyanskiy, K.J. Sung, M.D. Trevithick, A. Vainsencher, B. Villalonga, T. White, Z.J. Yao, P. Yeh, A. Zalcman, H. Neven, and J.M. Martinis. 2019. Quantum supremacy using a programmable superconducting processor. *Nature* 574: 505–510.

Assche, V.K., V. Valentinov, and G. Verschraegen. 2019. Ludwig von Bertalanffy and his enduring relevance: Celebrating 50 years general system theory. *Systems Research and Behavioral Science* 36 (3).

Bacon, D. 2006. Operator quantum error-correcting subsystems for selfcorrecting quantum memories. *Physical Review A* 73: 012340.

Benioff, P. 1980. The computer as a physical system: A microscopic quantum mechanical Hamiltonian model of computers as represented by Turing machines. *Journal of Statistical Physics* 22: 563–591.

Bennett, C.H., E. Bernstein, G. Brassard, and U. Vazirani. 1997. Strengths and weaknesses of quantum computing. *SIAM Journal on Computing* 26: 1510–1523.

Bennewitz, E.R., F. Hopfmueller, B. Kulchytskyy, J. Carrasquilla, and P. Ronagh. 2022. Neural error mitigation of near-term quantum simulations. *Nature Machine Intelligence* 4: 618–624.

Biamonte, J., P. Wittek, N. Pancotti, P. Rebentrost, N. Wiebe, and S. Lloyd. 2017. Quantum machine learning. *Nature* 549: 195–202.

Bloor, M.I.G., and M.J. Wilson. 1989. Generating blend surfaces using partial differential equations. *Computer-Aided Design* 21 (3): 165–171.

Bloor, M.I.G., and M.J. Wilson. 1990. Using partial differential equations to generate free-form surfaces. *Computer-Aided Design* 22 (4): 202–212.

Bloor, M.I.G., and M.J. Wilson. 1996. Spectral approximations to PDE surfaces. *Computer-Aided Design* 28 (2): 145–152.

Bravo-Prieto, C., R. LaRose, M. Cerezo, Y. Subasi, L. Cincio, and P.J. Coles. 2019. *Variational quantum linear solver*. https://doi.org/10.48550/arXiv.1909.05820.

Bush, A.B.G., M.P. Bishop, D. Huo, Z.H. Chi, and U. Tiwari. 2020. *Issues in climate analysis and modeling for understanding mountain erosion dynamics*. Amsterdam: Elsevier Inc.

Brill, K.F., L.W. Uccellini, J. Manobianco, P. Kocin, and J.H. Homan. 1991. The use of successive dynamic initialization by nudging to simulate cyclogenesis during GALE IOP 1. *Meteorology and Atmospheric Physics* 45: 15–40.

Butt, N., K. Epps, H. Overman, T. Iwamura, and J.M.V. Fragoso. 2015. Assessing carbon stocks using indigenous peoples' field measurements in Amazonian Guyana. *Forest Ecological Management* 338: 191–199.

Carmo, M.P. 2006. *Differential geometry of curves and surfaces*. Beijing: China Machine Press.

Caro, M.C., H.Y. Huang, M. Cerezo, K. Sharma, A. Sornborger, L. Cincio, and P.J. Coles. 2022. Generalization in quantum machine learning from few training data. *Nature Communications* 13: 4919.

Chalumuri, A., R. Kune, and B.S. Manoj. 2020. Training an artificial neural network using qubits as artificial neurons: A quantum computing approach. *Procedia Comput Sci* 171: 568–575.

Chambers, J.Q., G.P. Asner, D.C. Morton, L.O. Anderson, S.S. Saatchi, F.D.B. Jr Espirito-Santo, and M.P.C. Souza. 2007. Regional ecosystem structure and function: Ecological insights from remote sensing of tropical forests. *Trends in Ecology & Evolution* 22: 414–423.

Chen, C.F., and T.X. Yue. 2010. A method of DEM construction and related error analysis. *Computers & Geosciences* 36: 717–725.

Chen, C.F., Y.Y. Li, and T.X. Yue. 2013a. Surface modeling of DEMs based on a sequential adjustment method. *International Journal of Geographical Information Science* 27(7): 1272–1291.

Chen, C. F., T. X. Yue, H. L. Dai, and M. Y. Tian. 2013b. The smoothness of HASM. *International Journal of Geographical Information Science* 27 (8): 1651–1667.

Chiesi, M., L. Fibbi, L. Genesio, B. Gioli, R. Magno, F. Maselli, M. Moriondo, and F.P. Vaccari. 2011. Integration of ground and satellite data to model Mediterranean forest processes. *International Journal of Applied Earth Observation and Geoinformation* 13: 504–515.

Cohen, L.Z., I.H. Kim, S.D. Bartlett, and B.J. Brown. 2022. Low-overhead fault tolerant quantum computing using long-range connectivity. Science Advances 8: eabn1717.

Daley, A.J., I. Bloch, C. Kokail, S. Flannigan, N. Pearson, M. Troyer, and P. Zoller. 2022. Practical quantum advantage in quantum simulation. *Nature* 607: 667–676.

Davis, M. 1958. *Computability and unsolvability*. New York: Dover Publications.

Deshpande, A., A. Mehta, T. Vincent, N. Quesada, M. Hinsche, M. Ioannou, L. Madsen, J. Lavoie, H. Qi, J. Eisert, D. Hangleiter, B. Fefferman, and I. Dhand. 2022. Quantum computational advantage via high-dimensional Gaussian boson sampling. Sciences Advances 8: eabi7894.

Deutsch, D. 1985. Quantum theory, the Church-Turing principle and the universal quantum computer. *Proceedings of the Royal Society of London, Series A: Mathematical and Physical Sciences* 400: 97–117.

Djaferis, T.E., and I.C. Schick. 2000. *System theory: Modeling, analysis, and control. Boston*, MA: Kluwer.

Dobermann, A., and J.L. Ping. 2004. Geostatistical integration of yield monitor data and remote sensing improves yield maps. *Agronomy Journal* 96: 285–297.

Drack, M., and G. Schwarz. 2010. Recent developments in general system theory. *Systems Research and Behavioral Science* 27: 601–610.

Emili, E., C. Popp, S. Wunderle, M. Zebisch, and M. Petitta. 2011. Mapping particulate matter in alpine regions with satellite and ground-based measurements: An exploratory study for data assimilation. *Atmospheric Environment* 45: 4344–4353.

Evans, I.S. 1980. An integrated system of terrain analysis and slope mapping. *Z Geomorphol Suppl Bd* 36: 274–295.

Farhi, E., and S. Gutmann. 1998. Quantum computation and decision trees. *Physical Review A* 58: 915–928.

Fawzi, A., M. Balog, A. Huang, T. Hubert, B. Romera-Paredes, M. Barekatain, A. Novikov, R. Ruiz, F.J, Schrittwieser, G. Swirszcz, D. Silver, D. Hassabis, and P. Kohli. 2022. Discovering faster matrix multiplication algorithms with reinforcement learning. Nature 610: 47–53.

Feynman, R.P. 1982. Simulating physics with computers. *International Journal of Theoretical Physics* 21: 467–488.

Fritz, S., I. McCallum, C. Schill, C. Perger, L. See, D. Schepaschenko, M. van der Velde, F. Kraxner, and M. Obersteiner. 2012. Geo-Wiki: An online platform for improving global land cover. *Environmental Modelling & Software* 31: 110–123.

Grover, L.K. 1997. Quantum mechanics helps in searching for a needle in a haystack. *Physical Review Letters* 79: 325–328.

Gupta, M., K.K. Mohanty, D. Kumar, and R. Banerjee. 2014. Monitoring surface elevation changes in Jharia coal field, India using synthetic aperture radar interferometry. *Environment and Earth Science* 71 (6): 2875–2883.

Haber, W. 2021. Eco-environmental surface modelling requires integration of both extrinsic and intrinsic informations. *Science China Earth Sciences* 64 (1): 185–186.

Harrow, A.W., A. Hassidim, and S. Lloyd. 2009. Quantum algorithm for linear systems of equations. *Physical Review Letters* 103: 150502.

Henderson, D.W. 1998. *Differential geometry*. London: Prentice-Hall Inc.

Hilbert, D., von J. Neumann, and L. Nordheim. 1928. Ueber die Grundlagen der Quantenmechanik. Mathematische Annalen 98: 1–30

Hull, D.J. 2003. Optimal control theory for applications. New York, NY: Springer.

IPBES. 2016. Methodological assessment of scenarios and models of biodiversity and ecosystem services. Bonn, Germany: Secretariat of the Intergovernmental Platform for Biodiversity and Ecosystem Services.

IPBES. 2019. Global assessment of scenarios and models of biodiversity and ecosystem services. Bonn, Germany: Secretariat of the Intergovernmental Platform for Biodiversity and Ecosystem Services.

Kempe, J. 2003. Quantum random walks: An introductory overview. *Contemporary Physics* 44: 307–327.

Kerimov, I.A. 2009. F-approximation of the Earth's surface topography. *Izvestiya, Physics of the Solid Earth* 45: 719–729.

Khrennikov, A. 2021. Roots of quantum computing supremacy: Superposition, entanglement, or complementarity? *The European Physical Journal Special Topics* 230: 1053–1057.

Kirk, D.E. 2004. *Optimal control theory: An introduction*. New York: Dover Publications Inc.

Kohler, W. 1920. Die physischen Gestalten in Ruhe und im stationaeren Zustand. Erlangen.

Koehler, W. 1927. Zum Problem der Regulation. *Wilhelm Roux' Archiv fur Entwicklungsmechanik der Organismen* 112 (1): 315–332.

Laszlo, E. 1975. The meaning and significance of general system theory. *Behavioral Science* 20: 9–24.

Lee, E.B., and L. Markus. 1986. *Foundations of optimal control theory*. New York: John Wily and Sons Inc.

Liseikin, V.D. 2004. *A computational differential geometry approach to grid generation*. Berlin: Springer.

Li, L., K. Wu, and D.W. Zhou. 2014. Extraction algorithm of mining subsidence information on water area based on support vector machine. *Environment and Earth Science* 72 (10): 3991–4000.

Liu, Z.P., M.A. Shao, and Y.Q. Wang. 2013. Large-scale spatial interpolation of soil pH across the Loess Plateau, China. *Environmental Earth Sciences* 69 (8): 2731–2741.

Lotka, A.J. 1925. *Elements of physical biology*. New York: Dover Publications.

Mandayam, P., and H.K. Ng. 2012. Towards a unified framework for approximate quantum error correction. *Physical Review A* 86: 012335.

Marin, V.H. 1997. General system theory and the ecosystem concept. *Bulletin of the Ecological Society of America* 78: 102–104.

Medvidović, M., and G. Carleo. 2021. Classical variational simulation of the quantum approximate optimization algorithm. *Npj Quantum Information* 7: 101. https://doi.org/10.1038/s41534-021-00440-z.

Modani, M., M. Sharan, and S.C.S. Rao. 2008. Numerical solution of elliptic partial differential equations on parallel systems. *Applied Mathematics and Computation* 195: 162–182.

Mooij, H. 2005. The road to quantum computing. *Science* 307: 1210–1211.

Morrell, H.J. Jr., and H.Y. Wong. 2021. Step-by-step HHL algorithm walkthrough to enhance the understanding of critical quantum computing concepts. https://doi.org/10.48550/arXiv.2108.09004.

Mueller, N., A. Tarasov, and R. Venugopalan. 2021. Computing real time correlation functions on a hybrid classical/quantum computer. *Nuclear Physics A* 1005: 121889.

Nielsen, M.A., and I.L. Chuang. 2010. *Quantum computation and quantum information*. New York: Cambridge University Press.

Nimbe, P., B.A. Weyori, and A.F. Adekoya. 2021. Models in quantum computing: A systematic review. *Quantum Information Processing* 20: 80.

Pasadas, M., and M.L. Rodríguez. 2009. Construction and approximation of surfaces by discrete PDE splines on a polygonal domain. *Journal of Computational and Applied Mathematics* 59 (1): 205–218.

Pereira, H.M., S. Ferrier, M. Walters, G.N. Geller, R.H.G. Jongman, R.J. Scholes, M.W. Bruford, N. Brummitt, S.H.M. Butchart, A.C. Cardoso, N.C. Coops, E. Dulloo, D.P. Faith, J. Freyhof, R.D. Gregory, C. Heip, R. Höft, G. Hurtt, W. Jetz, D.S. Karp, M.A. McGeoch, D. Obura, Y. Onoda, N. Pettorelli, B. Reyers, R. Sayre, J.P.W. Scharlemann, S.N. Stuart, E. Turak, M. Walpole, and M. Wegmann. 2013. Essential biodiversity variables. Science 339: 277–278.

Pereira, L.M., et al. 2020. Developing multiscale and integrative nature–people scenarios using the nature futures framework. *People and Nature* 2: 1172–1195.

Phillips, J.D. 2002. Global and local factors in earth surface systems. *Ecological Modelling* 149: 257–272.

Ponsar, S., P. Luyten, and V. Dulière. 2016. Data assimilation with the ensemble Kalman filter in a numerical model of the North Sea. *Ocean Dynamics* 66: 955–971.

Perelshtein, M.R., A.I. Pakhomchik, A.A. Melnikov, A.A. Novikov, A. Glatz, G.S. Paraoanu, V.M. Vinokur, and G.B. Lesovik. 2022. Solving large-scale linear systems of equations by a quantum Hybrid Algorithm. *Annalen der Physik* 534: 2200082.

Protopopescu, V., R.T. Santoro, and J. Dockery. 1989. Combat modeling with partial differential equations. *European Journal of Operational Research* 3 (2): 178–183.

Rebentrost, P., T.R. Bromley, C. Weedbrook, and S. Lloyd. 2018. Quantum Hopfield neural network. *Physical Review A* 98: 042308.

Rebentrost, P., M. Mohseni, and S. Lloyd. 2014. Quantum support vector machine for big data classification. *Physical Review Letters* 113: 130503.

Rønnow, T.F., Z. Wang, J. Job, S. Boixo, S.V. Isakov, D. Wecker, J.M. Martinis, D.A. Lidar, and M. Troyer. 2014. Defining and detecting quantum speedup. *Science* 345: 420–424.

Quarteroni A, Sacco R, and Saleri F. 2000. *Numerical Mathematics*. Springer, New York.

Seidl, D., and D. Schoeneborn. 2016. System theory. In: *The international encyclopedia of communication theory and philosophy*, ed. by K.B. Jensen, R.T. Craig, J.D. Pooley, and E.W. Rothenbuhler. New York: John Wiley & Sons, Inc.

Sethi, S.P. 2019. *Optimal control theory: Applications to management science and economics*, 3rd ed. Cham: Springer Nature Switzerland AG.

Shao, C. 2018. Reconsider HHL algorithm and its related quantum machine learning algorithms. https://doi.org/10.48550/arXiv.1803.01486.

Shen, T.L. 2017. Special issue on model-based optimal control theory and technology in automotive control. *Control Theory and Technology* 15 (2): 81–82.

Shi, W. J., T. X. Yue, and Z. M. Fan. 2009. Surface modeling of soil pH. *Geoderma*, 150: 113–119.

Shi, W. J., T. X. Yue, and Z. M. Fan. 2011. An overview of high accuracy surface modeling of soil properties. *Procedia Environ Sci*, 13: 2304-2309.

Shi, Y.G., P.M. Thompson, I. Dinov, Osher St, and A.W. Tog. 2007. Direct cortical mapping via solving partial differential equations on implicit surfaces. *Medical Image Analysis* 11: 207–223.

Shor, P.W. 1994. Algorithms for quantum computation: Discrete logarithms and factoring. In Proceedings 35th Annual Symposium on Foundations of Computer Science, 124–134. New York.

Silver, D., T. Hubert, J. Schrittwieser, I. Antonoglou, M. Lai, A. Guez, M. Lanctot, L. Sifre, D. Kumaran, T. Graepel, T. Lillicrap, K. Simonyan, and D. Hassabis. 2018. A general reinforcement learning algorithm that masters chess, shogi, and Go through self-play. *Science* 362: 1140–1144.

Somasundaram, D. 2005. Differential geometry. Harrow, UK: Alpha Science International

Spivak, M. 1979. *A comprehensive introduction to differential geometry*, vol. 3. Boston, MA: Publish or Peril Inc.

Srivastava, P.K., D.W. Han, M.A. Rico-Ramirez, M. Bray, T. Islam, M. Gupta, and Q. Dai. 2014. Estimation of land surface temperature from atmospherically corrected LANDSAT TM image using 6S and NCEP global reanalysis product. *Environment and Earth Science* 72 (12): 5183–5196.

Thompson, J.F., F.C. Thames, and C.W. Mastin. 1974. Automatic numerical generation of body-fitted curvilinear coordinate system for field containing any number of arbitrary two-dimensional bodies. *Journal of Computational Physics* 15 (3): 299–319.

Toponogov, V.A. 2006. *Differential geometry of curves and surfaces*. New York: Birkhaeuser Boston.

Turing, A.M. 1937. On computable numbers, with an application to the Entscheidungs-problem. *Proceedings of London Mathematical Social* 42: 230–265.

Von Bertalanffy, L. 1949. Zu einer allgemeinen Systemlehre. Biologia Generalis1 9:114–129.

Von Bertalanffy, L. 1969. *General system theory: Foundations, development, applications*. New York: George Braziller.

Waldrop, M.M. 2016. More than Moore. *Nature* 530: 145–147.

Walder, C., B. Schoelkopf, and O. Chapelle. 2006. Implicit surface modeling with a globally regularized basis of compact support. *EUROGRAPHICS* 25: 635–644.

Wiebe, N., D. Braun, and S. Lloyd. 2012. Quantum algorithm for data fitting. *Physical Review Letters* 109: 050505.

Xu, G.L., and Q. Zhang. 2008. A general framework for surface modeling using geometric partial differential equations. *Computer Aided Geometric Design* 25 (3): 181–202.

Yang, W., Z.M. Zheng, C. Zheng, K.H. Lu, D.W. Ding, and J.Y. Zhu. 2018. Temporal variations in a phytoplankton community in a subtropical reservoir: An interplay of extrinsic and intrinsic community effects. *Science of the Total Environment* 612: 720–727.

Yue, T.X. 2011. *Surface modelling: High accuracy and high speed methods*. Boca Raton, FL: CRC Press.

Yue, T.X., and N.S. Ai. 1990. The mathematic model of cirque morphology. *J Glaciol Geocryol* 12: 227–234.

Yue, T.X., and Y.J. Song. 2008. The YUE-HASM method. In: *Accuracy in geomatics*, ed. D. Li D, Y. Ge, and G.M. Foody, 148–153. Edgbaston, UK: World Academic Union.

Yue, T.X., and S.H. Wang. 2010. Adjustment computation of HASM: A high-accuracy and high-speed method. *Int J Geograph Inf Sci* 24: 1725–1743.

Yue, T.X., S.P. Chen, B. Xu, Q.S. Liu, H.G. Li, G.H. Liu, and Q.H. Ye. 2002. A curve-theorem based approach for change detection and its application to Yellow River Delta. *International Journal of Remote Sensing* 23: 2283–2292.

Yue, T.X., Z.P. Du, and J.Y. Liu. 2004. High accuracy surface modelling and its error analysis. *Progress in Natural Science* 14 (3): 300–306 (in Chinese).

Yue, T.X., Z.P. Du, D.J. Song, and Y. Gong. 2007. A new method of surface modeling and its application to DEM construction. *Geomorphology* 91: 161–172.

Yue, T.X., C.F. Chen, and B.L. Li. 2010a. An adaptive method of high accuracy surface modeling and its application to simulating elevation surfaces. *Transactions in GIS* 14: 615–630.

Yue, T.X., D.J. Song, Z.P. Du, and W. Wang. 2010b. High-accuracy surface modelling and its application to DEM generation. *International Journal of Remote Sensing* 31: 2205–2226.

Yue, T.X., C.F. Chen, and B.L. Li. 2012. A high accuracy method for filling SRTM voids and its verification. *International Journal of Remote Sensing* 33: 2815–2830.

Yue, T.X., N. Zhao, H. Yang, Y.J. Song, Z.P. Du, Z.M. Fan, and D.J. Song. 2013a. The multi-grid method of high accuracy surface modelling and its validation. *Transactions in GIS* 17: 943–952.

Yue, T.X., N. Zhao, R.D. Ramsey, C.L. Wang, Z.M. Fan, C.F. Chen, Y.M. Lu, and B.L. Li. 2013b. Climate change trend in China, with improved accuracy. *Climate Change* 120: 137–151.

Yue, T.X., L.L. Zhang, N. Zhao, M.W. Zhao, C.F. Chen, Z.P. Du, D.J. Song, Z.M. Fan, W.J. Shi, S.H. Wang, C.Q. Yan, Q.Q. Li, X.F. Sun, H. Yang, C.L. Wang, Y.F. Wang, J.P. Wilson, and B. Xu. 2015a. A review of recent developments in HASM. *Env Earth Sci* 74: 6541–6549.

Yue, T.X., M.W. Zhao, and X.Y. Zhang. 2015b. A high-accuracy method for filling voids on remotely sensed $XCO_2$ surfaces and its verification. *Journal of Cleaner Production* 103: 819–827.

Yue, T.X., Z.P. Du, M. Lu, Z.M. Fan, C.L. Wang, Y.Z. Tian, and B. Xu. 2015c. Surface modelling of ecosystem responses to climatic change. *Ecological Modelling* 306: 16–23.

Yue, T.X., Y. Liu, M.W. Zhao, Z.P. Du, and N. Zhao. 2016a. A fundamental theorem of Earth's surface modelling. *Env Earth Sci* 75 (9): 751.

Yue, T.X., N. Zhao, Z.M. Fan, et al. 2016b. CMIP5 Downscaling and its uncertainty in China. *Glob Planet Change* 146: 30–37.

Yue, T.X., Y.F. Wang, Z.P. Du, M.W. Zhao, L.L. Zhang, N. Zhao, M. Lou, G.R. Larocqued, and J.P. Wilson. 2016c. Analyzing the uncertainty of estimating forest carbon stocks in China. *Biogeosciences* 13: 3991–4004.

Yue, T.X., N. Zhao, Z.M. Fan, J. Li, C.F. Chen, Y.M. Lu, C.L. Wang, J. Gao, B. Xu, and J. Wilson. 2019. Methods for simulating climate scenarios with improved spatiaotemporal specificity and less uncertainty. Global and Planetary Change 181 (Article 102973): 1–18.

Yue, T.X., N. Zhao, Y. Liu, Y.F. Wang, B. Zhang, Z.P. Du, Z.M. Fan, W.J. Shi, C.F. Chen, M.W. Zhao, D.J. Song, Y.J. Song, C.Q. Yan, Q.Q. Li, X.F. Sun, L.L. Zhang, Y.Z. Tian, W. Wang, Y.A. Wang, S.N. Ma, H.S. Huang, Y.M. Lu, Q. Wang, C.L. Wang, Y.Z. Wang, M. Lu, W. Zhou, Y. Liu, X.Z. Yin, Z. Wang, Z.Y. Bao, M.M. Zhao, Y.P. Zhao, Y.M. Jiao, U. Naseer, B. Fan, S.B. Li, Y. Yang, and J.P. Wilson. 2020. A fundamental theorem for eco-environmental surface modelling and its applications. *Science China-Earth Sciences* 63 (8): 1092–1112.

Yue, T.X., Y. Liu, Z.P. Du, J. Wilson, D.Y. Zhao, Y. Wang, N. Zhao, W.J. Shi, Z.M. Fan, X.M. Zhao, Q. Zhang, H.S. Huang, Q.Y. Wu, W. Zhou, Y.M. Jiao, Z. Xu, S.B. Li, Y. Yang, and B.J. Fu. 2022. Quantum machine learning of eco-environmental surfaces. *Scientific Bulletin* 67: 1031–1033.

Yue, T.X., C.C. Wu, Y. Liu, Z.P. Du, N. Zhao, Y.M. Jiao, Z. Xu, and W.J. Shi. 2023. HASM quantum machine learning. *Science China-Earth Sciences* 66 (9): 1937–1945.

Zhang, J.J., and L.H. You. 2004. Fast surface modeling using a 6th order PDE. *Eurographics* 23: 311–320.

Zhang, L.L., T.X. Yue, J.P. Wilson, N. Zhao, Y.P. Zhao, Z.P. Du, and Y. Liu. 2017. A comparison of satellite observations with the XCO2 surface obtained by fusing TCCON measurements and GEOS-Chem model outputs. *Sci Total Env*, 601–602: 1575–90.

Zhao, M.W., X.Y. Zhang, T.X. Yue, C.L. Wang, L. Jiang, and J.L. Sun. 2017. A high-accuracy method for simulating the XCO2 global distribution using GOSAT retrieval data. *Science China Earth Sciences*, 60: 143–155.

Zhao, N., and T. X. Yue. 2014a. A modification of HASM for interpolating precipitation in China. *Theor Appl Climatol*, 116: 273–285.

Zhao, N., T. X. Yue T. 2014b. Sensitivity studies of a high accuracy surface modelling method. *Sci China-Earth Sci*, 57: 1–11.

Zhao, N., T. X. Yue, C. F. Chen, M. W. Zhao, and Z. P. Du. 2018. An improved HASM method for dealing with large spatial data sets. *Science China Earth Sciences*, 61, 1078–1087.

Zhong, H.S., H. Wang, Y.H. Deng, M.C. Chen, L.C. Peng, Y.H. Luo, J. Qin, D. Wu, X. Ding, Y. Hu, P. Hu, X.Y. Yang, W.J. Zhang, H. Li, Y. Li, X. Jiang, L. Gan, G. Yang, L. You, Z. Wang, L. Li, N.L. Liu, C.Y. Lu, and J.W. Pan. 2020. Quantum computational advantage using photons. *Science* 370: 1460–1463.

Zhu, Q., S. Cao, F. Chen, M.C. Chen, X. Chen, T.H. Chung, H. Deng, Y. Du, D. Fan, M. Gong, C. Guo, C. Guo, S. Guo, L. Han, L. Hong, H.L. Huang, Y.H. Huo, L. Li, N. Li, S. Li, Y. Li, F. Liang, C. Lin, J. Lin, H. Qian, D. Qiao, H. Rong, H. Su, L. Sun, L. Wang, S. Wang, D. Wu, Y. Wu, Y. Xu, K. Yan, W. Yang, Y. Yang, Y. Ye, J. Yin, C. Ying, J. Yu, C. Zha, C. Zhang, H. Zhang, K. Zhang, Y. Zhang, H. Zhao, Y. Zhao, L. Zhou, C.Y. Lu, C.Z. Peng, X. Zhu, and J.W. Pan. 2022. Quantum computational advantage via 60-qubit 24-cycle random circuit sampling. *Scientific Bulletin* 67: 240–245.

Zidan, M., H. Eleuch, and M. Abdel-Aty. 2021. Non-classical computing problems: Toward novel type of quantum computing problems. *Results Phys* 21: 103536.

# Chapter 3
# Spatial Interpolation

**Learning Objective**

- Corollary 1 of FTEEM: If an intrinsic information is available only, HASM can be employed to create an eco- environment surface with higher accuracy after the necessary extrinsic information have been extracted from the intrinsic information by geo-statistics.

## 3.1 Introduction

Spatial interpolation is defined as predicting the values of a primary variable at points within the same region of sampled locations in terms of spatial data in the form of discrete points or in the form of data partition (Li and Heap 2014). Spatial interpolation provides effective means to construct a continuous surface from the discrete data or a surface with missing data (Wang and Wang 2012; Yue et al. 2015). It takes advantage of limited observation data to estimate the reasonable spatial distribution by filling missing data.

Since early 1950s, various approaches to spatial interpolation have been developed. In 1951, Krige proposed his basic idea of Kriging interpolation, a linear least square regression. In 1960, Birkhoff and Garabeddian introduced a two-dimensional spline approach and a concept of smooth surface interpolation by developing spline approximation that was first proposed by Schoenberg in 1946. In 1962, de Boor constructed bicubic spline interpolation on two-dimensional grid. In 1964, Bengtsson and Nordbeck developed a linear interpolation method, triangular irregular network (TIN), which can be used for irregularly distributed points. In 1968, Shepard published his interpolation method, Inverse Distance Weighting (IDW), which is highly dependent on an appropriate weight function (Franke 1982). In 1971, Hardy proposed multiquadric surfaces to representing irregular surfaces by the summation

 
T. Yue et al., *Eco-Environmental Informatics*,
https://doi.org/10.1007/978-981-96-1977-1_3

of equations of quadric surfaces having unknown coefficients, which are located at significant points throughout the region. In 1972, Harder and Desmarais found the thin-plate spline (TPS), which was originally termed surface spline and based on the small deflection equation of an infinite plate. In 1977, Talmi and Gilat introduced the smooth interpolation, based on choosing the smoothest analytical curve that approximates the data points. In 1981, Sibson developed Natural neighbor interpolation on the basis of Voronoi tessellation of a discrete set of spatial points.

Since the mid-1980s, most studies paid attention to comparing and preferring various interpolation methods. For instance, Yakowitz and Szidarovsky (1985) concluded that Kriging interpolation as a parametric regression method performs worse comparing with nonparametric regression approach when data have some trend features. Wagner et al. (2012) found that regression-IDW performed best in data scarce regions after they comparatively evaluated performances of seven spatial interpolation schemes for daily rainfall in Mula and Mutha rivers of India. Wang et al. (2014) used six interpolation methods to generate spatial distribution of precipitation based on 12 meteorological stations over Ontario of Canada; a cross-validation showed that local polynomial interpolation (LPI) is the optimal method with the least error. Xu et al. (2015) demonstrated that CK is the optimal method for spatial interpolation of the daily rainfall by examining ordinary Kriging (OK), IDW and ordinary CoKriging (CK) in the rainy season of Sichuan. Pereira et al. (2016) tested 17 univariate and multivariate methods at 36 metereological stations distributed in Sierra Nevada and neighbouring areas of Spain; the test results indicated that regression-Kriging was the most accurate. Arowolo et al. (2017) found that TPS was better than Kriging with an External Drift (KED) for generating monthly total precipitation, and minimum and maximum temperatures surfaces in Nigeria. Adhikary et al. (2017) evaluated 5 interpolation methods for enhanced spatial interpolation of monthly rainfall in the Middle Yarra River catchment and the Ovens River catchment in Victoria of Australia; they found that OCK method is the best interpolator. Hadi and Tombul (2018) used 6 spatial interpolation methods to interpolate daily, monthly and annual mean precipitation in Seyhan Basin of Turkey, for which a validation demonstrated that IDW had the smallest errors.

We notice that all classical methods for spatial interpolation above rely on local weighting. However, the algorithm for spatial interpolation, based on the method for high accuracy surface modelling (HASM), termed HASM-SI, provides a machine learning approach (Yue et al. 2020).

## 3.2 Filling Voids on Remotely Sensed $XCO_2$ Surfaces

To accurately project climate change, various investigations have been conducted to understand sources and sinks of $CO_2$. For instances, it was demonstrated that the yearly average carbon stock change was a net increase of 1.71 Mt C/year in Thailand's total rubber planting area, while the yearly average greenhouse gas emission from cultivation of rubber tree was 0.32 Mt C/year (Petsri et al. 2013). A study indicated that total carbon emissions in coastal Jiangsu amounted to $822.17 \times 10^4$ t in 1985 and increased to $2931.52 \times 10^4$ t in 2010, an increasing of 2.57 times from 1985 to 2010 (Chuai et al. 2015). A finding indicated that the conservation tillage practice had a greater contribution to the atmospheric $CO_2$ reduction (Chen et al. 2015). An input–output based methodology was proposed to identify key $CO_2$ emission sectors and the optimized production structure with respect to emission reduction targets in an economy (Chang 2015). Effects of post-earthquake reconstruction on carbon cycle were investigated in Guangyuan, a mountainous city in China, by means of an integrated accounting framework for urban carbon cycle (Hao et al. 2014).

However, insufficient knowledge of $CO_2$ leads to large uncertainties in future climate predictions because the observation of $CO_2$ is spatially and temporally limited on the globe (Yoshida et al. 2011). Satellite measurement is one of the most effective approaches to monitoring the global distributions of greenhouse gases at high spatiotemporal resolution and is expected to improve the accuracy of source and sink estimates of these gases (Rayner and O'Brien 2001). Two satellites have first been designed specifically to measure the column-averaged dry air mole fraction of $CO_2$ ($XCO_2$). They are Japan's Greenhouse Gases Observing Satellite (GOSAT) and NASA's Orbiting Carbon Observatory (OCO). Both missions try to identify cloud-free scenes for their retrievals, since clouds can cause large errors in the retrieved $CO_2$ concentrations (Baker et al. 2010). GOSAT was successfully launched in January 2009. Although Orbiting Carbon Observatory 1 (OCO-1) was lost because of its launch failure in February 2009, Orbiting Carbon Observatory 2 (OCO-2) was successfully launched in July 2014, with the instrument that is almost an exact copy of the OCO-1 instrument with minor changes.

GOSAT carried a thermal and near infrared sensor for carbon observation-Fourier transform spectrometer (TANSO-FTS) for collecting information about the concentrations of $CO_2$ in the upper-troposphere (Chevallier et al. 2009). Column-averaged $CO_2$ was retrieved from spectra in the shortwave infrared (SWIR) band. The GOSAT observing strategy is to observe $CO_2$ column-averaged concentrations under cloud-free conditions, with little disturbance from aerosols. Bias and random error were simulated under assumption that the bias depended on the aerosol optical thickness (AOT) and surface albedo. The number of successfully retrieved GOSAT data was corrected under assuming a global mean clear-sky probability of 11% (Kadygrov et al. 2009).

Clouds and aerosols are major sources of disturbance in greenhouse gases observations from space because they strongly modify the equivalent optical path length (Mao and Kawa 2004). 3.2% of measurement scenes are suitable for retrieval analysis for July 2009 and 2.7% for January 2010. The monthly mean fraction of the measurement scenes suitable for retrieval analysis is about 3% (Yoshida et al. 2011). In other words, no valid observations were available for GOSAT retrieval analysis on 97% of the earth surface.

The scanning imaging absorption spectrometer for atmospheric chartography (SCIAMACHY) spectrometer on board of the European Space Agency ENVISAT-1 was successfully launched in March 2002. SCIAMACHY first showed the global distribution of the $CO_2$ column abundance from space, although it was not specialized for $CO_2$ observations. SCIAMACHY is a passive hyper-spectral spectrometer designed to investigate tropospheric and stratospheric composition and processes (Bovensmann et al. 1999). The near-infrared nadir spectra of reflected solar radiation measured by SCIAMACHY contain information on the vertical columns of these gases, which we retrieve using the scientific algorithm, weighting function modified differential optical absorption spectroscopy (WFM-DOAS) approach. The horizontal resolution of the nadir measurements depends on orbital position and spectral interval but is typically 60 km for $CO_2$.

Calculations for a few typical aerosol classes, such as mineral dust, soot, sea salt, and sulfate aerosol, and a broad range of atmospheric conditions show that the impact of aerosols on SCIAMACHY retrieved $CO_2$ reaches significant levels. The largest errors are found over the deserts, owing to the high surface albedo in combination with relatively large aerosol loads. For instance, 33% of the data have to be ignored over deserts and the remaining data are used to obtain improved flux estimates. In other words, the voids for SCIAMACHY account for 33% of the total data set under cloud free conditions over deserts. The aerosol induced errors in the measured $CO_2$ total column are too large to allow meaningful source and sink estimates (Houweling et al. 2005).

Any areas of missing data that exist in a $CO_2$ surface are defined as voids. Void filling is the process of filling all no-data areas using the best-performing algorithm available (Reuter et al. 2007). In this section, the method for high accuracy surface modelling (HASM), inverse distance weighting (IDW) and ordinary Kriging are comparatively used to fill voids on $XCO_2$ surfaces of GOSAT and SCIAMACHY in order to identify the best-performing method.

### 3.2.1 An Algorithm of HASM for Void Filling

The solution procedure of HASM, taking the sampled points as its constraints, can be transformed into solving the following linear equation set in terms of least squares principle (Yue 2011),

$$\left[\mathbf{A}^T\ \mathbf{B}^T\ \mathbf{C}^T\ \lambda\cdot\mathbf{S}^T\right]\begin{bmatrix}\mathbf{A}\\ \mathbf{B}\\ \mathbf{C}\\ \lambda\cdot\mathbf{S}\end{bmatrix}\mathbf{z}^{(n+1)}=\left[\mathbf{A}^T\ \mathbf{B}^T\ \mathbf{C}^T\ \lambda\cdot\mathbf{S}^T\right]\begin{bmatrix}\mathbf{d}^{(n)}\\ \mathbf{q}^{(n)}\\ \mathbf{p}^{(n)}\\ \lambda\cdot\mathbf{k}\end{bmatrix} \tag{1}$$

The parameter $\lambda$ is the weight of the sampling points and determines the contribution of the sampling points to the simulated surface. $\lambda$ could be a real number, which means all sampling points have the same weight, or a sector, which means every sampling point has its own weight. An area affected by a sampling point in a complex region is smaller than in a flat region. Therefore, a smaller value of λ is selected in a complex region and a bigger value of λ is selected in a flat region.

Let

$$\mathbf{W} = \begin{bmatrix} \mathbf{A}^T & \mathbf{B}^T & \mathbf{C}^T & \lambda \cdot \mathbf{S}^T \end{bmatrix} \begin{bmatrix} \mathbf{A} \\ \mathbf{B} \\ \mathbf{C} \\ \lambda \cdot \mathbf{S} \end{bmatrix} \tag{2}$$

and,

$$\mathbf{v}^{(n)} = \begin{bmatrix} \mathbf{A}^T & \mathbf{B}^T & \mathbf{C}^T & \lambda \cdot \mathbf{S}^T \end{bmatrix} \begin{bmatrix} \mathbf{d}^{(n)} \\ \mathbf{q}^{(n)} \\ \mathbf{p}^{(n)} \\ \lambda \cdot \mathbf{k} \end{bmatrix} \tag{3}$$

then HASM has the following formulation,

$$\mathbf{W} \cdot \mathbf{z}^{(n+1)} = \mathbf{v}^{(n)} \tag{4}$$

### 3.2.2 Verification

The voids on $XCO_2$ surfaces of GOSAT and SCIAMACHY can be classified into two types: (1) inner voids that are embraced by grids with available data, and boundary voids that are surrounded by both grids with available data and grids without data. We design a verification procedure for these two types of voids (Fig. 1) as follows:

(1) data sets of monthly mean $XCO_2$ from GOSAT on 2.5° latitude–longitude grid and from SCIAMACHY on 0.5° latitude–longitude grid were downloaded respectively from the websites of http://www.gosat.nies.go.jp and http://www.iup.uni-bremen.de/sciamachy;

(2) SCIAMACHY data are up-scaled into a dataset at a spatial resolution of 7.5° × 7.5° using a bilinear re-sampling approach for its comparability with the GOSAT dataset;

(3) according to the data availability, we selected the dataset of monthly mean $XCO_2$ from SCIAMACHY during the period from January to December of 2009 and the one from GOSAT during the period from July of 2009 to June of 2010 so that the two datasets have an overlap for 6 months to compare the performances of GOSAT data and SCIAMACHY data;

(4) nine inner voids in size of 7.5° × 7.5° are artificially made by cleaning out the downloaded data on land, where there are no voids, in the high and middle latitude belts of the northern hemisphere (30°–70°N), the lower latitude belts

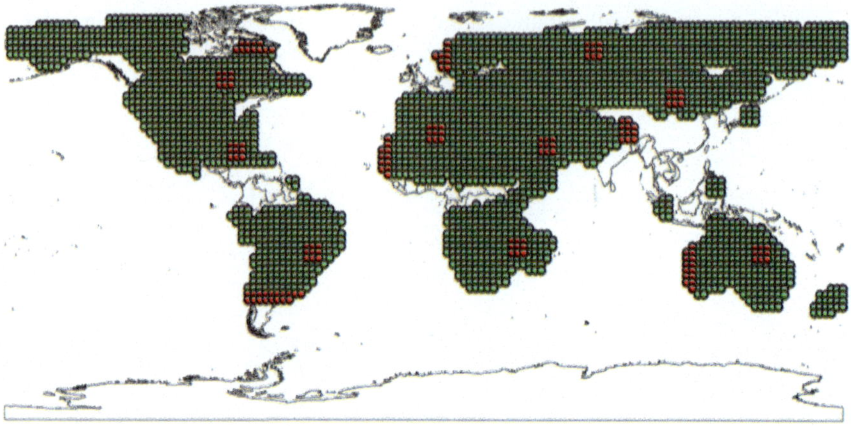


**Fig. 1** Location of the artificial voids

of the northern hemisphere (0°-30°N) and the southern hemisphere (0°–30°S) at random respectively; the cleaned-out data in the nine artificially made voids are selected as the 'true' values for verification;

(5) six boundary voids on land in different sizes, of which the maximum size is 22.5° × 5° and the minimum size is 7.5° × 5°, are artificially made by cleaning out the downloaded data in the three latitude belts and the cleaned-out data are selected as the 'true' values for verification;

(6) the voids on land are filled, taking available $XCO_2$ data over oceans as auxiliary data, by respectively using IDW, OK and HASM; errors of void filling results are comparatively analyzed in terms of mean absolute error (MAE) and mean relative error (MRE).

MAE and MRE are respectively formulated as,

$$MAE(t)=\frac{1}{n}\sum_{i}^{n}|o_i(t)-s_i(t)| \tag{5}$$

$$MRE(t)=100\% \times MAE(t)/\left(\frac{1}{n}\sum_{\mathrm{i}}^{n}o_i(t)\right) \tag{6}$$

where $o_i$ represents the original value at time $t$ on grid $i$; $s_i$ the simulated value on grid $i$.

For the inner voids of GOSAT data set (Tables 1 and 2), MAEs of the inner void filling from IDW, OK and HASM are 0.269 ppmv, 0.23 ppmv and 0.189 ppmv respectively on the average; their MREs are respectively 0.071%, 0.063% and 0.05%.

The maximum MAE of IDW is 0.445 ppmv, appearing in June of 2010; the one of OK is 0.351 ppmv, happening in March of 2010; and the one of HASM is 0.301 ppmv, in June of 2010. The minimum MAE of IDW is 0.099 ppmv in November of 2009; the one of OK is 0.01 ppmv in November of 2009; and the one of HASM is 0.053 ppm in December of 2009.

**Table 1** Mean absolute errors of void filling on $XCO_2$ surfaces of GOSAT (unit: ppmv)

| Time | Errors of inner void filling | | | Errors of boundary void filling | | |
|---|---|---|---|---|---|---|
| | IDW | OK | HASM | IDW | OK | HASM |
| January in 2010 | 0.152 | 0.126 | 0.121 | 0.487 | 0.484 | 0.476 |
| February in 2010 | 0.288 | 0.275 | 0.255 | 0.515 | 0.496 | 0.488 |
| March in 2010 | 0.321 | 0.351 | 0.26 | 0.313 | 0.362 | 0.149 |
| April in 2010 | 0.329 | 0.36 | 0.215 | 0.362 | 0.411 | 0.343 |
| May in 2010 | 0.395 | 0.28 | 0.262 | 0.829 | 0.567 | 0.432 |
| June in 2010 | 0.445 | 0.309 | 0.301 | 0.752 | 0.64 | 0.508 |
| July in 2009 | 0.271 | 0.228 | 0.156 | 0.517 | 0.495 | 0.333 |
| August in 2009 | 0.351 | 0.301 | 0.22 | 0.746 | 0.531 | 0.485 |
| September in 2009 | 0.229 | 0.212 | 0.116 | 0.397 | 0.379 | 0.297 |
| October in 2009 | 0.214 | 0.219 | 0.209 | 0.399 | 0.436 | 0.361 |
| November in 2009 | 0.099 | 0.01 | 0.099 | 0.461 | 0.488 | 0.457 |
| December in 2009 | 0.139 | 0.083 | 0.053 | 0.513 | 0.491 | 0.336 |
| On an average | 0.269 | 0.230 | 0.189 | 0.524 | 0.482 | 0.389 |

**Table 2** Mean relative errors of void filling on $XCO_2$ surfaces of GOSAT (unit: %)

| Time | Errors of inner void filling | | | Errors of boundary void filling | | |
|---|---|---|---|---|---|---|
| | IDW | OK | HASM | IDW | OK | HASM |
| January in 2010 | 0.04 | 0.033 | 0.031 | 0.128 | 0.129 | 0.125 |
| February in 2010 | 0.076 | 0.072 | 0.067 | 0.136 | 0.131 | 0.129 |
| March in 2010 | 0.085 | 0.092 | 0.068 | 0.082 | 0.094 | 0.039 |
| April in 2010 | 0.086 | 0.094 | 0.056 | 0.095 | 0.107 | 0.092 |
| May in 2010 | 0.104 | 0.074 | 0.068 | 0.217 | 0.149 | 0.116 |
| June in 2010 | 0.117 | 0.082 | 0.08 | 0.199 | 0.169 | 0.113 |
| July in 2009 | 0.072 | 0.061 | 0.041 | 0.136 | 0.13 | 0.088 |
| August in 2009 | 0.094 | 0.08 | 0.058 | 0.198 | 0.141 | 0.129 |
| September in 2009 | 0.061 | 0.056 | 0.031 | 0.105 | 0.1 | 0.079 |
| October in 2009 | 0.057 | 0.058 | 0.055 | 0.106 | 0.115 | 0.096 |
| November in 2009 | 0.026 | 0.026 | 0.026 | 0.122 | 0.128 | 0.121 |
| December in 2009 | 0.037 | 0.022 | 0.014 | 0.135 | 0.129 | 0.089 |
| On an average | 0.071 | 0.063 | 0.050 | 0.138 | 0.127 | 0.101 |

For the boundary voids of GOSAT data set (Tables 1 and 2), MAEs of IDW, OK and HASM are 0.524 ppmv, 0.482 ppmv and 0.389 ppmv respectively on average; their MREs are respectively 0.138%, 0.127% and 0.101%. The maximum MAE of IDW is 0.829 ppmv appearing in May of 2010; the minimum one is 0.313 ppmv in March of 2010. The maximum MAEs of OK and HASM are respectively 0.64 ppmv and 0.508 ppmv, both happening in June of 2010. The minimum MAEs of IDW, OK and HASM are respectively 0.313 ppmv, 0.362 ppmv and 0.149 ppmv, all happening in March of 2010.

For inner voids on SCIAMACHY surfaces of $XCO_2$ (Tables 3 and 4), MAEs of IDW, OK and HASM are respectively 1.552 ppmv, 1.618 ppmv and 1.466 ppmv on the average; their MREs are respectively 0.4%, 0.417% and 0.378%. Their maximum MAEs are respectively 1.858 ppmv, 1.928 ppmv and 1.775 ppmv, all appearing in February. The minimum MAEs of IDW and HASM are respectively 1.152 ppmv and 1.135 ppmv, both happening in January; the one of OK is 1.232 ppmv in September.

According to the tests of filling boundary voids on SCIAMACHY surfaces of $XCO_2$ (Tables 3 and 4), the average MAEs of IDW, OK and HASM are respectively 3.675 ppmv, 3.802 ppmv and 3.256 ppmv in 2009; their MREs are respectively 0.947%, 0.977% and 0.846%. The biggest MAEs of IDW, OK and HASM are respectively 5.159 ppmv, 5.638 ppmv and 4.7 ppmv, appearing in July. The smallest MAEs of IDW and OK are respectively 2.643 ppmv and 2.975 ppmv, both appearing in April; the one of HASM is 2.19 ppmv in February.

The verification results indicated that HASM always has the highest accuracy compared with the classical methods of IDW and OK, whether voids are inner ones or boundary ones, data sets are from GOSAT or from SCIAMACHY, and the void areas are larger or smaller. For all the methods of IDW, OK and HASM,

**Table 3** Mean absolute errors of void filling on $XCO_2$ surfaces of SCIAMACHY

| Time | Errors of inner void filling | | | Errors of boundary void filling | | |
|---|---|---|---|---|---|---|
| | IDW | OK | HASM | IDW | OK | HASM |
| January in 2009 | 1.152 | 1.388 | 1.135 | 3.667 | 3.824 | 3.62 |
| February in 2009 | 1.858 | 1.928 | 1.775 | 3.889 | 3.906 | 2.19 |
| March in 2009 | 1.473 | 1.591 | 1.358 | 3.66 | 3.794 | 3.65 |
| April in 2009 | 1.669 | 1.708 | 1.593 | 2.643 | 2.975 | 2.4 |
| May in 2009 | 1.516 | 1.467 | 1.472 | 4.172 | 4.122 | 4.09 |
| June in 2009 | 1.573 | 1.495 | 1.433 | 3.57 | 3.644 | 2.82 |
| July in 2009 | 1.979 | 2.117 | 1.832 | 5.159 | 5.638 | 4.7 |
| August in 2009 | 1.651 | 1.768 | 1.583 | 3.816 | 3.824 | 2.85 |
| September in 2009 | 1.267 | 1.232 | 1.183 | 3.481 | 3.604 | 3.37 |
| October in 2009 | 1.376 | 1.336 | 1.333 | 3.032 | 3.001 | 2.92 |
| November in 2009 | 1.706 | 1.829 | 1.544 | 3.645 | 3.732 | 4.07 |
| December in 2009 | 1.401 | 1.56 | 1.356 | 3.363 | 3.565 | 2.39 |
| On an average | 1.552 | 1.618 | 1.466 | 3.675 | 3.802 | 3.256 |

**Table 4** Mean relative errors of void filling on $XCO_2$ surfaces of SCIAMACHY (unit: %)

| Time | Errors of inner void filling | | | Errors of boundary void filling | | |
|---|---|---|---|---|---|---|
| | IDW | OK | HASM | IDW | OK | HASM |
| January in 2009 | 0.297 | 0.358 | 0.293 | 0.948 | 0.987 | 0.936 |
| February in 2009 | 0.478 | 0.496 | 0.456 | 0.995 | 0.999 | 0.564 |
| March in 2009 | 0.379 | 0.408 | 0.348 | 0.935 | 0.969 | 0.935 |
| April in 2009 | 0.429 | 0.438 | 0.409 | 0.673 | 0.758 | 0.667 |
| May in 2009 | 0.391 | 0.378 | 0.379 | 1.065 | 1.052 | 1.051 |
| June in 2009 | 0.403 | 0.383 | 0.369 | 0.92 | 0.939 | 0.731 |
| July in 2009 | 0.51 | 0.545 | 0.471 | 1.32 | 1.439 | 1.224 |
| August in 2009 | 0.429 | 0.46 | 0.413 | 0.985 | 0.986 | 0.738 |
| September in 2009 | 0.328 | 0.319 | 0.306 | 0.907 | 0.939 | 0.874 |
| October in 2009 | 0.356 | 0.345 | 0.344 | 0.791 | 0.76 | 0.758 |
| November in 2009 | 0.44 | 0.472 | 0.397 | 0.953 | 0.976 | 1.055 |
| December in 2009 | 0.364 | 0.405 | 0.352 | 0.871 | 0.923 | 0.618 |
| On an average | 0.400 | 0.417 | 0.378 | 0.947 | 0.977 | 0.846 |

the void filling errors on $XCO_2$ surfaces of SCIAMACHY are much bigger than the ones on the surfaces of GOSAT, especially for the boundary voids, which may be mainly caused by the greater spatial variability of SCIAMACHY surfaces. In terms of HASM, the average MAE and MRE of GOSAT are 0.269 ppmv and 0.07% respectively under consideration of both inner voids and boundary voids, while the ones of SCIAMACHY are 2.567 ppmv and 0.663% in the 12 months.

### 3.2.3 *Results*

According to the original data, taking the surfaces in August of 2009 as an example (Figs. 2a and 3a), spatial distribution of voids on land surfaces of $XCO_2$ from GOSAT are different comparing with the ones from SCIAMACHY. After GOSAT data, the voids mainly distribute in east-middle Africa, northwestern South America, southern South America, eastern North America, northwestern North America, southern Asia and eastern Asia, but after SCIAMACHY, voids mainly scattered over northwestern South America, southern South America, north North America, western Tibet plateau, eastern Africa, northern Europe and Northern Asia.

The surfaces of $XCO_2$, in which voids have been completely filled, show that spatial variability of SCIAMACHY surface is much higher than GOSAT surface (Figs. 2b and 3b). Standard deviation of GOSAT surfaces was 2.094 ppmv while the one of SCIAMACHY was 4.825 ppmv during the period from July to December in 2009 (Tables 5 and 6). The results from subtracting GOSAT surface from SCIAMACHY surface of $XCO_2$ indicates that almost all grids on SCIAMACHY surfaces

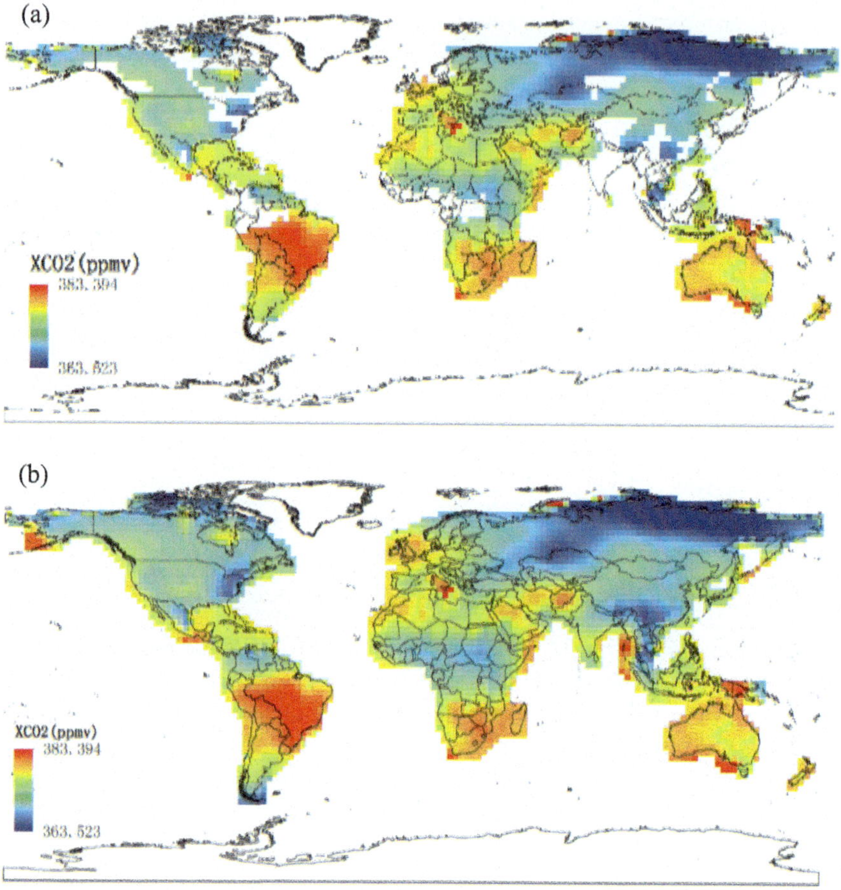


**Fig. 2** The land surfaces of $XCO_2$ from GOSAT in August of 2009, **a** the original surface with voids, **b** the surface on which voids have been filled

have bigger values of $XCO_2$ than the ones on GOSAT surfaces (Tables 5 and 6), except for a few grids on the boundaries of land (Fig. 4). The maximum difference is 28.255 ppmv. The grids, on which the difference is greater than 15 ppmv, distribute in middle Africa, India, western Tibet plateau, northern Russia and northern Alaska.

In terms of the GOSAT surface (Fig. 2b), $XCO_2$ on grids in southern hemisphere is generally much bigger than the ones on grids in northern hemisphere. The reason might be that southern hemisphere is in the season of winter and northern hemisphere in summer. However, this phenomenon could not be found from the SCIAMACHY surface. The difference of $XCO_2$ between SCIAMACHY and GOSAT is 11.463 ppmv on an average in Eurasian continent, 11.274 ppmv in African continent, 9.106 ppmv in the North American continent, 7.281 ppmv in mainland South America, and 6.861 ppmv in Oceania mainland (Table 7).

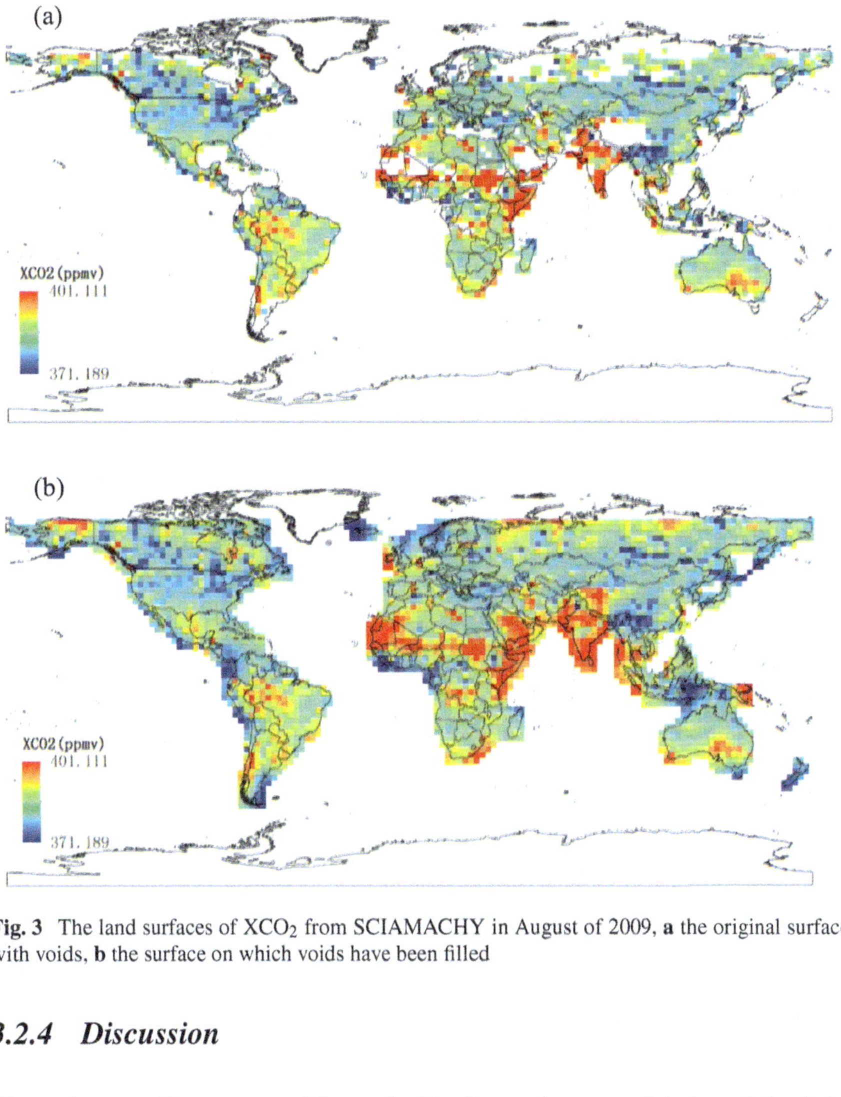


**Fig. 3** The land surfaces of $XCO_2$ from SCIAMACHY in August of 2009, **a** the original surface with voids, **b** the surface on which voids have been filled

### *3.2.4 Discussion*

The carbon satellites are providing valuable data to improve global models of the carbon cycle and to make more accurate predictions of global climate change. Measurements by the carbon satellites allow scientists to monitor the geographic distribution of $CO_2$ sources and to map the natural and human emissions. It becomes possible to simulate interaction mechanisms between cleaner production, the carbon cycle and climate change.

A simulation system for $XCO_2$ surfaces is being developed on the basis of HASM (simply termed as HASM system), which takes satellite remote sensing data as its driving field and ground observation data as its optimum control constraints. The $XCO_2$ surfaces from satellite remote sensing are difficult to be directly used as the

**Table 5** Effects of void filling on mean values and variances of $XCO_2$ surfaces of GOSAT by using HASM

| Time | Before void filling | | After void filling | |
|---|---|---|---|---|
| | Mean value (ppmv) | Standard deviation | Mean value (ppmv) | Standard deviation |
| January in 2010 | 378.745 | 1.63 | 379.14 | 1.659 |
| February in 2010 | 379.102 | 1.886 | 379.514 | 1.815 |
| March in 2010 | 380.384 | 1.91 | 380.597 | 1.622 |
| April in 2010 | 381.38 | 2.004 | 381.532 | 1.984 |
| May in 2010 | 381.708 | 1.956 | 381.349 | 1.933 |
| June in 2010 | 380.473 | 2.277 | 380.117 | 2.258 |
| July in 2009 | 377.649 | 2.623 | 377.22 | 2.629 |
| August in 2009 | 375.403 | 3.305 | 374.562 | 3.303 |
| September in 2009 | 376.096 | 2.093 | 375.566 | 2.299 |
| October in 2009 | 376.949 | 1.839 | 375.76 | 2.428 |
| November in 2009 | 377.905 | 1.281 | 377.256 | 1.913 |
| December in 2009 | 378.179 | 1.422 | 377.948 | 1.946 |
| On an average | 378.664 | 2.019 | 378.380 | 2.149 |

**Table 6** Effects of void filling on mean values and variances of $XCO_2$ surfaces of SCIAMACHY by using HASM

| Time | Before void filling | | After void filling | |
|---|---|---|---|---|
| | Mean value (ppmv) | Standard deviation | Mean value (ppmv) | Standard deviation |
| January in 2009 | 387.465 | 5.15 | 387.194 | 6.546 |
| February in 2009 | 388.728 | 6.072 | 390.249 | 7.031 |
| March in 2009 | 389.823 | 6.632 | 390.798 | 6.681 |
| April in 2009 | 389.883 | 6.486 | 390.391 | 5.988 |
| May in 2009 | 389.173 | 6.017 | 388.315 | 7.898 |
| June in 2009 | 386.352 | 5.481 | 386.761 | 4.754 |
| July in 2009 | 384.754 | 6.004 | 385.606 | 5.9 |
| August in 2009 | 384.887 | 4.956 | 385.158 | 4.115 |
| September in 2009 | 385.334 | 4.593 | 385.361 | 4.051 |
| October in 2009 | 386.639 | 4.451 | 386.097 | 3.594 |
| November in 2009 | 387.151 | 4.434 | 386.949 | 6.443 |
| December in 2009 | 387.519 | 4.509 | 386.052 | 5.702 |
| On an average | 387.309 | 5.399 | 387.411 | 5.725 |

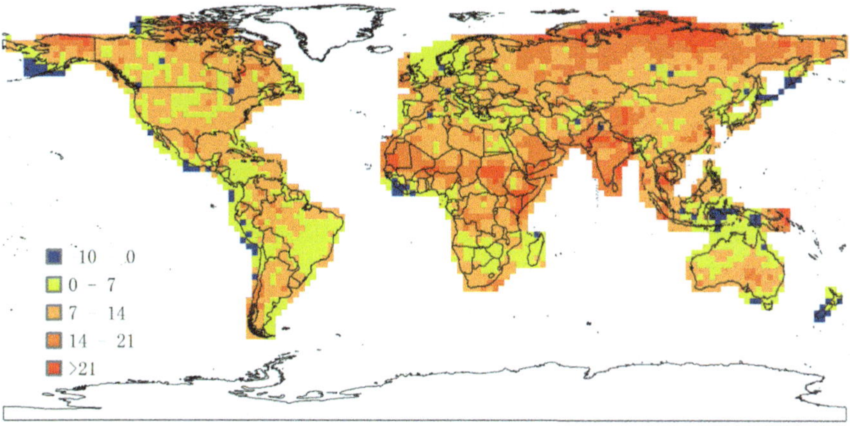


**Fig. 4** The difference map of $XCO_2$ surfaces from GOSAT and SCIAMACHY

**Table 7** Difference of $XCO_2$ from GOSAT and SCIAMACHY on the average in the five continents (unit: ppmv)

| | Eurasian continent | African continent | The North American continent | Mainland South America | Oceania mainland |
|---|---|---|---|---|---|
| GOSAT | 373.289 | 375.972 | 374.114 | 377.729 | 378.2 |
| SCIAMACHY | 384.752 | 387.246 | 383.22 | 385.01 | 385.061 |
| Difference | 11.463 | 11.274 | 9.106 | 7.281 | 6.861 |

driving fields because of a large number of voids caused by clouds, aerosols and the high surface albedo. It is necessary to conduct void filling for constructing the HASM system.

The verification results showed that HASM always has the highest accuracy compared with the classical methods of IDW and OK, whether voids are inner ones or boundary ones, data sets are from GOSAT or from SCIAMACHY, and the void areas are larger or smaller.

In terms of HASM, the MAEs on monthly $XCO_2$ surfaces of GOSAT are less than 0.6 ppmv for both inner voids and boundary voids. Although the void filling errors on the $XCO_2$ surfaces from SCIAMACHY are much bigger than the ones on the surfaces from GOSAT, the MREs on the SCIAMACHY surfaces are less than 0.9% on the average. HASM is an alternative approach to void filling for GOSAT, SCIAMACHY, Orbiting Carbon Observatory -2 (OCO-2) as well as TanSat, to be launched in 2015.

It is essential to validate the $XCO_2$ surfaces for understanding qualities of data sets from different satellites. The Total Carbon Column Observing Network (TCCON) has provided an essential validation resource for assessing the accuracy of $XCO_2$ surfaces from the satellites. However, the TCCON needs to be improved because the ground-based Fourier Transform Spectrometers are too sparsely scattered over the Earth's surface to give sufficiently believable conclusion, especially when TCCON would be also worked as the optimum control constraints for the simulation system for $XCO_2$ surfaces.

## 3.3 Spatial Interpolation of Mean Annual Temperature and Precipitation

Meteorological stations are primary sources for climatic data. However, sparsely distributed meteorological stations are often unable to satisfy the data requirements of most ecosystem change studies. One major problem is how to estimate values for locations where primary data is not available (Akinyemi and Adejuwon 2008).

GIS-based techniques have been widely used for interpolating observed point-based climatic data (Agnew and Palutikof 2000; Yue 2011). For instance, Ordinary kriging (OK) was used to interpolate the daily and monthly rainfall of Australia using ground-based observational data (Jeffrey et al. 2001). Lloyd (2005) compared the performance of different interpolation methods, including a moving window regression (MWR), inverse distance weighting (IDW) and Kriging, demonstrating that methods using elevation as secondary data performed better than others because of the relationships between climate factors, such as temperature, precipitation and evaporation, to elevation. Hancock and Hutchinson (2006) used thin plate smoothing splines (Spline) to interpolate mean annual temperature of the Australian and African continents.

Thiessen polygons (TP), IDW, Spline and OK were used to interpolate thirteen widely scattered rainfall stations and their daily time series into gridded rainfall surfaces over the 1950–1992 period in a West African catchment; assessment of the interpolation methods using reference point data indicated that interpolations using the IDW and OK were more efficient than TP and, to a lesser extent, Spline (Ruelland et al. 2008). Different interpolation models in a GIS environment were used to generate precipitation surfaces for the northwestern Himalayan Mountains and upper Indus plains of Pakistan at a spatial resolution of $250 \times 250$ m$^2$ for a baseline period (1960–1990). This precipitation simulation using a regional climate model (PRECIS) showed that OK, using elevation as secondary data, provided the best results especially for the monsoon months (Ashiq et al. 2010). Three interpolation approaches, IDW, Spline and Co-kriging, were used to interpolate monthly mean temperature, seasonal mean temperature, and annual mean temperature in the eastern part of India; it was found that Spline was preferred to Kriging and IDW because it was faster and easier to use (Samanta et al. 2012).

Many scholars are working to improve the interpolation of climate surfaces by using data from meteorological stations in China. For instance, Shang et al. (2001) employed IDW to interpolate mean annual precipitation from 1951 to 1980, including a digital elevation model (DEM) as secondary data; the interpolation had a mean absolute error of 102.23 mm. Lin et al. (2002) applied different interpolation techniques, OK and IDW, to estimate 10-day mean air temperature from 1951 to 1990; the results indicated that the mean absolute errors for OK and IDW were 2.15 °C and 1.9 °C respectively. Pan et al. (2004) interpolated mean annual temperature using 726 meteorological station observations in China from 1961 to 2000 using IDW with a mean absolute error of 1.51 °C. Hong et al. (2005) used climate data from 1971 to 2000 from meteorological stations in China to develop thin-plate smoothing spline surfaces for monthly mean temperature and precipitation for January, April, July and October. Their results showed interpolation errors for monthly temperatures varying from 0.42 to 0.83 °C and 8–13% for monthly precipitation.

A combination of interpolation methods applied through statistical transfer functions (STFs) is an efficient approach to improve the estimation error of climatic variables for locations where primary data is not available. It has been determined that the statistical relationship between mean annual temperature (MAT) and its environmental determinants is the same no matter where the measurement takes place (Yue 2011). But a simple 'global' model cannot explain the relationship between mean annual precipitation (MAP) and its environmental variables. The MAP relationship changes across space as a function of topographic structure across the landscape. In other words, MAT exhibits spatial stationarity and its statistical transfer function can be expressed by an Ordinary Least Squares regression (OLS), while MAP exhibits spatial non-stationarity and therefore its statistical transfer functions have to be formulated through Geographically Weighted Regression (GWR). In this section, OLS for MAT and GWR for MAP are combined with a high accuracy and high speed method for surface modeling (HASM) to produce surfaces of climatic change for the past 50 years in China at a spatial resolution of 1 km × 1 km.

### *3.3.1 The Algorithms for Spatial Interpolation*

10-day observation data for temperature and precipitation were collected from 752 meteorological stations scattered across China from 1961 to 2010 (National Meteorological Information Center of China: www.nmic.gov.cn) (Fig. 5). A digital terrain model (DTM) from the Shuttle Radar Topography Mission (SRTM) of China at a spatial resolution of 30 m × 30 m (http://srtm.csi.cgiar.org) was upscaled into a DTM at a spatial resolution of 1 km × 1 km using a bilinear resampling approach. The DTM was used as auxiliary data to develop the STFs for MAT and MAP. A variation coefficient is employed to analyze spatial stationarity of MAT and MAP. The variation coefficient can be formulated as:

$$\mathrm{CV} = \frac{\sigma}{\overline{Z}} \times 100\% \tag{7}$$

where $\sigma$ is the standard deviation and $\overline{Z}$ represents a mean property value.

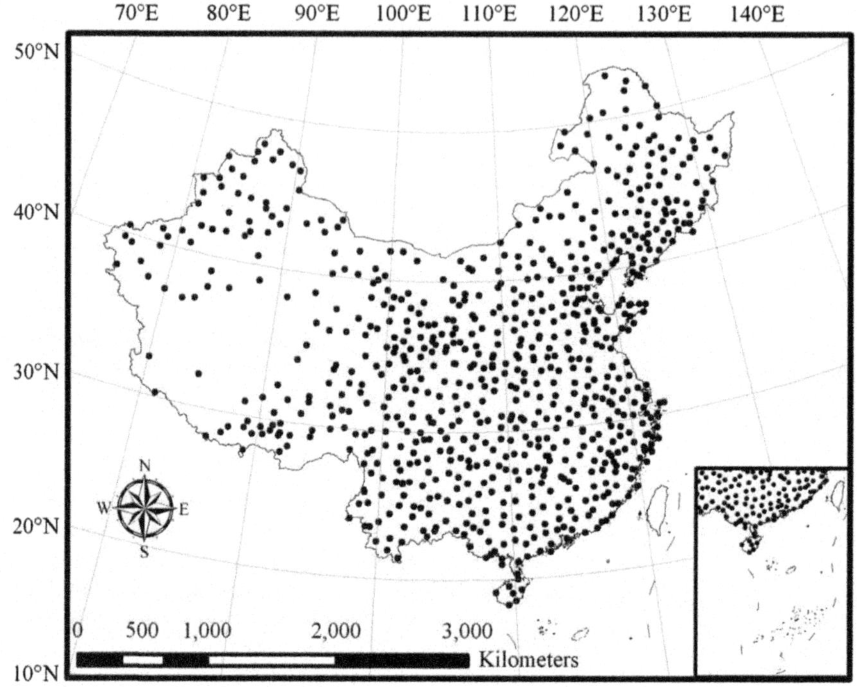


**Fig. 5** Spatial distribution of the meteorological stations in China

A statistical analysis of the time series data from the 752 meteorological stations during the 1961 to 2010 period, using the DEM as secondary data, indicates that MAT from 1961 to 2010 exhibits spatial stationarity, while MAP does exhibit spatial non-stationarity (Wang et and Wang 2012). MAT can be modeled as a linear function of a set of independent variables; local variations in rates of change are allowed for MAP so that the coefficients in the model are localized rather than using global estimates. In other words, regression coefficients for MAP are determined locally and not globally.

For the simulation of MAT from 1961 to 2010, HASM-OLS, a combination of HASM with the linear equation of OLS, was developed due to the spatial stationarity of MAT. Owing to the spatial non-stationarity of MAP during the 1961 to 2010 period, HASM-GWR-BC, an integration of HASM with the GWR equation of Box-Cox transformation of MAP, was developed.

#### 3.3.1.1 Annual Mean Temperature

In terms of the variation, mean annual temperature (MAT) exhibits spatial stationarity and trend surfaces of MAT can be simulated by ordinary least squares (OLS). A statistical transfer function (STF) of MAT was formulated using minimized residuals output from a high accuracy and high speed method for surface modeling (HASM) with an ordinary least squares (OLS) linear equation that used latitude and elevation as independent variables. The linear equation of an ordinary least squares (OLS) between observed and simulated temperatures from 1961 to 2010, with a correlation coefficient of 0.96, can be formulated as,

$$Ts = 38.552 - 0.705 \cdot La - 0.003 \cdot Ele + Te \tag{8}$$

where *Ts* represents the MAT simulated by OLS on a national level during 1961 to 2010 period; *La* = latitude; *Ele* = elevation, and *Te* is the residual.

The residuals of MAT from 1961 to 2010 were calculated by comparing the simulated values using the OLS Eq. (1) with the observed MAT from all meteorological stations. The residuals at all meteorological stations are used to drive HASM and interpolate residuals at every grid cell. The total number of 1 km × 1 km grid cells was 19,606,916 for a rectangle of computational domain covering China. The simulated MAT from 1961 to 2010 at every grid cell $i$, in which $i = 1, 2, \ldots, 19606916$, can be formulated as,

$$Ts_i(t) = \theta_{ols} \cdot \mathbf{x}^T + HASM\left(T_k(t) - \theta_{ols} \cdot \mathbf{x}^T\right) \tag{9}$$

where $\mathbf{x} = (1, Lat, Ele)$, $\theta_{ols} = (38.552, 0.705, 0.003)$, $T_k(t)$ is the MAT observed in the year of $t$ at the meteorological station $k$.

#### 3.3.1.2 Annual Mean Precipitation

Surfaces of latitude, longitude, elevation and aspect regression coefficients as well as intercept can be created by simulating the relationship of MAP from 752 meteorological stations scattered across China with latitude, longitude, elevation and aspect. The average value of the variation coefficients of the surfaces is 154% at the 1% significance level, which indicated that MAP is non-stationary in China. In other words, the regression coefficients of latitude, longitude, elevation and aspect for MAP are best determined locally instead of globally.

MAP values from the 752 meteorological stations were normalized by using the following formulation to stabilize the simulation processes:

$$P(x, y) = \frac{MAP(x, y)}{\mathrm{MAX}\{MAP(x, y)\}} \tag{10}$$

where $P(x, y)$ is a normalization value of MAP at grid cell $(x, y)$ and $MAX\{MAP(x, y)\}$ is the maximum value of MAP from the 752 meteorological stations.

To ensure the normality of GWR, a BOX-COX transformation (Box G E P, Cox D R. 1964) of $P(x, y)$ was conducted as follows:

$$\Psi_{0.4}(P(x,y)) = \left((P(x,y))^{0.4} - 1\right)/0.4 \tag{11}$$

A trend surface of the transformed $P(x, y)$ was then created by means of GWR:

$$\begin{aligned}\Psi_{0.4}(P(x,y)) = {} & \theta_0(x,y) + \theta_1(x,y)\cdot x + \theta_2(x,y)\cdot y \\ & + \theta_3(x,y)\cdot Ele(x,y) + \theta_4(x,y)\cdot ICA(x,y)\end{aligned} \tag{12}$$

where $\theta_1(x, y)$, $\theta_2(x, y)$, $\theta_3(x, y)$ and $\theta_4(x, y)$ as well as $\theta_0(x, y)$ are respectively the coefficients of the latitude, longitude, elevation, aspect, and the intercept.

The normalized MAP with the BOX-COX transformation at a spatial resolution of 1 km × 1 km can be formulated as follows:

$$\begin{aligned}\Psi_{0.4}(MAP(x,y)) &= \Psi_{0.4}(P(x,y)) + HASM\left\{\Psi_{0.4}(P_j(x,y)) - \Psi_{0.4}(P(x,y)\right\} \\ &= \Psi_{0.4}(P(x,y)) + HASM\left\{\varepsilon_j\right\}\end{aligned} \tag{13}$$

where $\varepsilon_j = \Psi_{0.4}(P_j(x,y)) - \Psi_{0.4}(P(x,y)$ is the residual of $\Psi_{0.4}(P(x,y))$ at meteorological station $j$, $HASM\left\{\varepsilon_j\right\}$ represents the spatial interpolation of the residuals, and $\Psi_{0.4}(P_j(x,y))$ is the normalized MAP at meteorological station $j$.

After an inverse BOX-COX transformation and inverse normalization, MAP was finally formulated as follows:

$$MAP(x,y) = MAX\{MAP(x,y)\}\cdot(0.4\cdot\Psi_{0.4}(P(x,y)) + 1)^{\frac{1}{0.4}} \tag{14}$$

The mean absolute and mean relative errors of the interpolated MAP surfaces were formulated as follows:

$$MAE = \frac{1}{n} \times \sum_{j=1}^{n} \left|\mathrm{SI}_j - MAP_j\right| \tag{15}$$

$$MRE = 100\% \times MAE / \left(\frac{1}{n}\sum_{j=1}^{n}\left|MAP_j\right|\right) \tag{16}$$

where $\mathrm{SI}_j$ is the interpolated MAP at the $j$th meteorological station; $MAP_j$ is the observed MAP at meteorological station $j$; and $MAE$ and $MRE$ are the mean absolute error and the mean relative error of the interpolated MAP, respectively.

### 3.3.2 Cross-Validation

Results of the HASM-OLS of MAT and HASM-GWR-BC of MAP were cross-validated using observational data from meteorological stations across China for the same period, for which mean absolute error (MAE) and mean relative error (MRE) of climate variables are calculated.

We used a consistent cross-validation methodology to test OK, IDW, Spline, and HASM results. Validation results of HASM-OLS for MAT during the 1961 to 2010 period were compared to the OK-OLS, IDW-OLS and Spline-OLS results. OK-OLS, IDW-OLS and Spline-OLS respectively refer to the combination of OK, IDW and Spline with the OLS. GWR-BC means that the result of a BOX-COX transformation of MAP during the 1961 to 2010 period at every meteorological station were used to operate the geographically weighted regression. HASM-GWR-BC, OK-GWR-BC, IDW-GWR-BC and Spline-GWR-BC respectively describe the interpolation processes of MAP by combining HASM, OK, IDW and Spline with GWR-BC.

#### 3.3.2.1 Cross-Validation in Space

Cross-validation in space was comprised of four steps: (i) 5% of the meteorological stations were removed for validation prior to model creation; (ii) MAT and MAP from 1961 to 2010 were simulated at a spatial resolution of 1 km × 1 km using the remaining 95% of meteorological stations, (iii) MAE and MRE were calculated using the 5% validation set, and (iv) the 5% validation set is returned to the pool of available station for the next iteration. This process is repeated until MAT and MAP at all meteorological stations have been simulated and the simulation error statistics for each station can be calculated.

Cross validation in space indicates that MAEs of MAT during the 1961 to 2010 period created by HASM-OLS, OK-OLS, IDW-OLS and Spline-OLS were 0.73 °C, 0.81 °C, 0.82 °C and 1.18 °C respectively. MAT accuracy of HASM-OLS was 11%, 12% and 62% higher than OK-OLS, IDW-OLS and Spline-OLS respectively. MAEs of MAP during the 1961 to 2010 period produced by HASM-GWR-BC, OK-GWR-BC, IDW-GWR-BC and Spline-GWR-BC are 48.78 mm, 65.29 mm, 65.10 mm and 124.14 mm respectively (Table 8). MAP accuracy of HASM-GWR-BC was 34%, 34% and 155% greater than the accuracies of OK-GWR-BC, IDW-GWR-BC and Spline-GWR-BC respectively. HASM was the best performer when compared to the more widely used classical methods.

**Table 8** Cross-validation in space

| Mean annual temperature (°C) | | | Mean annual precipitation (mm) | | |
|---|---|---|---|---|---|
| Method | MAE (°C) | MRE (%) | Methods | MAE (mm) | MRE (%) |
| HASM-OLS | 0.73 | 6 | HASM-GWR-BC | 48.78 | 8 |
| OK-OLS | 0.81 | 7 | OK-GWR-BC | 65.29 | 11 |
| IDW-OLS | 0.82 | 7 | IDW-GWR-BC | 65.10 | 11 |
| SPLINE-OLS | 1.18 | 10 | SPLINE-GWR-BC | 124.14 | 17 |
| IDW (Pan et al. 2004) | 1.51 | 13 | IDW (Shang et al. 2001) | 102.23 | 14 |

Results from Pan et al. (2004) and Shang et al. (2001) indicated that the interpolation results of MAT and MAP had MREs of 13% and 14% using IDW (Table 8). In other words, the accuracies of MAT and MAP interpolation were respectively increased by 6% and 3% because of the introduction of the STFs.

#### 3.3.2.2 Cross-Validation in Time

We used MAT and MAP from the 752 meteorological stations within 5 years, randomly selected from all years from the 50-year record, as the validation source, while the remaining data was used for model calibration. In other words, first, MAT and MAP were simulated at a spatial resolution of 1 km × 1 km using data from the 752 meteorological stations for the other 45 years; then, MAE and MRE were calculated using the 5-year validation data; the validation data was then returned to the pool of available MAT and MAP measurements. This process is repeated until MAT and MAP for all 50 years was simulated and error statistics for each simulation calculated (Table 9).

The cross-validation in time demonstrated that MAEs of MAT, created by HASM-OLS, OK-OLS, IDW-OLS and Spline-OLS, were respectively 0.75 °C, 0.95 °C, 0.79 °C and 0.81 °C. MRE of HASM-OLS was 3%, 2% and 2% lower than OK-OLS, IDW-OLS and Spline-OLS respectively. MAEs of MAP produced by HASM-GWR-BC, OK-GWR-BC, IDW-GWR-BC and Spline-GWR-BC are 89.29 mm, 98.08 mm, 93.17 mm and 143.23 mm respectively (Table 9). MRE of HASM-GWR-BC was 6%, 4% and 8% smaller than OK-GWR-BC, IDW-GWR-BC and Spline-GWR-BC respectively. HASM had the highest accuracy when compared to the widely used classic methods.

### *3.3.3 Results*

A zoning system dividing the land mass of China into 9 regions with similar temperature, precipitation and soil regimes was adopted to make it easier to analyze changes in precipitation and temperature from one place to another (Zhou et al. 1981). The 9 regions are respectively termed as $R_i$, i = 1, 2, …, 9 (Fig. 6).

**Table 9** Cross-validation in time

| Mean annual temperature (°C) | | | Mean annual precipitation (mm) | | |
|---|---|---|---|---|---|
| Method | MAE (°C) | MRE (%) | Methods | MAE (mm) | MRE (%) |
| HASM-OLS | 0.75 | 6 | HASM- GWR-BC | 89.29 | 14 |
| OK-OLS | 0.95 | 9 | OK- GWR-BC | 98.08 | 20 |
| IDW-OLS | 0.79 | 8 | IDW- GWR-BC | 93.17 | 18 |
| SPLINE-OLS | 0.81 | 8 | SPLINE- GWR-BC | 143.23 | 22 |

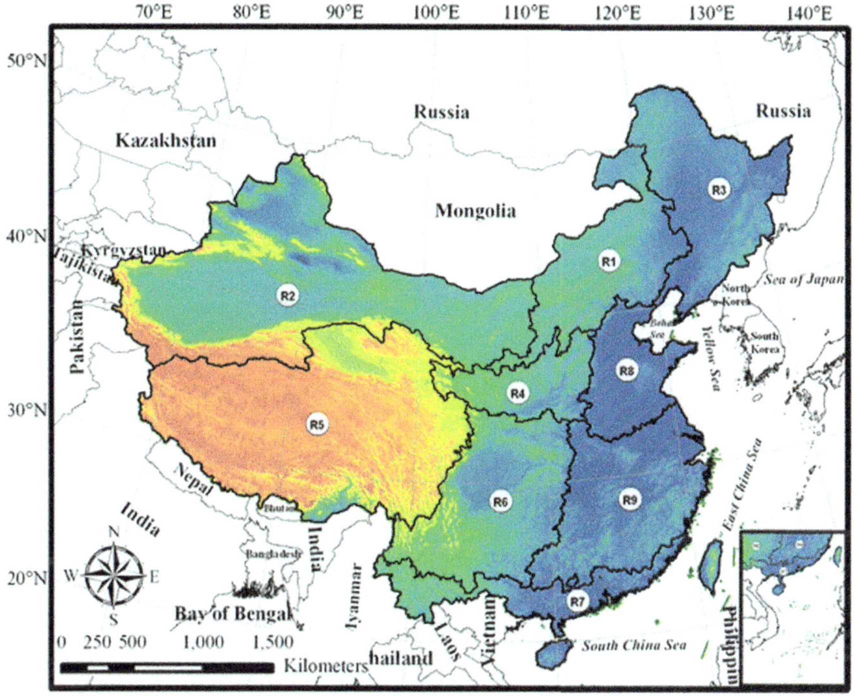


**Fig. 6** Digital elevation model and 9 regions of China

#### 3.3.3.1 Mean Annual Temperature

Annual temperature on average (ATOA) has generally shown an increasing trend in China from 1961 to 2010 with an average decadal increasing rate of 0.33 °C. The linear regression equation of the entire time series of ATOA, with correlation coefficient of 0.85 and significance level of 0.001, can be expressed as,

$$Tem(t) = 0.033t - 58.3 \tag{17}$$

where $Tem(t)$ is ATOA at the time of t; t = 1, 2,..., 49, 50 represents the year of 1961, 1962, ..., 2009 and 2010 respectively.

The significance level of 0.001 implies that there's only one chance in a thousand this could have happened by coincidence. The lower the significance level chosen, the stronger the evidence required. Smaller levels of significance level increase confidence in the determination of significance.

The period from 1961 to 2010 can be divided into 5 sub-periods: $C_1$ (from 1961 to 1970), $C_2$ (from 1971 to 1980), $C_3$ (from 1981 to 1990), $C_4$ (from 1991 to 2000) and $C_5$ (from 2001 to 2010). MATs during the sub-periods of $C_1$, $C_2$, $C_3$, $C_4$ and $C_5$ are respectively 6.91 °C, 7.15 °C, 7.33 °C, 7.82 °C and 8.35 °C. The changes of MATs are summarized in Table 10 where ΔC21 represents the result of subtracting the value of MAT in $C_1$ from the value of the MAT in $C_2$, ΔC32 represents the result of subtracting the value of the MAT in $C_2$ from the one in $C_3$, and so on. ΔC21, ΔC32, ΔC43 and ΔC54 are respectively 0.25 °C, 0.18 °C, 0.47 °C and 0.53 °C for all of China. Their increasing rates are respectively 4%, 3%, 6% and 7%.

The simulation results in terms of HASM-OLS indicate that on a cell-by-cell basis, each sub-period showed either warming or cooling. During the period from $C_1$ to

**Table 10** Change trends of MAT during the 1961 to 2010 period for the 9 regions of China

| Region | | $R_1$ | $R_2$ | $R_3$ | $R_4$ | $R_5$ | $R_6$ | $R_7$ | $R_8$ | $R_9$ | The whole land of China |
|---|---|---|---|---|---|---|---|---|---|---|---|
| From $C_1$ to $C_2$ | MAT in $C_1$ | 3.33 | 5.68 | 1.61 | 8.31 | −0.33 | 14.44 | 20.11 | 12.89 | 16.61 | 6.91 |
| | ΔC21 (°C) | 0.23 | 0.34 | 0.14 | 0.19 | 0.56 | 0.02 | 0.00 | 0.04 | −0.09 | 0.25 |
| | Changing rate (%) | 7 | 6 | 9 | 2 | 170 | 0 | 0 | 0 | −1 | 4 |
| | PGCIT (%) | 94 | 85 | 85 | 95 | 93 | 56 | 48 | 58 | 18 | 75 |
| From $C_2$ to $C_3$ | MAT in $C_2$ | 3.57 | 6.11 | 1.75 | 8.49 | 0.24 | 14.45 | 20.11 | 12.93 | 16.51 | 7.15 |
| | ΔC32 (°C) | 0.30 | 0.23 | 0.57 | 0.02 | 0.14 | −0.08 | 0.06 | 0.14 | 0.04 | 0.18 |
| | Changing rate (%) | 8 | 4 | 33 | 0 | 59 | −1 | 0 | 1 | 0 | 3 |
| | PGCIT (%) | 92 | 87 | 100 | 52 | 78 | 42 | 74 | 77 | 66 | 77 |
| From $C_3$ to $C_4$ | MAT in $C_3$ | 3.87 | 6.34 | 2.33 | 8.50 | 0.37 | 14.37 | 20.18 | 13.07 | 16.52 | 7.33 |
| | ΔC43 (°C) | 0.72 | 0.48 | 0.55 | 0.61 | 0.46 | 0.23 | 0.32 | 0.53 | 0.43 | 0.47 |
| | Changing rate (%) | 19 | 8 | 24 | 7 | 124 | 2 | 2 | 4 | 3 | 6 |
| | PGCIT (%) | 100 | 92 | 98 | 95 | 91 | 83 | 91 | 99 | 100 | 94 |
| From $C_4$ to $C_5$ | MAT in $C_4$ | 4.60 | 6.81 | 2.87 | 9.10 | 0.85 | 14.59 | 20.49 | 13.62 | 16.97 | 7.82 |
| | ΔC54 (°C) | 0.36 | 0.63 | 0.11 | 0.45 | 0.75 | 0.56 | 0.39 | 0.27 | 0.54 | 0.53 |
| | Changing rate (%) | 8 | 9 | 4 | 5 | 90 | 4 | 2 | 2 | 3 | 7 |
| | PGCIT (%) | 84 | 92 | 63 | 83 | 94 | 92 | 89 | 82 | 95 | 88 |

$C_2$, the percentage of 1 km grid cells that had an increasing trend of MAT (PGCIT) was 75% for the entire landmass of China. On a regional basis, all 9 regions had warming trends except for $R_9$ which had a cooling trend. $R_4$ had the biggest PGCIT of 95%, while $R_9$ had the smallest PGCIT of 18%. $R_5$ had the biggest increase of MAT from $C_1$ to $C_2$ on average (Fig. 7). From $C_2$ to $C_3$, PGCIT was 77% and MAT increased an average of 0.18 °C for the entire landmass of China. The increasing rate was 3%. Except for a cooling trend in $R_6$ for this period, all other regions had a warming trend. The highest warming rate was in $R_3$. From $C_3$ to $C_4$, PGCIT was 94%; the rise in MAT was 0.47 °C across China, with a rate of increase of 6%. $R_1$ had the largest PGCIT and the highest rise in MAT; $R_6$ had the smallest PGCIT and the lowest rise in MAT. From $C_4$ to $C_5$, PGCIT was 88%; MAT rose 0.53 °C with an average warming rate of 7% across China. $R_9$ had the largest PGCIT and $R_5$ had the highest increase in MAT.

Except the cooling trend in the sub-period from $C_1$ to $C_2$ in $R_9$ and from $C_2$ to $C_3$ in $R_6$, all other regions in China for all other sub-periods had a warming trend.

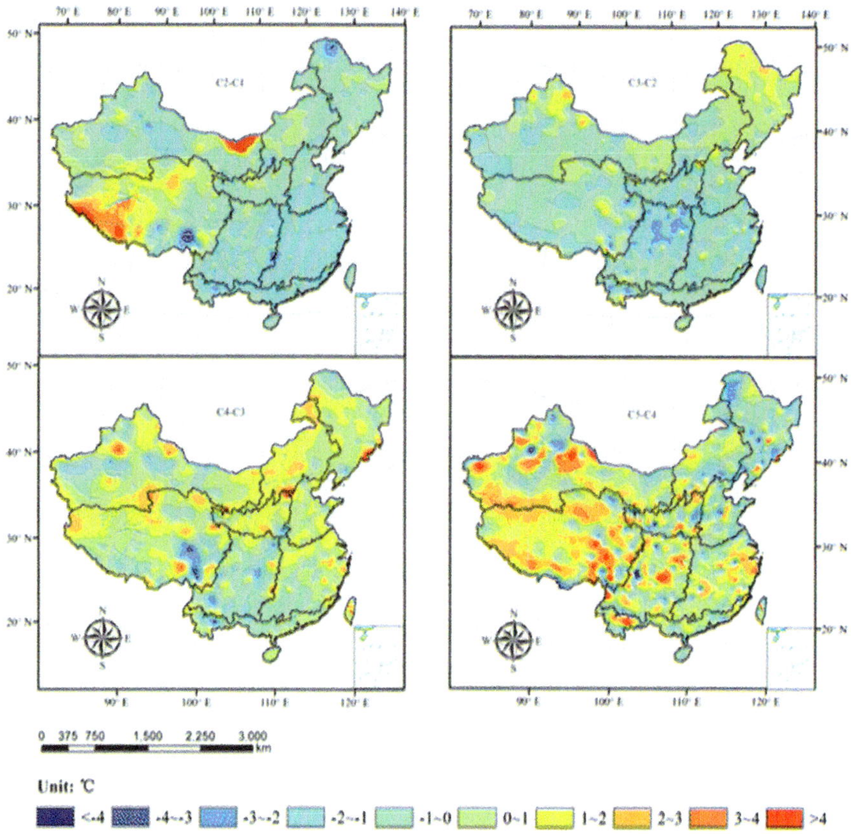


**Fig. 7** Changes in MAT from 1960 to 2000s

Especially in $R_5$, Qinghai-Xizang plateau, MAT increased by 1.91 °C with a warming rate of 443% in last 50 years. MAT in regions of $R_1$, $R_2$, $R_3$ and $R_4$ increased by 1.61, 1.68, 1.37 and 1.27 °C with rising rates of 42%, 27%, 79% and 14% respectively, while there were lower warming rates in the regions of $R_6$, $R_7$, $R_8$ and $R_9$. MAT in these regions were higher 0.73, 0.77, 0.98 and 0.92 °C in the last 50 years, with warming rates of 5%, 4%, 7% and 5% respectively.

#### 3.3.3.2 Mean Annual Precipitation

MAP averaged 583.78, 585.16, 591.08, 593.03 and 585.78 mm for the decadal sub-periods of $C_1$, $C_2$, $C_3$, $C_4$ and $C_5$ respectively across China. MAP increased by 1.39, 5.93 and 1.93 mm from $C_1$ to $C_2$, from $C_2$ to $C_3$ and from $C_3$ to $C_4$ respectively, while decreasing by 7.25 mm from $C_4$ to $C_5$. In Table 11, ΔC21 represents the result of subtracting the value of MAP in $C_1$ from the value of the MAP in $C_2$, ΔC32 represents the result of subtracting the value of the MAP in $C_2$ from the one in $C_3$, and so on.

Simulation results of HASM-GWR-BC indicate that portions of all 9 regions and all 5 sub-periods showed both increased and decreased precipitation. From $C_1$ to $C_2$, 55% of China showed an increasing trend in MAP (PGCIP); MAP increased by 1.39 mm across China on average and the increasing rate approached to 0. $R_5$ had the largest PGCIP and the highest increasing rate. $R_4$ the smallest PGCIP and the highest decrease in MAP. $R_7$ had the highest increase in MAP (Fig. 8). From $C_2$ to $C_3$, PGCIP was 59% and MAP increased by 5.93 mm at an average rate of 1% across China. $R_3$ had the biggest PGCIP, the highest increase in MAP and the largest increasing rate. $R_8$ exhibited the smallest PGCIP, the largest decrease of MAP and the biggest decreasing rate. From $C_3$ to $C_4$, PGCIP was 53%; MAP increased by 1.93 mm with an average rate approaching 0 across China. $R_9$ had the biggest PGCIP and the highest rise in MAP. $R_2$ had the largest increasing rate. The smallest PGCIP appeared in $R_4$. $R_4$ exhibited the deepest drop in MAP and the largest decreasing rate. From $C_4$ to $C_5$, PGCIP was 53% and MAP decreased by −7.25 mm, at a rate of −1%, across China. $R_4$ had the largest PGCIP. $R_8$ had the biggest increase in MAP. $R_1$ had the smallest PGCIP and the deepest dropping rate. $R_9$ had the deepest drop in MAP.

Since 1960s, MAP in $R_5$ and $R_2$ has had a continuously increasing trend; MAP increased by 52.55 and 20.86 mm with the increasing rates of 13% and 17% in recent 50 years respectively. In other 7 regions, MAP has been variably changed and increasing MAP alternated with decreasing MAP. On average, MAP in the regions of $R_1$, $R_4$, $R_6$ and $R_8$ decreased by 41.78, 58.14, 62.05 and 20.68 mm, with changing rates of −12%, −11%, −6% and −3% respectively, in recent 50 years; in $R_3$, $R_7$ and $R_9$, MAP increased by 4.5, 25.83 and 0.61 mm, with rising rates of 2%, 1% and 0.

**Table 11** Change trends of MAP during the 1961 to 2010 period for the 9 regions of China

| Region | | $R_1$ | $R_2$ | $R_3$ | $R_4$ | $R_5$ | $R_6$ | $R_7$ | $R_8$ | $R_9$ | The whole land of China |
|---|---|---|---|---|---|---|---|---|---|---|---|
| From $C_1$ to $C_2$ | MAP (mm) in $C_1$ | 325 | 117 | 547 | 530 | 372 | 1038 | 1465 | 735 | 1415 | 584 |
| | ΔC21 (mm) | −3 | 3 | −25 | −44 | 20 | −23 | 47 | −19 | 15 | 1 |
| | Changing rate (%) | −1 | 3 | −5 | −8 | 5 | −2 | 3 | −3 | 1 | 0 |
| | PGCIP (%) | 55 | 71 | 15 | 5 | 86 | 18 | 64 | 11 | 58 | 55 |
| From $C_2$ to $C_3$ | MAP (mm) in $C_2$ | 321 | 121 | 522 | 486 | 391 | 1015 | 1513 | 716 | 1430 | 585 |
| | ΔC32 (mm) | 6 | 4 | 92 | 15 | 5 | −5 | −14 | −47 | 3 | 6 |
| | Changing rate (%) | 2 | 3 | 18 | 3 | 1 | −1 | −1 | −7 | 0 | 1 |
| | PGCIP (%) | 45 | 61 | 98 | 68 | 59 | 57 | 44 | 14 | 58 | 59 |
| From $C_3$ to $C_4$ | MAP (mm) in $C_3$ | 328 | 124 | 614 | 502 | 397 | 1010 | 1499 | 670 | 1434 | 591 |
| | ΔC43 (mm) | 4 | 9 | −30 | −54 | 3 | −40 | 17 | 7 | 72 | 2 |
| | Changing rate (%) | 1 | 7 | −5 | −11 | 1 | −4 | 1 | 1 | 5 | 0 |
| | PGCIP (%) | 57 | 76 | 7 | 4 | 45 | 37 | 72 | 58 | 83 | 53 |
| From $C_4$ to – $C_5$ | MAP (mm) in $C_4$ | 331 | 133 | 549 | 447 | 399 | 970 | 1516 | 676 | 1505 | 593 |
| | ΔC54 (mm) | −48 | 5 | −32 | 24 | 25 | 5 | −25 | 38 | −90 | −7 |
| | Changing rate (%) | −14 | 4 | −6 | 5 | 6 | 1 | −2 | 6 | −6 | −1 |
| | PGCIP (%) | 8 | 60 | 15 | 88 | 87 | 62 | 27 | 79 | 11 | 53 |

### 3.3.4 Discussion

Topography of land in China from Qinghai-Xizang plateau eastward, is broadly arranged into three great steps (Zhao 1986). The first step is Qinghai-Xizang plateau, with mean elevation above 4000 m; the second step is from the eastern margin of

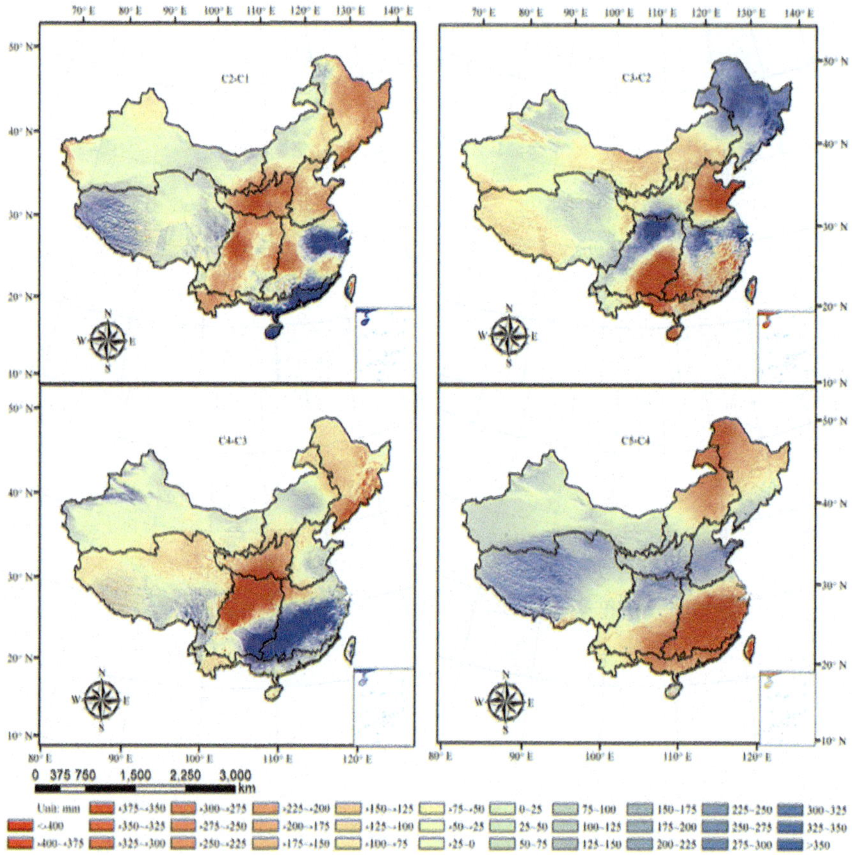


**Fig. 8** Changes in MAP from 1960 to 2000s

Qinghai-Xizang plateau eastward up to Da Hinggan-Taihang-Wushan mountains line, with elevations from 2000 to 1000 m; the third step is from the above-mentioned line eastward up to the coast, below 500 m in elevation (Fig. 6). The warming rates increase from south to north in the second and third steps of China. In particular, the 2100 °C $\cdots$ d contour line of annual accumulated temperature (AAT) of $\geq$ 10 °C in Heilongjiang province of $R_3$ shifted northwestward 255 km since the 1960s (Supplement 4). Areas suitable for planting early rice expanded about 3.8 million hectares because of the AAT change. The 2400 °C $\cdots$ d contour line of $\geq$ 10 °C AAT moved northward at least 167 km since 1960, which made the area suitable for planting early maize increase by about 5.5 million hectares in Heilongjiang province.

MAT had an increasing trend during the period from 1960 to 2010 in China, with an especially accelerating increase trend since the 1980s. MAT was respectively 6.91 °C, 7.15 °C、7.33 °C, 7.82 °C and 8.35 °C during the sub-periods of $C_1$, $C_2$, $C_3$, $C_4$ and $C_5$. The increasing rates of MAT were respectively 4%, 3%, 6% and 7% during the sub-periods from $C_1$ to $C_2$, from $C_2$ to $C_3$, from $C_3$ to $C_4$ and from $C_4$ to

$C_5$. The biggest values of MWRs and MCRs appeared in the regions of $R_1$, $R_2$, $R_3$ and $R_5$ in the four sub-periods. According to both changing rates and amplitudes, Qinghai-Xizang Plateau and Northern China had more extremes of temperature than Southern China since the 1960s.

MAP was 583.78 mm in $C_1$, 585.16 mm in $C_2$, 591.08 mm in $C_3$, 593.03 mm in $C_4$ and 585.78 mm in $C_5$. MAP increased by 1.39 mm from $C_1$ to $C_2$, by 5.93 mm from $C_2$ to $C_3$, and by 1.93 mm from $C_3$ to $C_4$, while MAP decreased by 7.254 mm from $C_4$ to $C_5$. On average, China became wetter during the period from 1960 to 1990s, but much drier from 1990 to 2000s. Since the 1960s, MAP in Qinghai-Xizang plateau and in arid region has had a continuously increasing trend, with the increasing rates of 13% and 17% in recent 50 years respectively. In other regions, increasing MAP alternated with decreasing MAP decade by decade.

In terms of changing rates, north China and Qinghai-Xizang plateau had more extremes of precipitation in the last five decades. According to changing amplitudes, South China and Sichuan basin had more extremes of precipitation.

## 3.4 Summary

Spatial interpolation consists of two kinds of algorithms. One is to estimate values for locations where primary data is not available by using sparsely distributed primary point-data. Another one is to construct a continuous surface on the basis of a surface with data voids.

Observation data for temperature and precipitation, collected from 752 meteorological stations scattered across China, are interpolated to form surfaces of MAT and MAP on spatial resolution of 1 km $\times$ 1 km. To do so, a DTM of China is used as auxiliary data to develop the STFs for MAT and MAP. Cross validations in space and in time demonstrate that HASM-based interpolation performs best when compared to the more widely used classical methods.

We tried to develop a simulation system for $XCO_2$ surfaces by taking satellite remote sensing data as its driving field and ground observation data as its optimum control constraints. However, the $XCO_2$ surfaces from satellite remote sensing are difficult to be directly used as the driving fields because of a large number of voids caused by clouds, aerosols and the high surface albedo. HASM, IDW and OK are comparatively used to fill voids on $XCO_2$ surfaces of GOSAT and SCIAMACHY. The verification results show that HASM always has the highest accuracy compared with the classical methods of IDW and OK.

## 3.5 Further Reading

(1) Yue TX, Zhao MW, Zhang XY. 2015. A high-accuracy method for filling voids on remotely sensed $XCO_2$ surfaces and its verification. Journal of Cleaner Production 103: 819–827.
(2) Yue TX, Zhao N, Ramsey D, Wang CL, Fan ZM, Chen CF, Lu YM, Li BL. 2013. Climate change trend in China, with improved accuracy. Climatic Change 120: 137–151.

### Exercises

(1) Define the terms "spatial interpolation".
(2) What is the meaning and advantages of HASM-based spatial interpolation (HASM-SI)?
(3) What are the logical relations between the fundamental theorem for eco-environmental surface modelling (FTEEM) and HASM-SI?

## References


Adhikary, S.K., N. Muttil, and A.G. Yilmaz. 2017. Cokriging for enhanced spatial interpolation of rainfall in two Australian catchments. *Hydrological Processes* 31: 2143–2161.

Agnew, M.D., and J.P. Palutikof. 2000. GIS-based construction of baseline climatologies for the Mediterranean using terrain variables. *Climatic Research Unit* 14: 115–127.

Akinyemi, F.O., and J.O. Adejuwon. 2008. A GIS-based procedure for downscaling climate data for west Africa. *Transactions in GIS* 12 (5): 613–631.

Arowolo, A.O., A.K. Bhowmik, W. Qi, and X.Z. Deng. 2017. Comparison of spatial interpolation techniques to generate high-resolution climate surfaces for Nigeria. *International Journal of Climatology* 37: 179–192.

Ashiq, M.W., C.Y. Zhao, J. Ni, and M. Akhtar. 2010. GIS-based high-resolution spatial interpolation of precipitation in mountain–plain areas of Upper Pakistan for regional climate change impact studies. *Theoretical and Applied Climatology* 99: 239–253.

Baker, D.F., H. Bosch, S.C. Doney, D. O'Brien, and D.S. Schimel. 2010. Carbon source/sink information provided by measurements from the Orbiting Carbon Observatory. *Atmospheric Chemistry and Physics* 10: 4145–4165.

Bengtsson, B.E., and S. Nordbeck. 1964. Construction of is arithms and is arithmic maps by computers. *BIT* 4: 87–105.

Birkhoff, G., and H. Garabedian. 1960. Smooth surface interpolation. *Journal of Mathematics and Physics* 39: 258–326.

Bovensmann, H., J.P. Burrows, M. Buchwitz, J. Frerick, S. Noeel, V. Rozanov, K.V. Chance, and A. Goede. 1999. SCIAMACHY–mission objectives and measurement modes. *Journal of the Atmospheric Sciences* 56: 127–150.

Buchwitz, M., R. de Beek, S. Noel, J.P. Burrows, H. Bovensmann, O. Schneising, I. Khlystova, M. Bruns, H. Bremer, P. Bergamaschi, S. Koerner, and M. Heimann. 2006. Atmospheric carbon gases retrieved from SCIAMACHY by WFM-DOAS: Version 0.5 CO and CH4 and impact of calibration improvements on $CO_2$ retrieval. *Atmospheric Chemistry and Physics* 6: 2727–2751.

Chang, N. 2015. Changing industrial structure to reduce carbon dioxide emissions: A Chinese application. *Journal of Cleaner Production* 103: 40–48.

Chen, Z.D., S.B. Dikgwatlhe, J.F. Xue, H.L. Zhang, F. Chen, and X.P. Xiao. 2015. Tillage impacts on net carbon flux in paddy soil of the Southern China. *Journal of Cleaner Production* 103: 70–76.

Chevallier, F., S. Maksyutov, P. Bousquet, F.M. Breon, R. Saito, Y. Yoshida, and T. Yokota. 2009. On the accuracy of the $CO_2$ surface fluxes to be estimated from the GOSAT observations. *Geophysical Research Letters* 36: L19807. https://doi.org/10.1029/2009GL040108.

Chuai, X.W., X.J. Huang, W.J. Wang, R.Q. Zhao, M. Zhang, C.Y. Wu. 2015. Land use, total carbon emissions change and low carbon land management in Coastal Jiangsu, China. Journal of Cleaner Production, 77–86.

De Boor, C. 1962. Bicubic spline interpolation. *Journal of Mathematics and Physics* 41: 212–218.

Franke, R. 1982. Scattered data interpolation: Tests of some methods. *Mathematics of Computation* 38: 181–200.

Hadi, S.J., and M. Tombul. 2018. Comparison of spatial interpolation methods of precipitation and temperature using multiple integration periods. *Journal of the Indian Society of Remote Sensing* 46 (7): 1187–1199.

Hancock, P.A., and M.F. Hutchinson. 2006. Spatial interpolation of large climate data sets using bivariate thin plate smoothing splines. *Environmental Modelling and Software* 21: 1684–1694.

Hao, Y., M.R. Su, L.X. Zhang, Y.P. Cai, and Z.F. Yang. 2014. Integrated accounting of urban carbon cycle in Guangyuan, a mountainous city of China: The impacts of earthquake and reconstruction. *Journal of Cleaner Production* 103: 231–240.

Harder, R.L., and R.N. Desmarais. 1972. Interpolation using surface Splines. *Journal of Aircraft* 9: 189–191.

Hardy, R.L. 1971. Multiquadric equations of topography and other irregular surfaces. *Journal of Geophysical Research* 76: 1905–1915.

Henderson, D.W. 1998. *Differential geometry*. London: Prentice-Hall Inc.

Hong, Y., H.A. Nix, M.F. Hutchinson, and T.H. Booth. 2005. Spatial interpolation of monthly mean climate data for China. *International Journal of Climatology* 25: 1369–1379.

Houweling, S., W. Hartmann, I. Aben, H. Schrijver, J. Skidmore, G.J. Roelofs, and F.M. Breon. 2005. Evidence of systematic errors in SCIAMACHY-observed $CO_2$ due to aerosols. *Atmospheric Chemistry and Physics* 5: 3003–3013.

Jeffrey, S.J., J.O. Carter, K.B. Moodie, and A.R. Beswick. 2001. Using spatial interpolation to construct a comprehensive archive of Australian climate data. *Environmental Modelling and Software* 16: 309–330.

Kadygrov, N., S. Maksyutov, N. Eguchi, T. Aoki, T. Nakazawa, T. Yokota, and G. Inoue. 2009. Role of simulated GOSAT total column $CO_2$ observations in surface $CO_2$ flux uncertainty reduction. *Journal of Geophysical Research* 114: D21208. https://doi.org/10.1029/2008JD011597.

Krige, D.G. 1951. A statistical approach to some basic mine valuation problems on the Witwatersrand. *Journal of the Chemical, Metallurgical and Mining Society of South Africa* 52 (6): 119–139.

Li, J., and A.D. Heap. 2014. Spatial interpolation methods applied in the environmental sciences: A review. *Environmental Modelling & Software* 53: 173–189.

Lin, Z.H., X.G. Muo, H.X. Li, and H.B. Li. 2002. Comparison of three spatial interpolation methods for climate variables in China. *Acta Geographica Sinica* 57: 47–56 (in Chinese).

Liseikin, V.D. 2004. *A computational differential geometry approach to grid generation*. Berlin: Springer.

Lloyd, C.D. 2005. Assessing the effect of integrating elevation data into the estimation of monthly precipitation in Great Britain. *Journal of Hydrology* 308: 128–150.

Makarieva, A.M., V.G. Gorshkov, D. Sheil, A.D. Nobre, P. Bunyard, and B.L. Li. 2014. Why does air passage over forest yield more rain? Examining the coupling between rainfall, pressure, and atmospheric moisture content. *Journal of Hydrometeorology* 15: 411–426.

Mao, J., and S.R. Kawa. 2004. Sensitivity studies for space-based measurement of atmospheric total column carbon dioxide by reflected sunlight. *Applied Optics* 43: 914–927.

Meinshausen, M., N. Meinshausen, W. Hare, S. Raper, K. Frieler, R. Knutti, D. Frame, and M. Allen. 2009. Greenhouse gas emission targets for limiting global warming to 2 °C. *Nature* 458: 1158–1163.

Mimikou, M.A., and E.A. Baltas. 2013. Assessment of climate change impacts in Greece: A general overview. *American Journal of Climate Change* 3: 46–56.

Moss, R., M. Babiker, S. Brinkman, E. Calvo, T. Carter, J. Edmonds, I. Elgizouli, S. Emori, L. Erda, K. Hibbard, R. Jones, M. Kainuma, J. Kelleher, J.F. Lamarque, M. Manning, B. Matthews, J. Meehl, L. Meyer, J. Mitchell, N. Nakicenovic, B. O'Neill, R. Pichs, K. Riahi, S. Rose, P. Runci, R. Stouffer, D. van Vuuren, J. Weyant, T. Wilbanks, J.P. van Ypersele, and M. Zurek. 2008. Towards new scenarios for analysis of emissions, climate change, impacts, and response strategies. Technical Summary. Intergovernmental Panel on Climate Change, Geneva.

Oberheitmann, A. 2010. A new post-Kyoto climate regime based on per-capita cumulative emissions rights—Rationale, architecture and quantitative assessment of the implication for the $CO_2$-emissions from China, India and the Annex-I countries by 2050. *Mitigation and Adapta-Tion Strategies for Global Change* 15 (2): 137–168.

Oberheitmann, A. 2013. Some remarks on the individual contribution to climate change. *American Journal of Climate Change* 2: 198–202.

Pan, Y.Z., D.Y. Gong, L. Deng, J. Li, and J. Gao. 2004. Smart distance searching based and DEM informed interpolation of surface air temperature in China. *Acta Geographica Sinica* 59: 366–374 (in Chinese).

Pereira, P., M. Oliva, and I. Misiune. 2016. Spatial interpolation of precipitation indexes in Sierra Nevada (Spain): Comparing the performance of some interpolation methods. *Theoretical and Applied Climatology* 126: 683–698.

Petsri, S., A. Chidthaisong, N. Pumijumnong, and C. Wachrinrat. 2013. Greenhouse gas emissions and carbon stock changes in rubber tree plantations in Thailand from 1990 to 2004. *Journal of Cleaner Production* 52: 61–70.

Quarteroni, A., R. Sacco, and F. Saleri. 2000. *Numerical mathematics*. New York: Springer.

Raisanen, J. 2007. How reliable are climate models? *Tellus* 59A: 2–29.

Rayner, P.J., and D.M. O'Brien. 2001. The utility of remotely sensed $CO_2$ concentration data in surface source inversions. *Geophysical Research Letters* 28: 175–178.

Reuter, H.I., A. Nelson, and A. Jarvis. 2007. An evaluation of void-filling interpolation methods for SRTM data. *International Journal of Geographical Information Science* 21: 983–1008.

Ruelland, D., S. Ardoin-Bardin, G. Billen, and E. Servat. 2008. Sensitivity of a lumped and semi-distributed hydrological model to several methods of rainfall interpolation on a large basin in West Africa. *Journal of Hydrology* 361: 96–117.

Samanta, S., D.K. Pal, D. Lohar, and B. Pal. 2012. Interpolation of climate variables and temperature modeling. *Theoretical and Applied Climatology* 107: 35–45.

Schoenberg, I.J. 1946. Contributions to the problem of approximation of equidistant data by analytic functions. *Quarterly of Applied Mathematics* 4 (45–99): 112–141.

Shang, Z.B., Q. Guo, and D.A. Yang. 2001. Spatial pattern analysis of annual precipitation with climate information system of China. *Acta Ecologica Sinica* 21: 689–694 (in Chinese).

Shepard, D. 1968. A two-dimensional interpolation function for irregularly-spaced data. Proceedings of the 1968 23rd ACM national conference. New York: Association for Computing Machinery.

Shi, W.J., J.Y. Liu, Y.J. Song, Z.P. Du, C.F. Chen, and T.X. Yue. 2009. Surface modelling of soil pH. *Geoderma* 150 (1–2): 113–119.

Shi, W.J., J.Y. Liu, Z.P. Du, A. Stein, and T.X. Yue. 2011. Surface modelling of soil properties based on land use information. *Geoderma* 162 (3–4): 347–357.

Sibson, R. 1981. A brief description of natural neighbor interpolation. In *Interpolating multivariate data*, ed. by V. Barnett, 21–36. Chichester: John Wiley.

Talmi, A., and G. Gilat. 1977. Method for smooth approximation of data. *Journal of Computational Physics* 23: 93–123.

Toponogov, V.A. 2006. *Differential geometry of curves and surfaces*. New York: Birkhaeuser Boston.

Wagner, P.D., P. Fiener, F. Wilken, S. Kumar, and K. Schneider. 2012. Comparison and evaluation of spatial interpolation schemes for daily rainfall in data scarce regions. *Journal of Hydrology* 464–465: 388–400.

Wang, H.W., and Y.H. Wang. 2012. Analysis of spatial interpolation methods. *Lecture Notes in Electrical Engineering* 129: 507–512.

Wang, S., G.H. Huang, Q.G. Lin, Z. Li, H. Zhang, and Y.R. Fan. 2014. Comparison of interpolation methods for estimating spatial distribution of precipitation in Ontario, Canada. *International Journal of Climatology* 34: 3745–3751.

West, T.O., and G. Marland. 2002. A synthesis of carbon sequestration, carbon emissions, and net carbon flux in agriculture: Comparing tillage practices in the United States. *Agriculture Ecosystems & Environment* 91: 217–232.

Xu, W.B., Y.J. Zou, G.P. Zhang, and M. Linderman. 2015. A comparison among spatial interpolation techniques for daily rainfall data in Sichuan Province, China. *International Journal of Climatology* 35: 2898–2907.

Yakowitz, S.J., and F. Szidarovsky. 1985. A comparison of Kriging with nonparametric regression methods. *Journal of Multivariate Analysis* 16 (1): 21–53.

Yoshida, Y., Y. Ota, N. Eguchi, N. Kikuchi, K. Nobuta, H. Tran, I. Morino, and T. Yokota. 2011. Retrieval algorithm for $CO_2$ and $CH_4$ column abundances from short-wavelength infrared spectral observations by the Greenhouse gases observing satellite. *Atmospheric Measurement Techniques* 4: 717–734.

Yue, T.X. 2011. *Surface modelling: High accuracy and high speed methods*. New York: CRC Press.

Yue, T.X., and S.H. Wang. 2010. Adjustment computation of HASM: A high-accuracy and high-speed method. *International Journal of Geographical Information Science* 24 (11): 1725–1743.

Yue, T.X., Z.P. Du, D.J. Song, and Y. Gong. 2007. A new method of surface modeling and its application to DEM construction. *Geomorphology* 91 (1–2): 161–172.

Yue, T.X., D.J. Song, Z.P. Du, and W. Wang. 2010a. High-accuracy surface modelling and its application to DEM generation. *International Journal of Remote Sensing* 31 (8): 2205–2226.

Yue, T.X., C.F. Chen, and B.L. Li. 2010b. An adaptive method of high accuracy surface modeling and its application to simulating elevation surfaces. *Transactions in GIS* 14 (5): 615–630.

Yue, T.X., C.F. Chen, and B.L. Li. 2012. A high accuracy method for filling SRTM voids and its verification. *International Journal of Remote Sensing* 33 (9): 2815–2830.

Yue, T.X., N. Zhao, H. Yang, Y.J. Song, Z.P. Du, Z.M. Fan, and D.J. Song. 2013a. The multi-grid method of high accuracy surface modelling and its validation. *Transactions in GIS* 17 (6): 943–952.

Yue, T.X., N. Zhao, D. Ramsey, C.L. Wang, Z.M. Fan, C.F. Chen, Y.M. Lu, and B.L. Li. 2013b. Climate change trend in China, with improved accuracy. *Climatic Change* 120: 137–151.

Yue, T.X., M.W. Zhao, and X.Y. Zhang. 2015. A high-accuracy method for filling voids on remotely sensed $XCO_2$ surfaces and its verification. *Journal of Cleaner Production* 103: 819–827.

Yue, T.X., N. Zhao, Y. Liu, Y.F. Wang, B. Zhang, Z.P. Du, Z.M. Fan, W.J. Shi, C.F. Chen, M.W. Zhao, D.J. Song, Y.J. Song, C.Q. Yan, Q.Q. Li, X.F. Sun, L.L. Zhang, Y.Z. Tian, W. Wang, Y.A. Wang, S.N. Ma, H.S. Huang, Y.M. Lu, Q. Wang, C.L. Wang, Y.Z. Wang, M. Lu, W. Zhou, Y. Liu, X.Z. Yin, Z. Wang, Z.Y. Bao, M.M. Zhao, Y.P. Zhao, Y.M. Jiao, U. Naseer, B. Fan, S.B. Li, Y. Yang, and J.P. Wilson. 2020. A fundamental theorem for eco-environmental surface modelling and its applications. *Science China-Earth Sciences* 63 (8): 1092–1112.

Zhao, S.Q. 1986. *Physical geography of China*. New York: John Wiley & Sons.

Zhao, N., and T.X. Yue. 2014. A modification of HASM for interpolating precipitation in China. *Theoretical and Applied Climatology* 116: 273–285.

Zhou, L.S., H. Sun, Y.Q. Shen, J.Z. Deng, and Y.L. Shi. 1981. *Comprehensive agricultural planning of China*. Beijing (in Chinese): China Agricultural Press.

# Chapter 4
# Upscaling and Downscaling

**Learning Objectives**

- Corollary 2 of FTEEM: If a surface on a finer spatial resolution is transferred to a one on a coarser spatial resolution, ground-based observations are necessary to operate HASM for obtaining higher accuracy.
- Corollary 3 of FTEEM: If a surface at a coarse spatial resolution and ground-based observations are available, HASM can be used to obtain a corresponding surface at finer spatial resolution with higher accuracy.

## 4.1 Introduction

The scale issue is an inherent part of eco-environmental informatics. From the early 1950s to the early 1970s, many ecologists (Greig-Smith 1952; Gould 1966; Hutchinson 1971) tried to incorporate scale into environmental biology. From the early 1970s to the 1980s, many ecologists focused increasing attention on the problem of spatial scale (Allen and Starr 1982; Golley 1989; Gosz and Sharpe 1989; Wiens 1989). In the 1990s, the problem of scale becomes a central issue in ecology, for unifying population biology and ecosystem science, and marrying basic ecology and applied ecology (Levin 1992). The explosion of interest in scale has created many methods for scale. For instance, the multiple-variables scaling method simultaneously examines each variable at different scales (Holling 1992). In the 1990s, spatially explicit models were developed to demonstrate scale-sensitive issues (Holt et al. 1995), and fractal geometry is now used to treat the dependence of various phenomena on scales (Mandelbrot 1982). Resampling techniques are used to frame samples within a hierarchical framework to assess how scale affects ecosystem characteristics (Cressie 1993) and neural models are used to test scale effects resulting from changes in grain size and spatial structure (Milne 1992). The hierarchy theory is

 
T. Yue et al., *Eco-Environmental Informatics*,
https://doi.org/10.1007/978-981-96-1977-1_4

also employed to address issues of spatial scale; this theory implies that an ecosystem is composed of interacting components and is itself a component of a larger system (O'Neill et al. 1989).

In the early 1980s, the multiscale issue was recognized as a fundamental problem in GIS (Li 2007). In 1983, a small group of leading scientists was gathered by National Aeronautics and Space Administration (NASA) to define research priorities in GIS and multiscale representation was identified as one of them. Since the early 1990s, multiscale representation has become a popular research topic in geographical information science. In 1996, the University Consortium for Geographic Information Science (UCGIS) selected multiscale representation as one of the 10 research priorities. By the late 1990s, multiscale representation gained a boost with the funding of the project of Automated Generalization New Technology (AGENT) by the European Union. In 2000, the International Society for Photogrammetry and Remote Sensing established a working group on multiscale representation. By 2003 this was adopted as one of the 14-longterm research priorities of UCGIS (Usery and McMaster 2005). Since then, special sessions on multiscale representation have been organized at conferences, and special issues on this topic have been published in various journals. Upscaling and downscaling are two essential aspects of the scale issues.

### *4.1.1 Upscaling*

Upscaling refers to the transfer of knowledge from a finer resolution to a coarser resolution in order to mostly reduce computational costs (Misaghian et al. 2018). There has been considerable number of upscaling studies in recent 10 years. For instance, Wood (2009) examined the difference between the mathematical process of averaging and the process of upscaling by introducing 6 scaling laws. It was found that upscaling via simple spatial averaging of soil moisture data acquired from randomly distributed measurement sites results in unacceptable upscaling errors (Crow et al. 2012). Soil organic carbon (SOC) storage in periglacial terrain (European Russian Arctic) was spatially upscaled by using point measurements and thematic maps (Hugelius 2012). The ensemble Kalman filter (EnKF) was coupled with upscaling to build an aquifer model at a coarser scale because an aquifer model at the support scale of observations would imply numerical models with many millions of cells, which is usually impractical (Li et al. 2012). Emissions of carbon dioxide from lakes to the atmosphere on national level were estimated by means of statistical upscaling procedures that aggregate data from 5140 Swedish lakes (Seekell et al. 2014). A support vector regression (SVR) was employed to upscale eddy flux observation data for estimates of continental $CO_2$ flux (Kondo et al. 2015). Regional groundwater of heterogeneous aquifers in Northern Ireland was simulated by coupling ground and airborne geophysical data through upscaling techniques (Dickson et al. 2014). Soil properties were upscaled by averaging plot-scale soil properties collected at two experimental fields at Michigan State University (Fry et al. 2017). A spatial upscaling algorithm was developed to estimate green aboveground biomass in the desert steppe of Inner Mongolia by establishing a statistical relation between the estimation on fine resolution and the one on coarse resolution (Xue et al. 2017). An upscaling method based on discrete wavelet transform was developed to create balance between accuracy and computation speed of the fine-grid model and the coarsened model (Misaghian et al. 2018).

It should be noticed that upscaling concept was sometimes confused with spatial interpolation, such as estimating the species-area relationship from small samples (Kunin et al. 2018), from field to atlas (van Bussel et al. 2015) and from point to area (Schlummer et al. 2014) as well as soil moisture values from station to area (Qin et al. 2015).

### 4.1.2 Downscaling

Spatial resolutions of many models or data are sometimes too coarse to be used for analyses on regional or local scales. To overcome this problem, downscaling approaches are developed to obtain information at finer resolution from the coarser resolution models and data (Yue et al. 2016).

Methods for spatial downscaling have been mostly introduced to climatic factors. Monte Carlo methods were used to downscale rainfall and temperature scenarios at multiple sites in two Mediterranean catchments (Palutikof et al. 2002). Dynamic downscaling was employed to improve the simulations of intensity of Atlantic Hurricanes (Bender et al. 2010). A stochastic weather generator was employed to reproduce daily precipitation characteristics in Korean peninsula (Min et al. 2011). K-nearest neighbor (KNN) method was used to downscale the outputs of global circulation models (GCM) for generating meteorological series for 60 years in Nakdong River Basin in Korea (Eum et al. 2010). Support vector machine (SVM) were applied to downscale the outputs of five atmosphere-ocean GCMs to reproduce local daily precipitations from large-scale weather variables (Yang et al. 2011). Statistical downscaling method (SDSM) was used to represent local climate conditions for the Upper Thames River Basin in Ontario of Canada by modifying outputs from Atmosphere–Ocean coupled Global Climate Models (AOGCMs) (King et al. 2012). A hybrid downscaling model was proposed to downscale daily rainfall series in Mahanadi basin of India by combining KNN and SV (Devak et al. 2015). High accuracy surface modeling (HASM) was developed to considerably improve accuracy of downscaled climate projections by identifying stationarity and non-stationarity of meteorological variables (Yue et al. 2013, 2016; Sinha et al. 2017). A method of random forest for machine learning was used to downscale MODIS evapotranspiration on spatial resolution of 1 km to 30 m on basis of Landsat-derived vegetation indices (Vis) and land surface temperature (LST) (Ke et al. 2017). A downscaling approach taking geopotential heights and relative humidity as predictors has been set up to downscale the time series of standardized precipitation index (SPI) in the different Mediterranean regions (Hertig and Tramblay 2017). The regional spectral model (RSM) with different physical schemes was employed to evaluate the 25-year (1983–2007) summer monsoon climate characteristics of precipitation and surface air temperature (Zong et al. 2017). A two-step statistical method for downscaling was adapted to simulate twenty-first-century climate scenarios for the Gulf of Fonseca on the basis of Coupled Model Intercomparison Project (CMIP5) climate models (Ribalaygua et al. 2018). A statistical downscaling technique was applied to ENSEMBLES global climate simulations for projecting future changes of air temperature over Italian agricultural areas (Tomozeiu et al. 2018).

In addition to climate, downscaling has been also employed in other fields. For instance, a spatial downscaling method based on Gaussian random fields was used to downscale an Australian national extent soil carbon map on spatial resolution of 100 m to 10 m for a farm in E.J. Holtsbaum Research Station, Australia (Malone et al. 2017). A participatory approach to downscaling socioeconomic scenarios was combined with land cover change modeling to assess future changes in mountain hazards in the French Pyrenees (Houet et al. 2017). A model of equilibrium moisture from topography, vegetation, and soil was generalized to downscale soil moisture from multiple coarse resolution grid cells (Hoehn et al. 2017).

It is an exploring spot to combine downscaling with bias correction in recent years. Statistical downscaling with bias correction was employed to project climate changes in the Indus River Basin by building a statistical relationship between outputs of model simulation and observed datasets (Su et al. 2016). Dynamical downscaling was combined with bias correction to reduce the simulated uncertainty of precipitation and sea-surface conditions (Sahai et al. 2017; Meyer and Jin 2017; Hernández-Díaz et al. 2017). A new approach to constructing the forcing data for the second phase of the WRF-based downscaling was developed by combining the transient weather signal from one global climate model with the monthly mean climate states from the multi-model ensemble mean for the present and future periods, together with a bias correction term (Dai et al. 2018).

Downscaling appears in two forms that are dynamical and statistical downscaling (Laflamme et al. 2016). Dynamical downscaling is computationally intensive, which makes use of the lateral boundary conditions combined with regional-scale forcing to produce regional climate models (RCMs) from a GCM. Statistical downscaling (SD) is a computationally less demanding alternative by establishing statistical relationships between local climate variables and large-scale predictors. Simulations of drought episodes over Greece showed that the drought characteristics were well reproduced by the regional climate models for long term drought indices, while a statistical downscaling technique based on artificial neural network are better for the short-term drought indices (Anagnostopoulou 2017). The rainy-season rainfall in Heihe River Basin of China simulated by dynamical downscaling was better than that by statistical downscaling in the upstream, while the one simulated by statistical downscaling was better than that by dynamical downscaling in the midstream (Su et al. 2017). LeRoux et al. (2018) demonstrated that for an equivalent sample size it is unnecessary to dynamically downscale the simulated temperatures because statistical downscaling provides results very close to those of dynamical downscaling, while requiring less computer resources. Mishra et al. (2018) found that dynamical downscaling led to an improvement in the spatial pattern of surface air temperature, but the errors in magnitude after downscaling were even worse in some cases in India, compared to the driving GCMs.

## 4.2 Spatial Upscaling of the Elevation Surface

### 4.2.1 The HASM Based Upscaling Algorithm

Two grids of $\Omega(H)$ and $\Omega(h)$ with different spatial resolutions are defined for the flight test area, where $\Omega(H) = \{(I, J)|(I \times H, J \times H)\}$, $(0 < I \leq I_H, 0 < J \leq J_H)$ and $\Omega(h) = \{(i, j)|(i \times h, j \times h)\}$, $(0 < i \leq I_h, 0 < j \leq J_h)$. $H$ and $h$ are the spatial resolutions of $\Omega(H)$ and $\Omega(h)$, respectively. $H > h$, $H \times I_H = h \times I_h$ and $H \times J_H = h \times J_h$, in which $I_H$ and $J_H$ are the row and column numbers of the grid $\Omega(H)$ with coarser resolution, and $I_h$ and $J_h$ are the row and column numbers of the grid $\Omega(h)$ with finer resolution, respectively. The DTMs, corresponding to the grids $\Omega(H)$ and $\Omega(h)$, are respectively expressed as $\{f_H\} = \{f_{I,J} = f(I, J), 0 < I \leq I_H, 0 < J \leq J_H\}$ and $\{f_h\} = \{f_{i,j} = f(i, j), 0 < i \leq I_h, 0 < j \leq J_h\}$.

For spatial resolution of $H$ and ground observation data $\Pi$, HASM can be formulated as:

$$\begin{cases} \min \left\| \mathbf{A}_H \mathbf{x}_H^{(n+1)} - \mathbf{b}_H^{(n)} \right\|_2 \\ s.t\ \mathbf{C}_H \mathbf{x}_H^{(n+1)} = \mathbf{d}_H^{(n)}|_\Pi \end{cases} \tag{1}$$

where $\mathbf{A}_H$ represents the coefficient matrix of the HASM master equation set under grid $\Omega(H)$; $\mathbf{x}_H^{(n+1)}$ is the unknown variables; $\mathbf{b}_H$ is the right-hand vector of the HASM master equation set; $\mathbf{C}_H$ is the coefficient matrix of equations for optimal constraints $\Pi$ and $\mathbf{d}_H^{(n)}$ is the right-hand vector of the equations for the optimal constraints $\Pi$ under grid $\Omega(H)$.

On the basis of optimal constraints of $\Pi$ , DTM $\{f_h\}$ at a spatial resolution of $h$ can be upscaled into DTM $\{f_H\}$ at a spatial resolution of $H$. During the upscaling process, a trend surface can be formulated as:

$$\bar{x}_H(I, J) = \begin{pmatrix} x_h(i-1, j+1) + x_h(i-1, j) + x_h(i-1, j-1) + x_h(i-1, j-2) \\ +x_h(i, j+1) + x_h(i, j) + x_h(i, j-1) + x_h(i, j-2) \\ +x_h(i+1, j+1) + x_h(i+1, j) + x_h(i+1, j-1) + x_h(i+1, j-2) \\ +x_h(i+2, j+1) + x_h(i+2, j) + x_h(i+2, j-1) + x_h(i+2, j-2) \end{pmatrix} /16 \tag{2}$$

A residual surface can be created by calculating the difference between the trend surface $\bar{x}_H$ and the optimal constraints $\Pi$ . The upscaled DTM can be generated by additive sum of the trend and the residual surface, of which the algorithm can be formulated as (Yue et al. 2020):

$$\begin{cases} \min \left\| \mathbf{A}_H(\mathbf{x}_H^{(n+1)} - \overline{x}_H) - \mathbf{b}_H^{(n)} \right\|_2 \\ s.t\ \mathbf{C}_H(\mathbf{x}_H^{(n+1)} - \overline{x}_H) = \mathbf{d}_H^{(n)}|_\Pi - \mathbf{C}_H\overline{x}_H \end{cases} \tag{3}$$

### *4.2.2 A Case Study*

Upscaling via simple spatial averaging frequently results in unacceptable errors (Crow et al. 2012; Fry et al. 2017). In this section, we take upscaling BeiJing-TianJin-HeBei (BTH) region as a case study to demonstrate the essential role of extrinsic and intrinsic invariants in. In other words, both completeness of data and appropriateness of algorithm are critical for upscaling.

We upscale the elevation surface of the BTH region from a spatial resolution of 90 to by combining data from the Shuttle Radar Topography Mission (SRTM) (http://srtm.csi.cgiar.org/) with ground observations using global positioning system (GPS) based on continuously operating reference stations (CORS).

HASM-US uses both intrinsic and extrinsic information as well as the appropriate method, HASM, for integrating them. A total of 560 points with information on altitude and geographic coordinates (Fig. 1), observed using GPS-based from CORS, was divided into 10 data sets. Nine data sets were employed as the optimal control constraints of HASM-US and the remaining data set was used for cross-validation. The maximum positive absolute error (MPAE), maximum negative absolute error (MNAE) and root-mean-square error (RMSE) of HASM-US were calculated and compared with the same metrics of two widely used upscaling algorithms, bilinear resampling (BR) and adjacent point resampling (APR).

The cross-validation results indicated that the MPAE, MNAE and RMSE of HASM-US are 59, −70 and 19 m respectively, whereas those for BR are 72, −107 and 28 m and those for ARP are 77, −113 and 28 m respectively (Table 1). The maximum amplitudes of HASM-US, BR and ARP are 129, 179 and 190 m respectively and taken as a whole, the aforementioned metrics indicate that HASM-US has the highest accuracy and the most stable performance of the three methods considered.

In terms of HASM-US, the BTH region is divided into two sub-regions using 1000 m as the cutoff. The lower-elevation sub-region accounts for 77.46% of BTH, about 166698 km$^2$ and the middle-elevation sub-region accounts for 22.54%, or about 48510 km$^2$. The lower-elevation sub-region is classified into six steps and the middle-elevation sub-region with a highest altitude of 2824 m is classified into five steps (Fig. 2 and Table 2). The first step accounting for 48.39% of BTH region, lower than 50 m dominates the lower-elevation sub-region and the eighth step accounting for 51.48% of BTH region, between 1200 and 1500 m stands out in the middle-elevation sub-region.

Relief amplitude (RA) is calculated by subtracting the smallest altitude from the highest one at a resolution of 90 m in each of the 1 km × 1 km grid cells. The

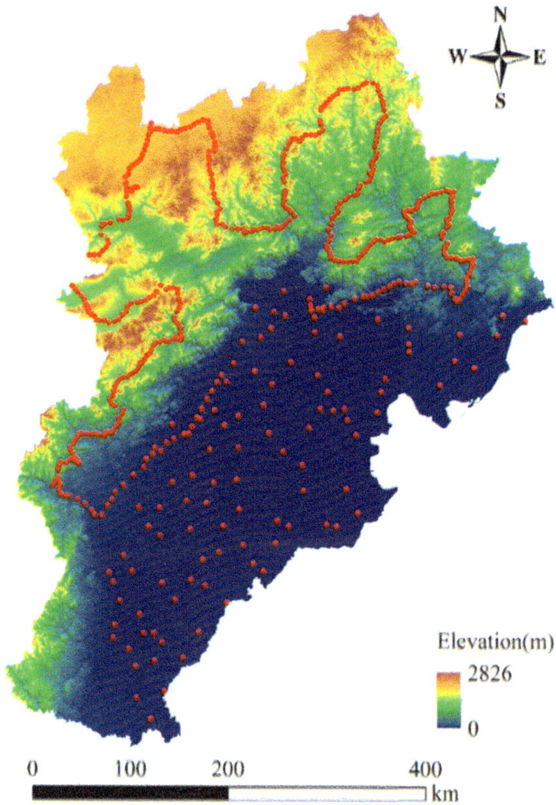


**Fig. 1** Sample points of elevation using GPS-based from CORS

**Table 1** Cross-validation results of HASM-US, BR and APR

| Data sets used for cross-validation | HASM-US (m) | | | BR (m) | | | APR (m) | | |
|---|---|---|---|---|---|---|---|---|---|
| | MPAE | MNAE | RMSE | MPAE | MNAE | RMSE | MPAE | MNAE | RMSE |
| 1 | 71 | −67 | 21 | 151 | −159 | 40 | 159 | −156 | 40 |
| 2 | 103 | −65 | 19 | 14 | −211 | 37 | 15 | −231 | 39 |
| 3 | 70 | −61 | 21 | 21 | −147 | 25 | 25 | −147 | 25 |
| 4 | 28 | −54 | 14 | 16 | −47 | 16 | 22 | −61 | 19 |
| 5 | 81 | −101 | 29 | 20 | −78 | 18 | 27 | −93 | 19 |
| 6 | 48 | −76 | 17 | 24 | −79 | 24 | 29 | −86 | 24 |
| 7 | 37 | −101 | 19 | 30 | −64 | 21 | 30 | −68 | 22 |
| 8 | 41 | −53 | 15 | 146 | −131 | 32 | 154 | −135 | 34 |
| 9 | 8 | −59 | 16 | 151 | −56 | 26 | 159 | −56 | 27 |
| 10 | 101 | −64 | 24 | 144 | −95 | 35 | 152 | −99 | 36 |
| Mean errors | 59 | −70 | 19 | 72 | −107 | 28 | 77 | −113 | 28 |

**Fig. 2** Elevation surface of the BTH region

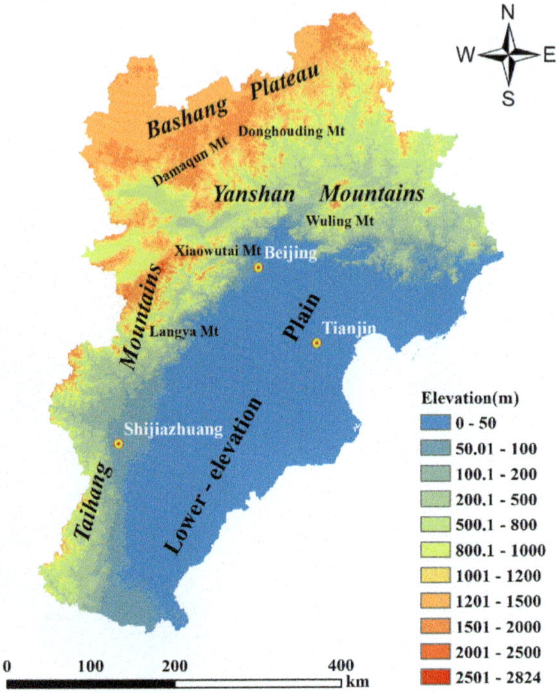


**Table 2** Elevation steps in the BTH region

| Lower-elevation sub-region | | | | Middle-elevation sub-region | | | |
|---|---|---|---|---|---|---|---|
| Step | Elevation range | Area ($km^2$) | Proportion (%) | Step | Elevation range | Area ($km^2$) | Proportion (%) |
| 1 | < 50 m | 80,660 | 37.48 | 7 | 1000–1200 | 13,490 | 6.27 |
| 2 | 50–100 | 14,325 | 6.66 | 8 | 1200–1500 | 24,972 | 11.60 |
| 3 | 100–200 | 9544 | 4.43 | 9 | 1500–2000 | 9764 | 4.54 |
| 4 | 200–500 | 21,737 | 10.10 | 10 | 2000–2500 | 267 | 0.12 |
| 5 | 500–800 | 24,804 | 11.53 | 11 | > 2500 | 17 | 0.01 |
| 6 | 800–1000 | 15,628 | 7.26 | Subtotal | | 48,510 | 22.54 |
| Subtotal | | 166,698 | 77.46 | | | | |

RA in the BTH region ranges from 0 to 1592 m and was classified into six levels corresponding to the standard used for the Geomorphologic Atlas of the People's Republic of China (Cheng et al. 2011) (Table 3 and Fig. 3).

Most of the flat areas are found in the lower-elevation plain, the Bashang plateau, and intermontane basins of northwestern Hebei province, and these taken as a whole, cover 40.28% of the BTH region. The next two classes (> 75 m relief) are found

**Table 3** Relief amplitude of the BTH region

| Level | | RA (m) | Area (km$^2$) | Proportion (%) |
|---|---|---|---|---|
| 1 | Flat | < 30 | 86,692 | 40.28 |
| 2 | Platform | 30–75 | 14,526 | 6.75 |
| 3 | Hill | 75–200 | 24,622 | 11.44 |
| 4 | Low-relief | 200–500 | 65,218 | 30.30 |
| 5 | Middle-relief | 500–1000 | 23,578 | 10.96 |
| 6 | High-relief | > 1000 | 572 | 0.27 |
| Total | | | 215,208 | 100 |

**Fig. 3** Relief amplitude surface of the BTH region

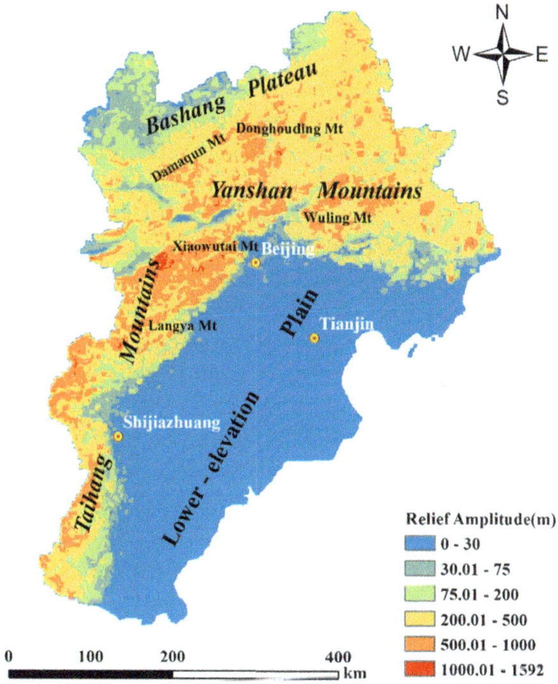


around the margins of the Bashang plateau, in transitional areas from the lower-elevation plain to the Taihang and Yanshan mountains, and along the borders of the flat areas in the intermontane basins of northwestern Hebei province. Together, these areas total 6.75% of the BTH region. The hills that mark the borders of the lower-elevation plain, the valley zones, and front belt of the Damaqu mountains, cover 11.44% of the BTH region. The low- and middle-relief areas found in the Taihang and Yanshan mountainous regions account for about 41.26% of the BTH region. The high-relief areas mark the locations of Xiaowutai, Wuling, and Donghouding mountains.

The landform pattern of the BTH region can be described as lower-elevation plain in the southeast and mountainous areas in the western and northern parts (Fig. 3). In the south, the altitude rapidly decreases from the Taihang mountains in the west to the lower-elevation plain in the east. In the north, intermontane basins alternate with high undulating mountains. In the Yanshan mountains, the altitude of the undulating terrain gradually diminishes from north to south.

## 4.3 Spatial Downscaling of CMIP5 Scenarios

GCMs are an available source for climate scenarios. However, although GCMs can give a good overview of the future climate on global level it is difficult to be used to assess impacts of climate change on ecosystem changes at local level because of the coarse spatial resolution of GCMs (Xue et al. 2007). It was presented that the range of temperature and precipitation changes at finer spatial scales is much broader from simulation of different models (Grotch and MacCracken 1991). It was found that GCM simulations were questionable on a regional level (von Storch et al. 1993). It was observed that making the spatial resolution of GCMs finer would improve the climate simulation and hence increase confidence in the GCM uses (Ciret and Sellers 1998; Akinyemi and Adejuwon 2008). Raisanen (2007) observed that many processes on finer spatial resolutions could not be simulated correctly in GCMs. Prudhomme and Davies (2009) stated that different GCMs often provided different climate outputs, especially on regional scales, from the same oceanic and atmospheric drivers.

The models of CMIP5 did not well simulate precipitation over eastern China, while the version on higher spatial resolutions better represented the frequency distribution of precipitation (Huang et al. 2013). The CMIP5 models failed to describe extremes, the diurnal cycle, and the dynamic forcing of precipitation. These characteristics of precipitation were improved by downscaling outputs of the CMIP5 models into 30 km × 30 km (Harding et al. 2013).

For improving the poor GCM performance, many statistical downscaling approaches (SDAs) and dynamic downscaling methods (DDMs) have been developed on local and regional scales. For instance, a fine computational grid over a limited domain was nested within the coarse grid of a GCM in Limited Area Models (LAMs) (Xu 1999; Lapris 2008). A regional climate model (RCM) on a spatial resolution of 50 km × 50 km was developed by using a one-way nesting method, which was driven by the output from a GCM (Jones et al. 1995). A weather research and forecasting model (WRF) was nested into the climate forecast system (CFS) of the coupled atmosphere–ocean general circulation model (CGCM) to downscale precipitation in winter over China between 1982 and 2008. A comparison of DDMs and SDAs indicated that a DDM based on non-linear artificial neural networks was the best at modelling the inter-annual rainfall variability (Haylock et al. 2006). These DDMs have many disadvantages such as costly computing requirements, too coarse resolutions that are not suitable for studies on local scale, and stationarity assumptions (Khalili et al. 2013).

SDAs employs statistical transfer functions (STFs) to represent relationships between larger-scale atmospheric variables and local quantities (Zorita and von Storch 1999; Pielke and Wilby 2012). The gap between large and local scales can be bridged by using statistical approaches (Hewitson and Crane 1996; Charles et al. 1999; Wilby et al. 2002; Gachon and Dibike 2007; Dibiki et al. 2008).

Downscaling GCMs using STFs have emerged as a popular approach to increase spatial resolutions and therefore makes GCMs more applicable to regional and local analysis in China. For instance, an STF was evaluated with data of daily mean temperature, precipitation, and pan-evaporation from 11 weather stations in the Haihe River basin. The results indicated that determination coefficients between downscaled and observed mean temperature, precipitation, and pan-evaporation were respectively 99%, 93%, and 73% (Chu et al. 2010). A daily time-series of temperature and precipitation estimates during the period from 1961 to 2099 was created in Yellow River Basin by the using an STF to downscale outputs from the third version of the Hadley Center coupled model (HadCM3) (Liu et al. 2011a). Two STFs, the nonhomogeneous hidden Markov model (NHMM) and the statistical down-scaling model (SDSM), were evaluated on basis of observed precipitation over the Tarim River basin in northwestern China. The results showed that there was a little difference between NHMM and SDSM in model performance for dry- and wet-spell length, but NHMM presented better performance than SDSM for wet-day precipitation estimates (Liu et al. 2011b). An STF developed for downscaling precipitation in the Qaidam Basin was presented by a relationship between precipitation and other environmental factors such as topography and vegetation (Jia et al. 2011).

An STF from the HadCM3 model under the A2 and B2 emission scenarios was constructed to develop scenarios of extreme daily temperature and precipitation in the Dongjiang River basin of China in the twenty-first century (Yang et al. 2012). An STF was developed to spatially downscale extreme precipitation and temperature of GCMs under three emission scenarios (A1B, A2, B1) to estimate their potential spatio-temporal changes on the Loess Plateau of China during the twenty-first century. The results demonstrated that the present change trend of longer heat-wave duration, more frequent and intense precipitation, and longer drought duration would continue during the twenty-first century (Li et al. 2012). An STF based on GCM outputs from a project of the Development of a European Multi-model Ensemble System for Seasonal to Interannual Prediction (DEMETER) and observation data was used to predict autumn precipitation in China (Liu and Fan 2013).

In this section, STFs for mean annual temperature (MAT) and mean annual precipitation (MAP) are developed by analyzing spatial stationarity. The STFs are constructed by a combination of trend surfaces with a method for high accuracy surface modeling (HASM) (Yue 2011; Yue et al. 2013b).

## 4.3.1 Materials and Methods

### 4.3.1.1 Data Preparation

The CMIP5 datasets of monthly mean temperature and precipitation were provided by the National Climate Center of China. They include historical simulations and three RCP scenarios: RCP2.6, RCP4.5 and RCP8.5. They were created by using the simple average of the outputs from the 21 CMIP5 models for which the spatial resolutions are unified into $1° \times 1°$ cells using the ANUSPLIN interpolation method (Fig. 4a).

Ten-day observation data of temperature and precipitation were collected from 735 meteorological stations scattered across China from 1961 to 2010 (National Meteorological Information Center of China; http://www.nmic.gov.cn) (Fig. 4b). A digital terrain model (DTM) of China acquired from the Shuttle Radar Topography Mission (SRTM) at a spatial resolution of 90 m × 90 m (http://srtm.csi.cgiar.org) was upscaled into a DTM at a spatial resolution of 1 km × 1 km using a cubic convolution resampling approach (Fig. 4c). The DTM was used as auxiliary data to develop the statistical transfer function for mean annual temperature (MAT) and mean annual precipitation (MAP).

A zoning system dividing the land mass of China into nine regions in terms of matching temperature, precipitation and soil was adopted to make it easier to analyze the changes in precipitation and temperature from one place to another. The nine regions are respectively labeled $R_1$, $R_2$, …, $R_9$ on the DTM (Fig. 4c).

A rectangular region (60–149°E, 0.5–49.5°N) was extracted from the CMIP5 datasets, which covers the whole land mass of China. This is referred to as the China-CMIP5 dataset hereafter and $t_1$, $t_2$, $t_3$ and $t_4$ are used to represent the periods 1976–2005, 2011–2040, 2041–2070, and 2071–2100, respectively.

## 4.3.2 Algorithms

### 4.3.2.1 A variation coefficient

A variation coefficient can be formulated as:

$$\mathrm{CV} = \frac{\sigma}{\overline{Z}} \times 100\% \tag{1}$$

where $\sigma$ is the standard deviation and $\overline{Z}$ represents a mean property value.

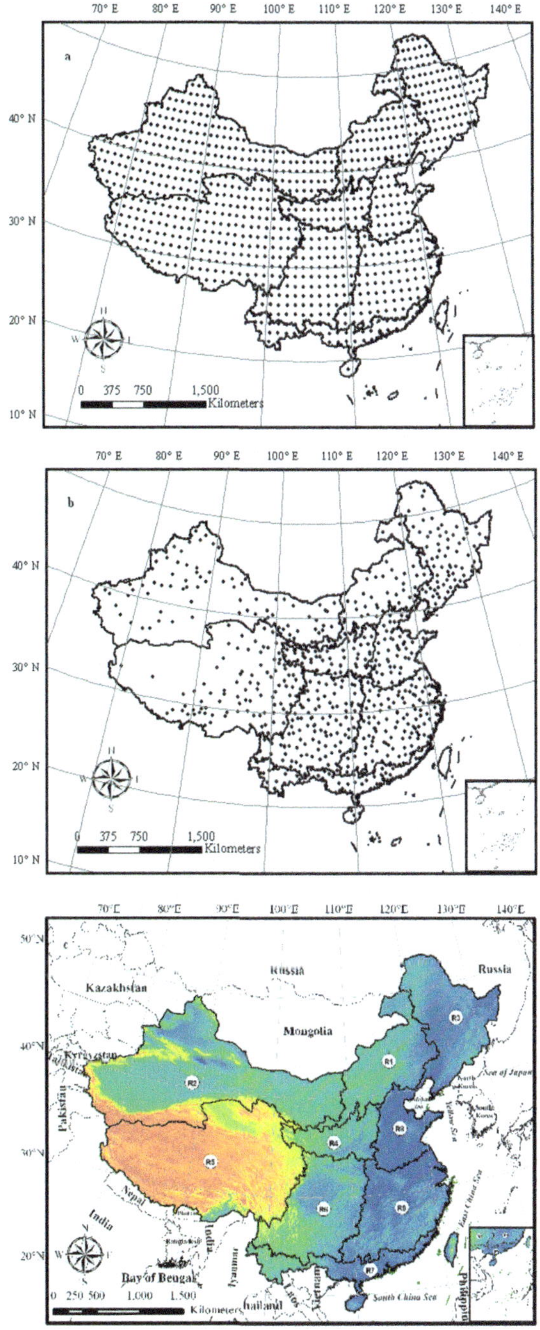


**Fig. 4** **a** The spatial distribution of the lattice centers of CMIP5 in China; **b** the spatial distribution of the 735 meteorological stations available in China; and **c** a digital elevation model China showing the elevation divided into nine ecoregions

#### 4.3.2.2 Algorithm for Downscaling of Annual Mean Temperature

Latitude and elevation were selected as explanatory variables for simulating MAT by means of geographically weighted regression (GWR). Then, we obtained three surfaces representing the intercept, latitude regression coefficient and elevation regression coefficient. The variation coefficients of the intercept, latitude regression coefficient and elevation regression coefficient were 24.66%, 23.36% and 21.30%, respectively, measured in terms of the variation coefficient formulation. They are all less than 25%. This means that MAT exhibits spatial stationarity and that trend surfaces of MAT can be simulated by ordinary least squares (OLS).

The OLS equation of MAT using the China-CMIP5 dataset and taking the DTM at the spatial resolution of 1 km × 1 km as auxiliary data can be expressed:

$$RMAT(x, y, t_1) = 41.585 - 10^{-5} \cdot (0.982y - 404.090Ele(x, y)) \qquad (2)$$

where $RMAT(x, y, t_1)$ is MAT from the OLS regression of the China-CMIP5 data during the period of $t_1$, $y$ represents latitude (m), and $Ele(x, y)$ is the elevation (m) at the location $(x, y)$.

The downscaling formulation of MAT can be expressed as:

$$DMAT(x, y, t_1) = RMAT(x, y, t_1) + HASM\,(MAT_j - RMAT_j) \qquad (3)$$

where $DMAT(x, y, t_1)$ represents the downscaled MAT at $t_1$, $MAT_j$ the observed MAT at weather station *j*, $HASM\,(MAT_j - RMAT_j)$ the residual surfaces of the downscaled MAT improved by HASM at the spatial resolution of 1 km × 1 km (Yue 2011; Yue et al. 2013a).

#### 4.3.2.3 Algorithm for Downscaling of Annual Mean Precipitation

Surfaces of latitude, longitude, elevation and aspect regression coefficients as well as intercept can be created by simulating the relationship of MAP from the China-CMIP5 dataset with latitude, longitude, elevation and aspect using GWR. The variation coefficients of the surfaces of the longitude and aspect regression coefficients exceeded 100%, which indicates strong variation; those for the latitude and elevation regression coefficients as well as the intercept are greater than 40%, indicating moderate variability. Latitude, longitude, elevation and aspect as well as intercept are non-stationary at the 1% significance level. It is better to simulate MAP by means of GWR. In other words, MAT can be modeled as a linear function of latitude and elevation, while regression coefficients of latitude, longitude, elevation and aspect for MAP are best determined locally instead of globally.

MAP values from the China-CMIP5 dataset were normalized by using the following formulation to stabilize the simulation processes:

$$P(x, y, t_1) = \frac{OMAP(x, y, t_1)}{\mathrm{MAX}\{OMAP(x, y, t_1)\}} \tag{4}$$

where $P(x, y, t_1)$ is a normalization value of MAP at grid cell $(x, y)$ during the period $t_1$,$OMAP(x, y, t_1)$ is the original MAP during the period $t_1$ from the China-CMIP5 dataset, and $MAX\{OMAP(x, y, t_1)\}$ is the maximum value of MAP during the period $t_1$ from the China-CMIP5 dataset.

To ensure the normality of GWR, a BOX-COX transformation of $P(x, y, t_1)$ (Box and Cox 1964) was conducted as follows:

$$\Psi_{0.4}(P(x, y, t_1)) = \left((P(x, y, t_1))^{0.4} - 1\right)/0.4 \tag{5}$$

A trend surface of the transformed $P(x, y, t_1)$ was then created by means of GWR:

$$\begin{aligned}\Psi_{0.4}(P(x, y, t_1)) &= \theta_0(x, y) + \theta_1(x, y) \cdot x + \theta_2(x, y) \cdot y \\ &\quad + \theta_3(x, y) \cdot Ele(x, y) + \theta_4(x, y) \cdot ICA(x, y)\end{aligned} \tag{6}$$

where $\theta_1(x, y)$, $\theta_2(x, y)$, $\theta_3(x, y)$ and $\theta_4(x, y)$ as well as $\theta_0(x, y)$ are respectively the coefficients of the latitude, longitude, elevation, aspect, and the intercept.

The MAP observations from weather stations were employed to improve the trend surface. The normalized downscaled-OMAP with the BOX-COX transformation on a spatial resolution of 1 km × 1 km can be formulated as follows:

$$\begin{aligned}\Psi_{0.4}(dMAP(x, y, t_1)) &= \Psi_{0.4}(P(x, y, t_1)) \\ &\quad + HASM\left\{\begin{array}{l}\Psi_{0.4}(MAP_j/MAX\{MAP_j(x, y, t_1)\}) \\ -\Psi_{0.4}(P(x, y, t_1)\end{array}\right\} \\ &= \Psi_{0.4}(P(x, y, t_1)) + HASM\left\{\varepsilon_j\right\}\end{aligned} \tag{7}$$

where $\varepsilon_j = \Psi_{0.4}(MAP_j/MAX\{MAP_j(x, y, t_1)\}) - \Psi_{0.4}(P(x, y, t_1)$ is the residue of $\Psi_{0.4}(P(x, y, t_1))$ at weather station $j$, $HASM\left\{\varepsilon_j\right\}$ represents the spatial interpolation of the residues at the weather stations, $\mathrm{MAP_j}$ is the observed MAP at weather station $j$, and $dMAP(x, y, t_1)$ is the normalized downscaled-OMAP at the location $(x, y)$ during the period $t_1$.

After an inverse BOX-COX transformation and inverse normalization, the downscaled-OMAP was finally formulated as follows:

$$DMAP(x, y, t_1) = MAX\{OMAP(x, y, t_1)\} \cdot (0.4 \cdot \Psi_{0.4}(dMAP(x, y, t_1)) + 1)^{\frac{1}{0.4}} \tag{8}$$

The procedure from Eqs. (7–11) is termed as HASM-GB. Similarly, we can get IDW-GB, OK-GB and Spline-GB when IDW, ordinary Kriging (OK) and Spline are used to interpolate the residues, respectively. The process of Eqs. (5) and (6) is termed HASM-OLS and IDW OLS, OK-OLS and Spline-OLS could be developed in similar ways.

### 4.3.3 Validation of CMIP5 Scenarios

The MATs and MAPs from the RCP2.6, RCP4.5 and RCP8.5 scenarios were downscaled using the algorithms above. The mean absolute errors (MAEs) of the downscaled MATs and MAPs from the RCP2.6, RCP4.5 and RCP8.5 scenarios during the period 2006–2010 were formulated as:

$$\mathrm{DT}AE_k(i)=\frac{1}{n_i}\times\sum_{j=1}^{n_i}\left|\mathrm{DMAT}_{k,j}(\mathrm{i})-\mathrm{MAT}_j(i)\right| \tag{9}$$

$$\mathrm{DP}AE_k(i)=\frac{1}{n_i}\times\sum_{j=1}^{n_i}\left|\mathrm{DMAP}_{k,j}(\mathrm{i})-\mathrm{MAP}_j(i)\right| \tag{10}$$

$$\mathrm{OT}AE_k(i)=\frac{1}{n_i}\times\sum_{j=1}^{n_i}\left|\mathrm{OMAT}_{k,j}(\mathrm{i})-\mathrm{MAT}_j(i)\right| \tag{11}$$

$$\mathrm{OP}AE_k(i)=\frac{1}{n_i}\times\sum_{j=1}^{n_i}\left|\mathrm{OMAP}_{k,j}(\mathrm{i})-\mathrm{MAP}_j(i)\right| \tag{12}$$

where $k = 1$, 2 and 3 represent the RCP2.6, RCP4.5 and RCP8.5s, respectively; $\mathrm{DMAT}_{\mathrm{k},j}(i)$ and $\mathrm{DMAP}_{k,j}(i)$ are the downscaled MAT and MAP of CMIP5 scenario $k$ at the $j$th meteorological station in region $i$;$\mathrm{OMAT}_{\mathrm{k},j}(i)$ and $\mathrm{OMAP}_{k,j}(i)$ are the original MAT and MAP of CMIP5 scenario $k$ at the $j$th meteorological station in region $i$;$\mathrm{MAT}_j(i)$ and $\mathrm{MAP}_j(i)$ are the observed MAT and MAP at the $j$th meteorological station in region $i$; $n_i$ is the total number of meteorological stations in region $i$; and $i = 1$ through 9 and represents the nine ecoregions shown in Fig. 4c.

In terms of the China-CMIP5 dataset, the MAEs of the three scenarios of OMAT were very similar during the period 2006–2010. The MAEs were 2.2 °C for the whole of China on average. The MAEs exceeded 2.2 °C in four regions: (1) the Tibet plateau ($R_5$); (2) the arid area ($R_2$); (3) the Loess plateau ($R_1$); and (4) the Sichuan basin plus the Yunnan-Guizhou plateau (R6) (Table 4). The MAEs for RCP2.6, RCP4.5 and RCP8.5 scenarios were 5.43, 5.32 and 5.37 °C for the Tibet plateau, 4.27, 4.24 and 4.14 °C for the arid area, 2.67, 2.64 and 2.56 °C for the Loess plateau, and 2.65, 2.62 and 2.61 °C for the Sichuan basin plus the Yunan-Guizhou plateau.

The CMIP5 scenarios mostly underestimated MAT as well. The proportion of the meteorological stations where OMAT was lower than MAT was 88% for the RCP2.6 and RCP4.5 scenarios and 87% for the RCP8.5 scenario in the whole of China. The proportion of stations where OMAT was lower than MAT in the Yangtze river basin was more than 92% under all three scenarios. In general, the RCP8.5 scenario performed better than the other two scenarios (Fig. 5).

After the downscaling process, all of the MAEs for the DMAT scenarios of RCP2.6, RCP4.5 and RCP8.5 were reduced to 0.61 °C for the whole land mass of China. The MAEs were improved for 72% of the meteorological stations in China

**Table 4** Mean absolute errors of mean annual temperature for the RCPs where the subscripts 1, 2 and 3 for OTAE, DTAE and RT represent the RCP2.6, RCP4.5 and RCP8.5 scenarios, respectively

| Region | $OTAE_1$ (°C) | $DTAE_1$ (°C) | $RT_1$ (%) | $OTAE_2$ (°C) | $DTAE_2$ (°C) | $RT_2$ (%) | $OTAE_3$ (°C) | $DTAE_3$ (°C) | $RT_3$ (%) |
|---|---|---|---|---|---|---|---|---|---|
| $R_1$ | 1.45 | 0.67 | 53.58 | 1.46 | 0.68 | 53.67 | 1.42 | 0.67 | 52.64 |
| $R_2$ | 4.27 | 0.86 | 79.92 | 4.24 | 0.85 | 79.84 | 4.14 | 0.86 | 79.31 |
| $R_3$ | 0.90 | 0.46 | 48.26 | 0.92 | 0.46 | 49.67 | 0.89 | 0.46 | 48.01 |
| $R_4$ | 2.67 | 0.98 | 63.26 | 2.64 | 0.98 | 62.89 | 2.56 | 0.98 | 61.78 |
| $R_5$ | 5.43 | 0.69 | 87.26 | 5.32 | 0.70 | 86.93 | 5.37 | 0.69 | 87.09 |
| $R_6$ | 2.65 | 0.85 | 67.97 | 2.62 | 0.85 | 67.59 | 2.61 | 0.85 | 67.51 |
| $R_7$ | 1.13 | 0.29 | 74.49 | 1.07 | 0.29 | 73.02 | 1.13 | 0.29 | 74.40 |
| $R_8$ | 1.58 | 0.32 | 79.91 | 1.55 | 0.32 | 79.55 | 1.52 | 0.32 | 79.21 |
| $R_9$ | 1.48 | 0.40 | 73.14 | 1.48 | 0.40 | 73.14 | 1.45 | 0.40 | 72.58 |
| China | 2.22 | 0.61 | 72.62 | 2.20 | 0.61 | 72.32 | 2.18 | 0.61 | 72.02 |

as a whole, for 87% of the stations on the Tibet plateau, 79% of the stations in the arid area, 62% of the stations on the Loess plateau, and 68% of the stations in the Sichuan basin plus the Yunan-Guizhou plateau.

The statistical analysis indicated that the MAEs of OMAP under the RCP2.6, RCP4.5 and RCP8.5 were 345.88, 338.18 and 340.93 mm, respectively, for the whole of China on average (Table 5). The Tibet plateau (R5), Loess plateau (R1), and the Sichuan basin plus the Yunan-Guizhou plateau (R6) had larger errors. The MAEs for the RCP2.6, RCP4.5 and RCP8.5 scenarios were 758.52, 765.54 and 765.01 mm, respectively, for the Tibet plateau, 583.55, 589.49 and 581.03 mm for the Loess plateau, and 400.94, 392.47 and 402.05 mm for the Sichuan basin plus the Yunan-Guizhou plateau. Precipitation was overestimated at 80% of the meteorological stations in all three CMIP5 scenarios. In the area covered by the Tibet, Loess and Inner Mongolia plateau, the estimates generated with the three scenarios exceeded the actual precipitation at all of the meteorological stations and, in general, the estimates generated with the three scenarios exceeded the actual precipitation at nearly all of the stations where MAP exceeded 1,500 mm. The RCP4.5 scenario performed better than the RCP2.6 and RCP8.5 scenarios (Fig. 6).

The downscaling process made the DMAP 70% more accurate than OMAP for the whole of China on average. The MAE ratios were improved at more than 90% of the meteorological stations in northeast Inner Mongolia, in the arid area, and the Tibet plateau. The MAEs, themselves, were decreased to 13.19, 12.91 and 13.82 mm under the RCP2.5, RCP4.6, and RCP8.5 scenarios, respectively in northeast Inner Mongolia, to 20.79, 20.43 and 20.46 mm for the three scenarios in the arid area, and to 72.89, 71.87 and 72.55 mm for the three scenarios on the Tibet plateau. However, the MAEs of the DMAP for the three scenarios were still high (165.27, 163.34 and 163.97 mm for the ECP2.5, RCP4.6, and RCP8.5 scenarios, respectively) in the middle and lower reaches of the Yangtze River.

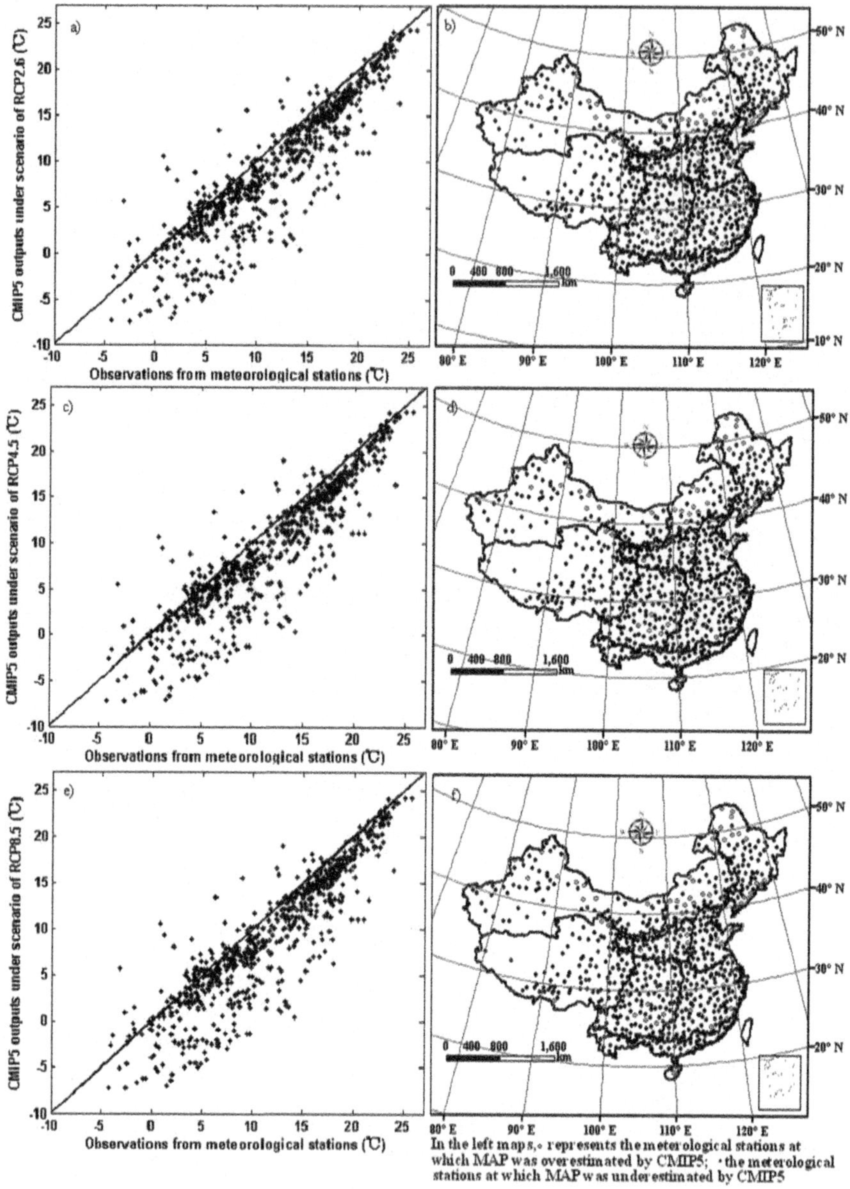


**Fig. 5** Comparison of OMAT with MAT: **a** observed and simulated values under the RCP2.6 scenario; **b** the locations of the meteorological stations whose values were under- and over-estimated with RCP2.6; **c** observed and simulated values under the RCP4.5 scenario; **d** the locations of the meteorological stations whose values were under- and over-estimated with RCP4.5; **e** observed and simulated values under the RCP8.5 scenario; and **f** the locations of the meteorological stations whose values were under- and over-estimated with RCP8.5. For the maps, ° represents the meteorological stations at which MAT was overestimated by CMIP5 and • represents the meteorological stations at which MAT was underestimated by CMIP5

**Table 5** Mean absolute errors of mean annual precipitation for the RCPs where the subscripts 1, 2 and 3 for OPAE, DPAE and RP represent the RCP2.6, RCP4.5 and RCP8.5 scenarios, respectively

| Region | $OPAE_1$ (mm) | $DPAE_1$ (mm) | $RP_1$ (%) | $OPAE_2$ (mm) | $DPAE_2$ (mm) | $RP_2$ (%) | $OPAE_3$ (mm) | $DPAE_3$ (mm) | $RP_3$ (%) |
|---|---|---|---|---|---|---|---|---|---|
| $R_1$ | 284.65 | 13.19 | 95.37 | 263.52 | 12.91 | 95.10 | 271.36 | 13.82 | 94.91 |
| $R_2$ | 229.66 | 20.79 | 90.95 | 228.15 | 20.43 | 91.05 | 220.40 | 20.46 | 90.72 |
| $R_3$ | 211.68 | 37.99 | 82.05 | 197.19 | 37.77 | 80.84 | 209.16 | 38.07 | 81.80 |
| $R_4$ | 583.55 | 67.74 | 88.39 | 589.49 | 66.39 | 88.74 | 581.03 | 68.01 | 88.29 |
| $R_5$ | 758.52 | 72.89 | 90.39 | 765.54 | 71.87 | 90.61 | 765.01 | 72.55 | 90.52 |
| $R_6$ | 400.94 | 59.90 | 85.06 | 392.47 | 56.39 | 85.63 | 402.05 | 56.55 | 85.93 |
| $R_7$ | 308.54 | 92.25 | 70.10 | 309.56 | 91.15 | 70.55 | 302.86 | 93.34 | 69.18 |
| $R_8$ | 252.95 | 70.60 | 72.09 | 230.29 | 70.66 | 69.32 | 246.40 | 70.89 | 71.23 |
| $R_9$ | 211.93 | 165.27 | 22.02 | 204.29 | 163.34 | 20.04 | 202.04 | 163.97 | 18.84 |
| China | 345.88 | 71.01 | 79.47 | 338.18 | 70.21 | 79.24 | 340.923 | 70.85 | 79.22 |

### 4.3.4 Climate Scenarios of China

The downscaling formulations for the CMIP5 scenarios can be expressed as follows:

$$\begin{aligned} SMAT(x, y, t_k) &= DMAT(x, y, t_1) \\ &\quad + HASM\,(OMAT(x, y, t_k) - OMAT(x, y, t_1)) \end{aligned} \tag{13}$$

$$\begin{aligned} SMAP(x, y, t_k) &= DMAP(x, y, t_1) \\ &\quad + HASM\,(OMAP(x, y, t_k) - OMAP(x, y, t_1)) \end{aligned} \tag{14}$$

where $k = 1, 2, 3$ and 4; $t_k$ represents, respectively, the periods 1976–2005, 2011–2040, 2041–2070, and 2071–2100; OMAT$(x, y, t_k)$ and OMAP$(x, y, t_k)$ represent the original CMIP5 MAT and MAP scenarios; $SMAT(x, y, t_k)$ and $SMAP(x, y, t_k)$ represent the downscaled OMAT and OMAP scenarios in the period $t_k$ ($k = 2, 3$ and 4) at a spatial resolution of 1 km × 1 km (Figs. 7 and 8); and $DMAT(x, y, t_1)$ and $DMAP(x, y, t_1)$ represent the downscaled OMAT and OMAP respectively during the period of $t_1$.

#### 4.3.4.1 Mean Annual Temperature

According to the downscaling formulation under scenario RCP2.6, almost all of China would have a warming trend during the period from $t_1$ to $t_2$ except for 2% of the grid cells on the Tibet plateau that would become cooler (Table 6 and Fig. 9). SMAT would be 1.41 °C warmer across the whole of China on average and the rise in SMAT would exceed 1.41 °C in regions $R_1$ (1.99 °C), $R_4$ (1.95 °C), $R_8$ (1.80 °C),

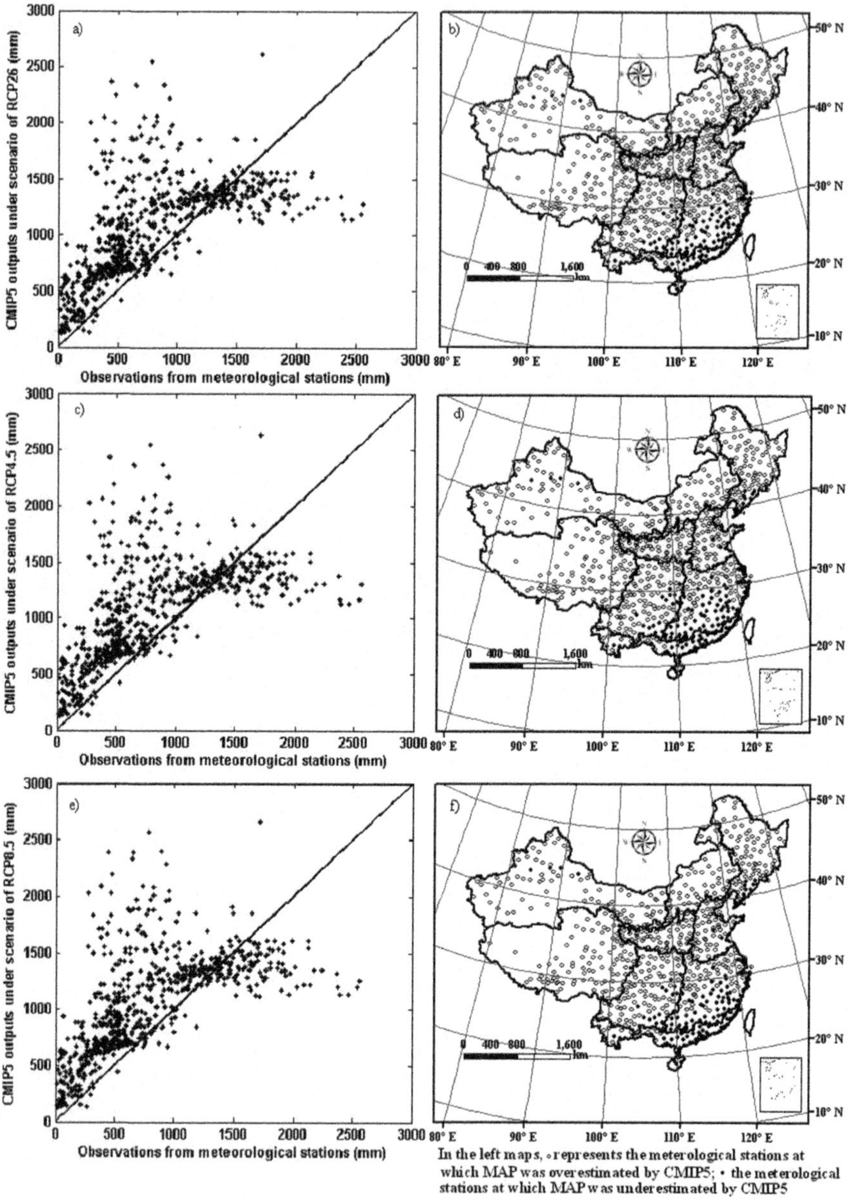


**Fig. 6** Comparison of OMAP with MAP: **a** observed and simulated values under the RCP2.6 scenario; **b** the locations of the meteorological stations whose values were under- and over-estimated with RCP2.6; **c** observed and simulated values under the RCP4.5 scenario; **d** the locations of the meteorological stations whose values were under- and over-estimated with RCP4.5; **e** observed and simulated values under the RCP8.5 scenario; and **f** the locations of the meteorological stations whose values were under- and over-estimated with RCP8.5. For the maps, ° represents the meteorological stations at which MAP was overestimated by CMIP5 and • represents the meteorological stations at which MAP was underestimated by CMIP5

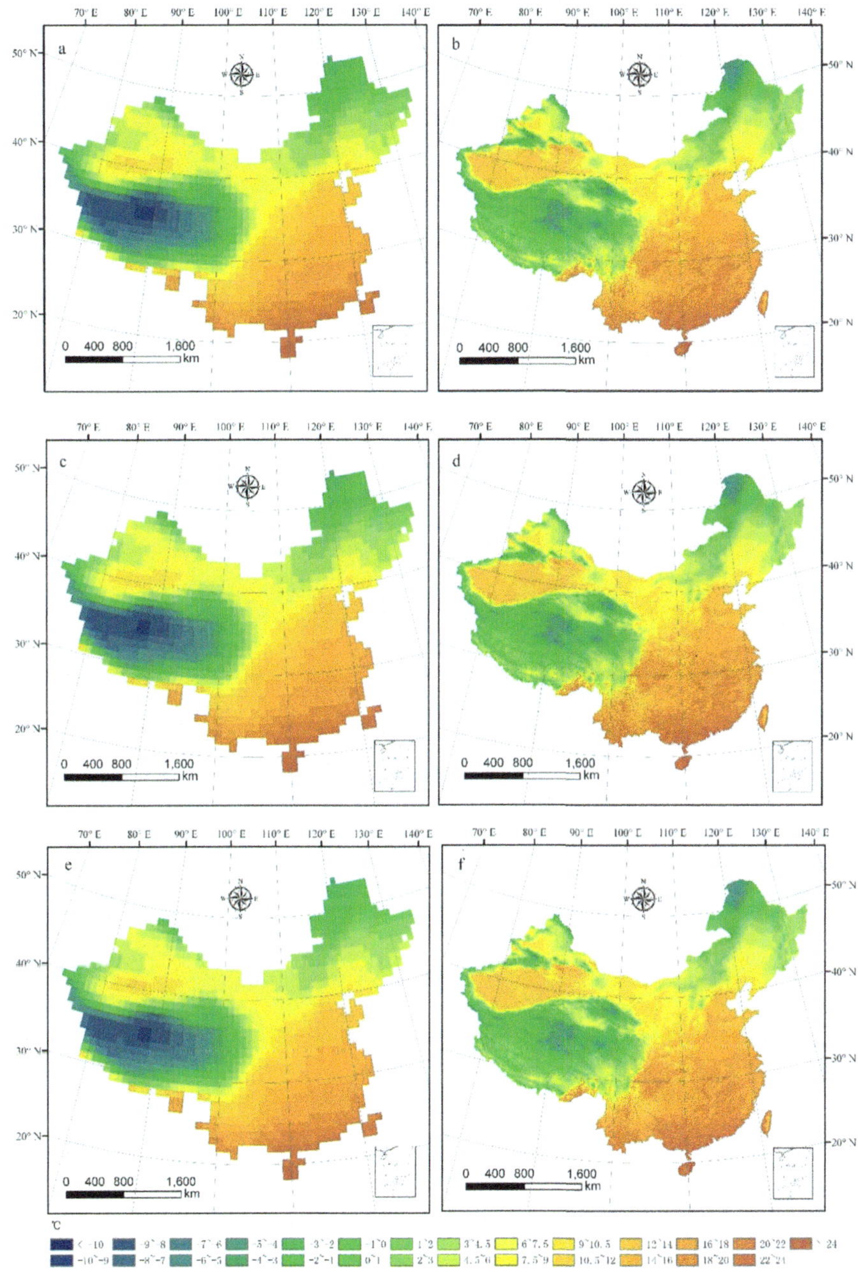


**Fig. 7** A comparison of the original and downscaled MAT RCP scenarios: **a** original RCP2.6; **b** downscaled RCP2.6; **c** original RCP4.5; **d** downscaled RCP4.5; **e** original RCP8.5; and **f** downscaled RCP8.5

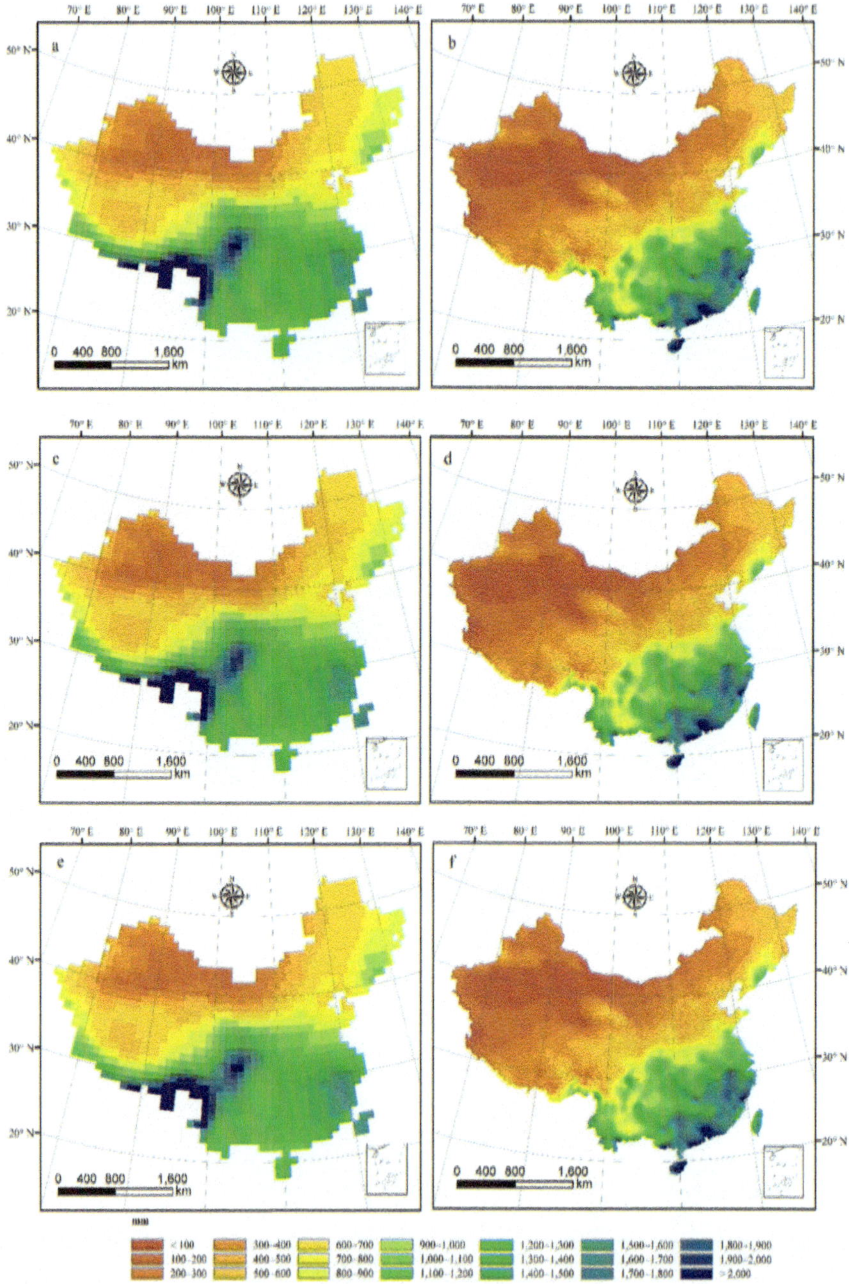

**Fig. 8** A comparison of the original and downscaled MAT RCP scenarios: **a** original RCP2.6; **b** downscaled RCP2.6; **c** original RCP4.5; **d** downscaled RCP4.5; **e** original RCP8.5; and **f** downscaled RCP8.5

**Table 6** Changes of mean annual temperature under scenario of RCP2.6

| Region | T2–T1 | | T3–T2 | | T4–T3 | |
|---|---|---|---|---|---|---|
| | Proportion of grid cells warmer (%) | ΔT21 (°C) | Proportion of grid cells warmer (%) | ΔT32 (°C) | Proportion of grid cells warmer (%) | ΔT43 (°C) |
| $R_1$ | 100 | 1.99 | 15.37 | −0.21 | 20.83 | −0.31 |
| $R_2$ | 100 | 1.75 | 21.46 | −0.35 | 93.08 | 0.51 |
| $R_3$ | 100 | 1.65 | 66.31 | 0.05 | 0.02 | −0.64 |
| $R_4$ | 100 | 1.95 | 0.00 | −0.52 | 89.09 | 0.49 |
| $R_5$ | 98 | 0.88 | 1.00 | −0.62 | 100.00 | 0.73 |
| $R_6$ | 100 | 1.18 | 2.01 | −0.54 | 100.00 | 0.98 |
| $R_7$ | 100 | 0.93 | 7.36 | −0.54 | 100.00 | 0.90 |
| $R_8$ | 100 | 1.80 | 9.76 | −0.13 | 66.99 | 0.19 |
| $R_9$ | 100 | 1.19 | 6.10 | −0.27 | 100.00 | 1.05 |
| China | 99 | 1.41 | 5.04 | −0.38 | 79.69 | 0.49 |

*Note* T1, T2, T3 and T4 represent downscaled MAT in the periods $t_1$, $t_2$, $t_3$ and $t_4$ respectively; and ΔTk(k−1) refers to the difference after subtracting downscaled MAT in period of $t_{k-1}$ from MAT in period $t_k$

$R_2$ (1.75 °C) and $R_3$ (1.65 °C). From $t_2$ to $t_3$, 94.96% of the grid cells across the whole of China would become cooler and SMAT would fall by 0.38 °C. The cooling would exceed the national average in regions $R_5$ (0.62 °C), $R_6$ (0.54 °C), $R_7$ (0.54 °C) and $R_4$ (0.52 °C). From $t_3$ to $t_4$, about 80% of the grid cells would become warmer and SMAT would increase by 0.49 °C across the whole of China on average; SMAT would increase more quickly in regions $R_9$ (1.05 °C), $R_6$ (0.98 °C), $R_7$ (0.90 °C), $R_5$ (0.73 °C), and $R_2$ (0.51 °C), while $R_1$ and $R_3$ would become cooler.

Under the RCP4.5 scenario (Table 7 and Fig. 10), all grid cells in China would be warmer and SMAT would increase by 2.03 °C from $t_1$ to $t_2$ across the whole of China on average. The warming in regions $R_1$ (2.98 °C), $R_3$ (2.82 °C), $R_4$ (2.42 °C), $R_2$ (2.37 °C) and $R_9$ (2.15 °C) would exceed the average for China as a whole. From $t_2$ to $t_3$, 71.7% of China would be warmer; including all of the Tibet plateau, and SMAT would increase by 1.72 °C on average; but regions $R_3$, $R_1$ and $R_8$ would be cooler. From $t_3$ to $t_4$, 71.06% of the whole of China would experience a warming trend, while 95% of the Tibet plateau would experience a cooling trend.

Under the RCP8.5 scenario (Table 8 and Fig. 11), all of the grid cells across the whole of China would become warmer during the periods from $t_1$ to $t_4$. From $t_1$ to $t_2$, SMAT would increase by 1.59 °C across the whole of China on average. The SMATs in regions $R_8$, $R_4$, $R_1$, $R_3$ and $R_2$ would warm the most, increasing by 2.10, 2.01, 1.92, 1.79 and 1.69 °C, respectively. From $t_2$ to $t_3$, SMAT would increase by 2.43 °C across China on average; led by increase in the arid area and on the Tibet plateau, which would be 3.02 and 2.57 °C warmer, respectively. From $t_3$ to $t_4$, northern China including regions $R_1$, $R_2$, $R_3$ and $R_8$ would warm the most; SMAT would increase by 3.77 °C in R2 (i.e., the arid area) for example; and China as a whole would be 1.88 °C warmer on average.

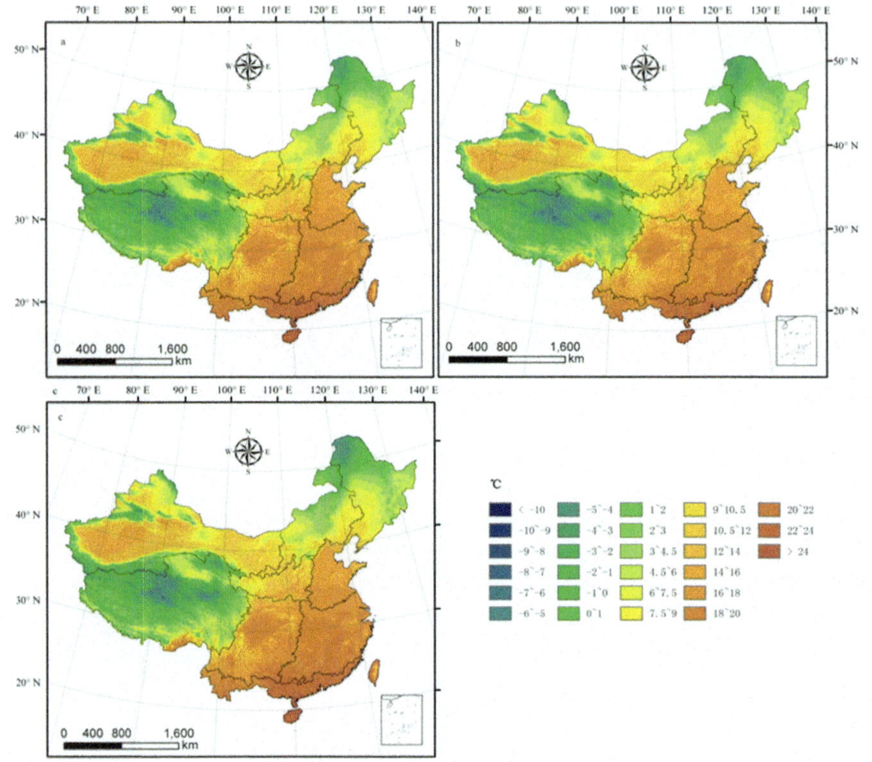


**Fig. 9** Changes of mean annual temperature under the RCP2.6 scenario

**Table 7** Changes in mean annual temperature under the RCP4.5 scenario

| Region | T2–T1 | | T3–T2 | | T4–T3 | |
|---|---|---|---|---|---|---|
| | Proportion of grid cells warmer (%) | ΔT21 (°C) | Proportion of grid cells warmer (%) | ΔT32 (°C) | Proportion of grid cells warmer (%) | ΔT43 (°C) |
| $R_1$ | 100.00 | 2.98 | 29.44 | –0.12 | 100.00 | 0.32 |
| $R_2$ | 100.00 | 2.37 | 88.66 | 0.45 | 78.89 | 0.26 |
| $R_3$ | 100.00 | 2.82 | 7.79 | –0.45 | 98.88 | 0.71 |
| $R_4$ | 100.00 | 2.42 | 31.09 | 0.14 | 84.28 | 0.49 |
| $R_5$ | 100.00 | 1.13 | 100.00 | 1.72 | 4.93 | –0.65 |
| $R_6$ | 100.00 | 1.64 | 76.22 | 0.38 | 98.36 | 0.57 |
| $R_7$ | 100.00 | 1.25 | 100.00 | 0.54 | 92.61 | 0.38 |
| $R_8$ | 100.00 | 1.79 | 40.67 | –0.01 | 98.54 | 0.59 |
| $R_9$ | 100.00 | 2.15 | 77.07 | 0.17 | 95.47 | 0.62 |
| China | 100.00 | 2.03 | 71.70 | 0.54 | 71.06 | 0.20 |

*Note* T1, T2, T3 and T4 represent downscaled MAT in the periods $t_1$, $t_2$, $t_3$ and $t_4$, respectively; and ΔTk(k–1) refers to the difference after subtracting downscaled MAT in period $t_{k-1}$ from MAT in the period $t_k$

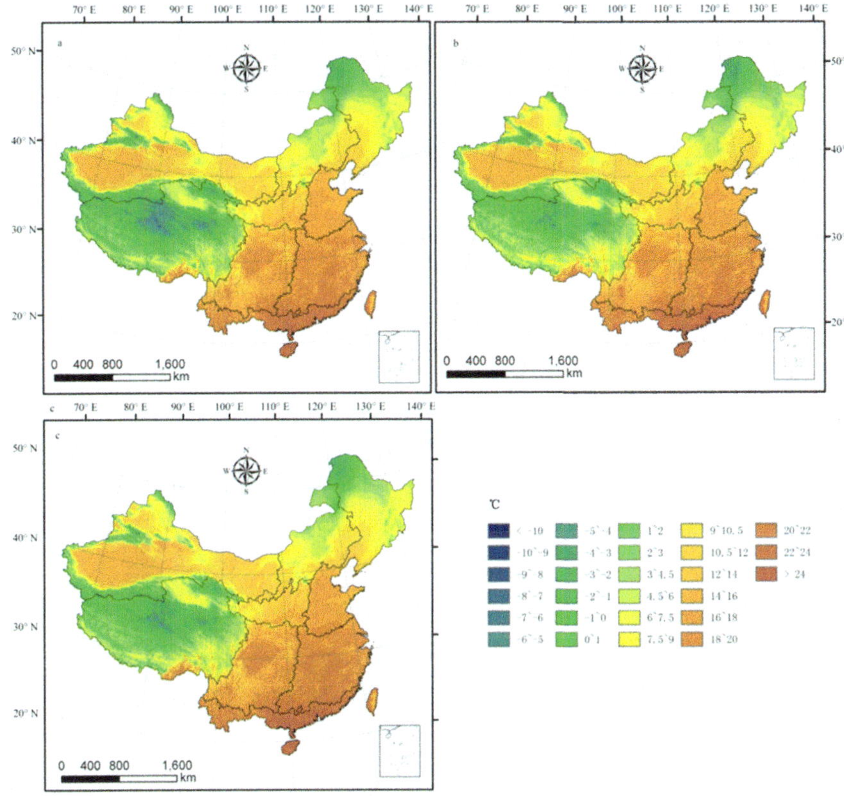


**Fig. 10** Changes of mean annual temperature under the RCP4.5 scenario

**Table 8** Changes in mean annual temperature under the RCP8.5 scenario

| Region | T2–T1 | | T3–T2 | | T4–T3 | |
|---|---|---|---|---|---|---|
| | Proportion of grid cells warmer (%) | ΔT21 (°C) | Proportion of grid cells warmer (%) | ΔT32 (°C) | Proportion of grid cells warmer (%) | ΔT43 (°C) |
| $R_1$ | 100.00 | 1.92 | 100.00 | 2.26 | 100.00 | 2.88 |
| $R_2$ | 100.00 | 1.69 | 100.00 | 3.02 | 100.00 | 1.94 |
| $R_3$ | 100.00 | 1.79 | 100.00 | 1.93 | 100.00 | 3.77 |
| $R_4$ | 100.00 | 2.01 | 100.00 | 2.42 | 100.00 | 1.33 |
| $R_5$ | 100.00 | 1.41 | 100.00 | 2.57 | 100.00 | 1.71 |

(continued)

**Table 8** (continued)

| Region | T2–T1 | | T3–T2 | | T4–T3 | |
|---|---|---|---|---|---|---|
| | Proportion of grid cells warmer (%) | ΔT21 (°C) | Proportion of grid cells warmer (%) | ΔT32 (°C) | Proportion of grid cells warmer (%) | ΔT43 (°C) |
| $R_6$ | 100.00 | 1.27 | 100.00 | 2.16 | 100.00 | 1.02 |
| $R_7$ | 100.00 | 1.04 | 100.00 | 1.89 | 100.00 | 0.71 |
| $R_8$ | 100.00 | 2.10 | 100.00 | 1.92 | 100.00 | 1.98 |
| $R_9$ | 100.00 | 1.53 | 100.00 | 2.18 | 100.00 | 1.11 |
| China | 100.00 | 1.59 | 100.00 | 2.43 | 100.00 | 1.88 |

*Note* T1, T2, T3 and T4 represent downscaled MAT in the periods $t_1$, $t_2$, $t_3$ and $t_4$, respectively; and ΔTk(k–1) refers to the difference after subtracting downscaled MAT in period $t_{k-1}$ from MAT in the period $t_k$

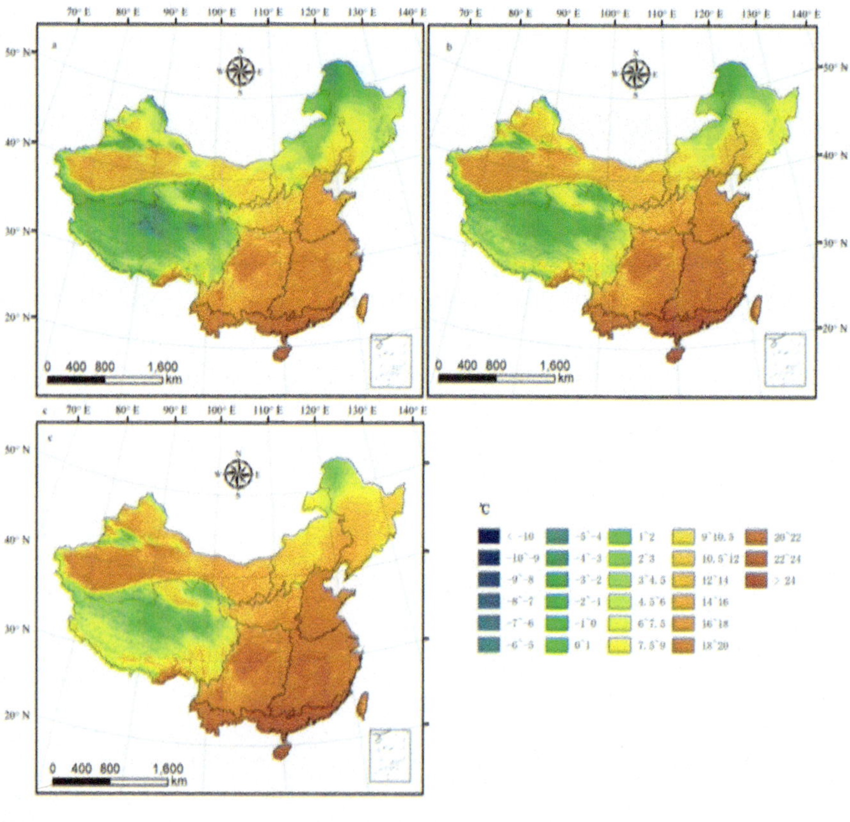


**Fig. 11** Changes of mean annual temperature under the RCP8.5 scenario

#### 4.3.4.2 Mean Annual Precipitation

Under the RCP2.6 scenario (Table 9 and Fig. 12), from $t_1$ to $t_2$, SMAP would increase in 88% of the grid cells and SMAP would increase by 24.69 mm over the whole of China on average. SMAP in regions $R_5$, $R_8$, $R_1$, $R_4$, $R_3$ and $R_2$ would increase by 37.88, 37.73, 30.37, 27.94, 27.37 and 24.70 mm, respectively, indicating that the Tibet plateau and northern China would experience the largest increases in SMAP compared to other regions. From $t_2$ to $t_3$, 95% of the grid cells would be wetter and SMAP would increase by 22.15 mm across the whole of China on average. SMAP would increase by 39.49, 37.99, 37.33, 26.95 and 23.91 mm in regions $R_9$, $R_6$, $R_7$, $R_5$ and $R_3$, respectively; showing that SMAP would experience the largest increases in southern China and on the Tibet plateau. From $t_3$ to $t_4$, 59% of the grid cells would be wetter and SMAP would increase by 2.1 mm on average. The SMAP in regions $R_6$, $R_5$, $R_9$ and $R_8$ would increase by 11.76, 5.76, 3.7 and 3.07 mm, respectively; but regions $R_7$, $R_3$, $R_1$ and $R_2$ would be drier and SMAP would decrease by 9.24, 2.93, 1.03 and 0.7 mm, respectively in the four regions.

**Table 9** Changes in mean annual precipitation under the RCP2.6 scenario

| Region | P2–P1 | | P3–P2 | | P4–P3 | |
|---|---|---|---|---|---|---|
| | Proportion of grid cells wetter (%) | ΔP21 (mm) | Proportion of grid cells wetter (%) | ΔP32 (mm) | Proportion of grid cells wetter (%) | ΔP43 (mm) |
| $R_1$ | 99.29 | 30.37 | 100.00 | 12.24 | 43.46 | −1.03 |
| $R_2$ | 100.00 | 24.70 | 80.49 | 3.33 | 42.03 | −0.70 |
| $R_3$ | 99.99 | 27.37 | 100.00 | 23.91 | 29.95 | −2.93 |
| $R_4$ | 97.76 | 27.94 | 100.00 | 18.33 | 52.60 | 0.81 |
| $R_5$ | 97.77 | 37.88 | 96.96 | 26.95 | 81.10 | 5.76 |
| $R_6$ | 48.96 | 5.07 | 100.00 | 37.99 | 87.76 | 11.76 |
| $R_7$ | 48.76 | 11.24 | 100.00 | 37.73 | 21.48 | −9.24 |
| $R_8$ | 94.93 | 37.73 | 100.00 | 19.58 | 82.47 | 3.07 |
| $R_9$ | 72.56 | 7.58 | 100.00 | 39.49 | 70.21 | 3.70 |
| China | 88.23 | 24.69 | 94.68 | 22.15 | 58.97 | 2.1 |

*Note* P1, P2, P3 and P4 represent downscaled MAP in the periods $t_1$, $t_2$, $t_3$ and $t_4$, respectively; and ΔPk(k−1) refers to the difference after subtracting downscaled MAP in period $t_{k-1}$ from MAP in the period $t_k$

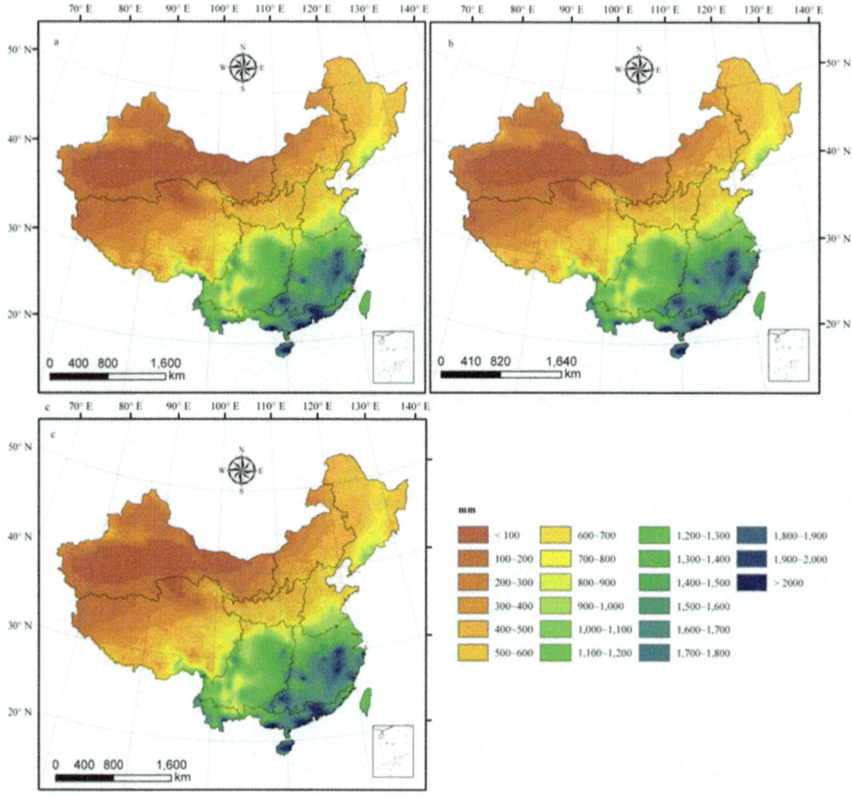


**Fig. 12** Changes of mean annual precipitation under the RCP2.6 scenario

Under the RCP4.5 scenario (Table 10 and Fig. 13), 86.55% of the grid cells would be wetter and SMAP would become 20.72 mm higher from $t_1$ to $t_2$ across the whole of China on average. The largest increases in SMAP would occur in regions $R_5$, $R_3$, $R_8$ and $R_1$ (33.31, 29.43, 28.24 and 22.69 mm, respectively), while SMAP would decrease in region $R_6$ by 2.02 mm. From $t_2$ to $t_3$, 98.1% of the whole of China would be wetter and SMAP would increase by 35.81 mm on average; regions $R_6$ (48.85 mm), $R_7$ (47.86 mm), $R_5$ (46.34 mm), $R_9$ (42.94 mm), $R_8$ (42.52 mm) and $R_4$ (41.39 mm) would experience the largest gains. From $t_3$ to $t_4$, 52.01% of China would be wetter and SMAP would increase by 21.5 mm on average led by 36.42, 34.18, 32.52, 26.67 and 24.77 mm increases in SMAP in regions $R_4$, $R_5$, $R_6$, $R_9$ and $R_8$, respectively.

**Table 10** Changes in mean annual precipitation under the RCP4.5 scenario

| Region | P2−P1 | | P3−P2 | | P4−P3 | |
|---|---|---|---|---|---|---|
| | Proportion of grid cells wetter (%) | ΔP21 (mm) | Proportion of grid cells wetter (%) | ΔP32 (mm) | Proportion of grid cells wetter (%) | ΔP43 (mm) |
| $R_1$ | 98.57 | 22.69 | 100.00 | 37.44 | 79.06 | 9.25 |
| $R_2$ | 100.00 | 19.93 | 96.33 | 9.86 | 98.59 | 11.24 |
| $R_3$ | 99.97 | 29.43 | 100.00 | 38.45 | 75.53 | 5.60 |
| $R_4$ | 94.26 | 18.38 | 100.00 | 41.39 | 100.00 | 36.42 |
| $R_5$ | 97.50 | 33.31 | 95.41 | 46.34 | 100.00 | 34.18 |
| $R_6$ | 34.41 | −2.02 | 100.00 | 48.85 | 100.00 | 32.52 |
| $R_7$ | 58.85 | 12.17 | 100.00 | 47.86 | 91.49 | 13.67 |
| $R_8$ | 94.28 | 28.24 | 100.00 | 42.52 | 100.00 | 24.77 |
| $R_9$ | 69.07 | 9.60 | 100.00 | 42.94 | 100.00 | 26.67 |
| China | 86.55 | 20.72 | 98.10 | 35.81 | 52.01 | 21.50 |

*Note* P1, P2, P3 and P4 represent downscaled MAP in the periods $t_1$, $t_2$, $t_3$ and $t_4$, respectively; and ΔPk(k−1) refers to the difference after subtracting downscaled MAP in period $t_{k-1}$ from MAP in the period $t_k$

Under the RCP8.5 scenario (Table 11 and Fig. 14), the wetting trend (i.e. SMAP) would increase from $t_1$ to $t_4$ across the whole of China. SMAP would increase by 18.88, 45.63 and 47.11 mm from $t_1$ to $t_2$, $t_2$ to $t_3$ and $t_3$ to $t_4$, respectively. Most of southern China would become drier, while the Tibet plateau and northern China would become wetter from $t_1$ to $t_2$. SMAP would increase across all of China from $t_2$ to $t_3$ with SMAP increasing by 67.69, 55.7, 54.06 and 53.26 mm in regions $R_5$, $R_4$, $R_8$ and $R_6$, respectively. From $t_3$ to $t_4$, 99.79% of the whole of China would be wetter; led by regions $R_5$, $R_3$, $R_6$, $R_8$ and $R_4$ for which SMAP would increase by 74.41, 57.51, 54,75, 50.58 and 49.8 mm, respectively.

In terms of HASM, China would become warmer and wetter under the RCP2.6, RCP4.5 and RCP8.5 scenarios from $t_1$ to $t_2$. The largest rise in SMAT is produced by RCP4.5 and the smallest one by RCP2.6; however, RCP2.6 shows the largest increase in SMAP and RCP8.5 shows the smallest increase. From $t_2$ to $t_3$, SMAT under RCP8.5 would have the largest rise and increase (2.43 °C), while SMAT under RCP2.6 would decrease by −0.38 °C; SMAPs under RCP2.6, RCP4.5 and RCP8.5 would all rise, led by RCP8.5 for which the increase would be 45.63 mm. From $t_3$ to $t_4$, most of the grid cells across China would be warmer and wetter under RCP8.5 (SMAP and SMAT would increase by 47.11 mm and 1.88 °C, respectively), but more than 40% of the land area in China would be drier and more than 20% of the land area in China would be cooler under RCP2.6 or RCP4.5.

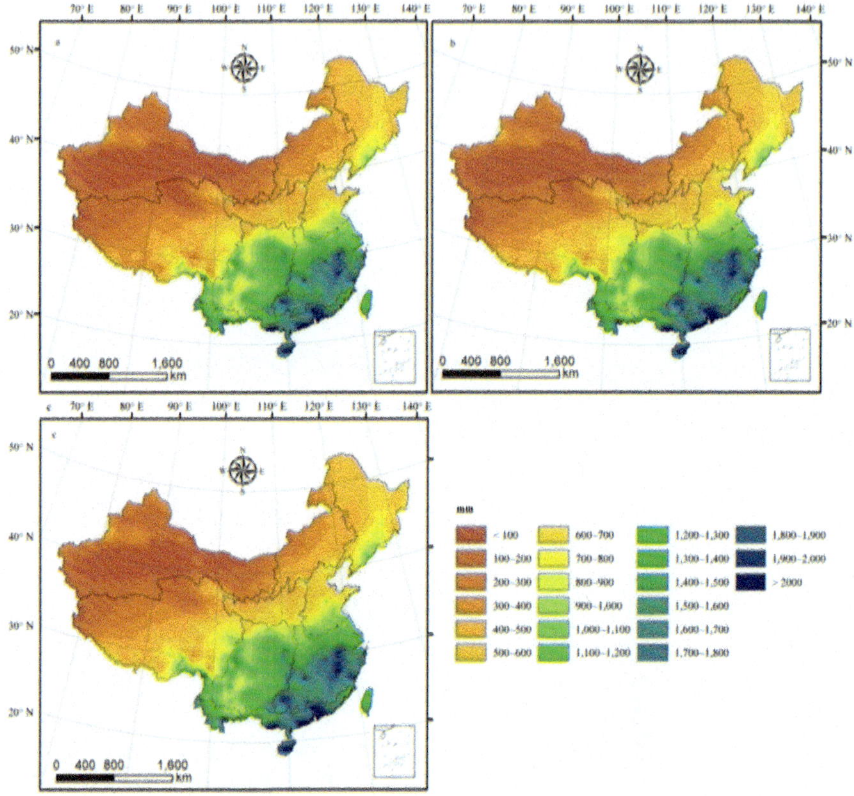


**Fig. 13** Changes of mean annual precipitation under the RCP4.5 scenario

**Table 11** Changes in mean annual precipitation under the RCP8.5 scenario

| Region | P2–P1 | | P3–P2 | | P4–P3 | |
|---|---|---|---|---|---|---|
| | Proportion of grid cells wetter (%) | ΔP21 (mm) | Proportion of grid cells wetter (%) | ΔP32 (mm) | Proportion of grid cells wetter (%) | ΔP43 (mm) |
| $R_1$ | 98.15 | 19.94 | 100.00 | 45.05 | 100.00 | 47.11 |
| $R_2$ | 100.00 | 23.88 | 100.00 | 18.72 | 98.97 | 18.10 |
| $R_3$ | 99.80 | 23.99 | 100.00 | 44.91 | 100.00 | 57.51 |
| $R_4$ | 90.65 | 21.25 | 100.00 | 55.70 | 100.00 | 49.80 |
| $R_5$ | 81.96 | 38.84 | 100.00 | 67.69 | 100.00 | 74.41 |

(continued)

**Table 11** (continued)

| Region | P2–P1 | | P3–P2 | | P4–P3 | |
|---|---|---|---|---|---|---|
| | Proportion of grid cells wetter (%) | ΔP21 (mm) | Proportion of grid cells wetter (%) | ΔP32 (mm) | Proportion of grid cells wetter (%) | ΔP43 (mm) |
| $R_6$ | 23.43 | −13.30 | 100.00 | 53.26 | 100.00 | 54.75 |
| $R_7$ | 38.30 | 4.11 | 100.00 | 45.05 | 99.77 | 33.89 |
| $R_8$ | 93.62 | 30.16 | 100.00 | 54.06 | 100.00 | 50.58 |
| $R_9$ | 33.29 | −8.97 | 100.00 | 43.14 | 100.00 | 37.56 |
| China | 80.54 | 18.88 | 100.00 | 45.63 | 99.79 | 47.11 |

*Note* P1, P2, P3 and P4 represent downscaled MAP in the periods $t_1$, $t_2$, $t_3$ and $t_4$, respectively; and ΔPk(k−1) refers to the difference after subtracting downscaled MAP in period $t_{k-1}$ from MAP in the period $t_k$

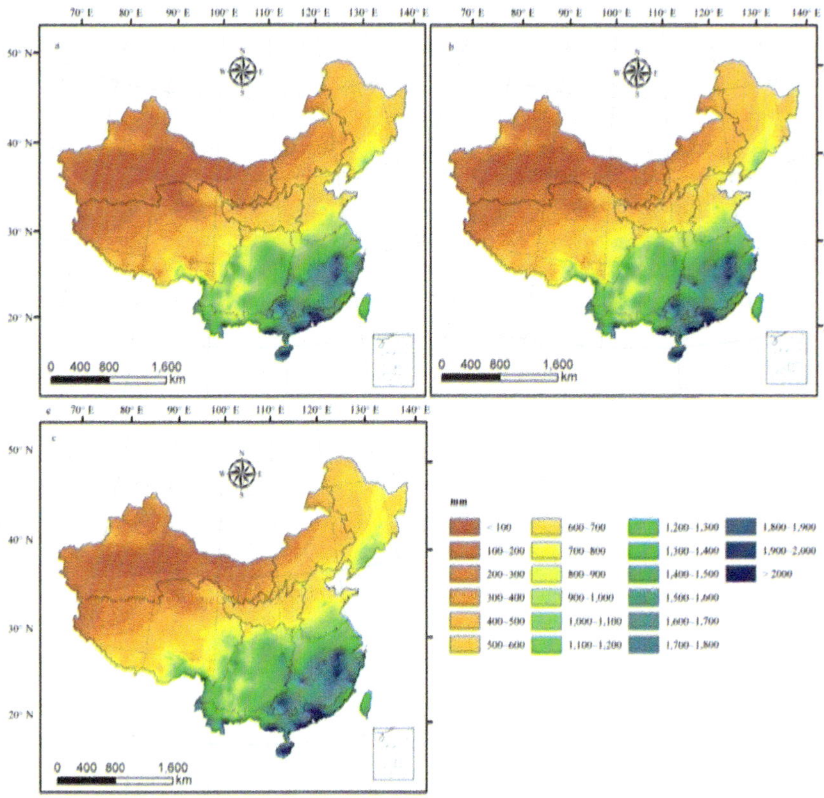


**Fig. 14** Changes of mean annual precipitation under the RCP8.5 scenario

### 4.3.5 Discussion

The CMIP5 scenarios incorporate too much uncertainty to be used for national level planning and decision-making in China. The MAE for annual mean temperature under all of the CMIP5 scenarios is as high as 2.2 °C for the whole of China on average, but 5 °C for Tibet plateau, 4 °C for the arid area, and 3 °C for the Loess plateau, Sichuan basin and the Yunan-Guizhou plateau due to the underestimation of CMIP5. For the whole land mass of China on average, the MAEs of OMAP under the RCP2.6, RCP4.5 and RCP8.5 scenarios were 345.88, 338.18 and 340.93 mm, respectively due to the considerable overestimation that occurred where annual mean precipitation was < 1,500 mm. The MAEs of OMAP were 758.52, 765.54 and 765.01 mm, respectively for the Tibet plateau, and 583.55, 589.49 and 581.03 mm, respectively for the Loess plateau for example.

The downscaling processes that are illustrated in this paper can considerably reduce the uncertainty because they introduce temperature and precipitation observations from meteorological stations and use HASM to predict the relevant surfaces. The MAE ratios of the temperature for all of China were greater than 72%, led by an improvement of 87% for the Tibet plateau and 79% for the arid area. The accuracy of the precipitation was 79% better for the whole of China on average, led again by a > 90% for the Tibet plateau and arid area.

In terms of the downscaled results, almost all China would become warmer under all the three scenarios and SMAT would be 1.41–2.03 °C higher in $t_2$ compared to $t_1$. In the twenty-first Century, both the 2,100°Cd contour line of annual accumulated temperature (AAT) of $\geq$ 10 °C and the 2,400°Cd contour line of $\geq$ 10 °C AAT would move considerably northward, greatly expanding the area suitable for planting early rice and maize.

SMAP across the whole of China would have an increasing trend on average, but the trends would vary under different scenarios, especially after the period $t_3$. Under scenario RCP2.6, SMAP would increase across the whole of China on average, but the increasing trend would gradually become weaker from $t_1$ to $t_4$, and the regions $R_1$, $R_2$, $R_3$ and $R_7$ would have a decreasing trend from period $t_3$ onwards. Under scenario RCP4.5, the largest increase in SMAP across the whole of China on average would occur from $t_2$ to $t_3$, and all nine regions would become wetter in all periods except $R_6$, which would be drier from $t_1$ to $t_2$. Under scenario RCP8.5, SMAP would rise markedly and the increasing trend would gradually become much stronger from $t_1$ to $t_4$ across the whole of China on average, except for regions $R_6$ and $R_9$, which would become drier from $t_1$ to $t_2$. The RCP8.5 predicts the largest increases in SMAP from $t_2$ to $t_3$ and from $t_3$ to $t_4$, compared with the RCP2.6 and RCP4.5 scenarios. In all of the periods, the Tibet plateau ($R_5$) and north China ($R_3$) would experience the largest increases in SMAP.

## 4.4 Summary

Multiscale issues are essential for eco-environmental informatics, especially upscaling and downscaling of eco-environmental surfaces. We develop an upscaling algorithm termed HASM-US by means of FTEEM Corollary 2 and take the elevation surface of BTH region as an example to exhibit the process of upscaling. The upscaling employed both extrinsic information such as satellite observation data and intrinsic information such as ground observation data as well as an appropriate approach. The results indicate that HASM-US has the highest accuracy and the most stable performance comparing with two classical methods.

The CMIP5 datasets of MAT and MAP are downscaled by means of FTEEM Corollary 3. Statistical transfer functions of MAT and MAP are respectively developed by combining data from 735 meteorological stations with DTM of China. The downscaling process made MAT and MAP of CMIP5 datasets respectively 87% and 70% more accurate.

## 4.5 Further Reading

(1) Yue TX, Liu Y, Zhao MW, Du ZP, Zhao N. 2016. A fundamental theorem of Earth's surface modelling. Environmental Earth Sciences 75(9): article 751 (pages 1–12).
(2) Yue TX, Zhao N, Fan ZM, et al. 2016. CMIP5 Downscaling and its uncertainty in China. Glob Planet Change 146: 30–37.
(3) Yue TX, Zhao N, Fan ZM, Li J, Chen CF, Lu YM, Wang CL, Gao J, Xu B, Wilson J. 2019. Methods for simulating climate scenarios with improved spatial-temporal specificity and less uncertainty. Global and Planetary Change 181 (Article 102973), 1–18.
(4) Yue TX, Zhao N, Liu Y, Wang YF, Zhang B, Du ZP, Fan ZM, Shi WJ, Chen CF, Zhao MW, Song DJ, Song YJ, Yan CQ, Li QQ, Sun XF, Zhang LL, Tian YZ, Wang W, Wang YA, Ma SN, Huang HS, Lu YM, Wang Q, Wang CL, Wang YZ, Lu M, Zhou W, Liu Y, Yin XZ, Wang Z, Bao ZY, Zhao MM, Zhao YP, Jiao YM, Naseer U, Fan B, Li SB, Yang Y, Wilson JP. 2020. A fundamental theorem for eco-environmental surface modelling and its applications. Science China-Earth Sciences 63(8): 1092–1112.

### Exercises

(1) Define the terms "upscaling" and "downscaling".
(2) Explain the Corollary 2 (Upscaling) and Corollary 3 (Downscaling) of FTEEM.

## References


Allen, T.F.H. and Starr, T.B. 1982. *Hierarchy: Perspectives for Ecological Complexity.* Chicago: University of Chicago Press.

Christensen, J.H., and F. Boberg. 2012. Temperature dependent climate projection deficiencies in CMIP5 models. *Geophysical Research Letters* 39: L24705.

Cressie, N.A.C. 1993. *Statistics for Spatial Data.* New York, NY: John Wiley & Sons.

Crow, W.T., A.A. Berg, M.H. Cosh, A. Loew, B.P. Mohanty, R. Panciera, P. de Rosnay, D. Ryu, and J.P. Walker. 2012. Upscaling sparse ground-based soil moisture observations for the validation of coarse-resolution satellite soil moisture products. Reviews of Geophysics 50: RG2002.

Dickson, N.E.M., J.C. Comte, J. McKinley, and U. Ofterdinger. 2014. Coupling ground and airborne geophysical data with upscaling techniques for regional groundwater modeling of heterogeneous aquifers: Case study of a sedimentary aquifer intruded by volcanic dykes in Northern Ireland. *Water Resources Research* 50: 7984–8001.

Elguindi, N., A. Grundstein, S. Bernardes, U. Turuncoglu, and J. Feddema. 2014. Assessment of CMIP5 global model simulations and climate change projections for the 21st century using a modified Thornthwaite climate classification. *Climatic Change* 122: 523–538.

Feng, S., Q. Hu, W. Huang, C.H. Ho, R.P. Li, and Z.H. Tang. 2014. Projected climate regime shift under future global warming from multi-model, multi-scenario CMIP5 simulations. *Global and Planetary Change* 112: 41–52.

Fry, J., A.K. Guber, M. Ladoni, J.D. Munoz, and A.N. Kravchenko. 2017. The effect of up-scaling soil properties and model parameters on predictive accuracy of DSSAT crop simulation model under variable weather conditions. *Geoderma* 287: 105–115.

Golley, F.B. 1989. Paradigm shift. *Landscape Ecology* 3(2): 65–66.

Gosz, J.R. and Sharpe, P.J.H. 1989. Broad-scale concepts for interaction of climate, topography, and biota at biome transitions. *Landscape Ecology* 3: 229–243.

Gould, S.J. 1966. Allometry and size in ontogeny and phylogeny. *Biological Review* 41: 587–640.

Greig-Smith, P. 1952. The use of random and contiguous quadrats in the study of the structure of plant communities. *Annals of Botany* 16: 293–316.

Holling, C.S. 1992. Cross-scale morphology, geometry, and dynamics of ecosystems. *Ecological Monographs* 62(4): 447–502.

Holt, R.D., Pacala, S.W., Smith, T.W., and Liu, J.G. 1995. Linking contemporary vegetation models with spatially explicit animal population-models. *Ecological Applications* 5(1): 20–27.

Hua, W.J., H.S. Chen, and S.L. Sun. 2014. Uncertainty in land surface temperature simulation over China by CMIP3/CMIP5 models. *Theoretical and Applied Climatology* 117: 463–474.

Hugelius, G. 2012. Spatial upscaling using thematic maps: An analysis of uncertainties in permafrost soil carbon estimates. *Global Biogeochemical Cycles* 26. https://doi.org/10.1029/2011gb004154.

Hutchinson, G.E. 1971. Banquet address: Scale effects in ecology. In Spatial Patterns and Statistical Distribution, eds., G.P. Patil, E.C. Pielou, and W.E. Waters, 17–22. Pennsylvania, PA: Pennsylvania State University Press.

Ji, M.X., J.P. Huang, Y.K. Xie, and J. Liu. 2015. Comparison of dryland climate change in observations and CMIP5 simulations. *Advances in Atmospheric Sciences* 32: 1565–1574.

Kondo, M., K. Ichii, H. Takagi, and M. Sasakawa. 2015. Comparison of the data-driven top-down and bottom-up global terrestrial $CO_2$ exchanges: GOSAT $CO_2$ inversion and empirical eddy flux upscaling. *Journal of Geophysical Research. Biogeosciences* 120: 1226–1245.

Kunin, W.E., J. Harte, F.L. He, C. Hui, R.T. Jobe, A. Ostling, C. Polce, A. Sizling, A.B. Smith, K. Smith, S.M. Smart, D. Storch, E. Tjorve, K.I. Ugland, W. Ulrich, and V. Varma. 2018. Upscaling biodiversity: Estimating the species-area relationship from small samples. *Ecological Monographs* 88 (2): 170–187.

Levin, S.A. 1992. The problem of pattern and scale in ecology. *Ecology* 73(6): 1943–1967.

Li, Z.L. 2007. Algorithmic Foundation of Multi-Scale Spatial Representation. London: CRC Press.

Li, L.P., H.Y. Zhou, H.J.H. Franssen, and J.J. Gomez-Hernandez. 2012. Modeling transient groundwater flow by coupling ensemble Kalman filtering and upscaling. *Water Resources Research* 48: W01537. https://doi.org/10.1029/2010WR010214.

Mandelbrot, B.B. 1982. The Fractal Geometry of Nature. San Francisco: W. H. Freeman.

Milne, B.T. 1992. Spatial aggregation and neutral models in fractal landscapes. American Naturalist 139 (1): 32–57.

Misaghian, N., M. Assareh, and M.T. Sadeghi. 2018. An upscaling approach using adaptive multi-resolution upgridding and automated relative permeability adjustment. *Computers & Geosciences* 22: 261–282.

O'Neill, R.V., Johnson, A.R., and King, A.W. 1989. A hierarchical framework for the analysis of scale. *Landscape Ecology* 3: 193–205.

Qin, J., L. Zhao, Y.Y. Chen, K. Yang, Y.P. Yang, Z.Q. Chen, and H. Lu. 2015. Inter-comparison of spatial upscaling methods for evaluation of satellite-based soil moisture. *Journal of Hydrology* 523: 170–178.

Scafetta, N. 2013. Solar and planetary oscillation control on climate change: Hind-cast, forecast and a comparison with the CMIP5 GCMs. *Energy and Environment* 24: 455–496.

Schlummer, M., H. Thomas, R. Dikau, M. Eickmeier, P. Fischer, R. Gerlach, J. Holzkämper, A.J. Kalis, I. Kretschmer, F. Lauer, A. Maier, J. Meesenburg, J. Meurers-Balke, U. Münc, S. Pätzoldg, F. Steininger, A. Stobbe, and A. Zimmermann. 2014. From point to area: Upscaling approaches for Late Quaternary archaeological and environmental data. *Earth-Science Reviews* 131: 22–48.

Seekell, D.A., J.A. Carr, C. Gudasz, and J. Karlsson. 2014. Upscaling carbon dioxide emissions from lakes. *Geophysical Research Letters* 41: 7555–7559.

Usery, E.L. and McMaster, R.B. 2005. Introduction to the UCGIS research agenda. In *A Research Agenda for Geographic Information Science*, eds., E.L. Usery and R.B. McMaster, 1–16. London: CRC Press.

van Bussel, L.G.J., P. Grassini, J.V. Wart, J. Wolf, L. Claessens, H.S. Yang, H. Boogaard, H. de Groot, K. Saito, K.G. Cassman, and M.K. van Ittersum. 2015. From field to atlas: Upscaling of location-specific yield gap estimates. *Field Crops Research* 177: 98–108.

Wang, C.L., T.X. Yue, Z.M. Fan, N. Zhao, and X.F. Sun. 2012. HASM-based climatic downscaling model over China. *Journal of Geo-Information Science* 14: 559–610 (In Chinese with English abstract).

Wiens, J. 1989. Spatial scaling in ecology. *Functional Ecology* 3: 385–397.

Wood, B.D. 2009. The role of scaling laws in upscaling. *Advances in Water Resources* 32: 723–736.

Xue, J.J., Y.H. Ge, and H.R. Ren. 2017. Spatial upscaling of green aboveground biomass derived from MODIS-based NDVI in arid and semiarid grasslands. *Advances in Space Research* 60 (9): 2001–2008.

Yue, T.X., N. Zhao, D. Ramsey, C.L. Wang, Z.M. Fan, C.F. Chen, Y.M. Lu, and B.L. Li. 2013. Climate change trend in China, with improved accuracy. *Climatic Change* 120: 137–151.

Yue, T.X., N. Zhao, Z.M. Fan, J. Li, C.F. Chen, Y.M. Lu, C.L. Wang, B. Xu, and J.P. Wilson. 2016. CMIP5 uncertainty in China. *Global and Planetary Change* 146: 30–37.

Yue, T.X., N. Zhao, Y. Liu, Y.F. Wang, B. Zhang, Z.P. Du, Z.M. Fan, W.J. Shi, C.F. Chen, M.W. Zhao, D.J. Song, Y.J. Song, C.Q. Yan, Q.Q. Li, X.F. Sun, L.L. Zhang, Y.Z. Tian, W. Wang, Y.A. Wang, S.N. Ma, H.S. Huang, Y.M. Lu, Q. Wang, C.L. Wang, Y.Z. Wang, M. Lu, W. Zhou, Y. Liu, X.Z. Yin, Z. Wang, Z.Y. Bao, M.M. Zhao, Y.P. Zhao, Y.M. Jiao, U. Naseer, B. Fan, S.B. Li, Y. Yang, and J.P. Wilson. 2020. A fundamental theorem for eco-environmental surface modelling and its applications. *Science China-Earth Sciences* 63 (8): 1092–1112.

# Chapter 5
# Data Fusion

**Learning Objectives**

- Corollary 4 of FTEEM: when remotely sensed data from satellites are available, ground measurements have to be obtained and incorporated before HASM can be used to generate a more accurate surface. This corollary emphasizes the necessity of information completeness for data fusion.
- Corollary 5 of FTEEM: when both remotely sensed data from satellites and ground measurements are available, HASM can be used to generate a surface that is more accurate than the one from either the satellite observations or the ground measurements. This corollary states the necessity of an appropriate method for data fusion with higher accuracy.

## 5.1 Introduction

Data fusion is the process of integration of multiple data and knowledge streams representing the same real-world object into a consistent, accurate, and useful representation. The aim of data fusion is to improve the quality of the information so that it is more accurate than the one if the data sources were used individually (Mitchell 2012). Since data from different sources have varying accuracy and coverage, the benefits of data fusion include improved system reliability, extended coverage and reduced uncertainty.

Methods for data fusion, which were mostly developed in the field of remote sensing, only gave consideration to the extrinsic property. For instance, Bayesian method for fusion of optical sensor data and radar data considerably improves the classification accuracy, compared to the methods based on the optical information alone (Park et al. 2002). Urban surface features are more distinguishable from each other using data fusion of hyperspectral data from Airborne Visible Infrared

 
T. Yue et al., *Eco-Environmental Informatics*,
https://doi.org/10.1007/978-981-96-1977-1_5

Imaging Spectrometer (AVIRIS) and the surface texture information from Topographic Synthetic Aperture Radar (TOPSAR) than using a single sensor data (Chen et al. 2003). Data fusion of remote sensing signal of imaging spectrometry and large footprint LiDAR was not only able to complement the canopy structure of forests and biophysical and -chemical canopy properties, but also helped to solve the ill-posed problem of the Radiative Transfer Model (RTM) inversion (Koetz et al. 2007). A spatial temporal adaptive algorithm for mapping reflectance change (STAARCH), a data fusion model of Landsat TM/ETM and MODIS reflectance data, was introduced to identify spatial and temporal changes in the landscape with a high level of detail in west-central Alberta, Canada (Hilker et al. 2009). A geostatistical technique for data fusion was used to exploit the complementary capabilities of multiangle (MISR) and multispectral (MODIS) sensors along with the ground-based Aerosol Robotic Network (AERONET) in better measuring aerosol optical thickness (AOT); cross-validation results over the contiguous United States demonstrated that AOT product was closer to the ground-based AOT observations than either of the individual satellite measurements (Chatterjee et al. 2010).

Fusing multispectral and LiDAR data allowed for a detailed characterization of fuel type distribution in the Natural Park of the Alto Tajo in central Spain, which exploited the spectral information provided by the optical data and the three-dimensional information provided by LiDAR data (García et al. 2011). Supervised maximum likelihood classification of land cover in Mecklenburg County of North Carolina, USA, by fusing LiDAR data with Landsat Thematic Mapper, increased total accuracy by 32% compared to LiDAR alone and by 8% over TM (Singh et al. 2012). Experimental analyses of three fields with different texture in the Netherlands demonstrated that soil property models based on fusion methods significantly improved the accuracy of predictions of soil properties measureable by both Vis–NIR spectrometer and EM38 sensor from those based on either of the individual sensor (Mahmood et al. 2012). An experimental analysis in the Southern Alps indicated that high density LiDAR data provided more information for tree species classification with respect to low density data, when fused with either hyperspectral or multispectral data (Dalponte et al. 2012).

A daily coverage of evapotranspiration on spatial resolution of 30 m was obtained by fusing daily MODIS (MODerate resolution Imaging Spectroradiometer) on spatial resolution of 1 km and biweekly Landsat LST imagery on spatial resolution of 30 m (Cammalleri et al. 2013). Data fusion of spaceborne LiDAR, synthetic aperture radar (SAR) and multispectral imagery demonstrated the ability of active and passive remote sensing for achieving the accuracy requirements of forest biomass mapping on global level (Montesano et al. 2013). Fusing multispectral with LiDAR data improved regression models for predicting both live and dead biomass and thus better predicted carbon stocks in tidal salt marshes in Texas of USA (Kulawardhana et al. 2014). Urban forest species in downtown Santa Barbara of California in USA was accurately mapped by fusing hyperspectral imagery with LiDAR data (Alonzo et al. 2014). An accuracy analysis of remote sensing products exposed that currently landcover maps are profiting from fusion of airborne laser scanning and imaging spectroscopy data (Torabzadeh et al. 2014). The Spatial and Temporal Adaptive

Reflectance Fusion Model (STARFM), to optimally exploit spatiotemporal characteristics of image datasets collected by the Landsat and Moderate resolution Imaging Spectroradiometer (MODIS) sensors, accurately reproduced the temporal dynamics of daily evapotranspiration for the experimental sites, and was able to capture the main differences between irrigated and rainfed fields at both daily and seasonal time scales (Cammalleri et al. 2014).

Applications of spaceborne hyperspectral and interferometric sensors to biomass estimation indicated that a data fusion provided higher accuracies than using solely one sensor type (Kattenborn et al. 2015). A lookup table (LUT)-based approach to data fusion between microwave and optical data was able to simultaneously describe spatial heterogeneity and track the temporal variability of inland waters by means of statistical regression (Mizuochi et al. 2017). Spatial data infrastructure (SDI) initiatives was proposed in order to fuse spatial data from the variety of existing, but distributed sources, especially on the Web under support of a number of open standards (Wiemann 2017). A framework for data fusion, based on a multivariate conditional autoregressive (MVCAR) model, was developed to integrate multiple data sources in species distribution in USA, which improved estimation of species distributions relative to single data set method (Pacifici et al. 2017). A fusion of LiDAR point cloud data and a hyperspectral image from Unmanned aerial vehicles (UAVs) provided more than 30% of improvement in species identification by leveraging differences in plant height, comparing with the classification of the hyperspectral data alone (Sankey et al. 2017). An approach to multisensor data fusion is developed to process the information from the entire sensor network in order to make a more accurate estimation of the environmental pollutions (Aiello et al. 2018).

In recent years, a number of methods for data fusion have paid attention to both intrinsic and extrinsic properties. For instance, directly measuring rainfall from rain gauges is the most accurate way to determine site-scale precipitation. However, rainfall estimations from ground-based measurements also contain obvious uncertainty in areas because of sparse networks of rainfall gauges, especially in many areas such as over the oceans, mountains, scattered islands and inaccessible land areas (Ragab and Prudhomme 2002; Alizadeh and Nikoo 2018). Compared with ground-based measurements, satellite-based methods provide rainfall estimations at much finer temporal and spatial resolutions, but satellite-based rainfall data cannot accurately represent local rainfall distributions (Lv and Zhou 2016). Fusing ground-based and satellite-based rainfall measurements could improve the accuracy of rainfall distribution measurements. Woodhouse et al. (2012) believed that combining ground truth reference data from field surveys with statistical predictors derived from remotely sensed information is a sound procedure to estimate forest biomass. Maack et al. (2016) used generalized additive models (GAM) to fuse a wall-to-wall air-borne LiDAR dataset, Landsat 7 satellite images, terrestrial inventory data derived from the National Forest Inventory (NFI) to estimate spatially explicit distribution of timber volume in forested areas in Baden-Württemberg of Germany. A data fusion approach, based on spatial correlation analysis, was developed for fusing observations from ambient monitors and simulated data from a community multiscale air quality (CMAQ) model (Huang et al. 2018).

In recent years, a few of studies paid attention to comparing data fusion methods. Seven fusion algorithms for satellite images were comparatively used for wildlife surveys, such as penguin guano, seals, and vegetation in Antarctica, which indicated that every algorithm had both its advantage and disadvantage (Witharana et al. 2016). Several regression methods for data fusion, including partial least squares regression (PLSR), support vector regression (SVR), and extreme learning machine based Regression (ELR), were developed to estimate crop biophysical and biochemical parameters; the comparative research indicated that ELR provided relatively accurate estimation of plant traits and valuable insight for plant stress assessment (Maimaitijiang et al. 2017).

This chapter, taking estimation of forest carbon stocks in China as an example, is to exhibit the procedure of HASM-based data fusion and its robustness comparing with other methods.

## 5.2 Data Fusion for Estimating Forest Carbon Stocks in China

Biomass dynamics reflects the potential of vegetation to act as a carbon sink over the long-term, as they integrate photosynthesis, autotrophic respiration and litter fall fluxes (Thurner et al. 2014). Forest ecosystems cover more than 41 million km$^2$ of the Earth's land surface and forests are thought to contain about half of the carbon in terrestrial biomes (Prentice et al. 2001). Forests play an important role in the active mitigation of atmospheric $CO_2$ through increased carbon stocks. The fixation of atmospheric $CO_2$ into plant tissue through photosynthesis is one of the most effective mechanisms for offsetting carbon emissions (Canadell and Raupach 2013; Gonzalez-Benecke et al. 2010).

Carbon sequestration by trees is the best way to store a large amount of terrestrial carbon over long durations (Jung et al. 2013). Estimation of carbon stocks at scales ranging from local to global is crucial for accurately predicting future changes in atmospheric carbon dioxide (Yu et al. 2014). However, substantial uncertainties remain in current model estimates of terrestrial carbon and there is an increasing need to quantify and reduce these uncertainties (Barman et al. 2014; Ahlstroem et al. 2012).

Several studies have estimated the forest carbon stocks in China. Piao et al. (2005), for example, used a satellite-based approach and estimated that: (1) the total forest biomass of China averaged 5.79 Pg C during the period 1981–1999, with an average biomass density of 4.531 kg C m$^{-2}$; and (2) the total forest biomass C stock increased from 5.62 Pg C in the early 1980s to 5.99 Pg C by the end of the 1990s, giving a total increase of 0.37 Pg C and an annual sequestration rate of 0.019 Pg C yr$^{-1}$. Zhang et al. (2013), on the other hand, analyzed seven forest inventories from 1973 to 2008 and suggested that the total biomass carbon stocks of all forest types increased by 65% during this period, reaching 8.12 Pg C in 2008.

Wang et al. (2007) used the Integrated Terrestrial Ecosystem C-budget model and estimated that China's forests were a source of 21.0 ± 7.8 Tg C $yr^{-1}$ due to human activities during the period 1901–1949 and that this flux increased to 122.3 ± 25.3 Tg C $yr^{-1}$ due to intensified human activities during the period 1950–1987. However, these forests became large sinks of 176.7 ± 44.8 Tg C $yr^{-1}$ during the period 1988–2001 owing to large-scale plantation and forest regrowth in previously disturbed areas (see the description of the Grain for Green Program below) as well as climatic warming, atmospheric $CO_2$ fertilization, and N deposition.

Yang and Guan (2008) utilized the continuous Biomass Expansion Factor (BEF) method with field measurements of forests plots in different age classes and forest inventory data, and showed that the carbon density of the forests in the Pearl River Delta increased by 14.3% from 19.08 to 21.81 kg C $m^{-2}$ during the period 1989–2003. Similarly, Piao et al. (2009) reported that China's terrestrial ecosystems were a net carbon sink of 0.19–0.26 Pg carbon per year and that they absorbed 28–37% of the fossil carbon emissions during the 1980s and 1990s. However, their results also showed that northeast China is a net source of $CO_2$ to the atmosphere due to the over-harvesting and degradation of forests, while southern China accounts for more than 65% of the carbon sink, which can be attributed to regional climate change, large-scale plantation programs initiated in the 1980s, and shrub recovery (Piao et al. 2009).

Guo et al. (2010) used three different approaches – the mean biomass density (MBD) method, the mean ratio (MR) method, and the continuous BEF method (CBM) – with forest inventory data to estimate China's forest biomass C stocks and their changes from 1984 to 2003. The MBD, MR, and CBM estimated that forest biomass C stocks increased from 5.7 to 7.7, 4.2 to 6.2, and 4.0 to 5.9 Pg C, respectively.

Deng et al. (2011) deployed a GIS approach and defined the vegetation carbon sink as the carbon sequestration from the atmosphere (1.63 × NPP), the vegetation carbon stock as the carbon content that aboveground vegetation holds, and the soil carbon stock as the carbon content that soil organic matter holds. These authors estimated vegetation and soil carbon stocks to 1.58 and 1.41 Pg C, respectively, in the forest ecosystems of China for the period 1981–2000.

Ni (2013) used available national-scale information to estimate that: (1) the mean vegetation carbon in China was 36.98 Pg and mean soil carbon was 100.75 Pg C; and (2) that the forest and grassland sectors supported mean carbon stocks of 5.49 and 1.41 Pg C, respectively.

The aforementioned studies show that the forest ecosystems of China store steadily increasing stocks of carbon and that these forest stands have great potential to absorb more biomass carbon in the future due to large fractions of young and middle-aged forests and programs to promote the conservation of soil and biological resources.

All these published results, which relied on either ground- or satellite-observation-based estimation, resulted in pronounced differences in carbon stock estimates. We try to evaluate if fusing these data sources could reduce the uncertainty associated with the final carbon stock estimates.

## 5.2.1 Materials and Methods

### 5.2.1.1 Ground Observation

The forest distribution data were created by combining the Vegetation Map of the People's Republic of China (Editorial Committee of Vegetation Map of China 2007) with a vegetation map produced from the forest inventory conducted during the period from 2004 to 2008 (State Forestry Administration of China 2009). The former provides a detailed classification of plant functional types and describes the phenological and regional character of forests in China, but it is not so exact. The latter shows the forest distribution in the period of the forest inventory, but its classification provides less information about plant functional types. The combination of the two kinds of maps retains the advantages of both. The forest distribution data covers 161 plant biomes, including five classes of deciduous needle-leaved trees, 57 classes of evergreen needle-leaved trees, 39 classes of deciduous broad-leaved trees, 56 classes of evergreen broad-leaved trees and four classes of mixed trees (Fig. 1). The national forestry inventory database (FID) for the period 2004–2008 includes 0,000 permanent sample plots and 90,000 temporary sample plots scattered across China.

The land mass of China was next divided into nine regions (Fig. 2) with similar temperature, precipitation and soil regimes to make it easier to analyze changes in

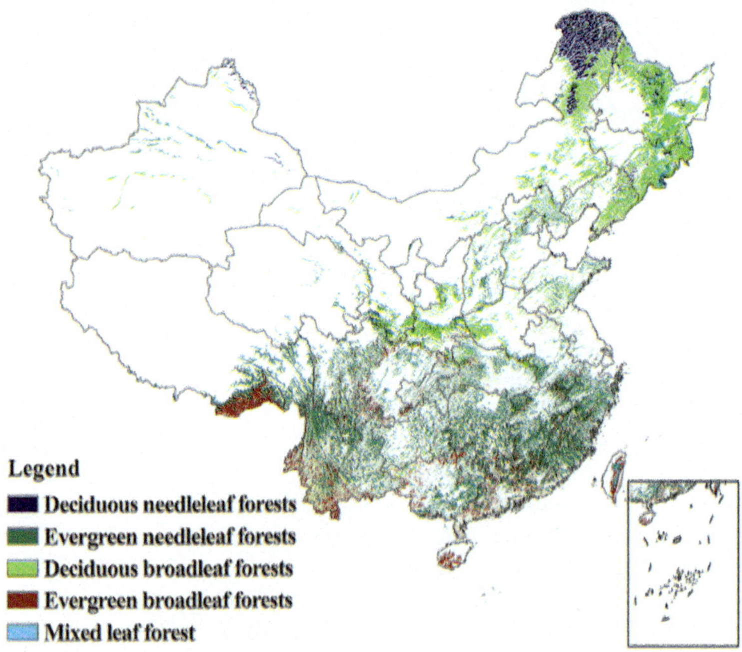


**Fig. 1** The forest cover map of China

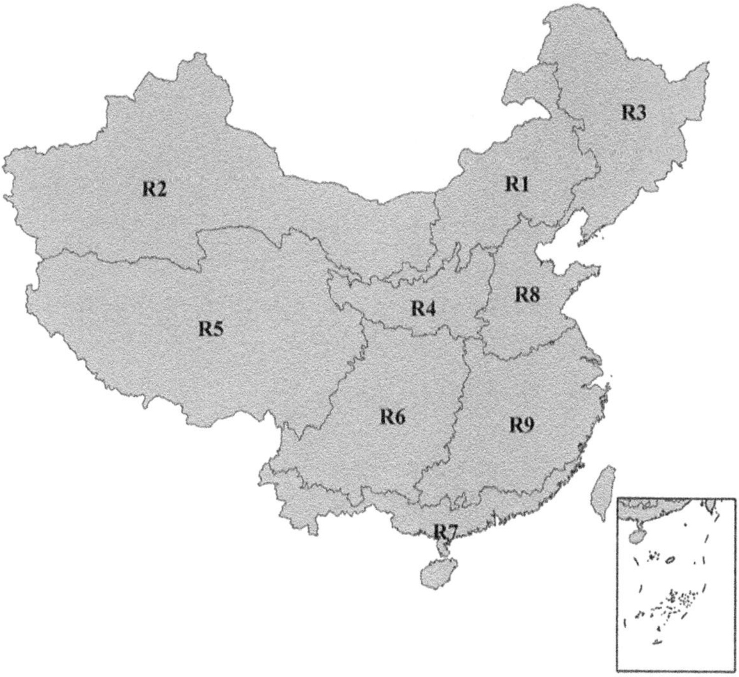


**Fig. 2** Map showing the nine regions of China used for detailed analysis

forest carbon storage from one place to another (Zhou et al. 1981). The nine regions are referred to as $R_k$ where k = 1 to 9 and we use periods 1, 2, 3, 4 and 5 to represent the periods 1984–1988, 1989–1993, 1994–1998, 1999–2003 and 2004–2008, respectively.

#### 5.2.1.2 Satellite Observation

The satellite observation (SO) used the normalized differential vegetation index (NDVI) at a temporal resolution of one month and at a spatial resolution of 1 km × 1 km from the Earth Observation System's moderate-resolution imaging spectroradiometer (EOS MODIS) (Piao et al. 2009). The biomass carbon density (BCD) from the FID data was matched with the NDVI data using the forest map of China reproduced in Fig. 1.

The BCD mirrored the latitude, longitude and maximum value of the monthly-averaged NDVI values during the Seventh National Forest Inventory conducted from 2004 to 2008:

$$\begin{aligned} BCD_j = {} & 93.351\ln(NDVI_j) - 2.96Lat_j - 21.388Lon_j \\ & + 0.047Lat_j^2 + 0.091Lon_j^2 + 1339.03 \end{aligned} \tag{1}$$

where $NDVI_j$ is the mean of the maximum values of the monthly-averaged NDVI values during the period 2004–2008 in the $j$th province and $Lat_j$ and $Lon_j$ refer to the latitude and longitude of the center of the $j$th province, respectively. The coefficient of correlation ($R = 0.91$) and significance ($P < 0.001$) show how latitude, longitude, and NDVI explained 83% of the variability in BCD.

#### 5.2.1.3 Algorithms

The HASM-based data fusion (HASM-DF) formulation of carbon stocks at various scales can be expressed as:

$$DF(x, y) = SO(x, y) + HASM\,(GO_k - SO_k) \tag{2}$$

where HASM($GO_k - SO_k$) represents the residual surface of carbon stocks from satellite observation; $GO_k$ is the carbon stocks from ground observation at the $k$th sample plot; $SO_k$ is the carbon stocks from satellite observation at the sample plot $k$; $SO(x, y)$ and $DF(x, y)$ represent the carbon stocks from satellite observation and the data-fusion results at the location $(x, y)$ respectively.

Forest carbon stocks and carbon densities were estimated by methods of spatial interpolation, satellite observation approach (SOA) and HASM-DF. The spatial interpolation provided an effective approach to construct a continuous surface from the FID by means of Kriging; it took advantage of limited observation data to estimate the most plausible spatial distribution by filling in missing data. The data fusion approach integrated the forest inventory and satellite data into a consistent, accurate and useful representation using HASM-DF; the aim of the data fusion was to improve the quality of the information so that it was more accurate than would be possible if the data sources had been used individually.

#### 5.2.1.4 Validation

The uncertainties of the carbon stock estimates reported in earlier studies relied on several different concepts and metrics. The same formula for absolute and relative error should be used to evaluate all estimates of carbon stocks so that the estimation results are comparable. We calculated the mean absolute errors (MAE) and mean relative errors (MRE), respectively, as:

$$\mathrm{MAE} = \frac{1}{n}\sum_{i=1}^{n} |GO_i - DF_i| \tag{3}$$

$$MRE = \frac{MAE}{\frac{1}{n}\sum_{i}^{n} |GO_i|} \tag{4}$$

where n is the total number of sample plots selected for validation, $GO_i$ is the carbon stocks from ground observation at the ith sample plot selected for validation; $DF_i$ is data-fusion result of HASM-DF at the ith sample plot.

Cross-validation is used to estimate how accurately a model performs, which is analyzed by removing certain data points in turn and summing the absolute value of the discrepancy of each removed data point from the simulated one at the same location (Hulme et al. 1995). It was comprised of four steps: (1) 5% of the sample plots from the national forest inventory were removed for validation; (2) the spatial distribution of average forest BCSs in China during the period 2004–2008 were simulated at a spatial resolution of 5 km × 5 km using the remaining 95% of the sample plots from the national forest inventory by means of the different methods; (3) the MAEs and MREs were calculated using the 5% validation set; and (4) the 5% validation set was returned to the pool for the next iteration, and another 5% validation set was removed. This final process was repeated until all the sample plots were used for validation at least one time and the simulation error statistics could be calculated for each sample plot.

### 5.2.2 Results

The three maps of mean annual carbon stocks during the period 2004–2008 reproduced with each of the aforementioned methods shows how the Kriging, SOA and HASM-DF methods were able to generate the same overall patterns based on the underlying forest cover, and how the estimates varied using each of these methods over large parts of the China (Fig. 3). This variability raises questions related to the reliability of the estimates produced with the three aforementioned approaches.

The cross-validation results indicated that Kriging and SOA had larger errors, with MREs of 50.12 and 48.77%, respectively. Kriging over-estimated the carbon stocks, while SOA under-estimated the carbon stocks (Table 1). Accuracy was considerably improved when the forest inventory and satellite data were fused by using HASM-S. The MRE of HASM-S was 22.71%.

The BCSs of all forest types estimated with HASM-S (the best approach) was 7.08 Pg in China during the period 2004–2008. The BCSs of coniferous, broadleaf and mixed forests were 2.74, 3.95 and 0.39 Pg, respectively (Table 1). The mean carbon densities (MBCDs) of the coniferous, broadleaf and mixed forests were 4.35, 4.74 and 4.20 kg/m$^2$, respectively.

The HASM-DF estimates showed that 89.9% of the MBCSs were found in the regions R5, R3, R6, R9 and R7 during the period P5, accounting for 28.61, 28.41, 14.48, 12.52 and 5.89% of the BCSs, respectively. The three largest BCDs occurred in R5 (Tibet plateau; 10.53 kg/m$^2$), R2 (arid area; 6.33 kg/m$^2$) and R3 (northeastern China; 4.44 kg/m$^2$) (Table 2 and Fig. 3c). The two smallest BCDs were predicted in the R8 (2.14 kg/m$^2$) and R9 (2.60 kg/m$^2$) regions.

The HASM-S estimates can be parsed by forest type as well (Table 3). Hence, the BCDs of evergreen broad-leaved and evergreen coniferous forests were 6.23 and

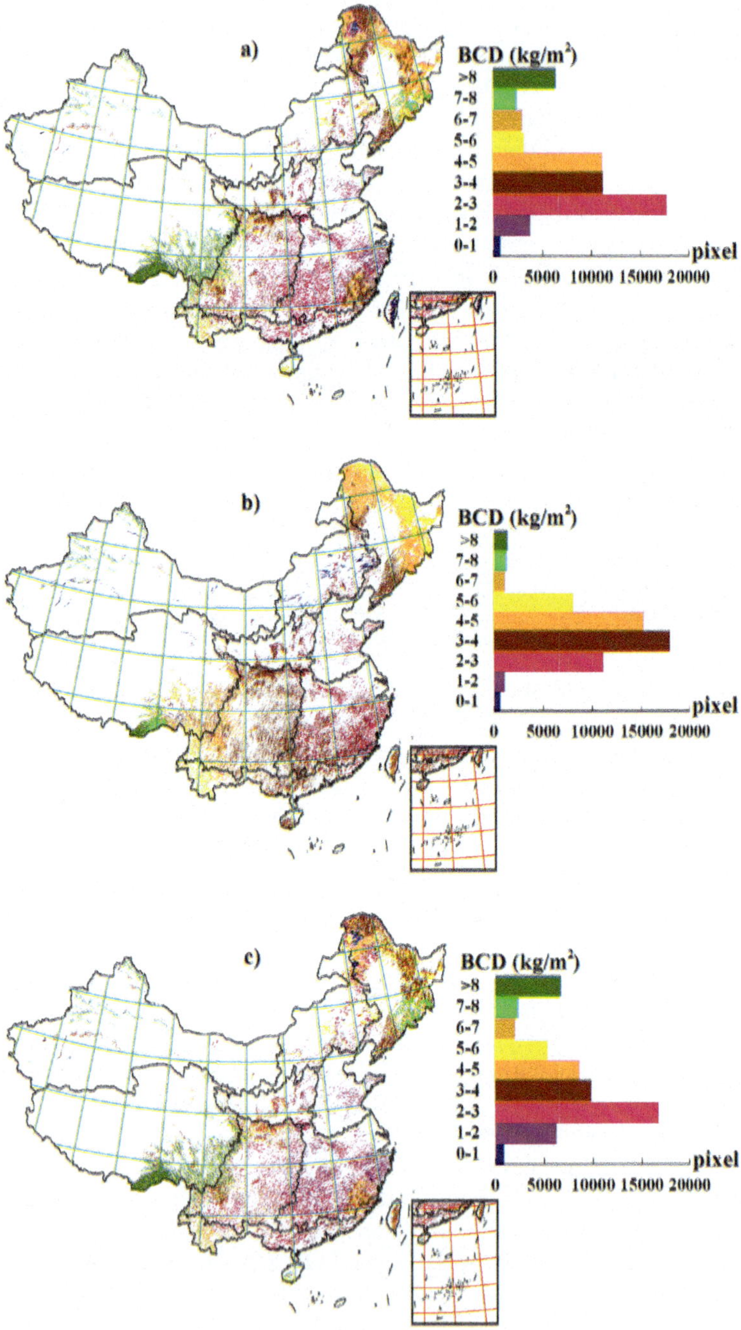


**Fig. 3** The spatial distribution of forest biomass BCDs estimated during the period 2004–2008 in China using: **a** Kriging; **b** SOA; and **c** HASM-DF

**Table 1** Biomass carbon stocks and biomass carbon densities as well as their errors produced by different methods

| Method | Calculated object | Coniferous forests | Mixed forests | Broadleaf forests | Total | MAE (kg $m^{-2}$) | MRE (%) |
|---|---|---|---|---|---|---|---|
| SOA | BCS (Pg) | 2.48 | 0.46 | 3.61 | 6.55 | 1.92 | 48.77 |
| | BCD (kg $m^{-2}$) | 3.94 | 4.93 | 4.34 | | | |
| Kriging | BCS (Pg) | 2.76 | 0.39 | 4.11 | 7.26 | 1.97 | 50.12 |
| | BCD (kg $m^{-2}$) | 4.38 | 4.24 | 4.94 | | | |
| HASM-DF | BCS (Pg) | 2.74 | 0.39 | 3.95 | 7.08 | 0.89 | 22.71 |
| | BCD (kg $m^{-2}$) | 4.35 | 4.2 | 4.74 | | | |

**Table 2** BCS and BCD of the forests in the nine regions of China during the periods 2004–2008 and 1984–1988

| Regions | P5 (from 2004 to 2008) | | | P1 (from 1984 to 1988) | | BCS accumulation rate ($Tg \cdot yr^{-1}$) |
|---|---|---|---|---|---|---|
| | BCD ($kg \cdot m^{-2}$) | BCS (Pg) | Percentage (%) | BCD ($kg \cdot m^{-2}$) | BCS (Pg) | |
| R1 | 3.710 | 0.28 | 3.94 | 2.666 | 0.16 | 6.2 |
| R2 | 6.330 | 0.20 | 2.80 | 6.358 | 0.15 | 2.3 |
| R3 | 4.445 | 2.01 | 28.41 | 4.493 | 1.59 | 21.3 |
| R4 | 3.274 | 0.19 | 2.68 | 3.035 | 0.13 | 2.8 |
| R5 | 10.525 | 2.03 | 28.61 | 6.718 | 0.99 | 52.0 |
| R6 | 3.671 | 1.03 | 14.48 | 3.734 | 0.82 | 10.4 |
| R7 | 3.693 | 0.42 | 5.89 | 3.643 | 0.36 | 2.9 |
| R8 | 2.138 | 0.06 | 0.83 | 1.515 | 0.03 | 1.3 |
| R9 | 2.598 | 0.87 | 12.26 | 2.358 | 0.62 | 12.7 |
| Total | | 7.08 | 100 | | 4.84 | 112 |

4.47 kg/$m^2$, respectively, while the BCDs for deciduous broad-leaved and deciduous coniferous forests were 3.93 and 3.77 kg/$m^2$, respectively in P5. The BCD of evergreen forests was 50% larger than that of deciduous forests, and the BCDs for broad-leaved forests were greater than those for both coniferous and deciduous forests. Turning next to the BCSs, the evergreen coniferous forests contributed the largest proportion, accounting for 33.05%, followed by deciduous broad-leaved forests (29.8%), and evergreen broad-leaved forests (25.99%). The deciduous coniferous and the broad-leaved and coniferous mixed forests accounted for the two smallest proportions of the total BCS, 5.65 and 5.51%, respectively.

The HASM-S estimates also indicate that BCSs rose from 4.84 Pg in period 1 to 7.08 Pg in P5 due to the increase of BCD and the expansion of forest area (Table 4). The BCD rose from 4.00 kg/$m^2$ in P1 to 4.55 kg/$m^2$ in P5 and the forest area grew

**Table 3** BCSs and BCDs for all forest types during the five periods estimated using HASM-DF

| Period | Calculation object | Deciduous coniferous forests | Evergreen coniferous forests | Broad-leaved and coniferous mixed forests | Deciduous broad-leaved forests | Evergreen broad-leaved forests |
|---|---|---|---|---|---|---|
| P1 | BCS (Pg) | 0.41 | 1.50 | 0.06 | 1.38 | 0.63 |
| | BCD (kg/m$^2$) | 4.35 | 3.81 | 3.08 | 3.75 | 4.35 |
| P2 | BCS (Pg) | 0.39 | 1.80 | 0.09 | 1.44 | 0.87 |
| | BCD (kg/m$^2$) | 4.28 | 4.13 | 3.75 | 3.77 | 5.65 |
| P3 | BCS (Pg) | 0.44 | 2.23 | 0.07 | 1.66 | 1.20 |
| | BCD (kg/m$^2$) | 4.20 | 4.09 | 3.03 | 3.87 | 6.35 |
| P4 | BCS (Pg) | 0.47 | 2.19 | 0.19 | 1.97 | 1.57 |
| | BCD (kg/m$^2$) | 4.37 | 4.40 | 5.18 | 3.89 | 7.49 |
| P5 | BCS (Pg) | 0.40 | 2.34 | 0.39 | 2.11 | 1.84 |
| | BCD (kg/m$^2$) | 3.77 | 4.47 | 4.20 | 3.93 | 6.22 |

from 1.21 million km$^2$ in P1 to 1.56 million km$^2$ in P5. The increasing trends of the BCS, BCD and forest area (FA) are captured by the following regression equations:

$$BCS(t) = 0.531t + 4.297 \quad R = 0.976 \tag{5}$$

$$BCD(t) = 0.125t + 3.958 \quad R = 0.943 \tag{6}$$

$$FA(t) = 0.083t + 1.1045 \quad R = 0.96 \tag{7}$$

where $t$ corresponds to periods 1, 2, 3, 4 and 5; $BCS(t)$, $BCD(t)$ and $FA(t)$ are BCS, BCD and FA, respectively in the period t; and $R$ represents the correlation coefficient for the corresponding regression equation.

Although BCS rose in all nine regions from period 1 to period 5, the spatial variability over China more or less mirrors the variability in the distribution of forests (Figs. 4 and 5, Table 4). For regions R1 and R8, for example, both BCS and BCD have continuously increased from period 1 to period 5. R8 has the smallest BCS, which only accounted for 0.83% of the BCS of the whole of China, and the smallest BCD of 2.14 kg/m$^2$ in P5 as well as the lowest BCS accumulation rate of 1.3Tg·yr$^{-1}$. In R1, BCS accounted for 3.94% of the total BCS of China for the period of P5 and the BCS accumulation rate has averaged 6.2 Tg·yr$^{-1}$ from P1 to P5.

The region R5 had the largest BCS, accounting for 28.61% of the total BCS of China in P5, along with the largest BCD of 10.53 kg/m$^2$ and the fastest BCS

**Table 4** Biomass carbon stocks and biomass carbon densities estimated by HASM-DF

| Regions | Calculation object | Period 1 | Period 2 | Period 3 | Period 4 | Period 5 |
|---|---|---|---|---|---|---|
| R1 | BCS (Pg) | 0.16 | 0.17 | 0.17 | 0.2 | 0.28 |
| | BCD (kg $m^{-2}$) | 2.67 | 2.71 | 2.88 | 2.98 | 3.71 |
| R2 | BCS (Pg) | 0.15 | 0.16 | 0.15 | 0.18 | 0.2 |
| | BCD (kg $m^{-2}$) | 6.36 | 6.27 | 6.25 | 6.23 | 6.33 |
| R3 | BCS (Pg) | 1.59 | 1.64 | 1.64 | 1.77 | 2.01 |
| | BCD (kg $m^{-2}$) | 4.49 | 4.42 | 4.5 | 4.43 | 4.44 |
| R4 | BCS (Pg) | 0.13 | 0.15 | 0.14 | 0.16 | 0.19 |
| | BCD (kg $m^{-2}$) | 3.04 | 3.23 | 3.13 | 3.15 | 3.27 |
| R5 | BCS (Pg) | 0.99 | 1.57 | 1.64 | 1.94 | 2.03 |
| | BCD (kg $m^{-2}$) | 6.72 | 10.15 | 10.83 | 11.49 | 10.53 |
| R6 | BCS (Pg) | 0.82 | 0.87 | 0.82 | 0.96 | 1.03 |
| | BCD (kg $m^{-2}$) | 3.73 | 3.78 | 3.66 | 3.88 | 3.67 |
| R7 | BCS (Pg) | 0.36 | 0.39 | 0.37 | 0.4 | 0.42 |
| | BCD (kg $m^{-2}$) | 3.64 | 3.79 | 3.66 | 3.54 | 3.69 |
| R8 | BCS (Pg) | 0.03 | 0.04 | 0.04 | 0.05 | 0.06 |
| | BCD (kg $m^{-2}$) | 1.52 | 1.64 | 1.89 | 1.89 | 2.14 |
| R9 | BCS (Pg) | 0.62 | 0.56 | 0.62 | 0.73 | 0.87 |
| | BCD (kg $m^{-2}$) | 2.36 | 2.05 | 2.3 | 2.51 | 2.6 |
| The whole of China | BCS (Pg) | 4.84 | 5.55 | 5.6 | 6.38 | 7.08 |
| | BCD (kg $m^{-2}$) | 4 | 4.32 | 4.33 | 4.47 | 4.55 |
| | Area (million $km^2$) | 1.2101 | 1.2864 | 1.292 | 1.4279 | 1.5559 |

accumulation rate of 52 Tg·$yr^{-1}$; the BCS has shown a monotonically increasing trend since P1. The second largest BCS occurred in R3 (Northeastern China). The BCS in R3 accounted for 28.41% of the total BCS of China. However, the BCD in R3 has declined since P3 following increases from P1 to P2 and from P2 to P3. The mean BCS accumulation rate in R3 was 21.3 Tg/yr.

In the regions R4 (Loess Plateau), R6 and R9, both BCS and BCD have increased since P3. The BCSs in R4, R6 and R9 accounted for 2.68, 14.48 and 12.26% of the BCS in the whole of China in P5, respectively. The average BCS accumulation rates were 2.8, 10.4 and 12.7 Tg·$yr^{-1}$, respectively in R4, R6, and R9. In R2 (an arid area), the BCS accounted for 2.80% of the total for China in the period P5. The BCS and BCD increased from P4 to P5, but the mean accumulation rate of BCS was only 2.3 Tg·$yr^{-1}$. The BCS accounted for 5.89% of the total for China in R7. The BCS and BCD both increased from P4 to P5 but like in R2, the mean BCS accumulation rate was relatively low at just 2.9 Pg·$yr^{-1}$.

In terms of forest types, evergreen broad-leaved forests had the fastest BCS accumulation rate and the largest BCD, while evergreen coniferous forests contributed the largest BCS. The BCSs of broad-leaved forests increased during all five periods.

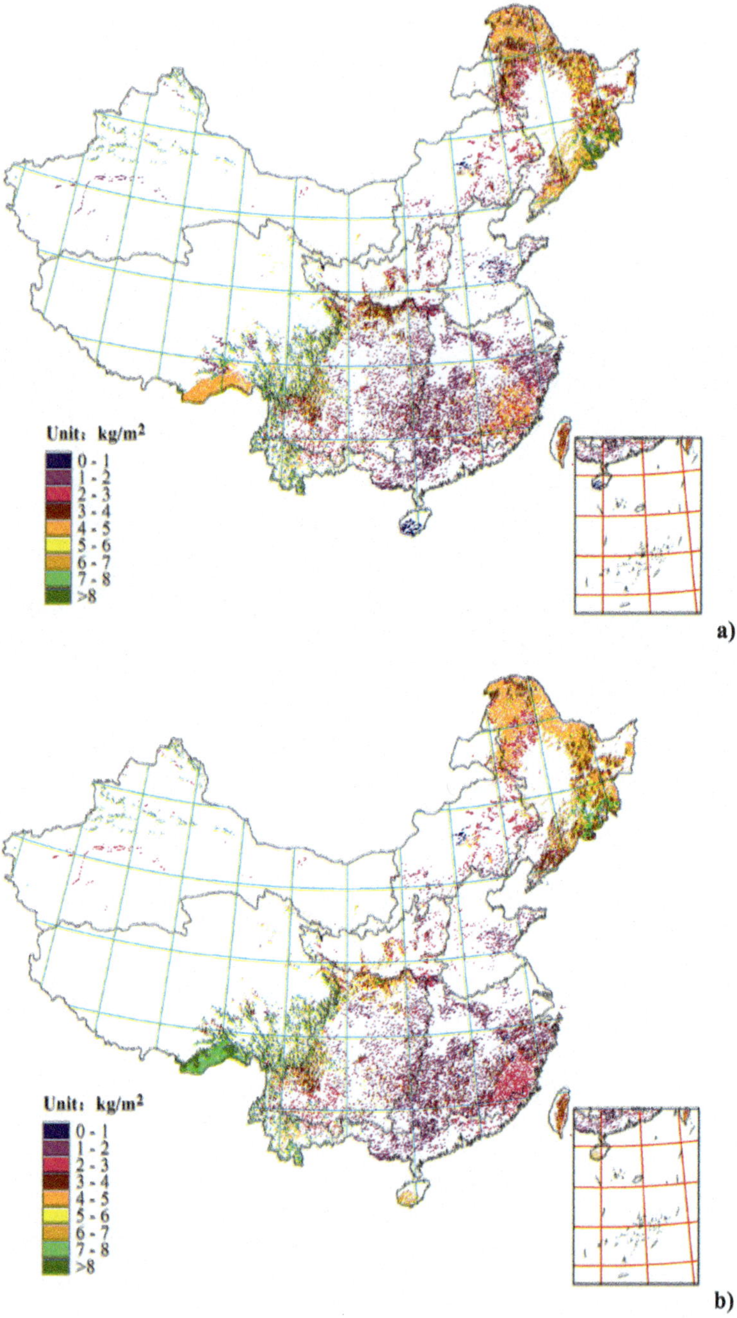


**Fig. 4** The spatial distribution of forest BCDs in China estimated during the five periods using HASM-S: **a** 1984–1988 (P1); **b** 1989–1993 (P2); **c** 1994–1998 (P3); **d** 1999–2003 (P4); and **e** 2004–2008 (P5)

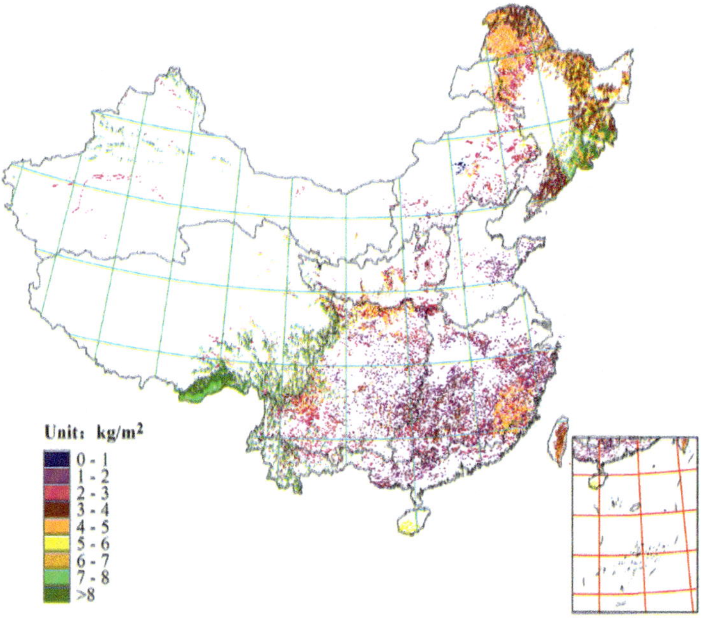


c)

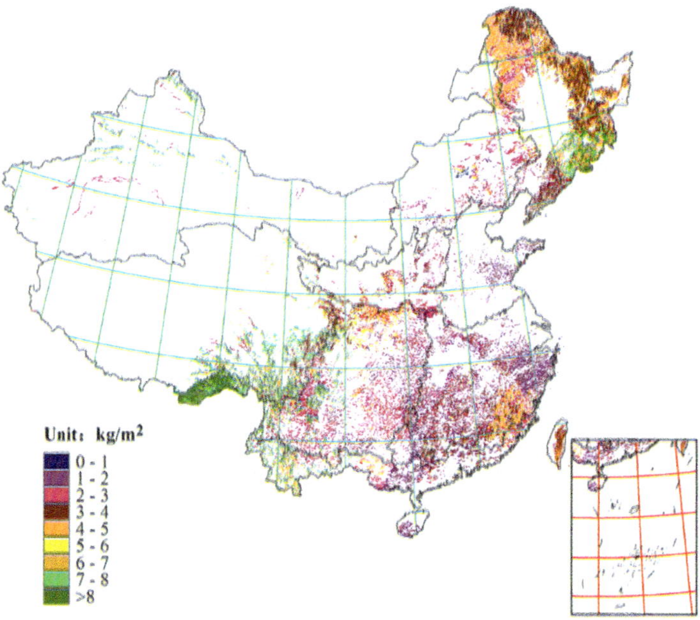


d)

**Fig. 4** (continued)

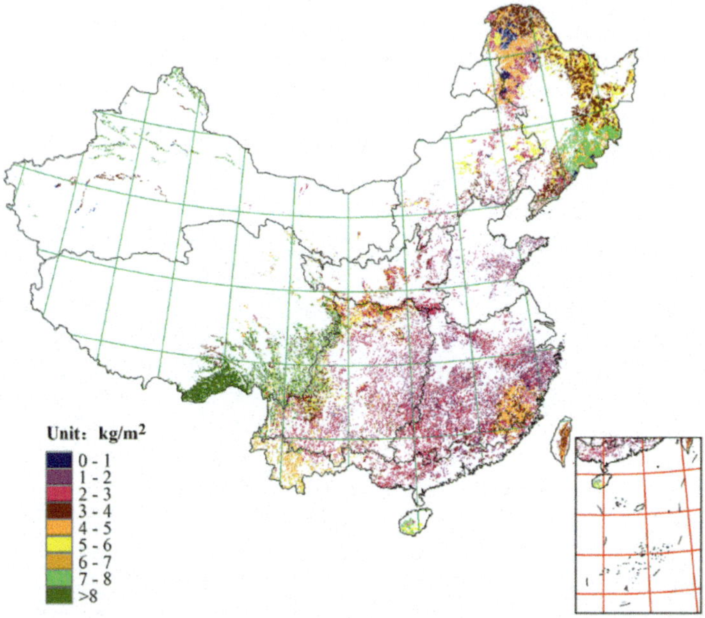


**Fig. 4** (continued)

The BCS of evergreen broad-leaved forests increased from 0.63 Pg in period 1 to 1.84 Pg in period 5, and the BCSs for deciduous broad-leaved forests rose from 1.38 Pg in period 1 to 2.11 Pg in period 5. These trends can be modeled with the following regression equations:

$$BCS_1(t) = 0.312t + 0.286, \quad R = 0.998 \tag{8}$$

$$BCS_2(t) = 0.199t + 1.115, \quad R = 0.981 \tag{9}$$

where $t$ corresponds to period t, $t = 1, 2, 3, 4$ and 5; $BCS_1(t)$ and $BCS_2(t)$ are respectively the BCSs of evergreen broad-leaved forests and deciduous broad-leaved forests in the period t and $R$ represents the correlation coefficient of the corresponding regression equation.

The BCSs of deciduous coniferous forests fluctuated from period to period. Evergreen coniferous forests and broad-leaved and coniferous mixed forests exhibited an increasing trend of BCS in general but declined in period 3. Their trends were modeled with the following regression equations:

$$BCS_3(t) = 0.207t + 1.391, \quad R = 0.932 \tag{10}$$

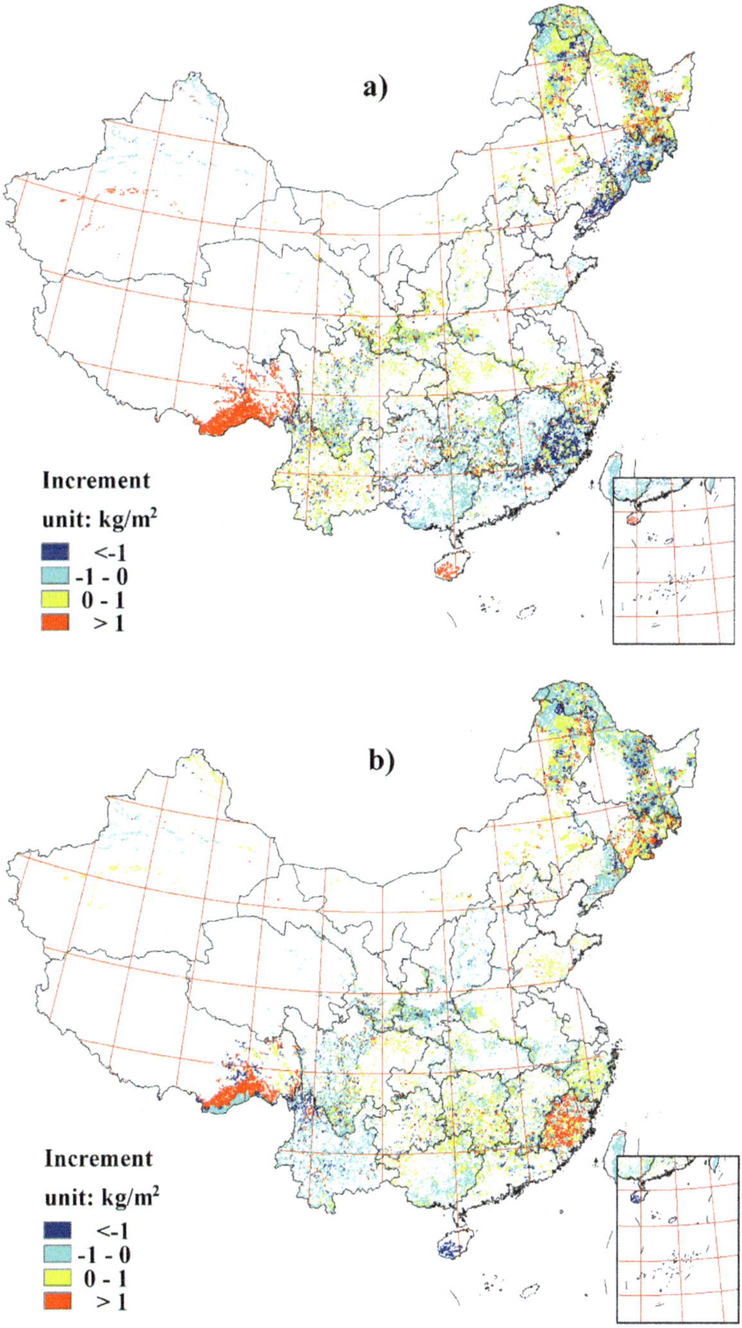


**Fig. 5** The increment maps of biomass carbon stocks from comparing two adjacent periods: **a** BCS in Period 2 minus BCS in Period 1, **b** BCS in Period 3 minus BCS in Period 2, **c** BCS in Period 4 minus BCS in Period 3, and **d** BCS in Period 5 minus BCS in Period 4

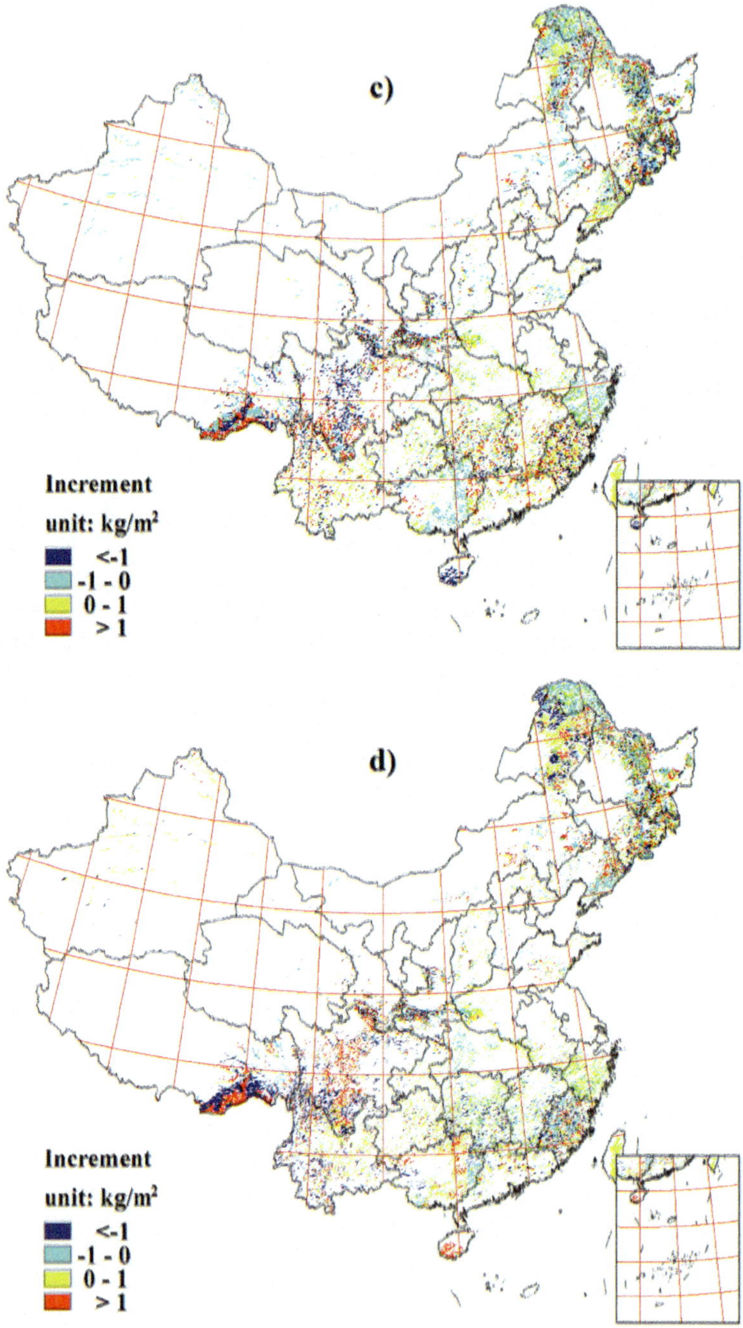


**Fig. 5** (continued)

$$BCS_4(t) = 0.076t - 0.068, \quad R = 0.867 \tag{11}$$

where $t$ corresponds to periods 1, 2, 3, 4 and 5; $BCS_3(t)$ represents the BCSs of evergreen coniferous forests in the period t; $BCS_4(t)$ refers to the broad-leaved and coniferous mixed forests; and $R$ represents the correlation coefficient of the corresponding regression equation.

Equations (5–7) indicate that increasing rates of BCS, BCD and FA were 0.531 Pg, 0.125 kg/m$^2$ and 0.083 million km$^2$, respectively, over a five-year period on average. According to Eqs. (8) and (9), the BCS of evergreen broad-leaved forests had a growth rate of 0.312 Pg over a five-year period, which was 0.113 Pg higher than that for deciduous broad-leaved forests because of the different increasing rates of BCD; the BCD of evergreen broad-leaved forests increased by 1.87 kg/m$^2$, while the BCD of deciduous broad-leaved forests increased by 0.18 kg/m$^2$ from period 1 to period 5. Equations (10) and (11) show that carbon stocks of evergreen coniferous forests grew much faster than those for broad-leaved and coniferous mixed forests. The former grew by 0.207 Pg but the latter by only 0.076 Pg in a five-year period on average because the area of evergreen coniferous forests was five times larger (0.50 $\times$ 10$^6$ km$^2$ compared to 0.09 $\times$ 10$^6$ km$^2$ for broad-leaved and coniferous mixed forests) and offset the lower BCD growth rate of the former (0.132 kg/m$^2$) compared to 0.56 kg/m$^2$ for broad-leaved and coniferous mixed forests per five-year period.

The results from Kriging and HASM-DF exhibit a similar spatial pattern on the national level, especially in southern Tibet, the Xiao Hinggan Mountains, the Changbai Mountains, and south China. Some differences occur in Taiwan and Xinjiang as well as the Qinling Mountains. Compared to HASM-S, Kriging is strongly influenced by sample-plot density. The spatial heterogeneity of the results from SOA is not so obvious because NDVI, a critical variable of SOA, is not so sensitive to a change of carbon stocks, especially in northeast China and near the lower reaches of Yangtze River.

The scatter diagrams of simulated BCD against observed BCD (Fig. 6) indicate that the BCD surface created by Kriging exhibits a higher correlation with observed BCD, $R^2 = 0.826$. But the Kriging interpolation generated large errors in the Xizang/Tibet and Xinjing regions, which have the highest BCD. The BCD was overestimated in Xizang and underestimated in Xinjiang. The BCD surface created by SOA has the lowest coefficient of determination, $R^2 = 0.627$, with the observed one. The SOA results show that BCDs in higher and lower latitudes are larger than those in middle latitudes, but for Xinjiang and Xizang. The BCD was overestimated in Xinjiang but underestimated in Xizang. The surface of BCDs generated by HASM-S has the best correlation with the observed one, $R^2 = 0.943$ and it generated the smallest errors in the regions of Xizang and Xinjiang, compared with the other methods.

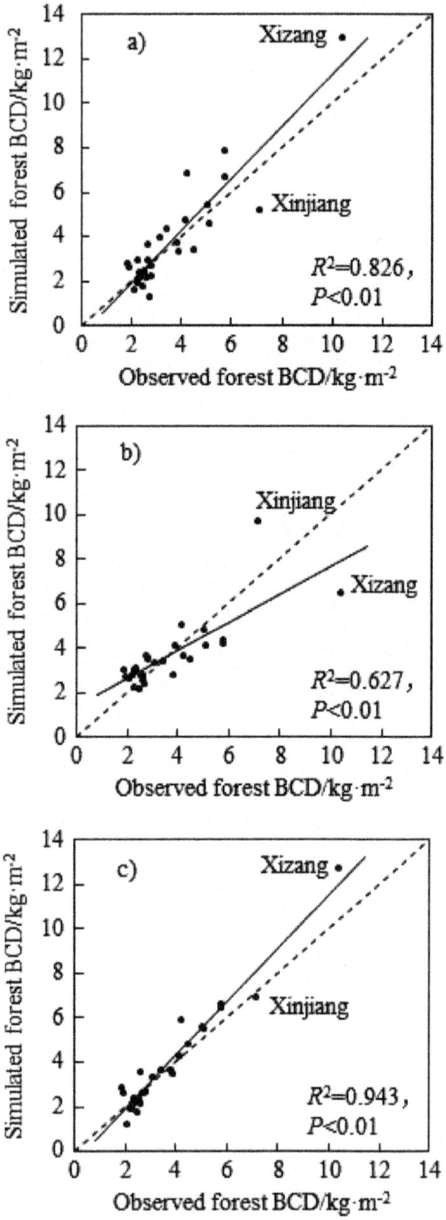


**Fig. 6** The scatter diagrams of simulated BCD against observed BCD: **a** Kriging, **b** SOA, and **c** HASM-DF

### 5.2.3 Discussion

HASM-DF overcame the shortcomings of both the ground-based national forest inventory and the satellite remote-sensing observations by fusing information about the details of the carbon stocks observed on the Earth's surface and the variability of the carbon surface observed from space. The cross-validation demonstrated that HASM-DF was 26.1% more accurate than the satellite-based approach and 28.4% more accurate than spatial interpolation of the sample plots. These findings suggest that China's forest biomass carbon stocks are more likely to be closer to our estimates than those generated by past efforts to estimate these same carbon stocks and their change over time.

Taken as a whole, the HASM-DF results show that the forest carbon stocks of China have increased by 2.24 Pg during the period 1984–2008 to a new high of 7.08 Pg C in 2008. These numbers fall in the middle of the previously published estimates. All of the estimates show forest biomass carbon stocks in China increasing from 1973 to 2008, notwithstanding the various methods used and the varying levels of uncertainty embedded in these different methods and the data sources used.

The results from HASM-DF are compare favorably well with those of other studies. For instance, the annual growth of total BCS from period 1 (circa 1986) to period 5 (circa 2006) in China was 0.112 $PgCyr^{-1}$. This estimate was higher than that of Zhang et al. (2013), which was 0.103 $PgCyr^{-1}$. However, our estimate of 0.148 $PgCyr^{-1}$ from period 3 (circa 1996) to period 5 (circa 2006) was lower than the 0.174 $PgCyr^{-1}$ estimated by Zhang et al. (2013). From period 4 (circa 2001) to period 5 (circa 2006), the 0.14 $PgCyr^{-1}$ estimated by HASM-S was twice that estimated by Liu et al. (2015).

In recent years, there are several different vegetation biomass maps for pan-tropical area (e.g., Avitabile et al. 2016; Baccini et al. 2012; Saatchi et al. 2011). They were based on combination of data from ground observations and satellite remote sensing. The spatial distribution pattern of the forest biomass carbon at low latitudes of China area simulated by HASM-S was in general consistent with the biomass carbon maps for pan-tropical area, i.e., south of Tibet plateau had the highest biomass carbon density, followed by southeast Tibet plateau, west of YunNan-GuiZhou plateau, the area of QinLing mountains and DaBa mountains; Southeastern China has the lowest forest biomass density except the area of Wuyi mountains. However, the biomass carbon densities from HASM-S were lower than the one from Baccini et al. (2012), especially in southeastern China.

The Grain for Green program, which was launched in 1999 and aims to restore the country's forests and grasslands to prevent soil erosion, has emerged as one of the key drivers of carbon sequestration in China. The increase in forest growth rate under climate change may also contribute to sequestering more carbon, but has relatively less impact than the restoration of forest lands or establishment of new forests (Forte 2009). This program targets land with slopes > 25° (Xu et al. 2006; Yue et al. 2016a, b) and has been implemented in four phases: (1) a pilot phase (1999–2001); (2) an initial construction phase (2002–2010); (3) a consolidation phase (2011–2013); and

(4) a second construction phase to be built around a new round of Grain for Green program expenditures (2014–2020).

The pilot program launched in 1999 focused on three provinces: Gansu, Shaanxi and Sichuan. Approximately 381,000 ha of farmland was converted into forestland and 66,000 ha of bare land was reforested. In 2000, the program was expanded to 17 provinces, and the converted farmland and reforested bare land totals grew to 410,000 and 449,000 ha, respectively. By 2001, 20 provinces were involved in the program and 420,000 and 563,000 ha of farm and bare land had been reforested, respectively (Table 5). The national Grain for Green program was launched in China in 2002 and by the end of 2010, 14.667 million ha of farmland had been converted to forest or grassland and 17.333 million ha of bare land had been reforested. During the consolidation phase from 2011 to 2013, scientific monitoring and management of the converted and reforested lands was strengthened to help sustain the aforementioned achievements of the Grain for Green program over the long-term.

To grow and consolidate these gains, the potential for farmland conversion at the county level during the period 2014–2020 was estimated in 2014 by counting up farmers' voluntary applications to determine how large an area could be converted to forest or grassland. By 2020, 2.827 million ha of farmland could be converted, which includes 1.449 million ha of farmland with slopes > 25°, 1.133 million ha of cultivated land threatened by desertification, and 247,000 ha of farmland with slopes between 15 and 25° around the Danjiangkou and Three Gorges reservoirs.

The results from this latest phase of the Grain for Green program are encouraging. Participating farmers can choose whether farmland is to be converted to forest or grassland, and which species will be planted, and they will receive a 22,500 RMB subsidy for every hectare of farmland converted to forest or grassland. In 2014,

**Table 5** Converted farmland and reforested bare land included in China's Grain for Green Program (millions hectares)

| Year | Converted farmland | Afforestation on bare land | Total |
|---|---|---|---|
| 1999 | 0.381 | 0.066 | 0.448 |
| 2000 | 0.405 | 0.468 | 0.872 |
| 2001 | 0.42 | 0.563 | 0.983 |
| 2002 | 2.647 | 3.082 | 5.729 |
| 2003 | 3.367 | 3.767 | 7.133 |
| 2004 | 0.667 | 3.333 | 4 |
| 2005 | 1.114 | 1.321 | 2.435 |
| 2006–2010 | 5.666 | 4.733 | 10.4 |
| 2014 | 0.333 | | 0.333 |
| 2015 | 0.667 | | 0.667 |
| 1999–2015 | 15.667 | 17.333 | 33 |
| 2016–2020 | 1.827 | | 1.827 |

Office of Converting Farmland to Forestry, State Forestry Administration of China (2014, 2016)

322,000 ha were converted to forest and 11,000 ha were converted to grassland, and in 2015 another 667,000 ha of farmland will be converted to either forest or grassland.

The BCS growth was improved by already existing forests and newly planted forests. The former accounted for about 55% while the latter about 40%. The BCS in existing forests had a growth rate of about 0.55 kg/m$^2$ per five-year period and the BCS in newly planted forests grew at 1.8 kg/m$^2$ per five-year period.

Methodologically, the fusion of forest inventory data with satellite observations achieved with HASM-DF provided much more accurate estimates of forest biomass carbon stocks and their changes. This kind of method can increase our understanding of the role of forests in the carbon cycle, and help to support greenhouse gas inventories and terrestrial carbon accounting projects (Muukkonen and Heiskanen 2007).

## 5.3 Summary

Data fusion approaches have limitations such as: (1) it can only simulate a surface at a specific time or the average condition for a period when there are ground observations; in other words, it is difficult to use these approaches to estimate changes at higher temporal resolutions; and (2) it cannot generate and capture the likely impacts of future interventions and scenarios. Development of model-data assimilation is an alternative way to find solutions to the limitations.

Future research should pay attention to improving uncertainty estimates by deriving confidence intervals by applying the basic principles of sampling theory and/or using models that contains the code for uncertainty analysis. Uncertainty estimates can allow decision makers to better appreciate the amplitude of the errors. For researchers, uncertainty estimates may contribute to improving methodologies.

## 5.4 Further Reading

(1) Yue TX, Wang YF, Du ZP, Zhao MW, Zhang LL, Zhao N, Lou M, Larocqued GR, Wilson JP. 2016. Analyzing the uncertainty of estimating forest carbon stocks in China. Biogeosciences 13: 3991–4004.

(2) Yue TX, Zhao N, Liu Y, Wang YF, Zhang B, Du ZP, Fan ZM, Shi WJ, Chen CF, Zhao MW, Song DJ, Song YJ, Yan CQ, Li QQ, Sun XF, Zhang LL, Tian YZ, Wang W, Wang YA, Ma SN, Huang HS, Lu YM, Wang Q, Wang CL, Wang YZ, Lu M, Zhou W, Liu Y, Yin XZ, Wang Z, Bao ZY, Zhao MM, Zhao YP, Jiao YM, Naseer U, Fan B, Li SB, Yang Y, Wilson JP. 2020. A fundamental theorem for eco-environmental surface modelling and its applications. Science China-Earth Sciences 63(8): 1092–1112.

## Exercises

(1) Define the term "data fusion".
(2) Explain Corollary 4 and Corollary 5 of FTEEM.

# References


Ahlstroem, A., G. Schurgers, A. Arneth, and B. Smith. 2012. Robustness and uncertainty in terrestrial ecosystem carbon response to CMIP5 climate change projections. *Environmental Research Letters* 7: 044008.

Aiello, G., I. Giovino, M. Vallone, P. Catania, and A. Argento. 2018. A decision support system based on multisensor data fusion for sustainable greenhouse management. *Journal of Cleaner Production* 172: 4057–4065.

Alizadeh, M.R., and M.R. Nikoo. 2018. A fusion-based methodology for meteorological drought estimation using remote sensing data. *Remote Sensing of Environment* 211: 229–247.

Alonzo, M., B. Bookhagen, and D.A. Roberts. 2014. Urban tree species mapping using hyperspectral and lidar data fusion. *Remote Sensing of Environment* 148: 70–83.

Avitabile, V., et al. 2016. An integrated pan-tropical biomass maps using multiple reference datasets. *Global Change Biology* 22: 1406–1420.

Baccini, A., et al. 2012. Estimated carbon dioxide emissions from tropical deforestation improved by carbon-density maps. *Nature Climate Change* 2: 182–185.

Barman, R., A.K. Jain, and M.L. Liang. 2014. Climate-driven uncertainties in modeling terrestrial gross primary production: A site level to global-scale analysis. *Global Change Biology* 20: 1394–1411.

Cammalleri, C., M.C. Anderson, F. Gao, C.R. Hain, and W.P. Kustas. 2013. A data fusion approach for mapping daily evapotranspiration at field scale. *Water Resources Research* 49: 4672–4686.

Cammalleri, C., M.C. Anderson, F. Gao, C.R. Hain, and W.P. Kustas. 2014. Mapping daily evapotranspiration at field scales over rainfed and irrigated agricultural areas using remote sensing data fusion. *Agricultural and Forest Meteorology* 186: 1–11.

Canadell, J.G., and M.R. Raupach. 2013. Managing forests for climate change mitigation. *Science* 20: 1456–1457.

Chatterjee, A., A.M. Michalak, R.A. Kahn, S.R. Paradise, A.J. Braverman, and C.E. Miller. 2010. A geostatistical data fusion technique for merging remote sensing and ground-based observations of aerosol optical thickness. *Journal of Geophysical Research* 115: D20207. https://doi.org/10.1029/2009JD013765.

Chen, C.M., G.F. Hepner, and R.R. Forster. 2003. Fusion of hyperspectral and radar data using the HIS transformation to enhance urban surface features. *ISPRS Journal of Photogrammetry & Remote Sensing* 58: 19–30.

Dalponte, M., L. Bruzzone, and D. Gianelle. 2012. Tree species classification in the Southern Alps based on the fusion of very high geometrical resolution multispectral/hyperspectral images and LiDAR data. *Remote Sensing of Environment* 123: 258–270.

Deng, S.H., Y.Q. Shi, Y. Jin, and L.H. Wang. 2011. A GIS-based approach for quantifying and mapping carbon sink and stock values of forest ecosystem: A case study. *Energy Procedia* 5: 1535–1545.

Editorial Committee of Vegetation Map of China. 2007. *Vegetation map of the People's Republic of China*. Beijing (in Chinese): Geological Publishing House.

Forte, R.W. 2009. Carbon sequestration in forests. *Congressional Research Service Report for Congress* 7–5700: 23p.

García, M., D. Riano, E. Chuvieco, J. Salas, and F.M. Danson. 2011. Multispectral and LiDAR data fusion for fuel type mapping using Support Vector Machine and decision rules. *Remote Sensing of Environment* 115: 1369–1379.

Gonzalez-Benecke, C.A., T.A. Martin, W.P. Cropper Jr., and R. Bracho. 2010. Forest management effects on in situ and ex situ slash pine forest carbon balance. *Forest Ecology and Management* 260: 795–805.

Guo, Z.D., J.Y. Fang, Y.D. Pan, and R. Birdsey. 2010. Inventory-based estimates of forest biomass carbon stocks in China: A comparison of three methods. *Forest Ecology and Management* 259: 1225–1231.

Hilker, T., M.A. Wulder, N.C. Coops, J. Linke, G. McDermid, J.G. Masek, F. Gao, and J.C. White. 2009. A new data fusion model for high spatial- and temporal-resolution mapping of forest disturbance based on Landsat and MODIS. *Remote Sensing of Environment* 113: 1613–1627.

Huang, R., X.X. Zhai, C.E. Ivey, M.D. Friberg, X.F. Hu, Y. Liu, Q. Di, J. Schwartz, J.A. Mulholland, and A.G. Russell. 2018. Air pollutant exposure field modeling using air quality model-data fusion methods and comparison with satellite AOD-drived fields: Application over North Carolina, USA. *Air Qual Aymos Health* 11: 11–22.

Hulme, M., D. Conway, P.D. Jones, T. Jiang, E.M. Barrow, and C. Turney. 1995. Construction of a 1961–1990 European climatology for climate change modelling and impact applications. *International Journal of Climatology* 15: 1333–1363.

Jung, J., S. Kim, S. Hong, K. Kim, E. Kim, J. Im, and J. Heo. 2013. Effects of national forest inventory plot location error on forest carbon stock estimation using k-nearest neighbor algorithm. *ISPRS Journal of Photogrammetry and Remote Sensing* 81: 82–92.

Kattenborn, T., J. Maack, F. Fanacht, F. Enlea, J. Ermert, and B. Koch. 2015. Mapping forest biomass from space—Fusion of hyperspectral EO1-hyperion data and Tandem-X and WorldView-2 canopy height models. *International Journal of Applied Earth Observation and Geoinformation* 35: 359–367.

Koetz, B., G.Q. Sun, F. Morsdorf, K.J. Ranson, M. Kneubühler, K. Itten, and B. Allgoewer. 2007. Fusion of imaging spectrometer and LIDAR data over combined radiative transfer models for forest canopy characterization. *Remote Sensing of Environment* 106: 449–459.

Kulawardhana, R.W., S.C. Popescu, and R.A. Feagin. 2014. Fusion of Lidar and multispectral data to quantify salt marsh carbon stocks. *Remote Sensing of Environment* 154: 345–357.

Liu, Y.Y., A.I.J.M. van Dijk, R.A.M. de Jeu, J.G. Canadell, M.F. McCabe, J.P. Evans, and G.J. Wang. 2015. Recent reversal in loss of global terrestrial biomass. *Nature Climate Change* 5: 470–474.

Lv, A.F., and L. Zhou. 2016. A rainfall model based on a geographically weighted regression algorithm for rainfall estimations over the Arid Qaidam Basin in China. *Remote Sensing* 8 (4): 311.

Maack, J., M. Lingenfelder, H. Weinacker, and B. Koch. 2016. Modelling the standing timber volume of Baden-Württemberg—A large-scale approach using a fusion of Landsat, airborne LiDAR and National Forest Inventory data. *International Journal of Applied Earth Observation and Geoinformation* 49: 107–116.

Maimaitijiang, M., A. Ghulam, P. Sidike, S. Hartling, M. Maimaitiyiming, K. Peterson, E. Shavers, J. Fishman, J. Peterson, S. Kadam, J. Burken, and F. Fritschi. 2017. Unmanned Aerial System (UAS)-based phenotyping of soybean using multi-sensor data fusion and extreme learning machine. *ISPRS Journal of Photogrammetry and Remote Sensing* 134: 43–58.

Mizuochi, H., T. Hiyama, T. Ohta, Y. Fujioka, J.R. Kambatuku, M. Iijima, and K.N. Nasahara. 2017. Development and evaluation of a lookup-table-based approach to data fusion for seasonal wetlands monitoring: An integrated use of AMSR series, MODIS, and Landsat. *Remote Sensing of Environment* 199: 370–388.

Montesano, P.M., B.D. Cook, G. Sun, M. Simard, R.F. Nelson, K.J. Ranson, Z. Zhang, and S. Luthcke. 2013. Achieving accuracy requirements for forest biomass mapping: A spaceborne data fusion method for estimating forest biomass and LiDAR sampling error. *Remote Sensing of Environment* 130: 153–170.

Mahmood, H.S., W.B. Hoogmoed, and E.J. van Henten. 2012. Sensor data fusion to predict multiple soil properties. *Precision Agriculture* 13: 628–645.

Mitchell, H.B. 2012. *Data fusion: Concepts and ideas*. Berlin: Springer-Verlag.

Muukkonen, P., and J. Heiskanen. 2007. Biomass estimation over a large area based on standwise forest inventory data and ASTER and MODIS satellite data: A possibility to verify carbon inventories. *Remote Sensing of Environment* 107: 617–624.

Ni, J., 2013. Carbon storage in Chinese terrestrial ecosystems: Approaching a more accurate estimate, *Climatic Change* 119: 905–917.

Office of Converting Farmland to Forestry. 2014. State Forestry Administration of China: A new round of the overall concept of returning farmland to forest and grass. *Newsletter for Converting Farmland to Forestry* 194: 1–2 (in Chinese).

Office of Converting Farmland to Forestry. 2016. State forestry administration of China: New year's speech. *Newsletter for Converting Farmland to Forestry* 198: 1–2 (in Chinese).

Pacifici, K., B.J. Reich, D.A.W. Miller, B. Gardner, G. Stauffer, S. Singh, A. McKerrow, and J.A. Collazo. 2017. Integrating multiple data sources in species distribution modeling: A framework for data fusion. *Ecology* 98 (3): 840–850.

Park, N.W., W.M. Moon, K.H. Chi, and B.D. Kwon. 2002. Multi-sensor data fusion for supervised land-cover classification using Bayesian and geostatistical techniques. *Geosciences Journal* 6 (3): 193–202.

Piao, S.L., J.Y. Fang, B. Zhu, and K. Tan. 2005. Forest biomass carbon stocks in China over the past 2 decades: Estimation based on integrated inventory and satellite data. *Journal of Geophysical Research* 110: G01006.

Piao, S.L., J.Y. Fang, P. Ciais, P. Peylin, Y. Huang, S. Sitch, and T. Wang. 2009. The carbon balance of terrestrial ecosystems in China. *Nature* 458: 1009–1014.

Prentice, I.C., et al. 2001. The carbon cycle and atmospheric carbon dioxide. In *Climate change 2001: The scientific basis*, ed. J.T. Houghton, Y. Ding, D.J. Griggs, M. Noguer, P.J. Van Der Linden, X. Dai, K. Maskell, and C.A. Johnson, 183–237. Cambridge: Cambridge University Press.

Ragab, R., and C. Prudhomme. 2002. Sw-soil and water: Climate change and water resources management in arid and semiarid regions: Prospective and challenges for the 21st century. *Biosystems Engineering* 81 (1): 3–34.

Saatchi, S.S., et al. 2011. Benchmark map of forest carbon stocks in tropical regions across three continents. *Proceedings of the National Academy of Sciences* 108: 9899–9904.

Sankey, T.T., J. McVay, T.L. Swetnam, M.P. McClaran, P. Heilman, and M. Nichols. 2017. UAV hyperspectral and lidar data and their fusion for arid and semi-arid land vegetation monitoring. *Remote Sensing in Ecology and Conservation* 1: 1–14.

Singh, K.K., J.B. Vogler, D.A. Shoemaker, and R.K. Meentemeyer. 2012. LiDAR-Landsat data fusion for large-area assessment of urban land cover: Balancing spatial resolution, data volume and mapping accuracy. *ISPRS Journal of Photogrammetry and Remote Sensing* 74: 110–121.

State Forestry Administration of China. 2009. *National forest report (2004–2008)*. Beijing (in Chinese): China Forestry Publishing House.

Thurner, M., C. Beer, M. Santoro, N. Carvalhais, T. Wutzler, D. Schepaschenko, A. Shvidenko, E. Kompter, B. Ahrens, S.R. Levick, and C. Schmullius. 2014. Carbon stock and density of northern boreal and temperate forests. *Global Ecology and Biogeography* 23: 297–310.

Torabzadeh, H., F. Morsdorf, and M.E. Schaepman. 2014. Fusion of imaging spectroscopy and airborne laser scanning data for characterization of forest ecosystems—A review. *ISPRS Journal of Photogrammetry and Remote Sensing* 97: 25–35.

Wang, S., J.M. Chen, W.M. Ju, X. Feng, M. Chen, P. Chen, and G. Yu. 2007. Carbon sinks and sources in China's forests during 1901–2001. *Journal of Environmental Management* 85: 524–537.

Wiemann, S. 2017. Formalization and web-based implementation of spatial data fusion. *Computers & Geosciences* 99: 107–115.

Witharana, C., M.A. LaRue, and H.J. Lynch. 2016. Benchmarking of data fusion algorithms in support of earth observation based Antarctic wildlife monitoring. *ISPRS Journal of Photogrammetry and Remote Sensing* 113: 124–143.

Woodhouse, I.H., E.T. Mitchard, M. Brolly, D. Maniatis, and C.M. Ryan. 2012. Radar backscatter is not a direct measure of forest biomass. *Nat. Clim. Change* 2 (8): 556–557.

Xu, Z.G., J.T. Xu, X.Z. Deng, J.K. Huang, E. Uchida, and S. Rozelle. 2006. Grain for Green versus grain: Conflict between food security and conservation set-aside in China. *World Development* 34: 130–148.

Yang, K., and D.S. Guan. 2008. Changes in forest biomass carbon stock in the Pearl River Delta between 1989 and 2003. *Journal of Environmental Sciences* 20: 1439–1444.

Yu, G.R., Z. Chen, S.L. Piao, C.H. Peng, P. Ciais, Q.F. Wang, X.R. Li, and X.J. Zhu. 2014. High carbon dioxide uptake by subtropical forest ecosystems in the East Asian monsoon region. *Proceedings of the National Academy of the Sciences of the USA* 111: 4910–4915.

Yue, T.X., and S.H. Wang. 2010. Adjustment computation of HASM: A high-accuracy and high-speed method. *Int J Geograph Inf Sci* 24: 1725–1743.

Yue TX, Liu Y, Zhao MW, Du ZP, Zhao N. 2016a. A fundamental theorem of Earth's surface modelling. *Environmental Earth Sciences* 75(9): article 751, 1–12.

Yue, T.X., Y.F. Wang, Z. P. Du, M. W. Zhao, L. L. Zhang, N. Zhao, M. Lou, G.R. Larocqued, J. Wilson. 2016b. Analyzing the uncertainty of estimating forest carbon stocks in China. *Biogeosciences* 13: 3991–4004.

Zhang, C.H., W.M. Ju, J.M. Chen, M. Zan, D.Q. Li, Y.L. Zhou, and X.Q. Wang. 2013. China's forest biomass carbon sink based on seven inventories from 1973 to 2008. *Climatic Change* 118: 933–948.

Zhou, L.S., H. Sun, Y.Q. Shen, J.Z. Deng, and Y.L. Shi. 1981. *Comprehensive agricultural planning of China*. Beijing (in Chinese): China Agricultural Press.

# Chapter 6
# Model-Data Assimilation

**Learning Objectives**

- Corollary 6 of FTEEM: When a spatial model is available, a more accurate surface can be produced when ground observations are incorporated into the spatial model. This corollary emphasizes the importance of intrinsic information for the model-data assimilation because extrinsic information can be obtained from output of the spatial model.
- Corollary 7 of FTEEM: When both a spatial model and ground observations are available, a surface can be produced by using HASM to incorporate the ground observations into the spatial model, which is more accurate than the one either from the spatial model or from the ground observations. This corollary pays attention to the significance of an appropriate method for the model-data assimilation.

## 6.1 Introduction

Model–data assimilation (MDA) includes model inversion and data assimilation. Data assimilation tries to improve the performance of a model by either optimizing parameters or initial state variables according to a given data set (Peng et al. 2011). Model inversion can be used to estimate parameters that are related to the measured quantity, by using observed properties to constrain physical or biological processes.

MDA algorithms can be distinguished into variational algorithms and sequential algorithms (Ehrendorfer 1992). A variational algorithm is to obtain the assimilation product as a particular solution of the assimilating model according to a prespecified cost function under the constraint that the assimilating model equations are exactly satisfied. The optimality of the assimilating-model trajectory can be obtained only at the expense of several model integrations associated with a minimizing algorithm

 
T. Yue et al., *Eco-Environmental Informatics*,
https://doi.org/10.1007/978-981-96-1977-1_6

(Courtier and Talagrand 1990). In contrast, the sequential algorithm only made one integration of the assimilating model over the assimilation interval, during which each observation is used exactly once to improve successively the predicted state of the model (Ghil and Malanotte-Rizzoli 1991). Variational and sequential algorithms have been developed separately by various groups and each brings its own benefits to data assimilation; in recent years, various ways have been developed to combine these algorithms to maximize the benefits and eliminate the inadequacies of the separate methods (Bannister 2017).

There have been many achievements in atmospheric and oceanic fields. For examples, an optimal adaptive MDA algorithm was derived by using the maximum likelihood method based on a conditional Gaussian probability density function for the first-guess and direct observations of the atmospheric state variables (Frehlich 2006). MDA algorithms were applied to derive anomaly fields of sea surface height (Falkowski et al. 1998) and support weather forecasting (Hanssen et al. 1999). They increased physical understanding of considered systems (Wang et al. 2000). A multimodel Kalman smoother for data assimilation was implemented by iteratively performing a least-squares analysis to detect changes in ice sheet melt rates, which permits the estimate of a nonstationary signal in contrast to ordinary least squares techniques (Hay et al. 2013). High-resolution seabird and ship drift data were assimilated into numerical ocean models to improve the accuracy of the ocean current forecast in the north-eastern sea of Japan (Miyazawa et al. 2016).

Since 2010s, there have appeared a considerable number of MDA approaches to hydrological issues. An ensemble-based Kalman filter approach was introduced to assimilate Gravity Recovery and Climate Experiment (GRACE) satellite data into the WaterGAP Global Hydrology Model to provide a step toward an effective spatial and vertical disaggregation of GRACE information of terrestrial water storage in Mississippi river basin (Eicker et al. 2014). Data assimilation method provides an effective solution to address the challenges in the history matching problem of highly non-Gaussian fractured reservoirs by combination of mimetic finite differences and ensemble Kalman filter (Ping et al. 2017). An ensemble Kalman filter-based data assimilation system, being able to run in a cloud computing environment, was developed for fully-coupled and physically-based hydrological model (Kurtz et al. 2017). Simulating hydrological processes within Murray-Darling Basin of Australia demonstrated that assimilating remotely sensed data from GRACE mission into WaterGAP Global Hydrology Model (WGHM), does not only improve simulation of seasonality and trend of terrestrial water storage changes (TWSC), but also it improves the simulation of individual water storage components (Schumacher et al. 2018). CryoSat-2 data was assimilated into a hydrodynamic model to accurately and bias-free represent spatial–temporal variations of water level over the Assam valley of Brahmaputra River in South Asia (Schneider et al. 2018).

MDA algorithms have been applied into analyses of carbon sinks and sources. For instances, Sacks et al. (2007) found that MDA applied to long eddy covariance time series could illuminate the impact of climate on terrestrial net ecosystem $CO_2$ exchange, gross ecosystem $CO_2$ exchange and ecosystem respiration. Normalized difference vegetation index (NDVI) observations were assimilated into a zero-order model of leaf area index and carbon uptake, in which frequent coarse-resolution observations were combined with temporally sparse fine-resolution measurements (Hill et al. 2011). A 4-D data assimilation system was constructed to simultaneously estimate atmospheric concentrations and surface fluxes of $CO_2$ at spatial of 2.8°and temporal resolution of 1.5 days by using various platform observation data sets (Miyazaki et al. 2011). A probabilistic inversion approach was developed to assimilate the biometric data sets to reduce uncertainties in forest carbon sink forecasting (Weng et al. 2011).

Results from assimilating Orbiting Carbon Observatory-2 (OCO-2) observations into a high-resolution global model indicated that the incorporation of OCO-2 observations corrected errors in the model's prediction of atmospheric $CO_2$ concentrations, and the model provided additional information about the vertical distribution of the gas and fills gaps in cloudy, data-sparse regions (Eldering et al. 2017). MDA was employed to analyze the growing season ecosystem $CO_2$ strength and dynamics in the Swiss central Alps (Scholz et al. 2018). An MDA approach within a Bayesian framework was used to improve estimation accuracy of gross primary productivity and ecosystem respiration by running terrestrial regional ecosystem exchange simulator against observations of evapotranspiration and net ecosystem exchange (Mitra et al. 2018). Eddy flux and meteorological data was assimilated into a simplified photosynthesis and evapotranspiration (SIPNET) model to improve uncertainties of estimating carbon and water exchanges between ecosystems and atmosphere over broad-leaved Korean pine mixed forest of Changbaishan research station in China (Ren and Hartnett 2017).

We can find many MDA methods for other issues such as soil, ecosystems, cholera and Venus. The ground-based microwave radiometer (GBMR) measurements from the soil moisture experiments in Iowa in 2002 were assimilated into a land data assimilation scheme (LDAS) via a radiative transfer model (RTM), which improved the results of LDAS (Pathmathevan et al. 2003). MDA methods were applied in spatially explicit ecosystem models in early 2000s to simulate the most likely past conditions (DeYoung et al. 2004; Wood 2001; Willebrand et al. 2001). An MDA system for a Venusian general circulation model was developed on the base of the local ensemble transform Kalman filter by integrating observations made by the Venus Monitoring Camera (VMC) onboard the Venus Express (Sugimoto et al. 2017). An MDA technique was used to update both state variables and parameters of a cholera model on the basis of observations for cholera outbreaks in Haiti (Pasetto et al. 2017). Jin et al. (2018) analyzed how to reduce the different errors of estimating canopy state variables and soil properties which were assimilated into crop models and discussed the impacts of different error sources on the different portions of the data assimilation chain.

Since 1990s, there have been many publications of assessing performances of MDA methods. For instance, Malanotte-Rizzoli and Young (1995) analyzed the effectiveness of data sets with different space coverage and time resolution in improving overall performance of an ocean circulation model, Semispectral Primitive Equation Model (SPEM). Numerical experiments demonstrated that adaptive observation strategy was a promising method for targeted observations in the context of 4D-Var data assimilation and its implementation was feasible for large scale atmospheric models (Daescu and Navon 2004). Rabier (2005) concluded that introduction of an adaptive component into a numerical weather-prediction system could bring additional improvement to the forecasting of high-impact weather. Cox and Stephenson (2007) concluded that MDA was common in numerical weather prediction, but was not yet used to constrain the climate projections of the Intergovernmental Panel on Climate Change (IPCC). Klimova (2008) argued that a practical implementation of the model-data assimilation algorithm based on the Kalman filter in its complete formulation was impossible due to high dimension of the associated equation sets and to nonlinearity of the predicted processes. Assimilation of satellite data into phenology model could be of great help to minimize the large differences between observed and predicted spatiotemporal phenological patterns (Peñuelas et al. 2009). Verbeeck et al. (2011) demonstrated that carbon models can be improved substantially by assimilating site level flux data over the tropics. Keenan et al. (2012) concluded that exogenous uncertainty dominates uncertainty in future projections from a data-informed process-based model. Numerical data assimilation can be successful only if an effective dimension is moderate or else one cannot reach reliable conclusions about the process one is modeling, even when the linear model is completely correct (Chorin and Morzfeld 2013). Khaki et al. (2017) evaluated performance of integrating data of terrestrial water storage into the world-wide water resources assessment (W3RA) model in Australia. Kim et al. (2018) assessed effects of MDA parameters on the estimation of surface $CO_2$ fluxes in Asia by means of an inverse modeling system based on the ensemble Kalman filter. A survey on data assimilation methods demonstrated that the quality of forecasts based on model-data assimilation is significantly better than the one from simple downscaling of larger-scale initial data (Gustafsson et al. 2018).

This chapter constructed a model-data assimilation algorithm on the basis of the high accuracy surface modelling (HASM) after the fundamental theorem for eco-environmental surface modelling. The algorithm is termed HASM-MD. The effects of HASM-MD on improving accuracy of the Dynamic Global Vegetation Model (LPJ-DGVM) are evaluated by taking carbon stocks in BeiJing-TianJin-HeBei (BTH) region as an example.

## 6.2 Methods

HASM is used to simulate a surface of a state variable in three dimensions (3-D) at a time (Yue 2011). The change of a state variable with time at a point is introduced into the HASM based model-data assimilation algorithm (HASM-MD) to realize 3-D spatial simulation in continuous time.

### 6.2.1 Formulation of HASM-MD

If $z$ is carbon stocks at location $(x, y)$, the surface of $z = f(x, y)$ at time $t$ can be formulated as (Yue 2011):

$$\begin{cases} f_{xx} = \Gamma_{11}^1 \cdot f_x + \Gamma_{11}^2 \cdot f_y + \frac{L}{\sqrt{E+G-1}} \\ f_{yy} = \Gamma_{22}^1 \cdot f_x + \Gamma_{22}^2 \cdot f_y + \frac{N}{\sqrt{E+G-1}} \end{cases} \quad (1)$$

where $E = 1 + f_x^2$, $F = f_x \cdot f_y$, $G = 1 + f_y^2$, $L = \frac{f_{xx}}{\sqrt{1+f_x^2+f_y^2}}$, $M = \frac{f_{xy}}{\sqrt{1+f_x^2+f_y^2}}$, $N = \frac{f_{yy}}{\sqrt{1+f_x^2+f_y^2}}$, $\Gamma_{11}^1 = \frac{1}{2}(GE_x - 2FF_x + FE_y)(EG - F^2)^{-1}$, $\Gamma_{11}^2 = \frac{1}{2}(2EF_x - EE_y - FE_x)(EG - F^2)^{-1}$, $\Gamma_{22}^1 = \frac{1}{2}(2GF_y - GG_x - FG_y)(EG - F^2)^{-1}$, and $\Gamma_{22}^2 = \frac{1}{2}(EG_y - 2FF_y + FG_x)(EG - F^2)^{-1}$.

If the steps of $\Delta x$ and $\Delta y$ are small enough, the first finite difference of $f_x(x, y, t)$ and $f_y(x, y, t)$ at time $t$ could be approximated to:

$$\begin{cases} f_x(x, y, t) \approx \frac{f(x+\Delta x,y,t)-f(x-\Delta x,y,t)}{2\Delta x} \\ f_y(x, y, t) \approx \frac{f(x,y+\Delta y,t)-f(x,y-\Delta y,t)}{2\Delta y} \end{cases} \quad (2)$$

The second order finite difference of $f_{xx}(x, y, t)$ and $f_{yy}(x, y, t)$ at time $t$ could approximately be expressed as:

$$\begin{cases} f_{xx}(x, y, t) \frac{f(x+\Delta x,y,t)-2f(x,y,t)+f(x-\Delta x,y,t)}{\Delta x^2} \\ f_{yy}(x, y, t) \approx \frac{f(x,y+\Delta y,t)-2f(x,y,t)+f(x,y-\Delta y,t)}{\Delta y^2} \end{cases} \quad (3)$$

For the state variable at location of $(x, y)$, the changing curve with time $t$ can be formulated as:

$$v(t) = f(x, y, t) \quad (4)$$

Then:

$$v_{tt} = \Gamma_{11}^1 v_t + \frac{L}{\sqrt{E}} \quad (5)$$

where $E = 1 + v_t^2$, $F = 0$, $G = 1$, $L = \frac{v_{tt}}{\sqrt{1+v_t^2}}$, $\Gamma_{11}^1 = \frac{E_t}{2E}$.

When the step of $\Delta t$ is small enough, the first-order finite difference of the changing curve of the state variable with time $t$ at location $(x, y)$ is expressed as:

$$v_t(t) \approx \frac{v(t + \Delta t) - v(t - \Delta t)}{2\Delta t} = \frac{f(x, y, t + \Delta t) - f(x, y, t - \Delta t)}{2\Delta t} \quad (6)$$

The second order finite difference of $v(t)$ is formulated as:

$$\begin{aligned} v_{tt}(t) &\approx \frac{v(t+\Delta t) - 2v(t) + v(t-\Delta t)}{\Delta t^2} \\ &= \frac{f(x,y,t+\Delta t) - 2f(x,y,t) + f(x,y,t-\Delta t)}{\Delta t^2} \end{aligned} \tag{7}$$

The (n + 1)th iteration of the HASM-MD finite difference can be expressed as:

$$\begin{cases} \dfrac{f_{i+1,j,t}^{(n+1)} - 2f_{i,j,t}^{(n+1)} + f_{i-1,j,t}^{(n+1)}}{\Delta x^2} \\ = (\Gamma_{11}^1)_{i,j,t}^{(n)} \dfrac{f_{i+1,j,t}^{(n)} - f_{i-1,j,t}^{(n)}}{2\Delta x} + (\Gamma_{11}^2)_{i,j,t}^{(n)} \dfrac{f_{i,j+1,t}^{(n)} - f_{i,j-1,t}^{(n)}}{2\Delta y} + \dfrac{L_{i,j,t}^{(n)}}{\sqrt{E_{i,j,t}^{(n)} + G_{i,j,t}^{(n)} - 1}} \\ \dfrac{f_{i,j+1,t}^{(n+1)} - 2f_{i,j,t}^{(n+1)} + f_{i,j-1,t}^{(n+1)}}{\Delta y^2} \\ = (\Gamma_{22}^1)_{i,j,t}^{(n)} \dfrac{f_{i+1,j,t}^{(n)} - f_{i-1,j,t}^{(n)}}{2\Delta x} + (\Gamma_{22}^2)_{i,j,t}^{(n)} \dfrac{f_{i,j+1,t}^{(n)} - f_{i,j-1,t}^{(n)}}{2\Delta y} + \dfrac{L_{i,j,t}^{(n)}}{\sqrt{E_{i,j,t}^{(n)} + G_{i,j,t}^{(n)} - 1}} \end{cases} \tag{8}$$

The iterative initial value $f_{i,j,t}^{(0)}$ on location $(i,j)$ at time $t$ can be calculated by:

$$\frac{f_{i,j,t}^{(0)} - 2f_{i,j,t-1} + f_{i,j,t-2}}{\Delta t^2} = (\Gamma_{11}^1)_{i,j,t-1} \frac{f_{i,j,t}^{(0)} - f_{i,j,t-2}}{2\Delta t} + \frac{L_{i,j,t-1}}{\sqrt{E_{i,j,t-1}}} \tag{9}$$

or

$$f_{i,j,t}^{(0)} = \frac{1}{2 - h_t(\Gamma_{11}^1)_{i,j,t-1}} \left( 4f_{i,j,t-1} - \left(2 + \Delta t(\Gamma_{11}^1)_{i,j,t-1}\right) f_{i,j,t-2} + \frac{2\Delta t^2 L_{i,j,t-1}}{\sqrt{E_{i,j,t-1}}} \right) \tag{10}$$

HASM-MD at time $t$ can be transformed into least squares under constraints:

$$\begin{cases} \mathbf{A} \cdot \mathbf{v}^{(n+1)} = \mathbf{d}^{(n)} \\ \mathbf{B} \cdot \mathbf{v}^{(n+1)} = \mathbf{q}^{(n)} \\ s.t. \mathbf{S} \cdot \mathbf{v}^{(n+1)} = \mathbf{k} \end{cases} \tag{11}$$

where $\mathbf{v}^{(n)} = (f_{1,1,t}^{(n)}, \cdots, f_{1,J,t}^{(n)}, \cdots, f_{I,1,t}^{(n)}, \cdots, f_{I,J,t}^{(n)})^{\mathrm{T}}$, $f_{i,j,t}^{(n)}$ is the value of $f(x,y,t)$ at the $n$th iterative step at location of $(x, y)$ in the year of $t$, $n$=1, 2, 3, ..., $t$=0, 1, 2, ...; **A** and **B** represent coefficients of the first and second equations, respectively; $\mathbf{d}^{(n)}$ and $\mathbf{q}^{(n)}$ are the right-hand vectors of the first and second equations. respectively; $\mathbf{d}^{(0)}$ and $\mathbf{q}^{(0)}$ are calculated in terms of $f_{i,j,t}^{(0)}$; **S** and **k** are, respectively, constraint points and corresponding values of $f(x, y, t)$ at the constraint points; if $\overline{f}_{i,j,t}$ is the value of $f(x, y, t)$ at the pth point $(x_i, y_j)$, then $s_{p,(i-1)\times I+j} = 1$, and $k_p = \overline{f}_{i,j,t}$.

For time window P, the matrix expression of HASM-MD could be:

$$\min_{f} \left\| \begin{bmatrix} \mathbf{A} \\ \mathbf{B} \\ \lambda \cdot \mathbf{S} \end{bmatrix} \mathbf{v}^{(n+1)} - \begin{bmatrix} \mathbf{d}^{(n)} \\ \mathbf{q}^{(n)} \\ \lambda \cdot \mathbf{k} \end{bmatrix} \right\| \tag{12}$$

A surface of carbon stocks could be simulated yearly by Eq. (12). The surface of carbon stocks in the year of $t + P$ could be employed as an initial surface in the next time window.

### 6.2.2 *Cross-Validation*

The cross-validation includes five steps: (1) 5% of the constraint points were selected randomly for validation, which did join the assimilation procedure; (2) the other 95% of the constraint points were used for the model-data assimilation; (3) the mean absolute error (MAE) and mean relative error (MRE) were calculated at the constraint points selected for validation; (4) the validation points were returned to the pool of available constraint points; and (5) the steps (1) to (4) were repeated until MAE and MRE at all constraint points were calculated.

MAE and MRE were formulated as follows:

$$MAE = \frac{1}{m}\sum_{i}^{m} |o_i - s_i| \tag{13}$$

$$MRE = \frac{MAE}{\frac{1}{m}\sum_{i}^{m} |o_i|} \tag{14}$$

where $m$ is the number of the constraint points for validation at each step; and $o_i$ and $o_i$ are the ground observed CS and HASM-MD simulated CS at the $i$th constraint point for validation, respectively.

### 6.2.3 *Trend Analysis*

The approach to nonparametric tests against trend was employed to analyze the changing trend of forest area and carbon stocks in the BTH region. The Mann–Kendall approach is formulated as (Mann 1945; Kendall 1948):

$$\beta = mean\left(\frac{x_j - x_i}{j - i}\right), \quad \forall j \rangle i \tag{15}$$

where $\beta$ represents the changing trend of forest area or forest carbon stocks; $i$ and $j$ are ordinal numbers of the time series; $x_i$ and $x_j$ are the forest area or carbon stocks at the time of $i$ and $j$, respectively, and $\beta > 0$ indicates the uptrend and $\beta < 0$ does downtrend.

## 6.3 Model-Data Assimilation of Carbon Stocks

In this section, a model-data assimilation algorithm is developed on the basis of HASM under the framework of the Lund-Potsdam-Jena General Ecosystem Simulator (LPJ-GUESS), an integrated model of the Lund-Potsdam-Jena dynamic global vegetation model (LPJ-DGVM) and the General Ecosystem Simulator (GUESS) (Smith et al. 2014). The HASM based model-data assimilation (HASM-MD) was used to support surface modelling of carbon stocks with a continuous time series by introducing changing curves of state variables with time into the HASM algorithm at three spatial dimensions.

The plant functional types (PFTs) in LPJ-GUESS are modified by taking the Vegetation Map of the People's Republic of China as a reference (Editorial Committee of Vegetation Map of China 2007). The modified PFTs in the BTH region include cold-temperate and temperate deciduous coniferous forests, cold-temperate and temperate mountainous needle-leaf forest, temperate needle-leaf forest, temperate broad-leaf deciduous forest and temperate grassland (C3). The climatic parameters were adjusted to match the modified PFTs in the BTH region, such as MMT in the coldest and warmest months as well as accumulated temperature for survival and establishment of the plant functional types (Table 1).

**Table 1** Climatic parameters distinguishing the five plant functional types (PFTs)

| PFT | $T_{c,min}$ | $T_{c,max}$ | $T_{w,max}$ | $T_{w\text{-}c,min}$ | $GDD_{min}$ |
|---|---|---|---|---|---|
| Cold-temperate and temperate deciduous coniferous forests | – | −13 | 23.5 | 35 | 350 |
| Cold-temperate and temperate mountainous needle-leaf forest | −32.5 | −13 | 23.5 | – | 600 |
| Temperate needle-leaf forest | −13 | 3 | – | – | 900 |
| Temperate broad-leaf deciduous forest | −32.5 | 3 | – | – | 1200 |
| Temperate grassland (C3) | – | 14 | – | – | 0 |

*Notes* $T_{c,min}$ and $T_{c,max}$ are the lower and upper limits of mean monthly temperature (MMT) in the coldest month, respectively; $T_{w,max}$ is the upper limit of MMT in the warmest month; $T_{w\text{-}c,min}$ is the lower limit of difference between MMT in the warmest and the coldest months; and $GDD_{min}$ is the lower limit of accumulated temperature for survival and establishment

Climatic data for driving the LPJ-GUESS model include mean monthly temperature, precipitation, sunshine percentage and number of rainy days, and the mean annual atmospheric $CO_2$ concentration. The point data came from 65 meteorological stations, selected from the 328 meteorological stations scattered across and around the BTH region (Fig. 1). The surface data were produced by using HASM with the point data.

Forest cover maps in 2003, 2008 and 2013 were developed on the basis of the national forestry inventory during the periods 1999–2003, 2004–2008, and 2009–2013 (Fig. 2). The carbon stocks data for validation was collected from 121 sample spots during the period 2013–2017. The measured parameters included diameter of a cross-section of a tree trunk at 1.3 m above the ground, tree species, tree height and

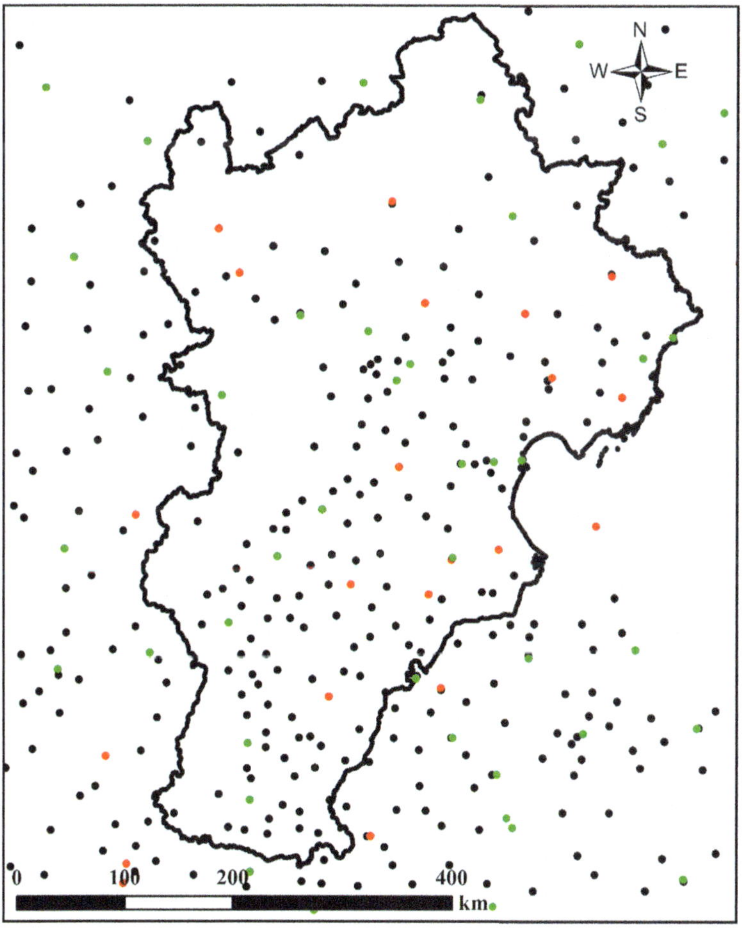


**Fig. 1** Meteorological stations scattered over and around the BTH region: red dots are used for model-data assimilation

the geographic coordinates of the tree. The biomass of the tree was calculated using the single-tree biomass equation. The carbon stocks of the tree were the product of the carbon content and biomass (Li and Lei 2010). The validation results indicated that the mean absolute and relative errors for the model were 2.084 kg·m$^{-2}$ and 68%, and HASM-MD provided estimates of were 0.848 kg·m$^{-2}$ and 28%, respectively. In other words, model-data assimilation improved the accuracy of LPJ-GUESS by 40% and using HASM-MD provided estimates of 37.14 km$^2$, 3.074 kg/m$^2$ and 114.17 Tg, for the forest area, carbon density and carbon stocks in the BTH region, respectively in 2013. The Mann–Kendall approach to trend analysis (Mann 1945; Kendall 1948) was employed to characterize the change trend of forest area (FA). The results showed that FA had an obvious increasing trend during the period from 2001 to 2015. The mean increment was 1.18 × 10$^5$ km$^2$ yr$^{-1}$ ($\beta$ = 1.325, p = 0.001). FA increased by 75% from 2001 to 2015. The largest increment occurred in 2008, following an average increment rate of 0.34 × 10$^5$ hm$^2$ yr$^{-1}$ ($\beta$ = 0.311, p = 0.005) prior to 2008 and a net decline ($\beta$ = −0.193, p = 0.1) after 2008.

Forest carbon stocks (CS), caused by afforestation, had a strong upward trend ($\beta$ = 4.325, p = 0.01) during the period 2001–2015. The CS increased by 90% from 2001 to 2015, growing by 20 Tg yr$^{-1}$ on average. The forest carbon density (CD) also displayed an upward trend ($\beta$ = 0.012, p = 0.01). The CD increased by 9% from 2001 to 2015, growing at the rate of 0.017 kg m$^{-2}$ yr$^{-1}$ on average. However, the net increment of CS was not as strong ($\beta$ = 0.003, p = 0.1) because the net increment (0.641 Tg yr$^{-1}$) for this metric displayed an upward trend prior to 2008 ($\beta$ = 0.603, p = 0.05) and an obvious downward trend ($\beta$ = −0.403, 0.1) after 2008.

The CS was divided between the original forest (OCS), which existed before 2001, and afforestation (ACS), the forest planted since 2001, in order to identify relationship between CS and forest cover change during the period 2001–2015 (Fig. 3). The carbon density of the original forest had an obvious upward trend ($\beta$ = 0.255, p = 0.01) such that the carbon density increased by 3.518 kg m$^{-2}$, but the OCS decreased by 10.035 Tg because the original forest cover declined during the period 2001–2015. The ACS had a continuous upward trend, and increased by 67.687 Tg during the same period ($\beta$ = 4.765, p = 0.01).

The net effect of forest cover change to carbon stocks changed from a negative to a positive contribution in 2006 and during the period 2011–2015, the CS increment of afforestation was 4.512 Tg yr$^{-1}$ and the CS loss due to deforestation and tree mortality was 4.147 Tg yr$^{-1}$. The contribution of forest cover change to CS, jointly caused by afforestation and deforestation, changed from −1.996 to 3.656 Tg yr$^{-1}$ from 2001 to 2015.

In short, the forest area increased at a rate of 1180 km$^2$ yr$^{-1}$ and the net increment was 16,550 km$^2$, in which forest area increased by 29,350 km$^2$ because of afforestation and decreased by 12,800 km$^2$ because of deforestation during the period 2001–2015. The CS had an increment rate of 4.020 Tg yr$^{-1}$. The total increase was 56.282 Tg over the 15 years, of which the original forest and afforestation contributed 48.807 and 66.317 Tg, respectively and deforestation removed 58.841Tg. Carbon density increased from 2.839 kg C m$^{-2}$ in 2001 to 3.079 kg C m$^{-2}$ in 2015.

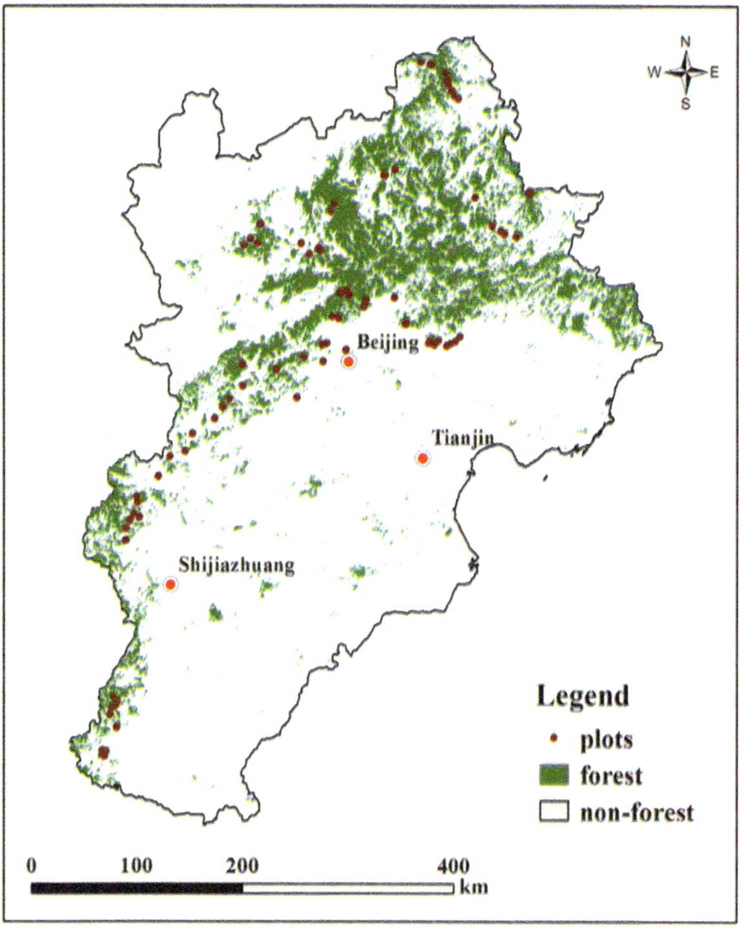


**Fig. 2** Forest cover and sample plots in the BTH region

## 6.4 Summary

System models and observations provide abundant pieces of information, but none of them provides the complete picture (Ponsar et al. 2016). A system model, if it is to be as accurate as possible, must supplement information from the currently available ground observations (Brill et al. 1991). MDA is the process by which measured observations are incorporated into a system model over a period of time to create an estimate of the system state. The aim of MDA is to use measured observations in combination with a system model in order to derive accurate estimates of the current and future states of the system, together with estimates of the uncertainty in the estimates (Nichols 2010). MDA can be used to match model simulation and ground observations by varying model structure, parameters and initial state variables (Wang

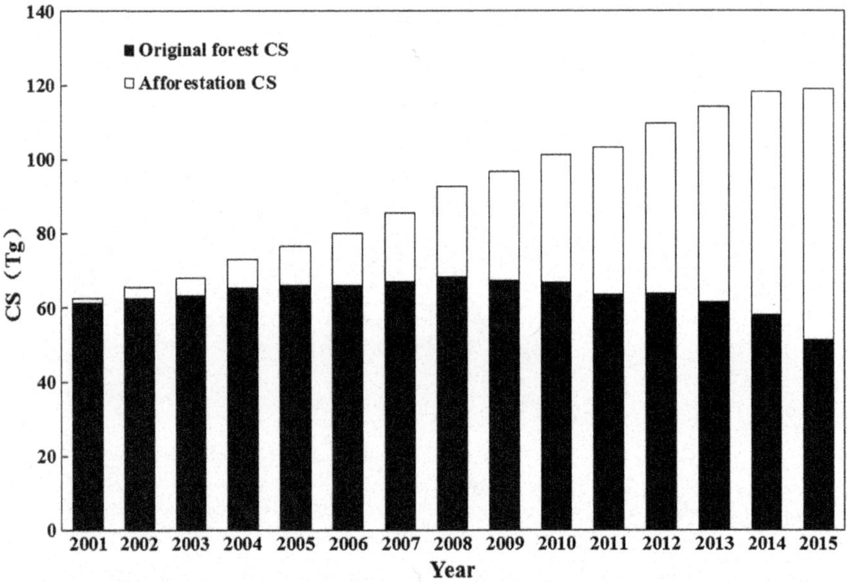


**Fig. 3** Contributions of original forest and afforestation to total carbon stocks by year

et al. 2009). MDA enables the optimal use of model and observational resources and offers the potential to generate forecasts that are more accurate than using a model alone (Smith et al. 2013). The MDA power is both in its ability to produce forecasts of the dynamic model, but also to make inferences in quantities that are otherwise hidden from observations (Li et al. 2014). Assimilating measurements into dynamic models makes the original model background states follow the observation trajectory, then uses it to provide more accurate forecasting information (Ren and Hartnett 2017).

After improvement of MDA, our next step is to develop robust algorithms for model coupling in terms of FTEEM. No single model can capture all of the dynamics of the Earth's surface at a high level of detail. The coupling of models may provide an alternative solution because the outputs of one model can serve as input to another model by loose or tight coupling, and thereby allow feedback (IPBES 2016). The choice of tight versus the loose coupling depends on the specific requirements of the system under consideration. A loosely coupled model approach allows for the identification of the role of system interactions, whereas a tight coupling approach will be required when feedback between the system components is important to the outcomes.

## 6.5 Further Reading

(1) Yue TX. 2011. Surface Modelling: High Accuracy and High Speed Methods. New York: CRC Press.
(2) Yue TX, Zhao N, Liu Y, Wang YF, Zhang B, Du ZP, Fan ZM, Shi WJ, Chen CF, Zhao MW, Song DJ, Song YJ, Yan CQ, Li QQ, Sun XF, Zhang LL, Tian YZ, Wang W, Wang YA, Ma SN, Huang HS, Lu YM, Wang Q, Wang CL, Wang YZ, Lu M, Zhou W, Liu Y, Yin XZ, Wang Z, Bao ZY, Zhao MM, Zhao YP, Jiao YM, Naseer U, Fan B, Li SB, Yang Y, Wilson JP. 2020. A fundamental theorem for eco-environmental surface modelling and its applications. Science China-Earth Sciences 63(8): 1092–1112.

### Exercises

(1) Define the terms "model-data assimilation".
(2) Explain Corollary 6 and Corollary 7 of FTEEM.

## References

Bannister, R.N. 2017. A review of operational methods of variational and ensemble-variational data assimilation. *Quarterly Journal Royal Meteorological Society* 143: 607–633.

Brill, K.F., L.W. Uccellini, J. Manobianco, P.J. Kocin, and J.H. Homan. 1991. The use of successive dynamic initialization by nudging to simulate cyclogenesis during GALE IOP 1. *Meteorology and Atmospheric Physics* 45: 15–40.

Chorin, A.J., and M. Morzfeld. 2013. Conditions for successful data assimilation. *Journal of Geophysical Research: Atmospheres* 118: 11522–11533.

Courtier, P., and O. Talagrand. 1990. Variational assimilation of meteorological observations with the direct and adjoint shallow-water equations. *Tellus* 42A: 531–549.

Cox, P., and D. Stephenson. 2007. A changing climate for prediction. *Science* 317: 207–208.

Daescu, D.N., and I.M. Navon. 2004. Adaptive observations in the context of 4D-Var data assimilation. *Meteorology and Atmospheric Physics* 85: 205–226.

deYoung, B., M. Heath, F. Werner, F. Chai, B. Megrey, and P. Monfray. 2004. Challenges of modeling ocean basin ecosystems. *Science* 304: 1463–1466.

Eicker, A., M. Schumacher, J. Kusche, P. Doell, and H.M. Schmied. 2014. Calibration/data assimilation approach for integrating GRACE data into the WaterGAP global hydrology model (WGHM) using an Ensemble Kalman filter: First results. *Surveys in Geophysics* 35: 1285–1309.

Ehrendorfer, M. 1992. Four-dimensional data assimilation: Comparison of variational and sequential algorithms. *Quarterly Journal Royal Meteorological Society* 118: 673–713.

Eldering, A., P.O. Wennberg, D. Crisp, D.S. Schimel, M.R. Gunson, A. Chatterjee, J. Liu, F.M. Schwandner, Y. Sun, C.W. O'Dell, C. Frankenberg, T. Taylor, B. Fisher, G.B. Osterman, D. Wunch, J. Hakkarainen, J. Tamminen, and B. Weir. 2017. The orbiting carbon observatory-2 early science investigations of regional carbon dioxide fluxes. *Science* 358: 1–8.

Falkowski, P.G., R.T. Barber, and V. Smetacek. 1998. Biogeochemical controls and feedbacks on ocean primary production. *Science* 281: 200–206.

Frehlich, R. 2006. Adaptive data assimilation including the effect of spatial variations in observation error. *Quarterly Journal Royal Meteorological Society* 132: 1225–1257.

Ghil, M., and P. Malanotte-Rizzoli. 1991. Data assimilation in meteorology and oceanography. *Advances in Geophysics* 33: 141–266.

Gustafsson, N., T. Janjic, C. Schraff, D. Leuenberger, M. Weissman, H. Reich, P. Brousseau, T. Montmerle, E. Wattrelot, A. Bucanek, M. Mile, R. Hamdi, M. Lindskog, J. Barkmeijer, M. Dahlbom, B. Macpherson, S. Ballard, G. Inverarity, J. Carley, C. Alexander, D. Dowell, S. Liu, Y. Ikuta, and T. Fujita. 2018. Survey of data assimilation methods for convective-scale numerical weather prediction at operational centres. *Quarterly Journal Royal Meteorological Society*. https://doi.org/10.1002/qj.3179.

Hanssen, R.F., T.M. Weckwerth, H.A. Zebker, and R. Klees. 1999. High-resolution water vapor mapping from interferometric radar measurements. *Science* 283: 1297–1299.

Hay, C.C., E. Morrow, R.E. Kopp, and J.X. Mitrovica. 2013. Estimating the sources of global sea level rise with data assimilation techniques. *PNAS* 110: 3692–3699.

Hill, T.C., T. Quaife, and M. Williams. 2011. A data assimilation method for using low-resolution Earth observation data in heterogeneous ecosystems. *Journal of Geophysical Research* 116: D08117. https://doi.org/10.1029/2010JD015268.

Hu, S., L.S. Shi, Y.Y. Zha, M. Williams, and L. Lin. 2017. Simultaneous state-parameter estimation supports the evaluation of data assimilation performance and measurement design for soil-water-atmosphere-plant system. *Journal of Hydrology* 555: 812–831.

IPBES. 2016. Methodological assessment of scenarios and models of biodiversity and ecosystem services. Bonn, Germany: Secretariat of the Intergovernmental Platform for Biodiversity and Ecosystem Services.

Jin, X.L., L. Kumar, Z.H. Li, H.K. Feng, X.G. Xu, G.J. Yang, and J.H. Wang. 2018. A review of data assimilation of remote sensing and crop models. *European Journal of Agronomy* 92: 141–152.

Keenan, T.F., E. Davidson, A.M. Moffat, W. Munger, and A.D. Richardson. 2012. Using model-data fusion to interpret past trends, and quantify uncertainties in future projections, of terrestrial ecosystem carbon cycling. *Global Change Biology* 18: 2555–2569.

Kendall, M.G. 1948. *Rank correlation methods*. London, UK: Griffin.

Khaki, M., I. Hoteit, M. Kuhn, J. Awange, E. Forootan, A.I.J.M. van Dijk, M. Schumacher, and C. Pattiaratchi. 2017. Assessing sequential data assimilation techniques for integrating GRACE data into a hydrological model. *Advances in Water Resources* 107: 301–316.

Kim, H., H.M. Kim, J. Kim, and C.H. Cho. 2018. Effect of data assimilation parameters on the optimized surface $CO_2$ flux in Asia. *Asia-Pacific Journal of Atmospheric Sciences* 54 (1): 1–17.

Klimova, E.G. 2008. A data assimilation technique based on the $\pi$-algorithm. *Russian Meteorology and Hydrology* 33 (3): 143–150.

Kurtz, W., A. Lapin, O.S. Schilling, Q. Tang, E. Schiller, T. Braun, D. Hunkeler, H. Vereecken, E. Sudicky, P. Kropf, H.J.H. Franssen, and P. Brunner. 2017. Integrating hydrological modelling, data assimilation and cloud computing for real-time management of water resources. *Environmental Modelling & Software* 93: 418–435.

Li, H.K., and C.Y. Lei. 2010. Estimation and Evaluation of Forest Biomass Carbon Storage in China, China Forestry Press, Beijing (in Chinese).

Li, K., A. Jackson, and P.W. Livermore. 2014. Variational data assimilation for a forced, inertia-free magnetohydrodynamic dynamo model. *Geophysical Journal International* 199: 1662–1676.

Malanotte-Rizzoli, P., and R.E. Young. 1995. Assimilation of global versus local data sets into a regional model of the Gulf Stream system. *Journal of Geophysical Research* 100(C12) 24: 773–796.

Mann, H.B. 1945. Nonparametric tests against trend. *Econometrica* 13: 245–259.

Mitra, B., D.S. Mackay, E. Pendall, B.E. Ewers, H. Kwon, M.B. Cleary, and K.J. Naithani. 2018. Model-data fusion approach to quantify evapotranspiration and net ecosystem exchange across the sagebrush ecosystem at different temporal resolutions. *Ecohydrology* 11: e1957.

Miyazaki, K., T. Maki, P. Patra, and T. Nakazawa. 2011. Assessing the impact of satellite, aircraft, and surface observations on $CO_2$ flux estimation using an ensemble-based 4-D data assimilation system. *Journal of Geophysical Research* 116: D16306. https://doi.org/10.1029/2010JD015366.

Miyazawa, Y., X.Y. Guo, S.M. Varlamov, T. Miyama, K. Yoda, K. Sato, T. Kano, and K. Sato. 2016. Assimilation of the seabird and ship drift data in the north-eastern sea of Japan into an operational ocean nowcast/forecast system. *Scientific Reports* 5: 17672. https://doi.org/10.1038/srep17672.

Nichols, N.K. 2010. Mathematical concepts of data assimilation. In *Data assimilation: Making sense of observations*, ed. W. Lahoz, B. Khattatov, and R. Menard, 13–39. Berlin: Springer-Verlag.

Pasetto, D., F. Finger, A. Rinaldo, and E. Bertuzzo. 2017. Real-time projections of cholera outbreaks through data assimilation and rainfall forecasting. *Advances in Water Resources* 108: 345–356.

Pathmathevan, M., T. Koike, X. Li, and H. Fujii. 2003. A simplified land data assimilation scheme and its application to soil moisture experiments in 2002 (SMEX02). *Water Resources Research* 39 (12): 1341. https://doi.org/10.1029/2003WR002124.

Peng, C.H., J. Guiot, H.B. Wu, H. Jiang, and Y.Q. Luo. 2011. Integrating models with data in ecology and palaeoecology: Advances towards a model–data fusion approach. *Ecology Letters* 14: 522–536.

Peñuelas, J., T. Rutishauser, and I. Filella. 2009. Phenology feedbacks on climate change. *Science* 324: 887–888.

Ping, J., O. Al-Hinai, and M.F. Wheeler. 2017. Data assimilation method for fractured reservoirs using mimetic finite differences and ensemble Kalman filter. *Computers & Geosciences* 21: 781–794.

Ponsar, S., P. Luyten, and V. Duli'ere. 2016. Data assimilation with the ensemble Kalman filter in a numerical model of the North Sea. *Ocean Dynamics* 66: 955–971.

Rabier, F. 2005. Overview of global data assimilation developments in numerical weather-prediction centres. *Quarterly Journal Royal Meteorological Society* 131: 3215–3233.

Ren, L., and M. Hartnett. 2017. Sensitivity analysis of a data assimilation technique for hindcasting and forecasting hydrodynamics of a complex coastal water body. *Computers & Geosciences* 99: 81–90.

Román-Cascón, C., T. Pellarin, F. Gibon, L. Brocca, E. Cosme, W. Crow, D. Fernández-Prieto, Y.H. Kerr, and C. Massari. 2017. Correcting satellite-based precipitation products through SMOS soil moisture data assimilation in two land-surface models of different complexity: API and SURFEX. *Remote Sensing of Environment* 200: 295–310.

Sacks, W.J., D.S. Schimel, and R.K. Monson. 2007. Coupling between carbon cycling and climate in a high-elevation, subalpine forest: A model-data fusion analysis. *Oecologia* 151: 54–68.

Schneider, R., M.E. Ridler, P.N. Godiksen, H. Madsen, and P. Bauer-Gottwein. 2018. A data assimilation system combining Cryo-Sat-2 data and hydrodynamic river models. *Journal of Hydrology* 557: 197–210.

Scholz, K., A. Hammerle, E. Hiltbrunner, and G. Wohlfahrt. 2018. Analyzing the effects of growing season length on the net ecosystem production of an Alpine Grassland using model-data fusion. *Ecosystems* 21: 982–999.

Schumacher, M., E. Forootan, A.I.J.M. van Dijk, H.M. Schmied, R.S. Crosbie, J. Kusche, and P. Doell. 2018. Improving drought simulations within the Murray-Darling Basin by combined calibration/assimilation of GRACE data into the WaterGAP global hydrology model. *Remote Sensing of Environment* 204: 212–228.

Smith, B., D. Wårlind,, A. Arneth, T. Hickler, P. Leadley, J. Siltberg, and S. Zaehle. 2014. Implications of incorporating N cycling and N limitations on primary production in an individual based dynamic vegetation model. Biogeosciences 11: 2027-2054, https://doi.org/10.5194/bg-11-2027-2014.

Smith, P.J., G.D. Thornhill, S.L. Dance, A.S. Lawless, D.C. Mason, and N.K. Nichol. 2013. Data assimilation for state and parameter estimation: Application to morphodynamic modelling. *Quarterly Journal Royal Meteorological Society* 139: 314–327.

Sugimoto, N., A. Yamazaki, T. Kouyama, H. Kashimura, T. Enomoto, and M. Takagi. 2017. Development of an ensemble Kalman filter data assimilation system for the Venusian atmosphere. *Scientific Reports* 7: 9321. https://doi.org/10.1038/s41598-017-09461-1.

Sun, M., X.Q. Yin, Y.Z. Yang, and K.J. Wu. 2017. An effective method based on dynamic sampling for data assimilation in a global wave model. *Ocean Dynamics* 67: 433–449.

Verbeeck, H., P. Peylin, C. Bacour, D. Bonal, K. Steppe, and P. Ciais. 2011. Seasonal patterns of $CO_2$ fluxes in Amazon forests: Fusion of eddy covariance data and the ORCHIDEE model. *Journal of Geophysical Research* 116: G02018. https://doi.org/10.1029/2010JG001544.

Wang, B., X.L. Zou, and J. Zhu. 2000. Data assimilation and its applications. *PNAS* 97 (21): 11143–11144.

Wang, Y.P., C.M. Trudinger, and I.G. Enting. 2009. A review of applications of model-data fusion to studies of terrestrial carbon fluxes at different scales. *Agricultural and Forest Meteorology* 149: 1829–1842.

Weng, E.S., Y.Q. Luo, C. Gao, and R. Oren. 2011. Uncertainty analysis of forest carbon sink forecast with varying measurement errors: A data assimilation approach. *Journal of Plant Ecology* 4 (3): 178–191.

Willebrand, J., B. Barnier, C. Boening, C. Dieterich, P.D. Killworth, C.L. Provost, Y.L. Jia, J.M. Molines, and A.L. New. 2001. Circulation characteristics in three eddy-permitting models of the North Atlantic. *Progress in Oceanography* 48: 123–161.

Wood, S.N. 2001. Partially Specified Ecological Models. *Ecological Monographs* 71: 1–25.

Yue, T.X. 2011. *Surface modelling: High accuracy and high speed methods*. Boca Raton, FL: CRC Press.

# Chapter 7
# Surface Modelling of Climatic Change

**Learning Objectives**

- The Methods and models of climate variables estimation.
- Apply the transfer function method-based method to estimate the climate variables in the past.
- Predict the future scenarios with high accuracy and high spatial resolution.
- The climate events and its' influence factors.
- The impacts of climate change on ecosystems.

## 7.1 Introduction

Climatology, an old field of scientific investigation in its own right, has many contacts with other fields. Climate influences agricultural, hydrology, human health and disease, forestry, transportation, and many other fields. Accurate information about the distribution of climate variables in space is the basis for scientifically understanding global or regional changes in processes involving water and its associated materials and energy, which is of great significance for the promotion of meteorological and hydrological monitoring and forecasting to enhance the capability to cope with natural disasters and optimize water resources management. Surface modeling of climatic change has been a vital scientific issue of common concern in many fields, such as meteorology, hydrology, ecology, geology, environment, and so on.

In recent years, numerous efforts on quantitative climate elements' spatial estimation have been made. At present, the number of spatial estimation methods available is relatively high, and new methods are still proposed continuously. These methods are based on special physical and mathematical principles and are adequate for different conditions and tempo-spatial scales. It is difficult to class the existing methods into different groups. However, from the information source perspective, it appears that

 
T. Yue et al., *Eco-Environmental Informatics*,
https://doi.org/10.1007/978-981-96-1977-1_7

the development of spatial estimation of climate variables has generally experienced three stages of development.

The first stage was spatial interpolation, based on the measurements of meteorological stations, which was the traditional way to obtain distributed climatic information. However, the density as well as the spacing required for this method restricts the results obtained by stations. The second stage was remote sensing retrieval and atmospheric model calculation, both of which are areal inherent and can obtain spatial continuous rainfall across a certain region. Since the 1980s, remote sensing retrieval and atmospheric model calculation have gradually become important methods for climatic spatial estimation. However, due to the influences of remote sensing instruments, atmospheric model capacity, and other factors, their uncertainty is relatively prominent. Since the 1990s, spatial estimation of climate variables has developed to the third stage, that is, the stage of multi-source data merging. Recently it has become one of the most important hotspots in meteorology and hydrology and can be used to obtain better climatic information by integrating various kinds of information, such as station observations, remote sensing retrieval, and atmospheric reanalysis estimates.

With the development and interaction of meteorology, hydrology, remote sensing, and spatial statistics, investigation on spatial estimation of climate variables has made significant progress in the three stages above and has produced a number of algorithm and climatic datasets. However, climate variables usually occur in a complex process of continuity and intermittence, with extremely intricate spatial variability. Therefore, accurate acquisition of spatial distribution of climatic variables still faces a series of challenges.

In order to solve the error problem that had long troubled the interpolation method, Yue (2011) developed a novel surface modeling method, High Accuracy Surface Modeling (HASM), based on the fundamental theorem of surface theory. The development of HASM can be divided into 4 stages. In the first stage from 1986 to 2001, a surface was regarded as a combination of its profiles that were fitted in terms of the fundamental theorem for curves (FTC). In this second stage from 2001 to 2007, we found solutions for error and multi-scale problems that plagued surface modelling prior to the development of HASM. Although the error and multi-scale problems were solved during the second stage, HASM could only be used with small areas because it must use the master equation set for simulating each lattice of a surface, which incurs a huge computation cost. In this third stage from 2008 to 2011, we solved the low computational speed and large memory requirement problems. In all of these applications, HASM produced more accurate results than the classical methods. During the fourth stage from 2011 to 2016, the fundamental theorem for earth surface system modeling (FTESM) was proposed on the basis of developing the HASM methods and their successful applications. Numerical tests have shown that HASM is much more accurate than the classical methods such as kriging, Inverse Distance Weighted (IDW) method, and spline (Zhao and Yue 2014; Yue et al. 2007). Surface modeling of Digital Elevation Model (DEM) and ecosystem change also indicate that HASM has increased interpolation accuracy (Yue et al. 2020).

## 7.2 Methods and Models

### 7.2.1 *Spatial Interpolation*

Spatial interpolation is the traditional way to transform point-wise rainfall into areal climatic information. It is a process of tapping and utilizing the spatial autocorrelation of climate variables and spatial intercorrelation between climate variables and related explanatory variables under a specific mathematical framework. Over the years, the evolution of spatial interpolation has occurred in two directions. One was the introduction of new mathematical statistical theories or methods to improve the utilization of spatial correlation information. Second was the integration of more explanatory variables into the quantitative estimation models to increase the amount of effective information for spatial interpolation. Spatial interpolation is defined as predicting the values of a primary variable at points within the same region of sampled locations, based on the station observations. This method is mainly based on the meteorological observations. Usually, the denser the station measurements, the higher the estimation accuracy for climate information and its application, such as hydrological modelling. Thus, a minimum density of station is needed for spatial estimation. However, when the density of a station exceeds a certain limit, the estimation accuracy will not change significantly when evaluated by usual statistical indices. On the other hand, the effect of introducing auxiliary information on spatial estimation accuracy depends on two factors. One is the correlation between auxiliary variables and climate variables—when they are strongly correlated, the introduced auxiliary information will produce a significant effect. Second is the richness of station measurements. When the density of observation network is low, the marginal effect of introducing auxiliary information is more obvious. The accuracy of spatial interpolation often varies with the study area, time scale, and climate elements. There is no optimal method for all circumstances, and in some cases, the accuracy of complex methods is not higher than, and may even be lower than, that of simple algorithms. Thus, any interpolation method has its own merits and demerits. It is highly recommended to select appreciate interpolation methods in terms of the application objective, the geographic and gauge conditions of the study area as well as the temporal and spatial scales.

### 7.2.2 *Remote Sensing Retrieval and Atmospheric Models*

Since the 1980s, remote sensing retrieval and atmospheric model calculation have gradually become important methods for spatial estimation, both of which are areal inherent and can obtain spatial continuous climate information across a certain region. One of the methods for obtaining spatial distribution of climate variables is the remote sensing retrieval. Satellite-borne sensors can detect climate information in a larger coverage area than ground-based stations and have the advantage of applicability to special situations, such as oceans, large lakes, high mountains, and deserts. Global satellite-based rainfall products are currently mainly based on passive microwave (PMW), calibrated infrared (IR), and PMW plus IR observations. IR sensors are on geostationary Earth orbit (GEO) satellites which provide precipitation estimates at high temporal resolutions, but the accuracy of IR-based rainfall estimates is generally not good due to the indirect linkage between IR signals and precipitation. PMW sensors are onboard low Earth orbiting (LEO) satellites and can provide precipitation estimation with a more direct link to rain and ice particles. There have been great numbers of multi-satellite precipitation products since the launch of TRMM in late 1997, such as Multi-satellite Precipitation Analysis (TMPA), Climate Prediction Center (CPC) MORPHing technique (CMORPH), Precipitation Estimation from Remotely Sensed Information using Artificial Neural Networks (PERSIANN) and Global Satellite Mapping of Precipitation (GSMaP), and so on. Table 1 lists the basic information of representative global or satellite precipitation datasets since the launch of TRMM. Datasets in this table are pure satellite remote sensing retrieval data, without calibration by surface gauge data.

Remote sensing data acquired from instruments aboard satellites require processing before the data are usable. Depending on differences in source data, the retrieval algorithms can generally be classified into four types: visible light and infrared, passive microwave, active microwave, and multi-sensor precipitation estimation. However, due to the influences of remote sensing instruments, the retrieval algorithm, the low spatial resolutions, and other factors, their uncertainty is relatively prominent. The performance of satellite precipitation datasets has been widely evaluated on different spatial and temporal scales. There is no single type of data that ubiquitously has the best performance.

At the same time, based on atmospheric physical process, various climate models were developed to produce the continuous climate information. Climate models are based on well-documented physical processes and use mathematical equations to simulate the transfer of energy and materials through the climate system. Building

**Table 1** Some representative global or quasi-global satellite precipitation datasets

| Names | Resolution/ frequency | Spatial coverage | Period | References |
|---|---|---|---|---|
| TRMA-3B42-RT | 0.25°/3 h | 50°S–50°N | 1998–2015 | Huffman et al. (2007) |
| CMORPH | 0.25°/3 h, 8 km/30 min | 60°S–60°N | 1998–present | Joyce et al. (2004) |
| PERSIANN | 0.25°/6 h | 60°S–60°N | 2000–present | Sorooshian et al. (2000) |
| IMERG | 0.1°/30 min, 3 h,1 d | 60°S–60°N | 2000–present | Hou et al. (2014) |
| GSMaP | 0.1°/1 h | 60°S–60°N | 2000–present | Ushio et al. (2009) |

and running a climate model is complex of identifying earth system processes, representing them with mathematical equations, setting variables to represent initial conditions and subsequent changes in climate forcing, and repeatedly solving the equations using powerful supercomputers. Generally, Climate models separate Earth's surface into a three-dimensional grid of cells. The results of processes modeled in each cell are passed to neighboring cells to model the exchange of matter and energy over time. Grid cell size defines the resolution of the model: the smaller the size of the grid cells, the higher the level of detail in the model. More detailed models have more grid cells, so they need more computing power.

The Weather Research and Forecasting (WRF) Model is a mesoscale numerical weather prediction system designed for both atmospheric research and operational forecasting applications (Fig. 1). It features two dynamical cores, a data assimilation system, and a software architecture supporting parallel computation and system extensibility. The model serves a wide range of meteorological applications across scales from tens of meters to thousands of kilometers. The effort to develop WRF began in the latter 1990's and was a collaborative partnership of the National Center for Atmospheric Research (NCAR), the National Oceanic and Atmospheric Administration (represented by the National Centers for Environmental Prediction (NCEP) and the Earth System Research Laboratory), the U.S. Air Force, the Naval Research Laboratory, the University of Oklahoma, and the Federal Aviation Administration (FAA). For researchers, WRF can produce simulations based on actual atmospheric conditions (i.e., from observations and analyses) or idealized conditions. WRF offers operational forecasting a flexible and computationally-efficient platform, while reflecting recent advances in physics, numerics, and data assimilation contributed by developers from the expansive research community. WRF is currently in operational use at NCEP and other national meteorological centers as well as in real-time forecasting configurations at laboratories, universities, and companies. The Mesoscale and Microscale Meteorology Division of NCAR is currently maintaining and supporting a subset of the overall WRF code (Version 4) that includes: WRF Software Framework (WSF); Advanced Research WRF (ARW) dynamic solver, including one-way, two-way nesting and moving nest; The WRF Preprocessing System (WPS); WRF Data Assimilation (WRF-DA) system which currently supports 3DVAR 4DVAR, and hybrid data assimilation capabilities. Numerous physics packages contributed by WRF partners and the research community; Several graphics programs and conversion programs for other graphics tools.

The second frequently used climate model is the Regional Climate Model system RegCM, originally developed at the National Center for Atmospheric Research (NCAR), is maintained in the Earth System Physics (ESP) section of the International Centre for Theoretical Physics (ICTP). The first version of the model, RegCM1, was developed in 1989 and since then it has undergone major updates in 1993 (RegCM2), 1999 (RegCM2.5), 2006 (RegCM3) and most recently 2010 (RegCM4). The latest version of the model, RegCM4, is now fully supported by the ESP, while previous versions are no longer available. This version includes major upgrades in the structure of the code and its pre- and post- processors, along with the inclusion of some new physics parameterizations. The model is flexible, portable and easy to use. It

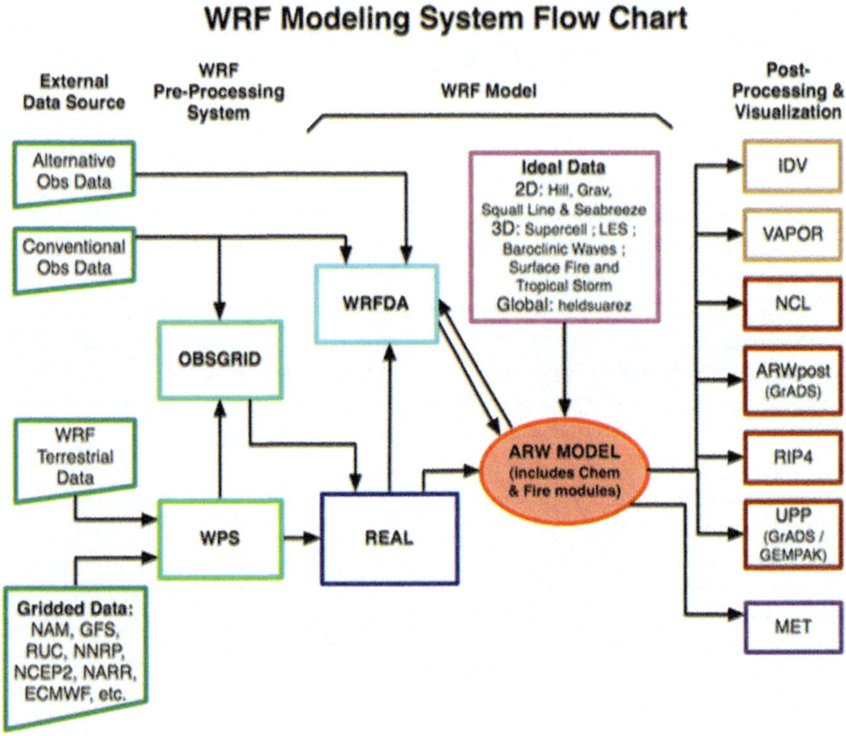


**Fig. 1** The WRF modeling system program components

can be applied to any region of the World, with grid spacing of up to about 10 km. The RegCM has been the first limited area model developed for long term regional climate simulation, it has participated to numerous regional model intercomparison projects, and it has been applied by a large community for a wide range of regional climate studies, from process studies to paleo-climate and future climate projections. Model improvements currently under way include the development of a new microphysical cloud scheme (to be released by the end of 2014), coupling with a regional ocean model, inclusion of full gas-phase chemistry, upgrades of some physics schemes (convection, PBL, cloud microphysics) and development of a non-hydrostatic dynamical core.

Despite the success of the climate modeling enterprise, the complexity of our Earth imposes important limitations on existing climate models. The limited ability of modern climate model is fairly accurate for large-scale organized synoptic systems but modeling skill quickly decreases for more localized events. Besides, the accuracy of the models' outputs are seriously restricted by the model initialization, the parameterization of physical processes and the definition of models.

**Table 2** Some representative reanalysis datasets

| Short name | Resolution \frequency | Spatial coverage | Period |
|---|---|---|---|
| NCEP/NCAR Reanalysis 1 | 2.5°/6 h | Global | 1948–present |
| NCEP/NCAR Reanalysis 2 | 0.5°/6 h | Global | 1979–present |
| NCEP-CFSR | 0.2°/1 h | Global | 2012–present |
| JRA-55 | 0.5625°/3 h | Global | 1958–present |
| ERA-Interim | 0.75°/6 h | Global | 1979–present |

With so many institutions developing and running climate models, there is a risk that each group approaches its modelling in a different way, reducing how comparable their results will be. This is where the Coupled Model Intercomparison Project ("CMIP") comes in. CMIP is a framework for climate model experiments, allowing scientists to analysis, validate and improve GCMs in a systematic way. The goal of CMIP is to generate a set of standard simulations that each model will run. One of the main sets of simulations run by models are future climate scenarios, where models are given a common set of future concentrations of greenhouse gases, aerosols and other climate forcings to project what might happen in the future. CMIP has been through several generations and each iteration becomes more sophisticated in the experiments that are being designed. A new generation comes round every 5–6 years. CMIP six is now underway, which will involve more than 30 modeling centres around the world.

Besides, for the model's outputs, Atmosphere reanalysis is the integration of irregular observation data and regular output of climate models. Over the last 40 years, the development of reanalysis datasets has experienced four generations, each with improved quality and diversity. Atmosphere reanalysis is the integration of irregular observation data and regular output of climate models. Some representative reanalysis datasets are listed in the Table 2.

Different methods of climate variables observation or estimation have different abilities to reflect the climate state in space, with varied reliability and error characteristics. Theoretically, station network can only obtain point wise climate information, and its ability is limited by the density and spacing of the stations. Remote sensing and reanalysis data feature strong spatial continuity and wide coverage but with prominent local errors, and cannot provide the details on very small spatial scales, making it difficult to directly apply them in practice currently.

### *7.2.3 Downscaling and Data Fusion*

At the third stage, a number of algorithms were proposed to meet the requirement of high accuracy with fine spatial and temporal scales, under a certain optimization criterion to integrate information from different sources with different levels of spatial and temporal resolution and precision. Downscaling and data fusion methods were thus developed. For bridging the gap between the scale of remote sensing products or climate models' output and required resolution for practical applications, downscaling provides climate change information at suitable spatial and temporal scales. Statistical and dynamical downscaling are two broad main types. The dynamical downscaling is mainly covered by Regional Climate Models (RCMs). The RCMs utilize large scale and lateral boundary conditions from Global Climate Models (GCMs) to produce higher resolution outputs that demands high computation time. Dynamical downscaling does not provide information at the point or station scale. The main problem with RCMs is that significant biases in the simulation of climate elements on large scales can be inherited from the driving GCM. The boundary conditions are derived from a specific GCM; use of different GCMs will result in different projections. The RCM outputs are still found to be sensitive to the numerical scheme and parameters. The discrepancies between areal average values and site-specific data are expected to remain a problem.

A variety of climate change assessment projects have been established to provided high-resolution climate change scenarios for specific regions. They are usually multi-country, multi-institutional, large-scale projects. They are an important source of regional projections as well as of additional information about RCMs, Methods, and even characteristics of current regional climate. Table 3 provides a summary about the projects.

Statistical downscaling is a two-step process consisting of the development of statistical relationships between local climate variables and large scale predictors; and the application of such relationship of the output of global climate model experiments to simulate local climate characteristics in the future. Statistical downscaling methods are computationally inexpensive in comparison to RCMs that require

**Table 3** Some climate change assessment projects

| Project | Region | Purpose | Method |
|---|---|---|---|
| PRUDENCE (prediction of regional scenarios and uncertainties for defining European climate change risks and effects) (2001–2004) | Europe | Provide high resolution climate change scenarios for twenty-first century Europe | Eight RCMs |
| ENSEMBLES (ENSEMBLE-based predictions of climate change and their impact) (2004–2009) | Europe | Develop an ensemble prediction system to construct integrated scenarios of future climate change for quantitative risk assessment | RCM ensemble |
| CLARIS (Climate change assessment and impact studies) (2008–present) | South America | Predict climate changes and their socio-economic impacts | Five RCMs |
| NARCCAP (North American regional climate change assessment program) (2006–present) | North America | Explore the separate and combined uncertainties in regional climate simulations that result from the use of different GCMs and RCMs | Six RCMs |

complex modeling of physical processes. Unlike RCMs, which produce downscaled projections at a spatial scale of 20–50 km, statistical methods can provide station-scale climate information. However, this approach relies on the critical assumption: The statistical relationship between the predictor and predict and does not change over time. Generally, statistical downscaling methods can be classified into three main categories: Linear methods; Weather classifications and Weather generators. Regression-based statistical downscaling approaches have gained popularity out of the above three categories owing to their simplicity in application. The regression techniques widely used in statistical downscaling include Multi Linear Regression (MLR), Generalized Linear Models (GLMs), Artificial Neural (ANNs), Support Vector Machine (SVM), Random Forest Regression (RF). Owing to the learning abilities from data and their use in computer algorithms, techniques such as ANN, SVM, and RF are often called machine learning techniques.

Choosing an appropriate downscaling method depends on the desired spatial and temporal resolution of the climate information, as well as resource and time constraints. A thorough investigation of these factors is required, and assistance from experts is recommended. In general, statistical methods are most appropriate if time and financial resources are a constraint. Large climate institutions and regional assessment projects primarily use RCMs to investigate climate over larger areas.

Earth observation big data faces the challenge that a huge amount of data must be collected, characterized, and analyzed, accounting for the multi-source, multi-variable, and multi-scale data with the different spatial and temporal attributes. The climate information is available now for different time periods and different spatial and temporal resolutions, from the multiple data sources, such as weather stations, satellite remote sensing, radar, and physical models. Data fusion is a necessary processing to bring all data together in coherent framework in space and time. Take precipitation for an example, the input information of precipitation merging can include ground surface observations, radar, satellite retrieval, and atmosphere reanalysis rainfall as well as other relevant auxiliary variables. Precipitation merging began in the 1970s with the associated precipitation estimation combining radar and rain gauge measurements. With the emergence of global satellite and atmosphere reanalysis precipitation datasets since the late 1990s, the research on gauge—satellite or gauge—reanalysis rainfall merging has become increasingly active.

At present, numerous rainfall merging algorithms have been proposed. These algorithms are roughly classified into three categories, which include the initial field correction mode, the interpolation mode with auxiliary information, and the optimal matching mode. Table 4 shows some representative algorithms of the three modes, their basic principles and technical characteristics.

### *7.2.4 HASM*

In terms of the fundamental theorem of surfaces, a surface is uniquely defined by the first and the second fundamental coefficients. The first fundamental coefficients of a

**Table 4** Some representative data fusion algorithms

| Method | Basic principle | Technical feature |
|---|---|---|
| OA | The initial rainfall field is generated with remote sensing or analysis data, and then corrected with the weighted average of the difference between the surface observation and the initial value | An empirical local correction method. The correction weights are decided subjectively and the analyzed field is not the optimal result |
| OI | The initial field is corrected with the weighted average of the difference between the surface observation and the initial value in a certain spatial neighborhood. The optimal correction weights are obtained based on the criterion of minimizing error variance | A local optimal estimation method, avoiding the subjectivity of weight selection. Error of the observation field and background field and their spatial correlation need to be inferred in advance |
| KED | Remote sensing, reanalysis rainfall or other auxiliary variables are used to describe the local variation trend of rainfall. The effect of space trend components on the estimated values is reflected by the constraint conditions of the KED equations | It is necessary to determine the spatial variation function of the residual components. An iterative method or other special treatment is needed to solve the KED equations |
| GWRR | A local variable coefficient regression model and extends the spatial estimation of rainfall to 3 D space–time domain. The spatial–temporal correlation between precipitation and related influences is viewed as non-stationary through the spatial variable coefficients | Combination of the GWR method and the ridge regression method, is proposed and implemented to generate improved spatial climate variable's estimates by overcoming the collinearity problem of the pure GWR method |
| BMA | BMA takes any data as a possible member of the real state ensemble and measure the importance of each member by the posterior probability density. The weighted average of all the member is finally calculated as the analysis result | BMA could blend data more than two kinds and provide the measurement for estimation uncertainty by the posterior variance. The key of this method is the iterative solution of weight coefficient, and usually the solution method of expectation maximization is adopted |
| PDM | The analyzed rainfall is expressed as the weighted average of different original rainfall data, so that its probability density distribution has the maximum overlap with the original data | Strict theoretical assumption is lacking for PDM. Iterative process is required to search the weighted for various original data. PDM does not directly provide uncertainty measurement indexes for spatial estimation |

surface yield information about some geometric properties of the surface, by which we can calculate the lengths of curves, the angles of tangent vectors, the areas of regions, and geodesics on the surface. The second fundamental coefficients reflect the local warping of the surface, namely its deviation from the tangent plane at the point under consideration.

Suppose a surface can be expressed as $z = f(x, y)$, the first fundamental coefficients.

$E, F, G$ are as follows,

$$E = 1 + f_x^2, \qquad F = f_x \cdot f_y, \qquad G = 1 + f_y^2 \tag{1}$$

The second fundamental coefficients can be formulated as,

$$L = \frac{f_{xx}}{\sqrt{1 + f_x^2 + f_y^2}} \quad M = \frac{f_{xy}}{\sqrt{1 + f_x^2 + f_y^2}} \quad N = \frac{f_{yy}}{\sqrt{1 + f_x^2 + f_y^2}} \tag{2}$$

Partial differential equations of the surface which E, F, G, L, M, N must satisfy are the following Gauss equations:

$$\begin{cases} f_{xx} = \Gamma_{11}^1 f_x + \Gamma_{11}^2 f_y + \frac{L}{\sqrt{E+G-1}} \\ f_{yy} = \Gamma_{22}^1 f_x + \Gamma_{22}^2 f_y + \frac{N}{\sqrt{E+G-1}} \\ f_{xy} = \Gamma_{12}^1 f_x + \Gamma_{12}^2 f_y + \frac{M}{\sqrt{E+G-1}} \end{cases} \tag{3}$$

The computation procedure of HASM, taking the climatic sampled values as its constraints, can be changed into solving the following equality-constrained least squares problem (LSE),

$$\begin{cases} min \left\| \begin{bmatrix} A \\ B \\ C \end{bmatrix} x^{n+1} - \begin{bmatrix} d \\ q \\ p \end{bmatrix}^n \right\| \\ s.t. \quad Sx^{n+1} = k \end{cases} \tag{4}$$

where $S(l, (i-1) \cdot J + j) = 1$, $k(l) = \overline{f}_{i,j}$, this means the sampled value is $\overline{f}_{i,j}$ at the $l$th sampling point $(x_{i,} y_j)$.

One advantage of HSAM is its extensibility, due to this, HASM can be performed as an interpolation method, a downscaling method also an upscaling method. For HASM-SI, The climate field is obtained by summing the trend information obtained by some geostatics, and its corrections interpolated by using HASM. The driving field $x^0$ zero in this equation is obtained by subtracting trend information from station observations. The trend information can be acquired using some regression methods such as ordinary least squares regression (OLS), geographical weighted regression (GWR) and Bayesian statistical method, by integrating the explanatory variables.

$$\cdot \begin{cases} \mathbf{min} \left\| \begin{bmatrix} A \\ B \\ C \end{bmatrix} x^{n+1} - \begin{bmatrix} d \\ q \\ p \end{bmatrix}^n \right\| \\ s.t. \quad Sx_{station}^{n+1} = k \end{cases} \tag{5}$$
$$F = f(trend) + HASM(station\ observations - f)$$

As a downscaling method of HASM, we modified the equations by adding a downscaling scheme f due to the extensibility of HASM. The driving field $x^0$ zero is obtained by integrating the explanatory variables and station observations using some downscaling methods f, such as geographical weighted regression (GWR), random forest and Artificial neural networks (ANN):

$$\begin{cases} min \left\| \begin{bmatrix} A \\ B \\ C \end{bmatrix} x^{n+1} - \begin{bmatrix} d \\ q \\ p \end{bmatrix}^n \right\| \\ s.t. \quad Sx_{station}^{n+1} = k \\ f\left(x^0{}_{low}, covriate\right) = x^0{}_{high} \end{cases} \tag{6}$$

As a data fusion method, the driving field $x^0$ of HASM is obtained from the remote sensing, the climate model's output, or the weighted summation from multiple. The ground observations are used to be as a constraint condition.

$$\begin{cases} \min \left\| \begin{bmatrix} A \\ B \\ C \end{bmatrix} x_{remotesensingorclimatemodel'soutput}^{n+1} - \begin{bmatrix} d \\ q \\ p \end{bmatrix}^n \right\| \\ s.t.Sx_{station}^{n+1} = k \end{cases} \tag{7}$$

## 7.3 Change Trends and Scenarios of Climate Variables

### *7.3.1 Methods for Simulating Temperature in Beijing-Tianjin-Hebei Region (BTH)*

A combination of interpolation methods applied through statistical transfer functions (STFs) is an efficient approach to improve the estimation error of climatic variables for locations where primary data is not available. The spatial distribution of climate elements has the characteristics of structure and randomness. The observed climate value at each point in space is regarded as the sum of the climatic trend value and its error value at that point.

$$f_i = \theta_0 + \sum_{j=1}^{J} x_{i,j}\theta_j + \varepsilon_i \tag{8}$$

$i = 1, 2, ..., n$ is the number of grid cells. $j = 1, 2, ..., J$ is the number of explanatory variables. $x_{i,j}$ is the jth explanatory in the ith cell; $\theta_j$ is the jth regression coefficient. The statistical transfer function f can be obtained by using the ordinary least squares regression method (OLS), the Bayesian statistical method (Baye) or other regression methods.

In statistics, ordinary least squares (OLS) is a type of linear least squares method for estimating the unknown parameters in a linear regression model. OLS chooses the parameters of a linear function of a set of explanatory variables by the principle of least squares: minimizing the sum of the squares of the differences between the observed dependent variable in the given dataset and those predicted by the linear function. For OLS, $\boldsymbol{\theta} = \left(\mathbf{X}^T\mathbf{X}\right)^{-1}\mathbf{X}^T\mathbf{f}$. The assumption of OLS that climate element and explanatory variables are correlated in a globally stationary way and that the residuals are subject to independent normal distribution does not comply with the reality.

To find the Bayesian solution, the parameter's prior distribution is usually specified as a conjugate prior distribution. Supposing that an isotropic Gaussian prior $p(\alpha|\beta) = G(\alpha; 0, \beta I)$ with mean zero is conjugate, the posterior distribution of $\alpha$ is also Gaussian and can be given as follows (Bishop 2006):

$$p(\alpha|tem\_sim, \beta, \sigma^2) = \frac{p(tem\_sim|\alpha, \sigma^2)p(\alpha|\beta)}{p(tem\_sim|\beta, \sigma^2)} \propto G(\alpha; m, S), \tag{9}$$

$$m = \sigma^2 S(tem\_obs)^T tem\_sim, \tag{10}$$

$$S = (\beta I + \sigma^{-2}(tem\_obs)^T tem\_obs)^{-1}. \tag{11}$$

where $p$ is the density function.$G$ is the Gaussian distribution.$m$ is the Bayesian estimation for weight vector $\alpha$.$S$ and $\beta$ are the covariance matrix and the variance of $\alpha$, respectively.$\sigma^2$ is the Gaussian noise variance (Qin et al. 2013).

To estimate $m$, we should solve for parameters $\beta$ and $\sigma^2$. In this research, we use the maximum likelihood estimation method (Lu et al. 2015). The marginal likelihood function is

$$\begin{aligned} p(tem\_sim|\beta, \sigma^2) &= \int p(tem\_sim|\alpha, \sigma^2)p(\alpha|\beta)d\alpha \\ &= (2\pi)^{-M/2}|C|^{-1/2}\exp(-\frac{1}{2}(tem\_sim)^T C^{-1} tem\_sim), \end{aligned} \tag{12}$$

where $C = \sigma^2 I + \beta^{-1}(tem\_obs)^T(tem\_obs)$.

Taking the logarithm base $e$, we find

$$\begin{aligned} \ln p(tem\_sim|\beta, \sigma^2) &= \frac{M}{2}\ln\beta + \frac{N}{2}\ln\sigma^2 - \frac{1}{2}\ln|S^{-1}| \\ &- \frac{\sum_{n=1}^{N}(tem\_sim_n - m^T tem\_obs_n)}{2\sigma^2} - \frac{\beta m^T m}{2} - \frac{N}{2}\ln(2\pi). \end{aligned} \tag{13}$$

From (13), we have

$$\frac{d\ln p(tem\_sim|\beta,\sigma^2)}{d\beta} = \frac{M}{2\beta} - \frac{1}{2}tr(S) - \frac{m^T m}{2} \quad (14)$$

Letting the left side of (14) be zero, we have

$$\beta = \frac{M}{m^T m + tr(S)}. \quad (15)$$

Similarly, setting $d\ln p(tem\_sim|\beta,\sigma^2)/d\sigma^2 = 0$, we find

$$\sigma^2 = \frac{N}{\sum_{n=1}^{N}(tem\_sim_n - m^T tem\_obs_n) + tr(tem\_obs^T tem\_obsS)}, \quad (16)$$

where $tr(.)$ is the trace of a matrix. To maximize this marginal likelihood function with respect to $\beta$ and $\sigma^2$, we apply the expectation maximization (EM) algorithm, iterating 1000 times to find the global maximum solution (Chen and Martin 2009).

To produce the temperature field, we use a mixed interpolation technique that combines multiple regression with residual modification by HASM. Several factors influence the climate for a specific place. The temperature in a region is mainly affected by the interaction of important natural controls, e.g., longitude, latitude, elevation, aspect, slope, and orographic factors, especially over a long time scale (Ninyerola et al. 2007). This method not only produces accurate results but also allows us to explore the relationship between temperature and climatic factors (Apaydin et al. 2011).

Including the above 14 stations, 329 stations are used to produce the annual mean temperature field in the BTH region. Eighty percent of the stations are used to interpolate the temperature surfaces, and the remaining 20% are used for validation. A multiple regression with residual correction method was used in this section.

We compared different combinations to find the optimal technique. The regression methods were the OLS and Bayesian statistical method (Bayes). The residual methods included HASM and Kriging, since Kriging remains one of the most preferred stochastic surface interpolation techniques. We first used a stepwise regression method to select the most influential factors from longitude, latitude, digital elevation model (DEM), the sky view factor (Svf), the modified aspect (Asp), and the distance to the coastline (Dis). The formulas for Svf and Asp can be found in Yue (2011). Following are the regression equations for the OLS and Bayes methods, respectively:

$$\begin{aligned} tem_{\mathrm{OLS}} = {} & 35.65 - 8.87 \times 10^{-6} \times lon - 3.40 \times 10^{-5} \times lat - 0.0057 \times DEM \\ & - 0.10 \times Asp + 2.98 \times 10^{-5} \times Dis \end{aligned}$$

$$\begin{aligned} tem_{\mathrm{Bayes}} = {} & 35.51 - 8.96 \times 10^{-6} \times lon - 3.43 \times 10^{-5} \times lat \\ & - 0.0057 \times DEM - 0.10 \times Asp + 3.02 \times 10^{-5} \times Dis \end{aligned}$$

**Table 5** Comparison of different interpolation methods

| Errors | OLS | OLS-Kriging | OLS-HASM | Bayes | Bayes-Kriging | Bayes-HASM |
|---|---|---|---|---|---|---|
| Bias | 0.16 | 0.06 | 0.19 | 0.14 | −0.02 | 0.13 |
| MAE | 0.75 | 0.64 | 0.41 | 0.71 | 0.58 | 0.37 |
| RMSE | 0.94 | 1.13 | 0.52 | 0.91 | 1.03 | 0.47 |
| r | 0.95 | 0.94 | 0.99 | 0.95 | 0.95 | 0.99 |

The goodnesses-of-fit measures for both methods were 95.57% and 95.61%, respectively, with a significance level of 0.05. The residuals produced by the two methods could then be modified by HASM or Kriging. The final results were obtained by summing the regression results and the modified residuals.

Table 5 lists the errors of the different methods. It can be seen that the trend surface produced by the Bayesian method is more accurate than that of the OLS method. The reason is that OLS depends only on the sample information, while Bayesian combines empirical knowledge with the sample information. The better results were obtained by the mixed interpolators. Bayes-Kriging had the lowest Bias. However, from other error indexes, Bayes-HASM produced the lowest errors, followed by OLS-HASM, which indicates the critical importance of the residual modification by HASM. From RMSE, OLS-Kriging and Bayes-Kriging have larger values than the other methods, which suggests that Kriging produces extreme values in this study.

Figure 2 displays the simulation results of the annual mean temperature for the period 1980–2014. Since the mixed interpolators produced better results than the regression method, we only display their results to provide detailed information on the annual mean temperature. It can be seen that the temperature fields are consistent with each other. The annual mean temperature in the BTH region has a remarkable change from less than −3 °C in the northwest to more than 17 °C in the southeast, which agrees well with other research (Domroes and Peng 1998). Since the Taihang and Yan mountains are in the northwest of BTH, the annual temperature in this area ranges from −3 °C to about 7 °C. All of the methods reflect the actual situation of the annual mean temperature in BTH region. However, more accurate information can be obtained from the Bayes-HASM method.

### *7.3.2 Methods for Simulating Future Precipitation in Poyang Lake Basin*

To investigate the future precipitation changes with high accuracy in Poyang Lake basin, a two-step downscaling method was designed. First, the CMIP6 outputs during the historical period of 1980–2014 were downscaled in the Poyang Lake basin by using the geographical weighted regression method (GWR) together with a residual correction technique using kriging with external drift (KED). Second, by combining a time transfer matrix, the downscaling method GWR-KED was used to downscale

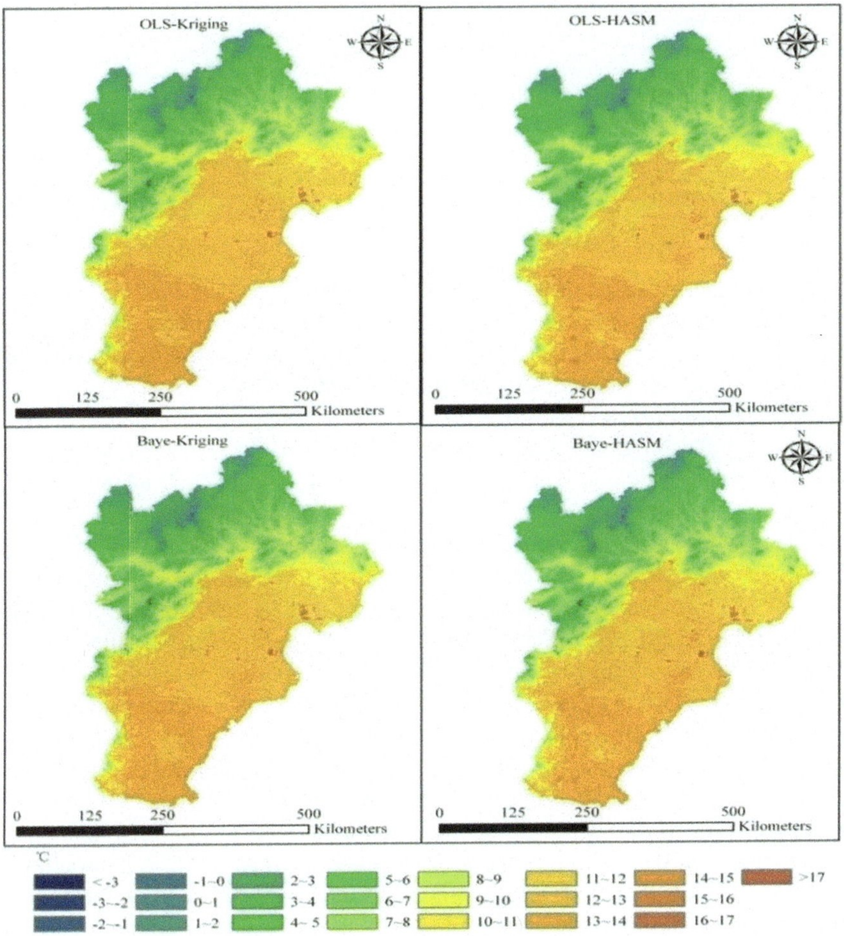


**Fig. 2** Annual mean temperature produced by different mixed interpolators

the precipitation in the future scenarios (SSP1-2.6, SSP2-4.5, SSP3-7.0, and SSP5-8.5) for the period of 2015–2019. This established scale-based downscaling method was validated by using meteorological observations for the period 2015–2019 and was finally applied to downscale precipitation for future projections (2020–2100).

For the statistical downscaling method, an accurate exploration of the relationship between precipitation and its influencing factors is crucial to obtain high accuracy downscaled results. Other than the atmosphere–ocean-land interactions, the topographic structure and the Poyang Lake shape a complex and regionally variable precipitation climate in the study area. A spatial variable relationship could be more suitable for downscaling precipitation in the Poyang Lake basin, which can be constructed as follows:

$$CMIP6_{T_0}(s,t) = \beta_o(s,t) + \beta_i(s,t) * f_i(s,t) + \varepsilon(s,t), \tag{17}$$

where $f_i(s,t)$ with $i = 1, 2, \cdots n$ represents the influencing factors at spatial location $s$ at time $t$(spring, summer, autumn, winter, and annual); $\beta_o(s,t)$ is the intercept; $\beta_i(s,t)$ is the corresponding coefficient of $f_i(s,t)$; and $\varepsilon(s,t)$ is the residual of the regression method. Relationship (17) can be implemented by applying the widely used GWR method, which is a local spatial statistical method designed to address spatial non-stationarity Brunsdon et al. 1996). Many studies have demonstrated that this regression method combined with the residual correction algorithm can obtain improved results, as part of the precipitation variability may not be explained by the regression method alone (Alexakis & Tsanis 2016; Zhao et al. 2018; Sharifi et al. 2019; Zhang et al. 2020). The bias of the GWR method $\varepsilon^{1^o}(u,t)$ was corrected by using the KED method with gauge observations in this study. In addition, it is critical to identify the most influential factors in the regression method. The coarse-resolution precipitation values in the central and adjacent grid cells are considered the most likely important predictors (He et al. 2016). Topographical and geographical factors, including altitude, slope, aspect, surface roughness and distance to the coastline (defined as the Euclidean distance of a given location to the nearest coastline), were also employed to explore their interactions with precipitation. In addition, as previous studies have shown, seasonal precipitation was significantly correlated with the normalized difference vegetation index (NDVI) (Kang et al. 2014; Chen et al. 2020). The most influential factors were selected in this study by using the stepwise regression method in each time period.

Based on the GWR-kriging downscaling method, we finally obtained high spatial resolution precipitation data in future scenarios by using a time transfer matrix, which can be constructed as follows:

$$C_{i,j} = \frac{\Pr e(T_i, S_j)}{\Pr e(T_0)} \tag{18}$$

where $C_{i,j}$ is the time transfer matrix at time $i$ in scenario $j$; $\Pr e(T_0)$ represents the CMIP6-GCM simulations in the historical period (1980–2014); $\Pr e(T_i, S_j)$ is the CMIP6 outputs at the future time period $i$ in scenario $j$; $i = 1,\ 2,\ 3,\ 4$ represents the periods of 2015–2019, 2020–2049, 2050–2079, 2080–2100, respectively; and $j = 1,\ 2,\ 3,\ 4$ represents the scenarios of SSP1-2.6, SSP2-4.5, SSP3-7.0, and SSP5-8.5, respectively.

After interpolating $C_{i,j}$ from 1° to 0.01° by using a simple bilinear method, the downscaling method for future scenarios can be given as:

$$CMIP6_{T_i} = C_{i,j} \cdot CMIP6_{T_0} \tag{19}$$

Similar to the process reported by Huffman et al. (2007), Eq. (19) was established based on the assumption that the ratio of future precipitation to historical precipitation remains the same before and after downscaling. Here, $CMIP6_{T_0}$ is the downscaled result yielded from Eq. (17) in the historical period $T_0$.

The GWR method is frequently used for downscaling coarse precipitation datasets due to its ability to deal with data with spatial non-stationarity. Moreover, several studies have shown that the residual corrections of GWR can yield improved results compared with the GWR method alone (Hutengs and Vohland 2016; Chen et al. 2018; Yue et al. 2019; Zhao 2021). To obtain the final results, we added the modified residuals back to the GWR results by using interpolators combined with the station values. The performance of different interpolators, including ordinary kriging, KED, inverse distance weighted method (IDW), and spline, were first compared based on the MAE, which is the common index used to measure overall accuracy (Table 6). Local evaluation, aspect, and slope were used as the auxiliary variables in KED. The results showed that using the residual corrections in the downscaling process improves the accuracy of seasonal and annual precipitation fields. Models generated with the help of auxiliary variables gave more accurate results, and methods, including GWR-OK, GWR-IDW, and GWR-Spline, which consider only the residual value at the stations performed worse than GWR-KED with secondary data. The inclusion of residual corrections with local auxiliary information improves precipitation predictions. Therefore, the GWR-KED was used as the primary downscaling method in this study.

The GWR-KED based statistical downscaling method was used to downscale the CMIP6 outputs to a 1 km resolution during the period of 1985–2014. Figure 3 compares the original CMIP6 products and the downscaling results, revealing that all the downscaling precipitation fields have similar spatial patterns and distributions to the original CMIP6 outputs, yet the latter has more detailed information than the former due to the inclusion of the fine-scale explanatory variables in the downscaling process. In addition, the precipitation characteristics in the Poyang Lake basin are shown in Fig. 3. Significant differences in the spatial patterns of precipitation can be observed in different time periods. The mean differences between the original CMIP6 outputs and the downscaling results in the Poyang Lake basin were 13.55, 34.81, 40.58, 13.47, and 4.72 mm at the annual, spring, summer, autumn, and winter, scales respectively. Table 7 further gives the accuracy of the downscaling results, which showed that the accuracy of the downscaling results was improved by 60.30, 60.69, 19.59, 52.01, and 75.30%, for spring, summer, autumn, winter and annually, respectively, based on MAE. The errors at the spring, summer, winter and annual scales notably improved, while in autumn, the MAE and RMSE values decreased by 19.59% and 23.37%, respectively.

**Table 6** Comparison of different downscaling methods based on MAE (mm)

| Downscaling methods | Spring | Summer | Autumn | Winter | Annual |
|---|---|---|---|---|---|
| GWR | 76.99 | 69.87 | 47.21 | 59.06 | 105.16 |
| GWR-OK | 63.15 | 62.00 | 38.98 | 49.33 | 100.01 |
| GWR-KED | 56.02 | 51.81 | 29.18 | 33.50 | 87.96 |
| GWR-IDW | 65.32 | 61.54 | 39.00 | 49.88 | 98.06 |
| GWR-SPLINE | 69.18 | 66.72 | 41.07 | 54.26 | 103.42 |

**Table 7** Accuracy of precipitation fields before and after downscaling

| Error (mm) | Precipitation | Spring | Summer | Autumn | Winter | Annual |
|---|---|---|---|---|---|---|
| MAE | CMIP6 | 141.10 | 131.79 | 36.29 | 69.80 | 356.07 |
| | Pre_down | 56.02 | 51.81 | 29.18 | 33.50 | 87.96 |
| RMSE | CMIP6 | 155.20 | 151.71 | 42.75 | 72.03 | 378.32 |
| | Pre_down | 62.22 | 58.51 | 32.76 | 36.02 | 94.31 |

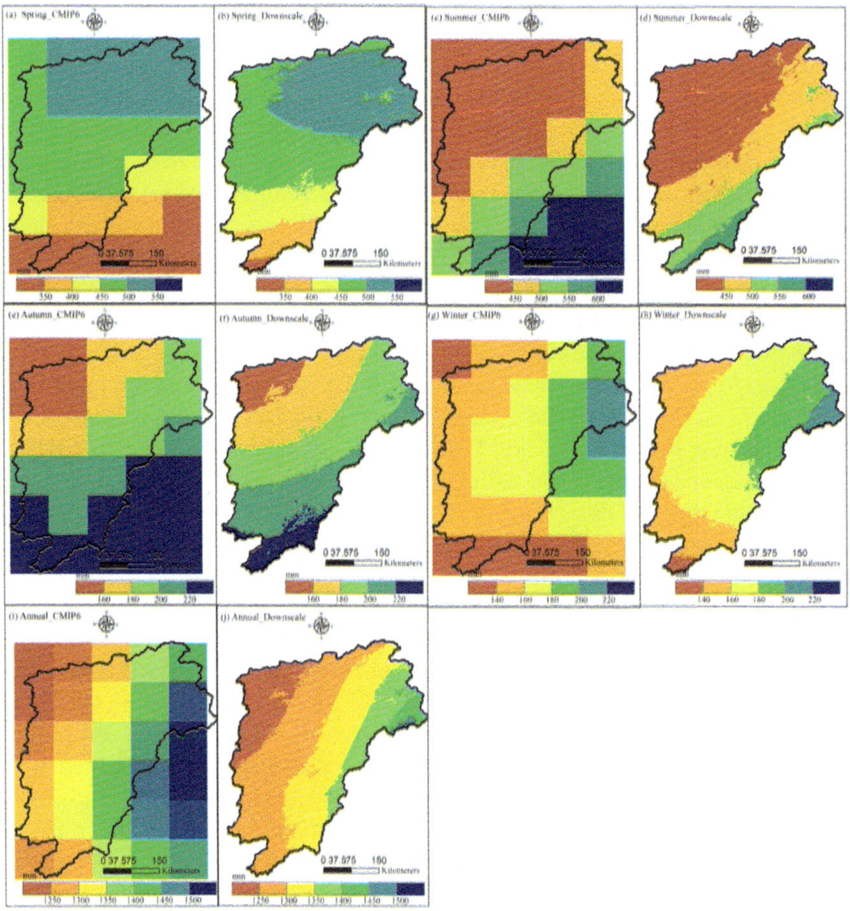

**Fig. 3** Comparison of the original CMIP6 products and the downscaling results during 1985–2014

The final established scale-based downscaling method was validated during the period 2015–2019 under the SSP1-2.6, SSP2-4.5, SSP3-7.0, and SSP5-8.5 scenarios. Figures 4 and 5 display the downscaling results in the SSP2-4.5 and SSP5-8.5 scenarios as examples. Similar to the downscaling results in the historical period, the downscaled precipitation fields exhibit similar spatial patterns to the original CMIP6 outputs but show more detailed information. Compared with the precipitation distribution in the period 1980–2014, significant differences in the spatial patterns of precipitation can be observed in the SSP2-4.5 and SSP3-7.0 scenarios, especially at the winter and annual scales. Precipitation at the annual scale generally exhibited decreasing trends from east to west in the period of 1980–2014, with values ranging from 1193.18 to 1651.27 mm, while precipitation showed north–south spatial patterns, with maximum precipitation occurring in the central region under the SSP2-4.5 and SSP3-7.0 scenarios, with annual precipitation amounts ranging from 988.11 to 1505.56 mm and 1087.00 mm to 1408.46 mm. Precipitation at the winter scale exhibited a west–east spatial pattern, changing from 129.39 to 236.07 mm, and 100.47 mm to 244.50 mm in the historical and validated periods in the SSP2-4.5 scenario, respectively, while precipitation in the SSP3-7.0 scenario shows a south-north distribution, with values changing from 112.26 to 202.21 mm. Although the north–south distributions were observed in both the historical and validated periods in autumn, precipitation in the SSP3-7.0 scenario displays a larger difference than that in the other two scenarios. The precipitation amounts in autumn range from 143.72 to 233.11 mm, 114.05 mm to 300.05 mm, 53.71 mm to 333.48 mm for historical and validated periods in the SSP2-4.5 and SSP3-7.0 scenarios, respectively. Precipitation in summer showed an increasing pattern from north to south in the historical period and validated periods in the SSP2-4.5 and SSP3-7.0 scenarios, while precipitation in the SSP3-7.0 scenario exhibited a larger difference, with most of the regions having values changing between 500 and 550 mm. In spring, precipitation in the SSP3-7.0 scenario also showed significant differences compared with the other two scenarios (Fig. 4a), and the rainfall belt had maximum values ranging from 400 to 450 mm, while the maximum rainfall belts changed from 450 to 500 mm in the SSP2-4.5 scenario and the historical period. Precipitation in the SSP3-7.0 scenario shows a larger difference than that in the other scenarios. These obvious differences provide a basis for assessing the variations in precipitation in the future, since the precipitation change analyses at fine scales were based on the scale-based downscaling method.

More importantly, the uncertainty of precipitation change in future scenarios is also based on the accuracy of the scale-based downscaling method established in the validated period under different scenarios. The accuracy of the scale-based downscaling method is listed in Table 7. The downscaling accuracy varies with different time scales. In spring, the downscaled precipitation field with the highest accuracy was observed under the SSP1-2.6 scenario, with MAE and RMSE values of 72.81 mm and 83.76 mm, respectively. According to the MAE values, the downscaling accuracy in spring improved by 57.22%, 61.39%, 59.39% and 59.30% under the SSP1-2.6, SSP2-4.5, SSP3-7.0 and SSP5-8.5 scenarios, respectively. The precipitation field with the highest accuracy was produced under the SSP3-7.0 scenario in summer, with MAE and RMSE values of 69.81 mm and 72.17 mm, respectively. However, in summer, the downscaled method performed the best under the SSP1-2.6 scenario, with MAE and RMSE reduced by 66.81% and 67.15%, respectively, compared with the original CMIP6 outputs. In addition, the accuracy of precipitation

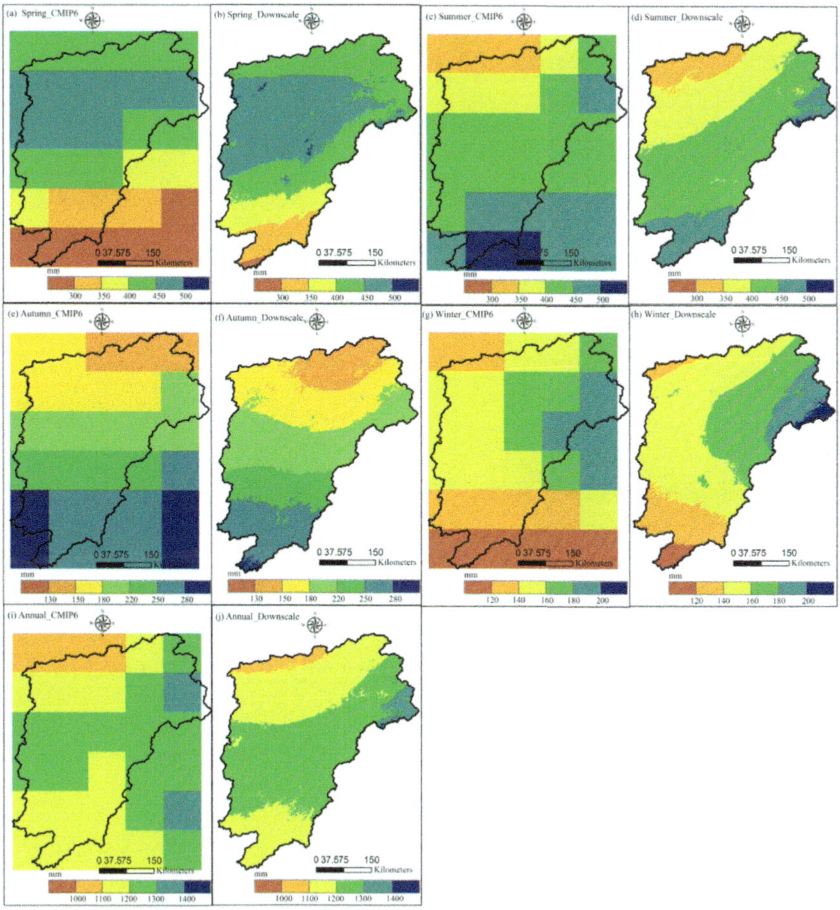


**Fig. 4** Downscaled precipitation under the SSP2-4.5 scenario in the validated period of 2015–2019

in summer improved by 62.96%, 55.26%, and 59.25% under the SSP2-4.5, SSP3-7.0, and SSP5-8.5 scenarios, respectively, based on MAE values. In autumn, the MAE values decreased by 42.96%, 42.71%, 37.49%, and 47.04%, under the SSP1-2.6, SSP2-4.5, SSP3-7.0 and SSP5-8.5 scenarios, respectively compared with the original CMIP6 products; the precipitation field with the highest accuracy occurred under the SSP3-7.0 scenario, with MAE and RMSE values of 47.10 mm and 51.96 mm, respectively. In winter, the accuracy of the downscaled precipitation field improved by 42.06%, 39.12%, 42.15%, and 45.15% under the SSP1-2.6, SSP2-4.5, SSP3-7.0, and SSP5-8.5 scenarios, respectively, compared with the CMIP6 datasets, and a high accuracy precipitation field was found in the SSP5-8.5 scenario, with MAE and RMSE values of 43.32 mm and 46.68 mm, respectively. In terms of the MAE value, the accuracy of annual precipitation improved by 75.20%, 74.95%, 75.08%,

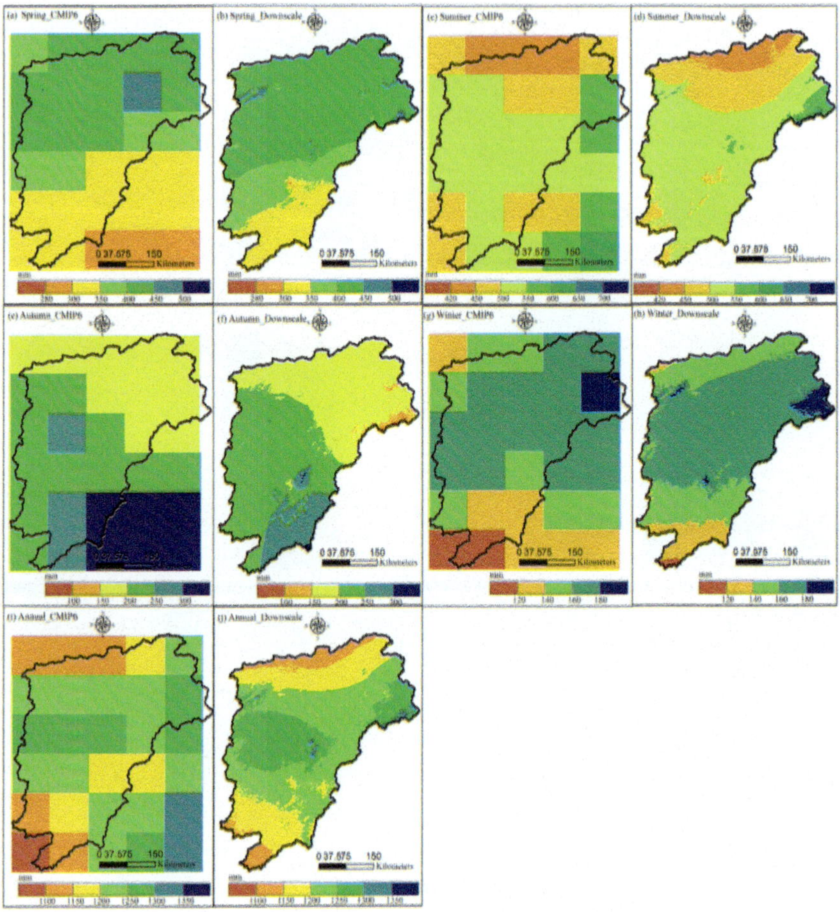

**Fig. 5** Downscaled precipitation under the SSP3-7.0 scenario in the validated period of 2015–2019

and 75.38% under SSP1-2.6, SSP2-4.5, SSP3-7.0, and SSP5-8.5 scenarios, respectively. In addition, compared with the seasonal scales, the accuracy of precipitation at annual scales improved the most. In other words, the scale-based downscaling method performs the best at the annual scale, followed by summer and spring, which both have relatively large amounts of precipitation. Moreover, the time transfer matrix did not introduce large errors in the final downscaling results based on Tables 7 and 8, particularly for the results in autumn, where the scale-based downscaling method performed better in the validated period than the GWR-KED method that was carried out in the historical period.

The scale-based downscaled method was then applied for the CMIP6 outputs under the SSP1-2.6, SSP2-4.5, SSP3-7.0, and SSP5-8.5 scenarios. Table 8 shows the error levels under different scenarios by using the scale-based downscaling method. Under these error levels, the precipitation changes in future scenarios are given for the

**Table 8** Accuracy of the downscaled precipitation under different scenarios

| Scenarios | Error (mm) | Precipitation | Spring | Summer | Autumn | Winter | Annual |
|---|---|---|---|---|---|---|---|
| SSP1-2.6 | MAE | CMIP6 | 170.23 | 300.65 | 115.96 | 119.02 | 638.02 |
| | | Pre_down | 72.81 | 99.77 | 66.14 | 68.96 | 158.25 |
| | RMSE | CMIP6 | 182.44 | 312.36 | 125.62 | 121.85 | 655.79 |
| | | Pre_down | 83.76 | 102.62 | 74.26 | 70.31 | 176.20 |
| SSP2-4.5 | MAE | CMIP6 | 197.03 | 248.62 | 84.96 | 104.72 | 628.14 |
| | | Pre_down | 76.07 | 92.10 | 48.67 | 63.75 | 157.36 |
| | RMSE | CMIP6 | 210.62 | 264.50 | 94.23 | 107.63 | 644.78 |
| | | Pre_down | 85.47 | 98.00 | 52.09 | 68.00 | 164.99 |
| SSP3-7.0 | MAE | CMIP6 | 226.22 | 156.05 | 75.35 | 99.32 | 610.02 |
| | | Pre_down | 91.87 | 69.81 | 47.10 | 57.46 | 152.00 |
| | RMSE | CMIP6 | 236.24 | 176.02 | 89.02 | 103.63 | 629.17 |
| | | Pre_down | 99.04 | 72.17 | 51.96 | 64.74 | 163.38 |
| SSP5-8.5 | MAE | CMIP6 | 179.62 | 181.32 | 128.50 | 78.97 | 597.33 |
| | | Pre_down | 73.11 | 73.88 | 68.05 | 43.32 | 147.09 |
| | RMSE | CMIP6 | 197.30 | 200.85 | 139.36 | 82.49 | 616.90 |
| | | Pre_down | 81.54 | 84.16 | 71.32 | 46.68 | 154.92 |

future period of 2020–2100. The change trends of precipitation at a spatial resolution of 1 km were obtained by using the well-known Mann–Kendall statistical test and Sen's estimator (Chatterjee et al. 2016).

Under the SSP1-2.6 scenario (Fig. 6), 72.68% of the Poyang Lake basin would be drier in spring, with a maximum decreasing change rate of 0.77 mm/year. The maximum change rate of precipitation in spring was 0.75 mm/year and occurred in the north, and the mean change rate was −0.15 mm/year for the whole study region (Fig. 6a). In summer, 82.93% of the study area would be wetter and the change rate of precipitation was 0.36 mm/year on average, with a changing trend of precipitation from −0.26 mm/year in the north to 1.07 mm/year in the south (Fig. 6b). In autumn, the basin would be wetter, and precipitation would change from 0.06 to 0.82 mm/year with a mean change rate of 0.38 mm/year (Fig. 6c). In winter, precipitation would increase by 0.47 mm/year on average, with a maximum increasing rate of 0.59 mm/year and a maximum decreasing rate of 0.20 mm/year. In the west and east, occupying by 36.13% of the Poyang Lake basin, precipitation changes occurred at rates of greater than 0.5 mm/year (Fig. 6d). The annual precipitation increases from 0.30 to 2.95 mm/year across the whole study area, with the maximum changing rate occurring in the south. The average changing rate of annual precipitation is 1.32 mm/year for the whole basin (Fig. 6e).

Under the SSP2-4.5 scenario, the entire Poyang Lake basin would be wetter in spring, with precipitation increasing at a rate of 0.85 mm/year on average. The minimum and maximum increasing rates were 0.44 mm/year and 1.00 mm/year, respectively, and 56.24% of the total area showed a wetting trend with a changing

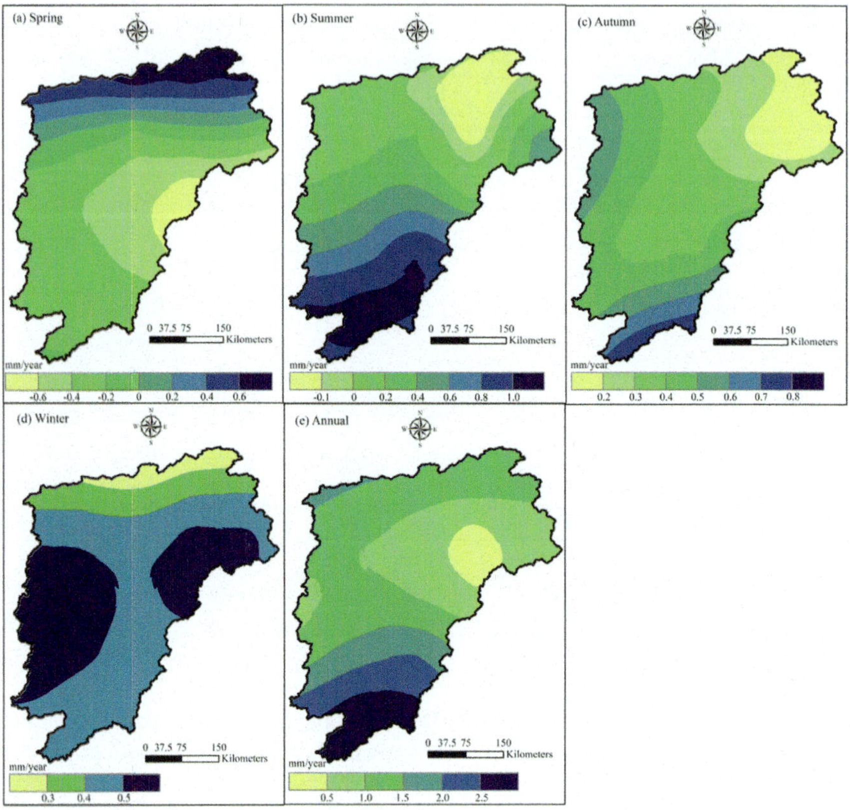


**Fig. 6** The change trend of precipitation under the SSP1-2.6 scenario during 2020–2100

rate greater than 0.85 mm/year (Fig. 7a). In summer, 62.32% of the Poyang Lake basin would be drier, and precipitation would decrease at a rate of 0.02 mm/year. The increasing trends mainly occurred in the south, with a maximum changing rate of 0.62 mm/year. In autumn, precipitation exhibits increasing trends in the south and decreasing trends in the west, with a maximum change rate of 0.89 mm/year and a minimum change rate of −0.21 mm/year. A total of 80.04% of the total study area would be wetter, and the regional average change rate was 0.25 mm/year (Fig. 7c). In winter, 69.56% of the total basin would be wetter. The change rate of precipitation ranges from −0.17 mm/year in the east to 0.13 mm/year in the west, and 66.18% of the region exhibits increasing trends with rates between 0 and 1 mm/year. The annual precipitation increases across the whole basin at rates of 1.14–1.64 mm per year (Fig. 7e).

Under the SSP3-7.0 scenario, the entire Poyang Lake basin would be wetter in spring, and precipitation would change at rates of 0.49–1.87 mm/year, with a mean change rate of 1.18 mm/year. The maximum change rate occurred in the north while the minimum change rate occurred in the south (Fig. 8a). In summer, precipitation

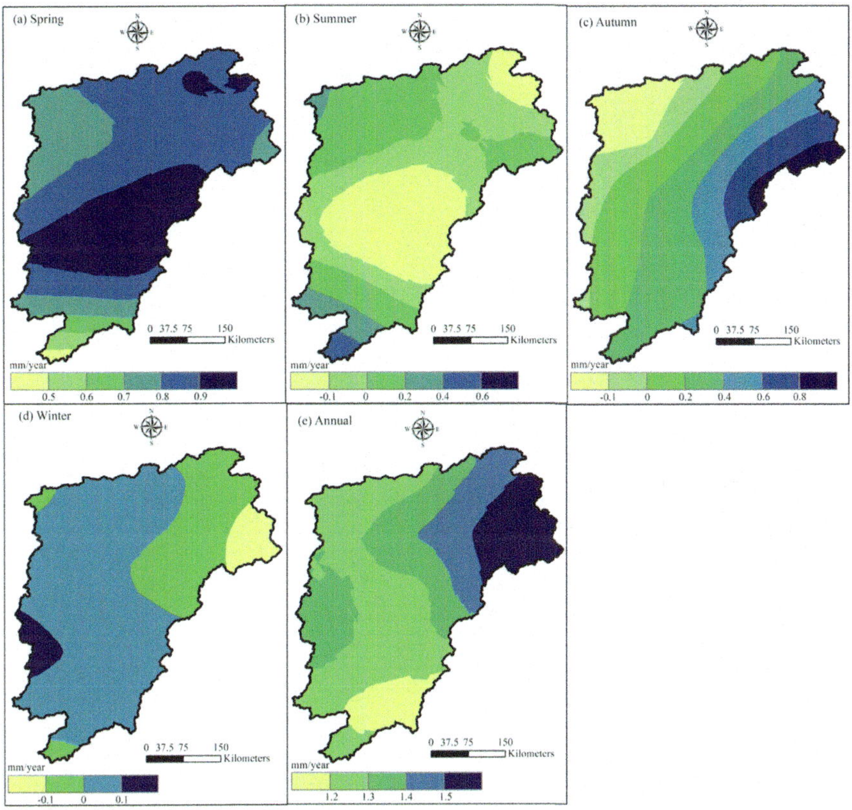


**Fig. 7** The change trend of precipitation under the SSP2-4.5 scenario during 2020–2100

changes at a rate of −0.17 to 1.54 mm/year, increases in the east, and decreases in the west. 97.18% of the total area would be wetter (Fig. 8b). In autumn, precipitation shows increasing trends across the whole study area, with a maximum changing rate of 1.10 mm/year occurring in the south, and a minimum changing rate of 0.25 mm/year occurring in the north (Fig. 8c). In winter, 47.47% of the Poyang Lake basin would be wetter, with a maximum increasing rate of 0.48 mm/year, which occurred in the north. Decreases in precipitation would mainly occur in the southeastern part of the region, with a maximum decreasing rate of 0.53 mm/year (Fig. 8d). The annual precipitation would increase at rates of 0.93 to 4.09 mm/year across the whole area, with a regional mean increasing rate of 2.25 mm/year and precipitation in the north would increase the most (Fig. 8e).

Under the SSP5-8.5 scenario, the Poyang Lake basin becomes wetter in spring, and precipitation increases at rates of 0.80 to 1.61 mm/year, with a regional mean increasing rate of 1.24 mm/year. Precipitation in the west increases the most (Fig. 9a). In summer, 52.58% of the total study area would be wetter, and precipitation in the central region would decrease with a maximum decreasing rate of 0.5 mm/year.

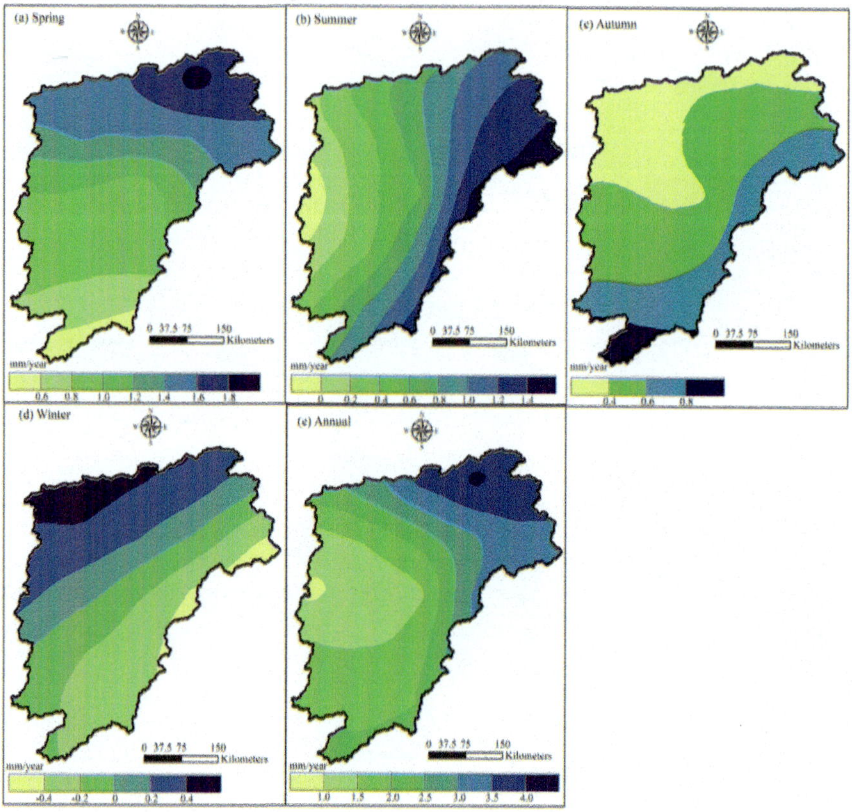


**Fig. 8** The change trend of precipitation under the SSP3-7.0 scenario during 2020–2100

Precipitation in the north would increase the most at a rate of 1.04 mm/year (Fig. 9b). In autumn, the entire basin would be wetter, and precipitation in the northwest would increase the most, with a maximum increasing rate of 0.93 mm/year (Fig. 9c). In winter, 63.37% of the Poyang Lake basin would be drier, mainly located in the north. Precipitation in the south shows increasing trends, with a maximum changing rate of 0.11 mm/year (Fig. 9d). The annual precipitation would increase across the whole region, with a maximum changing rate of 3.26 mm/year, a minimum increasing rate of 1.00 mm/year, and a regional average increasing rate of 2.11 mm/year (Fig. 9e).

In general, precipitation in spring would be increase over the entire basin under all future scenarios except SSP1-2.6, and the largest regional-average increasing trend occurs in SSP3-7.0, with an average changing rate of 1.18 mm/year. Parts of the Poyang Lake basin exhibit drier trends under the four scenarios, with the largest drying area occurring in the SSP2-4.5 scenario, covering 62.32% of the entire basin. In the SSP1-2.6 and SSP2-4.5 scenarios, the regions with drying trends occur in the northern part, while these regions occur in the western and central parts of the basin in SSP3-7.0 and SSP5-8.5, respectively. Except in the SSP2-4.5 scenario, precipitation

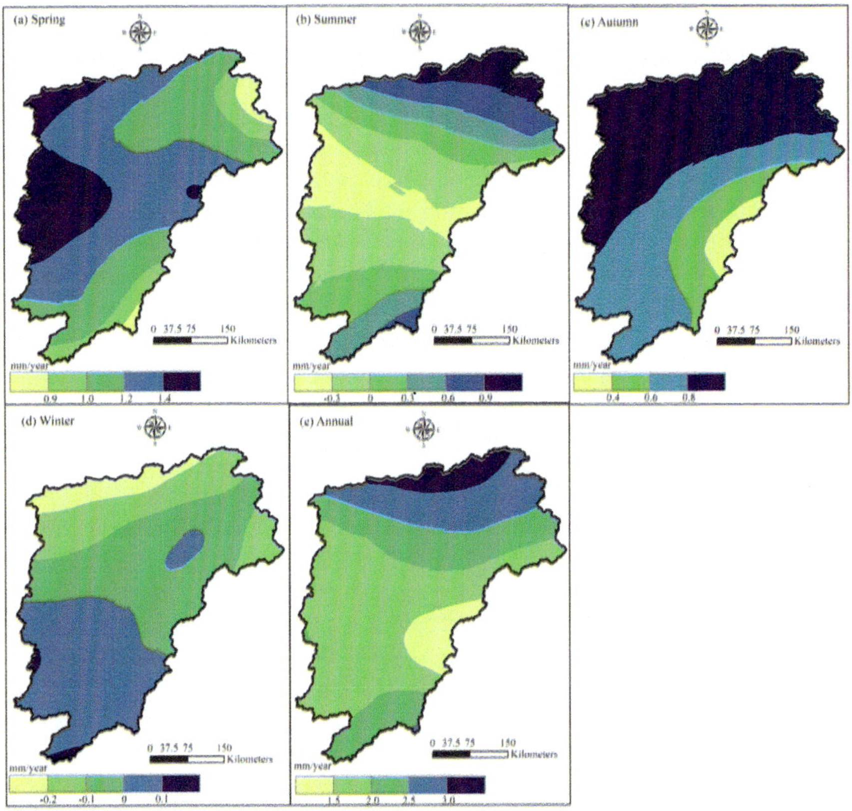


**Fig. 9** The change trend of precipitation under the SSP5-8.5 scenario during 2020–2100

in autumn has increasing trends in all scenarios across the study area, with the largest increasing rate being 0.75 mm/year on average. Precipitation in winter shows drying trends in some areas except in the SSP1-2.6 scenario, with the largest area occurring in SSP5-8.5, occupying 63.37% of the total basin. The annual precipitation exhibits increasing trends across the entire basin under the four scenarios, with the largest rising rate of 2.25 mm/year occurring in the SSP3-7.0 scenario.

## 7.4 Understanding Climate Events

An extreme event is one that is rare at a particular place and/or time of year. Definitions of 'rare' vary, but an extreme event would normally be as rare as or rarer than the 10th or 90th percentile of a probability density function estimated from observations. When a pattern of extreme weather persists for some time, such as a season, it may be classified as an extreme climate event, especially if it yields an average or total that is itself extreme (e.g., drought or heavy rainfall over a season). For some climate extremes such as drought, floods and heat waves, several factors such as duration and intensity need to be combined to produce an extreme event.

Different types of climate extremes are projected to intensify and become more frequent in a number of regions worldwide due to climate change. Extreme events, such as droughts and heat waves, can adversely impact agricultural production and have implications for the livelihoods and food security of communities. A recent example of the impacts of climate extremes on agricultural productivity is the European heat wave and drought in summer 2018, which lead to widespread harvest failures and shortages of fodder for livestock in many countries across the continent. Extreme weather and climate-related events affect human health by causing death, injury, and illness, as well as having large socioeconomic impacts. During 1980–2016, the annual average number of billion-dollar events in US was 5.5. For the most recent 5 years (2012–2016), the annual average is 10.6 events. The year 2005 was the costliest since 1980 due to the combined impacts of Hurricanes Katrina, Rita, Wilma, and Dennis, as losses exceed $200 billion.

27 major climate extreme indices are defined by the Expert Team on Climate Change Detection and Indices (ETCCDI). The ETCCDI has a mandate to facilitate the international coordination of the calculation of a suite of temperature and precipitation indices for the monitoring and assessment various aspects of changes in climate extremes. This team has focused on the following aspects including (1) the development of a set of indices and software that computes the indices, (2) capacity building through hands-on training of data analysis by organizing regional climate change workshops, (3) coordinating the calculation of indices for both observed and climate model simulated data, and (4) contributing to climate change science.

The 27 major climate extreme indices Annual count of days when precipitation is greater than or equal to "n mm", n is a user defined threshold (Rn mm); Annual count of days when precipitation is greater than or equal to 10 mm (R10 mm); Annual count of days when precipitation is greater than or equal to 20 mm (R20 mm); Annual total precipitation in wet days (PRCPTOT); Annual total precipitation when daily precipitation is greater than the 95th percentile (R95pTOT); Annual total precipitation when daily precipitation is greater than the 99th percentile (R99pTOT); Maximum length of dry spell, maximum number of consecutive days with daily precipitation less than 1 mm (CDD); Maximum length of wet spell, maximum number of consecutive days with daily precipitation greater than or equal to 1 mm (CWD); Maximum 1-day precipitation (Rx1day); Maximum consecutive 5-day precipitation (Rx5day); Simple precipitation intensity index, Annual precipitation divided by the number of wet days (SDII); Monthly maximum value of daily maximum temperature ($TX_x$); Monthly maximum value of daily minimum temperature ($TN_x$);Monthly minimum value of daily maximum temperature ($TX_n$); Monthly minimum value of daily minimum temperature ($TN_n$); Cold spell duration index (CSDI); Daily temperature range (DTR); Growing season length (GSL); Warm spell duration index (WSDI); Number of icing days (ID); Number of summer days (SU); Number of tropical nights (TR); Number of frost days (FD); Percentage of days when daily maximum temperature is greater than the 90th percentile (TX90p); Percentage of days when daily maximum temperature is less than the 10th percentile (TX10p); Percentage of days when daily minimum temperature is greater than the 90th percentile (TN90p); Percentage of days when daily minimum temperature is less than the 10th percentile (TN10p).

Besides, Karl et al. (1996) developed a combined index known as the Climate Extremes Index (CEI) to act as a communication tool that provides an initial indication of how combined extremes are changing over time. The CEI determines the areal extent of climate extremes for a specified region during a given year and is achieved by averaging several components, each identifying the fraction of an area experiencing a specific type of extreme associated with temperature or precipitation. The index is calculated as the arithmetic average of values from a specified set of extreme events, weighted by the percent area that experienced the defined conditions. The CEI components calculate the fraction of the area of a region experiencing extreme minimum and maximum temperatures, drought/moisture surplus, daily precipitation, and wet/dry days calculated on monthly or longer time scales. The CEI components identify extremes in both tails of the distribution, that is, hot and cold temperature extremes and wet and dry extremes.

Another example of the combined index is the Integrated Extreme Climatic Index (IECI). The index is composed of seven single extreme climatic indices, i.e. the country-average frequencies of high temperature days, low temperature days, intensive precipitation days, dust storm days, strong wind days, percentage of drought areas, and number of land falling tropical cyclones. Varied weights are assigned to the single extreme climatic indices based on the average economic losses and social impacts induced by the extreme weather and climatic events during recent five years. The largest weight, for example, is given to the frequency of intensive precipitation events because of the huge economic and life losses caused by the heavy rainfall and floods in eastern China. It is followed by the percentage of drought areas and number of land falling tropical cyclones.

The level of concern for the extreme events varies among different populations and regions. For example, people in northern Inland are generally not affected by typhoons, but they pay special attention to the extreme events such as drought, dust storms and high winds; People in Southern china give more attention to extreme events such as typhoons, heavy rainfall and floods, high temperatures and heat waves. Some extreme events happen at the same time. We know that High temperatures and drought often interact with each other; Drought can occur simultaneously with a heat wave. And you should know that One can define integrated extreme indicators according to the characteristics of his/her study area, but the first 27 indices are the generally and commonly used indicators.

Understanding of the changes in climate extremes and the underling drivers is important for human society, economies and ecosystems. Global warming can contribute to the intensity of heat waves by increasing the chances of very hot days and nights. Warming air also boosts evaporation, which can worsen drought. More drought creates dry fields and forests that are prone to catching fire, and increasing temperatures mean a longer wildfire season. Global warming also increases water vapor in the atmosphere, which can lead to more frequent heavy rain and snowstorms.

A warmer and moister atmosphere over the oceans makes it likely that the strongest hurricanes will be more intense, produce more rainfall, and possibly be larger. In addition, global warming causes sea level to rise, which increases the amount of seawater, along with more rainfall, that is pushed on to shore during coastal storms. That seawater, along with more rainfall, can result in destructive flooding. While global warming is likely making hurricanes more intense, scientists don't know yet if global warming is increasing the number of hurricanes each year. The effect of global warming on the frequency, intensity, size, and speed of hurricanes remains a subject of scientific research.

Extreme weather events are influenced by many factors in addition to global warming. Daily and seasonal weather patterns and natural climate patterns such as El Niño or La Niña affect when and where extreme weather events take place. The risk of a fire could depend on past forest management, natural climate variability, human activities, and other factors, in addition to human-caused climate change. Changes in greenhouse gases forcing lead to rise in TXx, TNx, TXn and TNn, increase in frequencies of SU and TR and decrease in frequencies of ID and FD over China through increased Greenhouse Effect with positive clear sky longwave radiation and play a dominant role in simulated changes of both hot and cold extremes over China.

The anthropogenic aerosol emissions forcing lead to warming over China during winter through cloud feedbacks related to aerosol induced cooling over tropical Indian Ocean and western tropical Pacific, and also induce changes in cold extremes the same sign as those induced by GHG, but with weak magnitude. Urbanization exacerbated the rainfall and flooding caused by hurricane harvey. Increased spatial variability and intensification of extreme monsoon rainfall due to Urbanization. A considerable impact of urbanization on microclimate is the urban heat island (UHI) effect, which generates higher temperature in urban areas than in suburban or rural areas.

### *7.4.1 Variations of Temperature Extremes Across China Based on Bayesian Hierarchical Spatial–Temporal Model*

During recent decades, studies on extreme temperature events are receiving increased attention throughout the world. With the global warming, spatial differences for the change patterns of temperature extremes existed due to the complex interaction between local and regional forcing factors and the non-linear climate system. Next, we investigate the Variations of temperature extremes across China based on Bayesian hierarchical spatial–temporal model.

Significant increasing warm extremes and decreasing cold extremes have been detected over the 70% of the global area, including Australia, Europe, South America, and so on; however, some areas, such as the northwestern America, had experienced decreasing warm extremes. Over China, temperature experienced significant changes in recent decades. The annual mean temperature increased by 1.4 °C at rates of 0.35 °C/10a, which is higher than the global average over the past 50 years. It is clear that regional differences were observed for extreme temperature indices in response to the background of the different areas, and it is of great importance to study the spatial differences by considering the complicated natural and human environment in China. Detection and attribution of changes in extremes to assess the relative contributions from external forcings remain a scientific challenge. The field of extreme event attribution has burgeoned recently.

There is now general consensus that natural forcing and human activity are the main driving forces of extreme events. Many researchers have relied on sophisticated climate models to study the effect of natural and anthropogenic climate forcings on the extreme events. However, numerical models have some disadvantages because of the numerical approximation and a series of parameterized procedures; other drawbacks are their high computational cost and their spatial resolution, which are too coarse for many applications.

Statistical methods for event attribution have several advantages over the numerical models, including low computational cost and strong robustness. However, most of these methods relied on a Gaussian assumption, which may not be conformed for most climate variables. Besides, very few studies in event attribution concerning spatial–temporal structure effect. New scientific approaches make it possible to determine how global warming affected individual extreme weather events. Bayesian spatial–temporal hierarchical model partitions the data into spatial–temporal patterns and their interaction terms. The left figure is the distribution of meteorological stations in china, 2272 stations are used. Figure 10 is thiessen polygon generated based on the stations. The temperature extremes were estimated for each of these polygon.

Poisson regression is a type of generalized linear models where the random component is specified by the Poisson distribution of the response variable which is a count. When all explanatory variables are discrete, log-linear model is equivalent to Poisson regression model. For more on Poisson regression models beyond to what is covered in this lesson. Poisson regression can be used for this data type (TX90p, TX10p, TN90p, TN10p).

A Thiessen polygon was generated based on the meteorological stations, creating 2272 regions for the whole study area, which are indexed by $j \in \{1, \cdots, 2272\}$. The temperature extremes were estimated for each of these regions for $t \in \{1992, \cdots, 2018\}$ time periods. Poisson regression was used for this data type. Specifically, let $Y_{jt}$ be the temperature event in the area $j \in \{1, \cdots, 2272\}$ at time $t \in \{1992, \cdots, 2018\}$, $Y_{jt} \sim Poisson(E_{jt}R_{jt})$, where $E_{jt}$ represents the expected level and $R_{jt}$ means the relative risk of a temperature extreme in the $j$th area and $t$th time. The Bayesian hierarchical spatial–temporal model was developed as follows (Rue et al. 2009):

$$\log(E_{jt}) = \beta_0 + S_j + T_t + C_{jt} + X_{jt}'\beta, \tag{20}$$

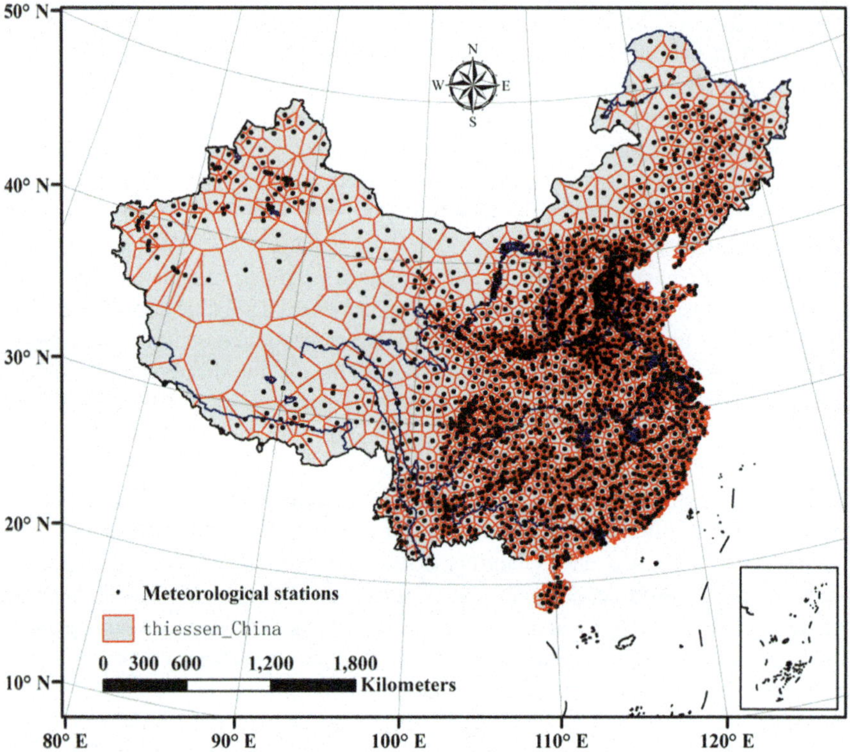


**Fig. 10** Meteorological stations and thiessen polygons

where $\beta_0$ is the intercept; $S_j$ is the spatial component accounting for existing spatial autocorrelation; $T_t$ is the time component meaning random time effects; $C_{jt}$ is the residual spatial–temporal interaction term providing spatial variations not accounted for by the main space and time effects; $X_{jt}$ represents the covariates of region $j$ and time $t$; and $\beta$ denotes the regression coefficients. The parameters of the model are computed through an Integrated Nested Laplace Approximation (INLA) in R software (Bivand et al. 2015).

Daily temperature observations from 2419 weather stations in China were obtained from the China Meteorological Administration (CMA). The dataset was quality controlled and homogeneity adjusted. RClimDex software was used for quality control. Daily temperature values are considered anomalous if the deviations from the mean value of temperature over the 1981–2010 reference period were greater than four standard deviations. The outliers are manually checked and edited by consulting the data centers. The in homogeneities were tested using RH test (Wang 2003). The in homogeneous were then excluded from the data series and not adjusted as suggested by some researchers. Finally, 2272 observations were used with a relatively complete sequence.

China is located in southeast Asia and the west of the Pacific with complicated and varied topography. The terrain descends from west to east, and the land extends out into the ocean in the east. The digital elevation model (DEM), obtained from http://srtm.csi.cgiar.org/SELECTION/inputCoord.asp, was used as one of the possible influencing factors for temperature extremes. EI Nino-Southern Oscillation (ENSO) phenomenon, the best known natural fluctuations inherent in the global climate system, has teleconnection signals throughout the global and has a direct and indirect impact on extreme events exerts with different magnitudes at different regions in the world (Horton et al. 2015; Qi and Wang 2012). The Niño 3.4 index, available from http://www.esrl.noaa.gov/psd/data/correlation/nina34 was used, which is one of the most commonly used indices to define EI Niño and La Niña events. The distances between the meteorological stations and the sea were computed by using ArcGIS to characterize the contribution of the sea on the temperature extreme changes. The time series of the nighttime light data (NTL) has been used for estimating the urbanization process, socio-economic activity and then the spatial heterogeneity in human activities from local, regional to global scales (Ma et al. 2012; Xie et al. 2019; Zhao et al. 2019). The nighttime light data from 1992 to 2013 at a resolution of 1 km were derived from the Defense Meteorological Satellite Program's (DMSP) Operational Linescan System (OLS).

The maximum value of TN10p was 39 days/year, and the high values were mainly located in the northeast, northwest and central part of China (Fig. 11). The low values were mainly located in Tibetan Plateau, Yunnan province and the North China Plain, with the minimum value of 25 days/year. The regionally average value of cool nights was 34 days/year over China with the standard deviation of 1.5 days/year. The cool nights were above the average value 34 days/year in most regions, taking up 58% of the total land area of China.

For the cool days, the high values were distributed in the northeast, northwest, Sichuan Basin and its south part, with the maximum value of 40 days/year. The low values were mainly located in the southwest of China, with minimum value of 29 days/year. The regionally average TX10p was 35 days/year, and the standard deviation was 1.9 days/year, which was larger than the cool nights, indicating the larger spatial variation of the cool days. Cool days with values greater than 35 days/year accounted for about 55% of the mainland of China.

The average value was 40 days/year, and the standard deviations were 1.6 days/year and 1.4 days/year for TN90p and TX90p, respectively (Fig. 12). The high values of TN90p were mainly observed in the Tibetan Plateau, Gansu province, the Inner Mongolia and eastern coastal area, while the low values were mainly distributed in the south part of China. For TX90p, the high values were mainly located in the Sichuan Basin, the Yangtze River Delta and the north part of the Inner Mongolia. The low values of TX90p were mainly distributed in Shandong province, and the west and southeast part of China. The values greater than 40 days/year accounted for 61% and 37% for TN90p and TX90p, respectively (Fig. 13).

The spatial variations of the four temperature extremes indicated the important role of the non-stationary of natural and human forcings. Compare with other regions, the south part in China showed more cold extremes and few warm extremes, while the

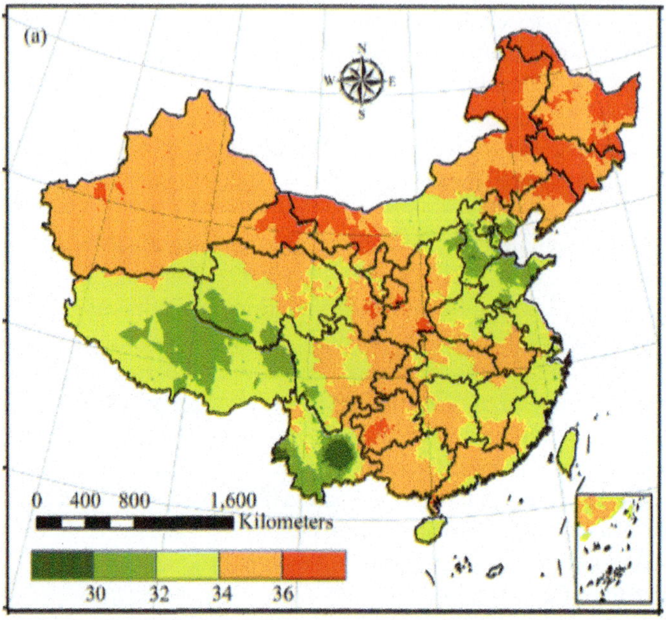


Fig. 11 Spatial distribution of TN10p

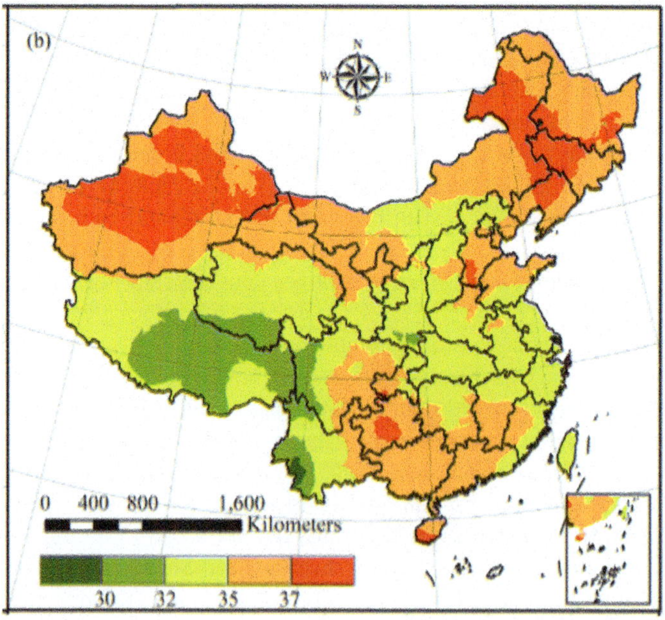


Fig. 12 Spatial distribution of TX10p

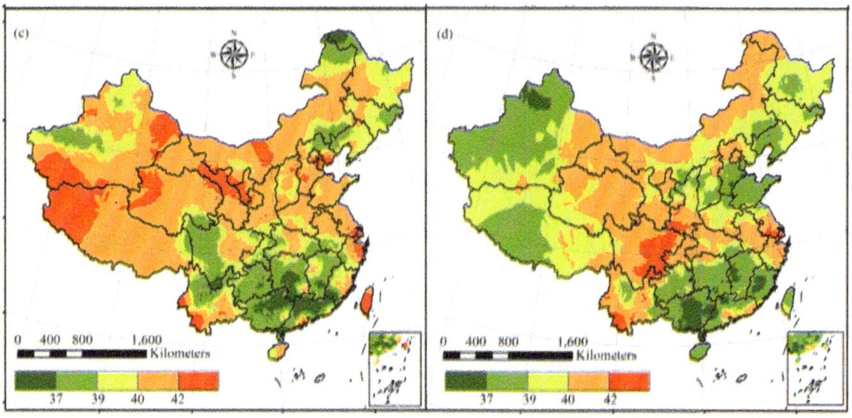


**Fig. 13** Spatial distribution of TN90p and TX90p

North China Plain had relatively few cold nights and warm days, and more cool days and warm nights. In general, Tibetan Plateau had experienced more warm extremes and less cold extremes. Compared with other regions in China, more cold and warm extreme events were both observed in Inner Mongolia and Gansu province, indicating people at high risk from temperature extremes in these regions.

A clear decrease of extreme cold nights and days (TN10p and TX10p) can be observed in the southwest part of China, especially in Tibetan Plateau, with maximum decreasing rates of 1.4 day/year and 2.5 day/year, respectively (Fig. 14). About 47% and 62% of the total land area showed positive trends for TN10p and TX10p, occurring at the northwest, southeast and northeast of China, with the maximum change value of 1.8 and 2.1 day/year, respectively. Cold nights decreased significantly at rates of 1.4 day/year and 1.2 day/year in Yunnan provinces and the east and southeast of Tibetan Plateau, respectively. Cold days increased significantly along the Taihang mountain and in northeast of China, with mean change rates of 1 day/year and 1.2 day/year, respectively, and decreased significantly in the southwest of China with change rate of 1.1 day/year.

TN90p increased in almost of China, occupying 99% of the land area of China, and in the middle and south of China, it increased significantly at a mean rate of 1.5 days/year, accounting for 30% of the total stations (Fig. 15). Approximately 97% of the total land area experienced increased trend for TX90p, with maximum rising rate of 2.9 days/year occurring at Yunnan province. Regions with significant increased warm days were mainly concentrated in the south of Yangtze River and the upper reach of the Yellow River; while TX90p changed significantly with a rising rate of 1.2 days/year at 38% of the total stations.

Detect the effects of four external forces on the temperature extreme events, including anthropogenic feature (nighttime light data), circulation pattern (Niño 3.4 index), and geographical and topographical factors (Elevation, distances between the meteorological stations and the sea), by using Bayesian hierarchical spatial–temporal

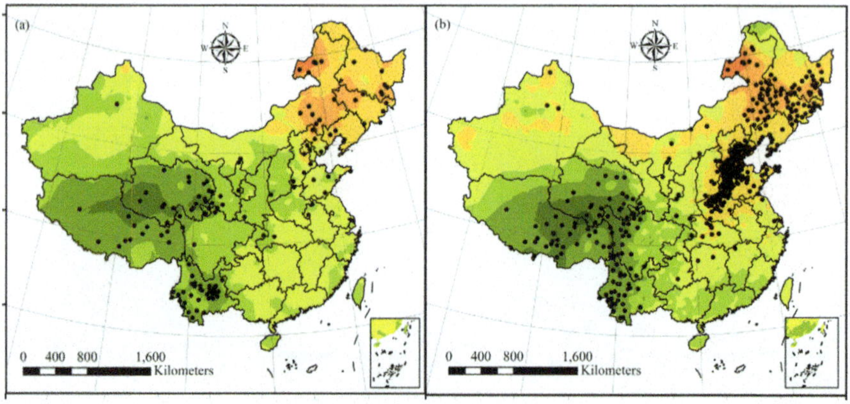


**Fig. 14** Change trends of TN10p and TX10p

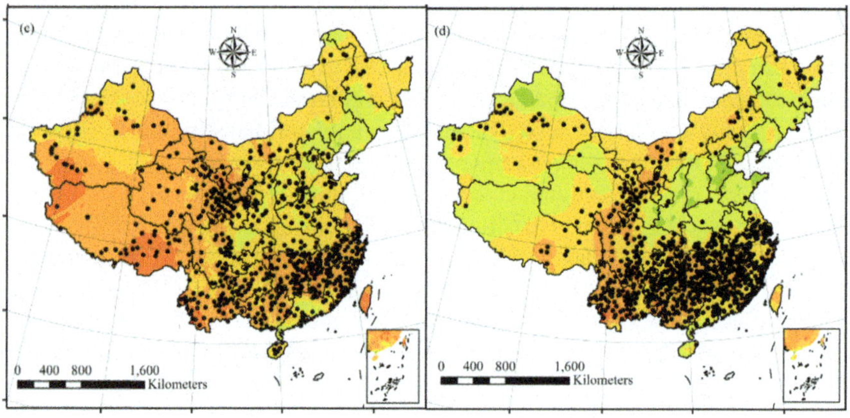


**Fig. 15** Change trends of TN90p and TX90p

model (INLA in R). The negative effects of ENSO event (N3.4) on TN10p and TN90p were detected with some uncertainties at 25% and 22% levels, respectively. The effect of ENSO phenomenon was not clear detected for TX10p and TX90p. It can be concluded that, compared to the day related indices (TX90p and TX10p), the ENSO played a stronger effect on the night-related indices (TN90p and TN10p). Distance to the sea (Dist.) had positive effects on TN10p, TX10p, and TX90p; while the effect on TN90p was undetectable, as the median of the 95% confidence interval closing to zero. The negative effect of DEM on TN10p and TX10p can be clearly detected; while DEM had weak positive effect on TX90p (82% possibility), and negative contribution on TN90p (at a 66% probability level). The strong positive effects of human activities on TX10p, TN90p, and TX90p and negative effect on TN10p can be found in Fig. 16, suggesting that the anthropogenic forcing might be the important factor for the four extreme changes.

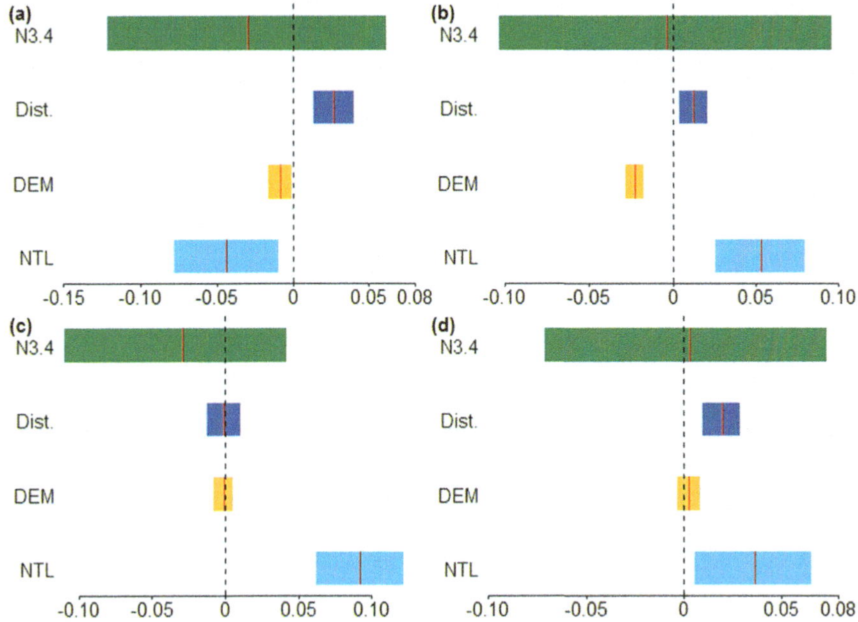


**Fig. 16** The effect of external and human forces on temperature extremes

### *7.4.2 Estimating the Effect of Urbanization on Extreme Climate Events in the Beijing-Tianjin-Hebei Region, China*

Urbanization, characterized by demographic transition, land expansion and surface thermal characteristic alterations, has caused profound influences on environment, biodiversity, biogeochemical cycles and energy flow. One important effect of urbanization that needs to be clearly understood is the changing climate of the considered region. A significant difference exists between the local climates of urban and nearby suburban regions. The urban heat island (UHI) effect is now well established, and it is well recognized that the replacement of green space with impervious surfaces in urban areas has decreased the evaporation and latent heat flux and partitioned more energy into sensible heat. Studies showed that urban areas have nearly 80% more precipitation than surrounding regions. Urbanization can lead to an increase in total precipitation in downtown and downwind areas in big cities.

China has experienced a rapid and profound urbanization process during the past forty years under the reform and opening-up policy proposed in the late 1970s. In 2010, the gap between China's urbanization process and the world average level gradually declined and was nearly eliminated, with the urbanization rate increasing from 17.9% in 1978 to 57.4% in 2016. As the main areas of urbanization, the metropolitan regions, including the capital city in developing countries, have experienced some environmental issues owing to ongoing massive migration from rural to urban areas and extensive human interferences.

Comparative studies of urbanization development in the three major economic areas in China (the Yangtze River Delta, the Pearl River Delta, and the BTH urban agglomeration) revealed that the BTH urban agglomeration, the largest city agglomeration in China, exhibited the fastest urban sprawl and growth from 1990 to 2010; the BTH region has a population of 109.22 million people in 2014 and an area of 216,756 km$^2$.

The impacts of urbanization on extreme temperature and precipitation indices in the BTH region were investigated by comparing meteorological records in urban regions with those in surrounding rural areas. The K-means clustering approach was used to identify large cities, small and medium cities, and rural areas by combining land use information, population data, gross domestic product (GDP) data, and satellite measurement of nighttime lights with Google Earth. The long-term trends of the frequently used extreme event series were estimated, and the urbanization-induced extreme climate changes during 1980–2015 were distinguished.

BTH agglomeration is located in the North China Plain and includes three provincial-level cities (Fig. 17): Beijing, Tianjin, and Shijiazhuang, and ten prefectural -level cities in Hebei province. Beijing is the national capital and one of the most populous cities in the world. Tianjin is the largest open coastal city in northern China. Hebei is an important agricultural and industrial province in China. As the largest socio-economic region in north China, the urban system of the BTH agglomeration area is in full development and boasts the well-known Jing-Jin-Tang manufacturing belt. The urban agglomeration of BTH region has experienced a fast urbanization process, with the most populated and rapidly changing land-cover during the past decades.

A total of 176 stations of country- and province-level political ideology were drawn from the National Meteorological Information Center for the time period 1980–2015 over the BTH region. The dataset was quality controlled and homogeneity adjusted. RClimDex software was used for quality control. Daily temperature values are considered anomalous if the deviations from the mean value of temperature over the 1981–2010 reference period were greater than four standard deviations. The outliers are manually checked and edited by consulting the data centers. The in homogeneities were tested using RH test.

Climate extreme indices (Table 9): 36 extreme indices obtained from daily temperature and precipitation observations over the period 1980–2015 were computed, including 20 extreme temperature events and 16 extreme precipitation events. These selected indices were mainly used for assessing some aspects of a changing local urban microclimate, and the changes in them could have significant societal impacts.

To measure city size, spatial land use data, night-time lights (NTL) data from the Defense Meteorological Satellite Program /Operational Linescan System (DMSP/ OLS), and socio-economic data are used.

Night-time DMSP/OLS satellite images of city lights from 1978 to 2013 at a resolution of 1 km were provided by the NOAA's National Centers for Environmental

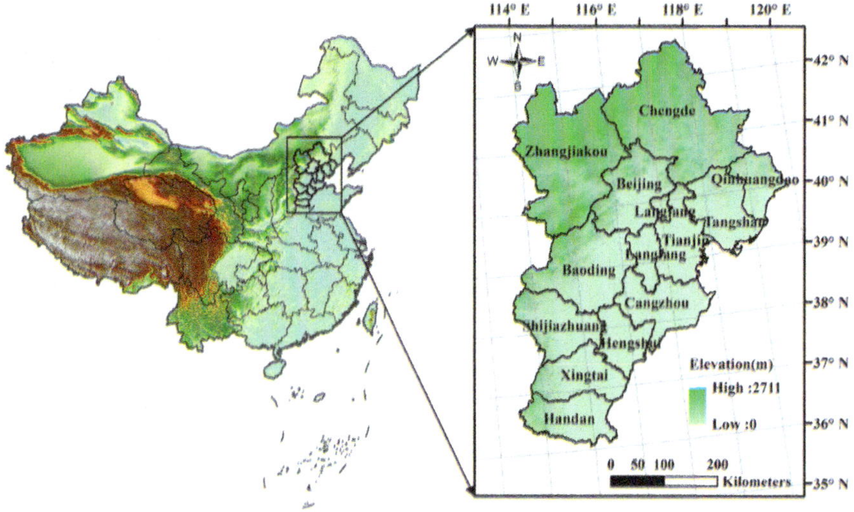


**Fig. 17** The location of BTH agglomeration

Information (NCEI), which are usually used to reflect the urban sprawl and socio-economic activity. The satellite- based land cover products acquired for every five years from 1980 to 2015 were provided by the Data Center for Resources and Environmental Sciences, Chinese Academy of Sciences. The GDP data were derived from the China Statistical Yearbook from 2000 to 2010. The resident populations near the meteorological stations were acquired from the year-2010 census report published by the Statistics Bureau of China. The size of the artificial buildings within a 2-km buffer zone of the meteorological sites and the Euclidean distances between the sites and the settlement centers were derived from Google Earth images.

Two different methods are typically used to investigate the influence of urbanization on observed extreme climate change: the method of physical modeling and statistical tools, including comparing the observations from urban stations with those from rural sites (Trusilova 2008; Wang et al. 2013; Feng et al. 2015). However, because of the existing uncertainties of the parameters and driving fields in the models, uncertainties probably remain in the simulation results from the modeling perspective. Using observations to directly reflect climate change is relatively reliable and realistic. In this case, meteorological stations should be grouped into urban and rural categories by using some objective criteria and other additional information. To get reliable results for the urbanization influence on extreme events, it is crucial to choose the optimum reference meteorological sites. The density and the spatial distribution of the reference stations should be considered because of the spatial heterogeneity of the extreme climate events influenced by regional factors such as land use/land cover change and pollutant particles beyond city scales.

Obtain the multidimensional attribute information for each station with 2-km radius buffer zones by using DMSP/OLS nighttime lights data, population data and

**Table 9** Temperature and precipitation extremes

| Type | Index | Definition | Unit |
|---|---|---|---|
| Extreme temperature indices | FD0 | Number of days with daily minimum temperature (TN) < 0 °C (number of frost days) | d |
| | ID0 | Annual count of days when daily maximum temperature (TX) < 0 °C (number of icing days) | d |
| | Ta10P | Number of days with daily average temperature (Ta) < 10th percentile | d |
| | Ta90P, Ta95P, Ta99P | Number of days with Ta > 90th/95th/99th percentiles (warm days) | d |
| | Tr10P | Number of days with diurnal temperature range < 10th percentile | d |
| | Tr90P, Tr95P, Tr99P | Number of days with diurnal temperature range (DTR) > 90th/95th/99th percentiles | d |
| | TN10 | Number of days with TN < 10th percentile | d |
| | TN90P, TN95P, TN99P | Days with TN > 90th/95th/99th percentiles of daily min temp (warm night) | d |
| | TX10 | Number of days with TX < 10th percentile (cool day) | d |
| | TX90P, TX95P, TX99P | Days with TX > 90th/95th/99th percentiles of daily max temp (warm day-times) | d |
| | TR20 | Number of days with TN > 20 °C (number of tropical nights) | d |
| | Su25 | The number of days with TX > 25 °C (number of summer days) | d |
| Extreme precipitation indices | CDD | Maximum number of consecutive days with daily precipitation amount (RR) < 1 mm (Maximum length of dry spell) | d |
| | CWD | Maximum number of consecutive days with RR ≥ 1 mm (maximum length of wet spell) | d |
| | R90P | Number of days with RR > 90th percentile (wet days) | d |
| | R90PTOT | Annual total precipitation amount with RR > 90th percentile | mm |
| | PR90P | Mean precipitation amount of R90P | mm |
| | R95P | Number of days with RR > 95th percentile (very wet days) | d |
| | R95PTOT | Annual total precipitation amount with RR > 95th percentile | mm |
| | PR95P | Mean precipitation amount of R95P | mm |
| | R99P | Number of days with RR > 99th percentile (extremely wet days) | d |

(continued)

**Table 9** (continued)

| Type | Index | Definition | Unit |
|---|---|---|---|
| | R99PTOT | Annual total precipitation amount with RR > 99th percentile | mm |
| | PR99P | Mean precipitation amount of R99P | mm |
| | SDII | Annual precipitation divided by the number of wet days | mm/d |
| | PRCPTOT | Annual precipitation amount in wet days | mm |
| | R10 | Number of days with RR > 10 mm | d |
| | R25 | Number of days with RR > 25 mm | d |
| | Rx1day | Maximum daily precipitation amount | mm |

GDP. K-means cluster analysis is employed to divide the observational stations into urban, suburban and rural groups (Jain 2010). K-means cluster method divides the observations into k mutually exclusive clusters, and the observations within one cluster are as close to each other and as far from observations in others clusters. It is a dynamic method, meaning that observations can change cluster and their members are closest to the centroid than to the mean of other clusters.

K-means cluster analysis was employed to divide the observational stations into urban, suburban and rural groups, which is one of the most popular and simple clustering algorithms. The K-means clustering method divides the observations into k mutually exclusive clusters, and the observations within one cluster are as close to each other and as far from observations in others clusters. It is a dynamic method, which means that observations can change cluster and that their members are closest to the centroid than to the mean of other clusters.

Let $Y = \{y_i\}, i = 1, \cdots, 174$ be the set of 174 4-dimensional points, which will be classified into $K = 3$ clusters: $C = \{c_k, k = 1, \cdots, K\}$. The goal of K-means method is to find a division such that the Euclidean distance of the empirical mean of a cluster from the points in the cluster is minimized. Let $v_k$ be the mean of cluster $c_k$. The Euclidean distance between $v_k$ and the points in $c_k$ is defined as

$$D(c_k) = \sum_{y_i \in c_k} \|y_i - v_k\|^2 \tag{21}$$

The aim of K-means method is to solve the following optimum problem for all k clusters:

$$D(C_k) = \min_k \sum_{k=1}^{K} \sum_{y_i \in c_k} \|y_i - v_k\|^2 \tag{22}$$

The K-means method first calculates the k centroid and then takes each point to the cluster that has minimum Euclidean distance from the centroid. The partition is generated and modified by recalculating the new centroid of each cluster and the new

Euclidean distance between each center and point. This process is finished until the cluster membership stabilizes.

Finally, Google earth was used to verify and modify the classification results of the stations based on the built-up area, vegetation distribution in the buffer area of each station, and the distance to the center of the city. The extreme temperature and precipitation differences between stations in different types were used to evaluate the urbanization effects on change extreme events.

The extreme temperature/precipitation differences between stations in different types are used to evaluate the urbanization effects on changing extreme events. The non-parameteric Mann–Kendall (MK) statistical test and Sen's slope estimator, which have been frequently used for trend examination in climatic and hydrologic data, were applied to detect the change trends of extreme temperature/precipitation time series.

The effect of the urbanization on temperature and precipitation extremes is defined as:

$$\Delta_{j1} = \beta_j - \beta_1, \quad j = 2, 3 \tag{23}$$

where $\beta_j$ denotes the change trend of $j$th urbanization level. The linear trend of extreme events averaged over the stations in rural areas, reflected by $\beta_1$, is defined as the reference trend in BTH region (Santer et al. 2000). $\Delta_{j1} > 0$ means the urbanization effect on extreme events is positive; $\Delta_{j1} = 0$ means that no urbanization influence on extreme events exists; $\Delta_{j1} < 0$ denotes the urbanization effect is negative.

The urbanization contribution is defined as the ratio of the urbanization effect $\Delta_{j1}$ to the average change trends $\beta_j$ at the urban stations:

$$r_j = \left| \frac{\Delta_{j1}}{\beta_j} \right| \times 100\% \tag{24}$$

Generally, the value of $r_j$ is less than 1. $r_j = 100\%$ means the change of the extreme event is caused entirely by urbanization; $r_j > 100\%$ means the extra trend might have been caused by other local factors not yet identified and it is regarded as 100% in this research. Moreover, the urbanization contribution is not considered once the urbanization effect is not statistically significant.

Figure 18 shows the classification of urbanization levels in BTN region by using K-means cluster method. The stations include the 47 rural stations level one in the figure, 108 sub-urban stations level 2, and 17 urban stations stand for level one.

Table 10 gives the trends in temperature extremes for the three different urbanization levels over the time period 1980–2015. The increasing trends were found for SU25, TR20, Ta90P, Ta95P, Ta99P, TN90P, TN95P, TN99P, TX90P, TX95P, TX99P, and Tr10P. Although the change trends for TX90P, TX95P and TX99P all rose for different urbanization levels, they were all statistically significant at the 0.05 level only in level 3. The positive urbanization effects on TR20, Ta90P, Ta95P, Ta99P,

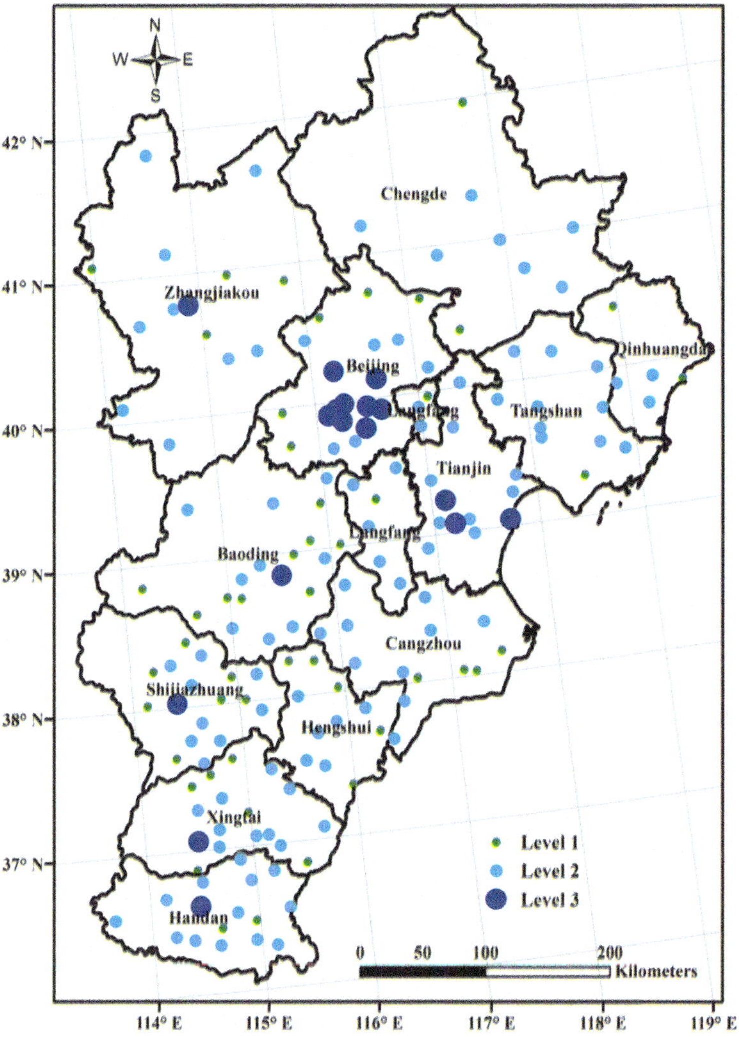


**Fig. 18** Classification of urbanization levels in BTH region

TN90P, TN95P, and TN99P for the period 1980–2015 were all statistically significant at the 95% confidence level. A significant negative urbanization effect was found for TN10P in both level 2 and level 3. The larger increasing trends were seen for TR20 and TN90P, with both reaching approximately 0.7 days/yr and having urbanization contributions of 15.71% and 21.43%, respectively, all related to the daily minimum temperature. The largest decreasing trend was seen for TN10P, with changing rate of −0.79 days/yr in large cities. We assume the trends for all extreme indices in level 1 were caused by the global climate change, the differences between the extreme indices in level 3 or 2 and level 1 indicated the urban-induced climate change. The

**Table 10** Trends in temperature extremes for different urbanization levels

| Index | Level1 | Level 2 | Level 3 | △21 | △31 | r21 (%) | r31 (%) |
|---|---|---|---|---|---|---|---|
| FD0 | −0.53** | −0.64** | −0.45** | −0.11 | 0.08 | 17.19 | 20.45 |
| ID0 | 0 | −0.02 | −0.09 | −0.02 | −0.09 | 100.00 | 100.00 |
| SU25 | 0.12 | 0.15 | 0.2 | 0.03 | 0.08 | 20.00 | 40.00 |
| TR20 | 0.59** | 0.7** | 0.7** | 0.11 | 0.11 | 15.71 | 15.71 |
| Ta10P | −0.4 | −0.46 | −0.42* | −0.06 | −0.02 | 13.04 | 4.76 |
| Ta90P | 0.39** | 0.44** | 0.6** | 0.05 | 0.21 | 11.36 | 35.00 |
| Ta95P | 0.09** | 0.09** | 0.11** | 0 | 0.02 | 0.00 | 18.18 |
| Ta99P | 0.09** | 0.09** | 0.11** | 0 | 0.02 | 0.00 | 18.18 |
| TN10P | −0.67** | −0.85** | −0.79** | −0.18 | −0.12 | 21.18 | 15.19 |
| TN90P | 0.55** | 0.7** | 0.7** | 0.15 | 0.15 | 21.43 | 21.43 |
| TN95P | 0.41** | 0.5** | 0.5** | 0.09 | 0.09 | 18.00 | 18.00 |
| TN99P | 0.1** | 0.13** | 0.17** | 0.03 | 0.07 | 23.08 | 41.18 |
| TX10P | 0 | −0.04 | −0.09 | −0.04 | −0.09 | 100.00 | 100.00 |
| TX90P | 0.26 | 0.24 | 0.6** | −0.02 | 0.34 | 8.33 | 56.67 |
| TX95P | 0.12 | 0.17 | 0.44** | 0.05 | 0.32 | 29.41 | 72.73 |
| TX99P | 0.05 | 0.03 | 0.09** | −0.02 | 0.04 | 66.67 | 44.44 |
| Tr10P | 0.44** | 0.5** | 0.33** | 0.06 | −0.11 | 12.00 | 33.33 |
| Tr90P | −0.57** | −0.85** | −0.42** | −0.28 | 0.15 | 32.94 | 35.71 |
| Tr95P | −0.35** | −0.5** | −0.2** | −0.15 | 0.15 | 30.00 | 75.00 |
| Tr99P | −0.07** | −0.1** | −0.06* | −0.03 | 0.01 | 30.00 | 16.67 |

relatively largest and significant positive urbanization effect was found for Ta90P in level 3, with an urbanization contribution as high as 35%, while the largest negative urbanization effects was found for Tr90P in level 2. The absolute larger positive urbanization effects were seen for TR20, Ta90P, TN90P, Tr90P, and Tr95P at the large city stations, reaching 0.1 days/yr, 0.21 days/yr, 0.15 days/yr, 0.15 days/yr and 0.15 days/yr, respectively. It was found that the slopes of the linear change trends for FD0, Ta10P, TN10P, Tr90P, Tr95P, and Tr99P were negative; however, the urbanization effects for FD0, Tr90P, Tr95P, and Tr99P were all positive at level 3 at the 95% significant confidence level, which highlights the impact of the local human activities. The negative and positive urbanization effects for FD0, Tr10P, Tr90P, Tr95P, and Tr99P in level 2 and level 3 indicate the different scale effects of urbanization on extreme events. Cities in level 2 were mainly focused on tourism and modern agriculture, with relatively lower per capita GDP, serving for the large cities in level 3. With the impact of strong governmental intervention on urban expansion, excessive resources have been concentrated in the cities in level 3. Since the enhanced UHI magnitude has obviously contributed to the overall increasing trends of mean temperature, UHI must have exerted some influence on trends of FD0, Tr90P, Tr95P, and Tr99P based on the daily minimum and maximum temperature. It was found that

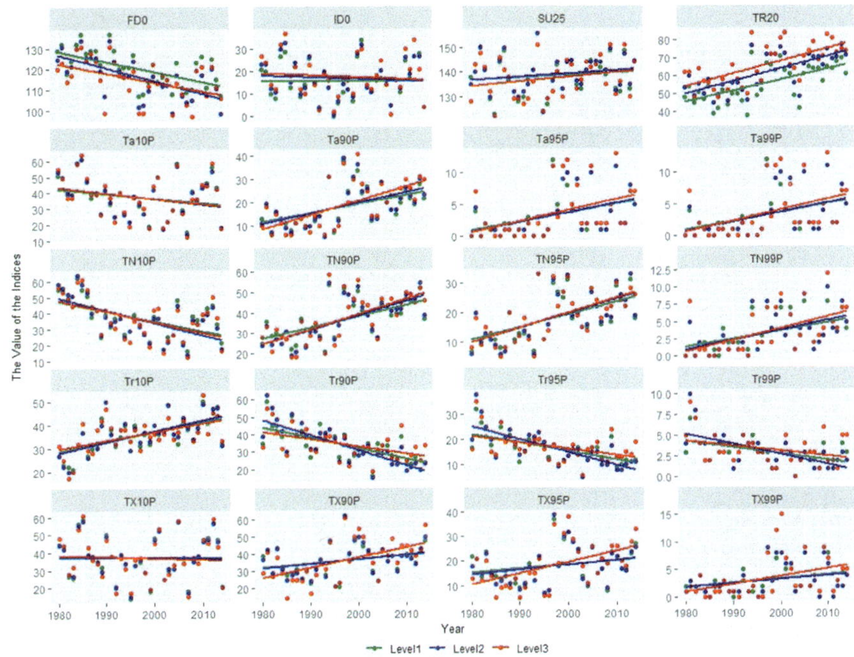


**Fig. 19** Changes in the temperature extremes in BTH region over the period 1980–2015

the urbanization contributions on Ta10P, TN10P, and Tr99P in level 2 are 13.04%, 21.18%, and 30%, respectively, which were higher than those in level 3, indicating the cooling effect of the relative larger vegetated areas for cities in level 2 (Fig. 19).

Figure 20 presents the time series of the 20 extreme temperature indices in three different urbanization levels over the period 1980–2015. Among the 20 extreme temperature indices, 8 indices showed decreasing trends, which are FD0, ID0, Ta10P, TN10P, Tr90P, Tr95P, Tr99P, and TX10P, and the others exhibited increasing trends. From the concept of the extreme climate indices, it was found that the decreased indices were all related to the low value part of the sequence, except for the diurnal temperature range series, while the increasing indices were related to the high value part of the extreme temperature series. This means that the values of the 20 extreme temperature indices increased over the period 1980–2015. Large differences of the urbanization effect were found for FD0, TR20, Tr90P, Tr95P and Tr99P, which indicates that urbanization processes might have exerted a significant influence upon the daily minimum temperature in the period analyzed. The indices SU25, Ta90P, Ta95P, Ta99P, TX90P, TX95P, and TX99P generally showed larger difference trends between level 3 and level 1 than those between level 2 and level 1, which might have been related to the intense human activities and the generally reinforced urban heat island.

The change trends in extreme precipitation events for the period 1980–2015 in the BTH region are complex for different urbanization levels. Positive trends were

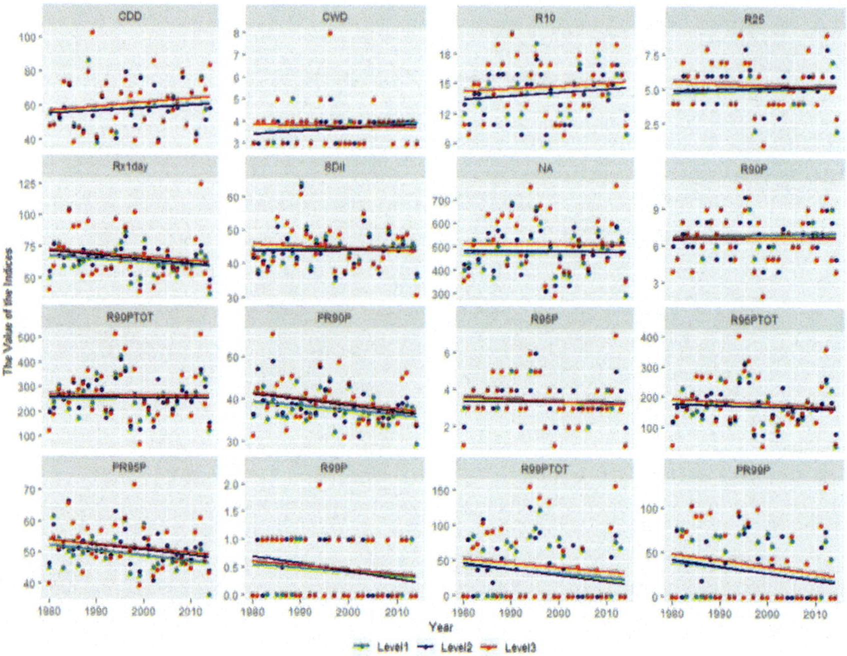

Fig. 20 Changes in the precipitation extremes in BTH region over the period 1980–2015

all detected for CDD, PRCPTOT, SDII, R10, and R90PTOT in levels 1 and 2, while four indices, including Rx1day, PR90P, R95PTOT, and PR95P, showed decreasing trends. However, it was found that, except for CDD in level 3 and PR95P in level 2, the urbanization effects on CDD, PRCPTOT, SDII, R10, R90PTOT, Rx1day, PR90P, R95PTOT, and PR95P were negative. Larger urbanization effects were found for PRCPTOT, R90PTOT and R95PTOT, reaching −1.21 mm/yr, −0.93 mm/yr and −0.83 mm/yr, respectively. From the percentile-based indices, it can be found that urbanization might decrease the extreme heavy precipitation events in the BTH region. Besides, the positive CDD urbanization effect in level 3 means that the urbanization lengthened the consecutive dry days in the studied area. No trends were detected for CWD, R25, R90P, R95P, R99P, R99PTOT, and PR99P. Figure 20 shows changes in the precipitation extremes in the BTH region over the period 1980–2015. The linear trends of the indices exhibit patterns of changes consistent with those shown in Table 11 the differences between the extreme precipitation events in levels 3 and 1 are slightly larger than those between level 2 and level 1, which indicates that urbanization processes might have exerted an influence on the extreme precipitation in the period analyzed. Researches have shown that cities with heavy pollution probably generate more intense rainfall events, and cities in level 3 witness more pollution than medium and small cities in the BTH region. However, in contrast to temperature extremes, almost no significant effect on extreme precipitation events has been detected. Previous studies have found that the East Asian summer monsoon

**Table 11** Trends in precipitation extremes for different urbanization levels

| Index | Level1 | Level 2 | Level 3 | Δ21 | Δ31 | r21 (%) | r31 (%) |
|---|---|---|---|---|---|---|---|
| CDD | 0.2 | 0.16 | 0.29 | −0.04 | 0.09 | 25.00 | 31.03 |
| CWD | 0* | 0 | 0 | 0 | 0 | 0.00 | 0.00 |
| PRCPTOT | 1.11 | 0.29 | −0.1 | −0.82 | −1.21 | 282.76 | 1210.00 |
| Rx1day | −0.15 | −0.23 | −0.31 | −0.08 | −0.16 | 34.78 | 51.61 |
| SDII | 0.07 | 0.05 | 0 | −0.02 | −0.07 | 40.00 | 0.00 |
| R10 | 0.06 | 0.04 | 0 | −0.02 | −0.06 | 50.00 | 0.00 |
| R25 | 0 | 0 | 0 | 0 | 0 | 0.00 | 0.00 |
| R90P | 0 | 0 | 0 | 0 | 0 | 0.00 | 0.00 |
| R90PTOT | 0.51 | 0.24 | −0.42 | −0.27 | −0.93 | 112.50 | 221.43 |
| PR90P | −0.07 | −0.11* | −0.11 | −0.04 | −0.04 | 36.36 | 36.36 |
| R95P | 0 | 0 | 0 | 0 | 0 | 0.00 | 0.00 |
| R95PTOT | −0.01 | −0.1 | −0.84 | −0.09 | −0.83 | 90.00 | 98.81 |
| PR95P | −0.13 | −0.12 | −0.16 | 0.01 | -0.03 | 8.33 | 18.75 |
| R99P | 0 | 0 | 0 | 0 | 0 | 0.00 | 0.00 |
| R99PTOT | 0 | 0 | 0 | 0 | 0 | 0.00 | 0.00 |
| PR99P | 0 | 0 | 0 | 0 | 0 | 0.00 | 0.00 |

(EASM) and tropical cyclones played a significant role in the precipitation extremes over the regions including the BTH area, which has been found useful to explain to some extent that non-significant urban effect on extreme precipitation events has been found in this region.

## 7.5 Impacts of Climate Change on Ecosystems

Changing climate affects ecosystems in a variety of ways. For instance, warming may force species to migrate to higher latitudes or higher elevations where temperatures are more conducive to their survival. Similarly, as sea level rises, saltwater intrusion into a freshwater system may force some key species to relocate or die, thus removing predators or prey that are critical in the existing food chain. The elevated atmospheric CO2 concentration, enhanced air temperature, and changed precipitation exert great influence on crop output, crop growth, diseases and pests, agricultural water resources, and structure and function of agro-ecosystem.

Climate influences the structure and function of forest ecosystems and plays an essential role in forest health. A changing climate may worsen many of the threats to forests, such as pest outbreaks, fires, human development, and drought. Global warming is melting the permafrost layer, releasing greenhouse gases into the environment. Permafrost thawing may produce microbially degradable rich organic matter, which, in turn, will cause massive greenhouse emissions and speed up global warming.

Permafrost is a permanently frozen layer on or under Earth's surface. It consists of soil, gravel, and sand, usually bound together by ice. Permafrost usually remains at or below 0 °C (32 °F) for at least two years. Permafrost can be found on land and below the ocean floor. It is found in areas where temperatures rarely rise above freezing. This means permafrost is often found in Arctic regions such as Greenland, the U.S. state of Alaska, Russia, China, and Eastern Europe. Permafrost thickness can range from one meter (about three feet) to more than 1000 m (about 3281 feet). Permafrost covers approximately 22.8 million $km^2$ (about 8.8 million square miles) in Earth's Northern Hemisphere. Frozen ground is not always the same as permafrost. A layer of soil that freezes for more than 15 days per year is called "seasonally frozen ground." A layer of soil that freezes between one and 15 days a year is called "intermittently frozen ground." Permafrost is frozen for two years or more.

Permafrost does not always form in one solid sheet. There are two major ways to describe its distribution: continuous and discontinuous. Continuous permafrost is just what it sounds like: a continuous sheet of frozen material. Continuous permafrost extends under all surfaces except large bodies of water in the area. The part of Russia known as Siberia has continuous permafrost. Discontinuous permafrost is broken up into separate areas. Some permafrost, in the shadow of a mountain or thick vegetation, stays all year. In other areas of discontinuous permafrost, the summer sun thaws the permafrost for several weeks or months. The land near the southern shore of Hudson Bay, Canada, has discontinuous permafrost.

Permafrost and seasonally frozen ground are key components of the Arctic and global climate system because of their influence on energy, water, and carbon cycles. The freeze–thaw status of the ground is a critical threshold in the terrestrial system that is closely linked to the timing and length of the vegetation growing season, boreal plant productivity, the seasonal evolution of land–atmosphere carbon dioxide and methane exchange, and the amplitude and timing of spring snowmelt river discharge peaks.

When the temperature of permafrost rises above 0 °C, it may thaw, in which case any ice it contains will undergo a phase change from solid to liquid (i.e., it will melt). Generally, the mineral and organic components of the former permafrost will remain solid. Therefore, in regions with more ground ice, thawing permafrost may result in more of a change to the land surface, and vice versa. Thawing permafrost can raise water levels in Earth's oceans and increase erosion. Erosion happens when permafrost thaws because soil and sediment are easily washed away without ice binding them together.

As the ground thaws and freezes, it contracts and expands, putting stress on foundations and twisting rail lines. In arctic areas where structures have been built over permafrost, insufficient insulation has led to the collapse of buildings and twisting of rail beds, due to the movement of thawing permafrost.

Scientists are worried about melting Arctic permafrost. When permanently frozen ground turns out to be not so permanent, structures built on permafrost can catastrophically collapse—and the bizarre, ranging from abandoned nuclear waste to anthrax, can be revealed. Now, there's another issue to be concerned about: A massive release of carbon dioxide. Permafrost is vital to the world's climate because it stores twice as much carbon as there is in the atmosphere. Under usual circumstances, the top layer of the frozen soil thaws during summer, when plants and microorganisms spring to life. The microbes munch on plant roots and inevitably emit greenhouse gases since they have to respirate to live.

Permafrost is the result of present climate. Many temperature profiles show, however, that permafrost is not in equilibrium with present climate at the sites of measurement. Some areas show, for example, that climatic warming since the last third of the nineteenth century has caused a warming of the permafrost to a depth of more than 100 m. In such areas much of the permafrost is a product of a colder, former climate. Ground temperatures have increased throughout Alaska since the late 1970s, and permafrost has already thawed in many places. As temperatures keep getting warmer, permafrost will continue to thaw. For example, the map on the right shows how permafrost in northwestern Alaska could change by the year 2100. Increased rainfall led to deeper thaw with an increase of 0.7 cm of thaw per cm of additional rain. Disturbed and wetland were the most vulnerable to rain-induced thaw with –1 cm of surface thaw per additional 1 cm of rain. Permafrost in tussock tundra, mixed forest, and conifer forest was less sensitive to rain-induced thaw.

The total area of permafrost in China is estimated at $-1.59 \times 10^6$ $km^2$(glacier and lakes excluded), and the area of seasonally frozen ground (excluding instantaneous frozen ground) is $-1.35 \times 10^6$ $km^2$,the area of mountain permafrost is $-0.30 \times 10^6$ $km^2$ and the area of plateau permafrost is $-1.05 \times 10^6$ $km^2$. The latitudinal permafrost is located in the northern part of northeastern China, and its area is $0.24 \times 10^6$ $km^2$. The development and distribution of permafrost in China is mainly dependent upon climatic, geologic, topographic, hydrologic and surface cover conditions. Among the major climatic parameters, mean annual air temperature and precipitation apparently control and influence permafrost. As a result, permafrost distribution displays distinct vertical and horizontal zonation (Fig. 21).

The distribution of permafrost tallies roughly with the variations of the temperature and precipitation in three dimensions. In northeastern China, the distribution of permafrost reveals a latitudinal zonation. Permafrost decreases in its mean annual ground temperature (MAGT) from 0 to − 1 °C at its southern limit down to − 2.4 °C (locally down to − 4.2 °C) in the north. Correspondingly, its thickness increases from several meters to more than 100 m. The continuity of permafrost varies northward from sporadic to continuous permafrost zones. In western China, which lies beyond the southern limit of Eurasian continental permafrost zone, permafrost is present only in high mountains and plateaus, revealing an obvious vertical zonation, i.e. permafrost decreases in temperature and increases in thickness and continuity with rising elevations.

With an average elevation of more than 4000 m, the Tibetan Plateau is the highest and the most extensive plateau in the world. Almost all of the Tibetan Plateau is underlain by permafrost and seasonally frozen ground because of its high elevation. The present permafrost on the Tibetan Plateau is a relfict of the late Pleistocene permafrost, which had degraded during the Holocene. In response to the

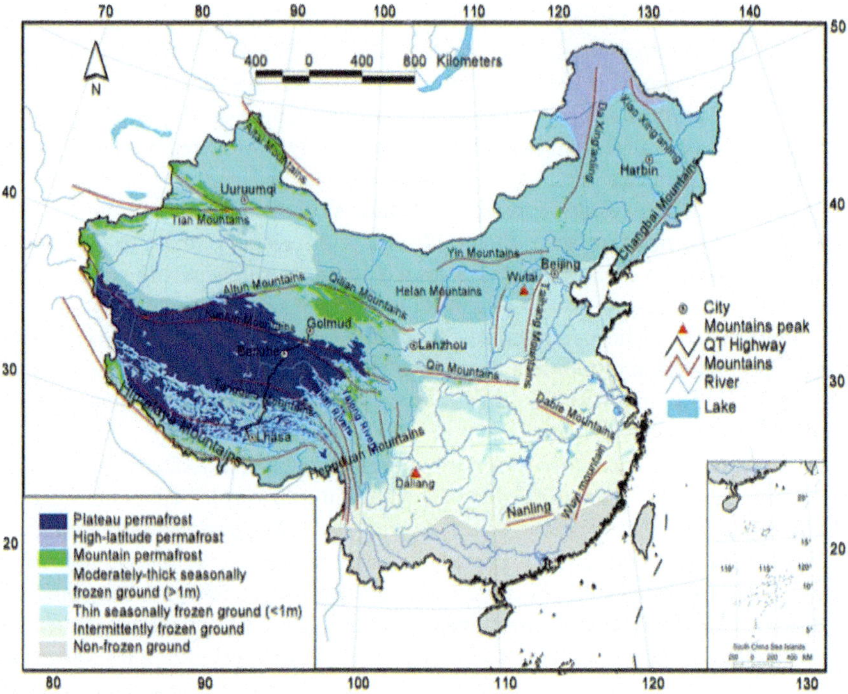


**Fig. 21** Permafrost distribution in China

Plateau warming of approximately 0.44 C/decade from 1981 to 2010, the near-surface permafrost area decreased at a rate of 9.20104 km$^2$/decade, and the area-mean active layer thickness increased by 0.15 m/decade. The area-mean maximum freezing depth of the seasonally frozen ground decreased by 0.34 m/decade.

Super-heavy precipitation (44 mm occurred on 2 July 2015) caused a sharp increase in soil water content and dramatic rises in soil temperatures by 0.3–0.5 °C at shallow depths and advancement thawing of the active layer by half a month. More summer precipitation amount tends to reduce the seasonal amplitude of soil temperatures, decrease mean annual soil temperatures and thawing indices and thin active layers.

Terrestrial permafrost monitoring is coordinated within the Global Terrestrial Network for Permafrost where ground temperatures are measured in > 1000 boreholes, but only at discrete points. Extrapolating these observations to larger regions is hampered by the considerable spatial variability of the ground thermal regime and an uneven distribution of sites that results in extensive unsampled areas.

The active layer of the permafrost, currently appears as a specific phenomenon related to the presence of permafrost. The active layer is a layer of ground that is subject to annual thawing and freezing in areas underlain by permafrost. The active layer is key feature in cold regions, particularly as most of the ecological, hydrological, biochemical and pedogenic processes take place inside it.

Dynamic Global Vegetation Models (DGVMs) provide a suitable modeling framework to also assess grassland dynamics, grassland productivity, and the impact on the biogeochemical cycles under 5 different grassland management schemes at the global scale. When the LPJ DGVM was expanded by agricultural activities, forming LPJmL. Include the integration of managed grassland and simulate plant physiological and ecological processes and includes a newly developed discrete layer energy balance permafrost module and a vertical carbon distribution within the soil layer. The model is able to reproduce the interactions between vegetation and soil carbon dynamics as well as to simulate dynamic permafrost changes resulting from changes in the climate.

By combining core features of global biogeographical and biogeochemical models developed in the 1990s, Dynamic Global Vegetation Models (DGVMs) emerged as the main tool to simulate processes underlying the dynamics of natural vegetation types (growth, mortality, resource competition, disturbances such as wildfires) and the associated carbon and water fluxes. In light of the strengthening human 40 interferences, DGVMs were further developed to integrate additional processes that are relevant to the original research quest of studying biogeography and biogeochemical cycles under climate change. This includes the incorporation of human land-use and the simulation of agricultural production systems, nutrient limitation, as well as hydrological modules and river routing 45 schemes.

The parts of LPJmL4 (Fig. 22) build on not only allow for quantifying changes in vegetation composition, the water cycle, the carbon cycle, and agricultural production, but also for explicitly simulating the dynamics and constraints within and among the modules, thereby providing a consistent and comprehensive representation of Earth's land surface processes.

Recent developments focused on an improved energy balance model able to estimate permafrost dynamics based on a vertical soil carbon distribution scheme and a new soil hydrological scheme. The permafrost module is based on a new calculation of the soil energy balance. Soil temperatures (Tsoil) for each layer are computed with an energy balance model, including one-dimensional heat conduction and convection of latent heat. Freezing and thawing has been added to better account for soil ice dynamics. Soil water dynamics are computed daily. The soil column is divided into six hydrological active layers of 0.2, 0.3, 0.5, 1, 1 and 10 m depth. Soil parameters for thermal diffusivity at wilting point, at 15% of water holding capacity, and at field capacity and for thermal conductivity at wilting point, and at saturation (for water and ice) are derived for each grid cell using soil texture from the Harmonized World Soil Database (HWSD) version.

The active layer thickness represents the depth of maximum thawing of the year. Freezing depth is calculated by assuming that the fraction of frozen water is congruent with the frozen soil bucket. The 0 °C–isotherm within a layer is estimated by assuming a linear temperature gradient within the layer and this fraction of heat is assumed to be used for the thawing respectively freezing process. Temperature represents the amount of thermal energy available, whereas heat transport represents the movement of thermal energy into the soil by rain and melt water. Precipitation and percolation energy and the amount of energy which arises from the temperature difference

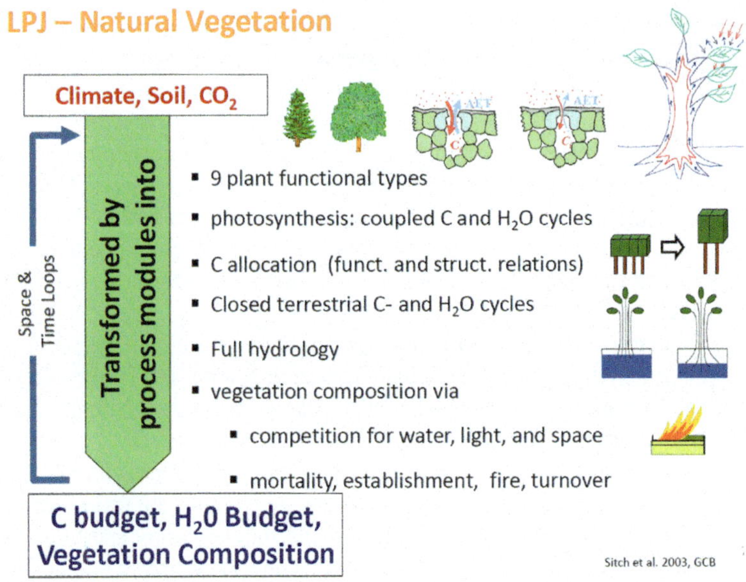


**Fig. 22** The main features of LPJml

between the temperature of the above layer (or the air temperature for the upper layer) and the temperature of the below layer, are assumed to be used for converting latent heat fluxes first. The residual energy is used to increase soil temperature. Tsoil is initialized at the beginning of the spinup simulation by the mean annual air temperature.

We use LPJml to simulate the permafrost in China-Mongola-Russian region. Figure 23 shows the simulated permafrost distribution in the year 2018. The lowest active-layer thickness is mainly located in northern Russia and Qinghai-Tibet Plateau region. The value of active-layer thickness is between 1 and 3 m in the south of Russia, the north of Mongolia, and the north east of China. The active-layer thickness exhibited observous Latitude zonality, and is significantly affected by elevation.

From T1 to T2 (Fig. 24), the thickness of permafrost activity layer in China, Mongolia and Russia increased by 0.07 m on average, with maximum increase value of 8.02 m and maximum decrease value of 8.67 m. From T2 to T3, the thickness of the permafrost activity layer increased by 0.04 m on average, with a maximum increase of 4.86 m and a maximum decrease of 11.7 m. The area where the thickness of the active layer decreased between 0 and 0.5 m accounted for 47.7% of the whole study area. From T3 to T4, the thickness of the permafrost activity layer in the study area increases by 0.33 m on average, with a maximum increase of 8.53 m and a maximum decrease of 8.35 m. The area with the increased value between 0 and 0.5 m accounts for 62.58%.

During the past 118 years, the maximum and minimum change value of the thickness of active layer are 69 mm/a and −117 mm/a, and the average change

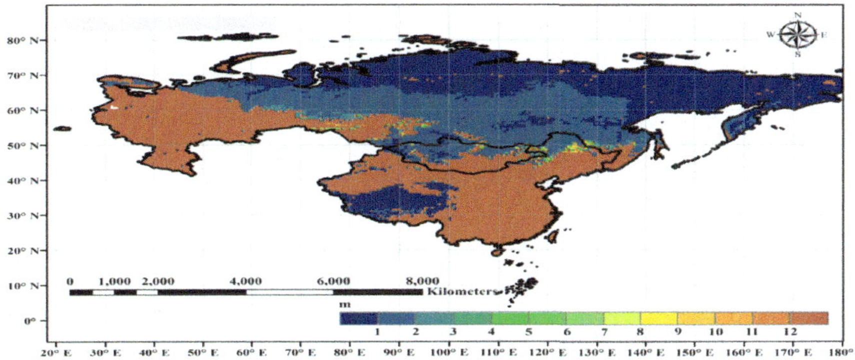


**Fig. 23** Simulated permafrost distribution in 2018

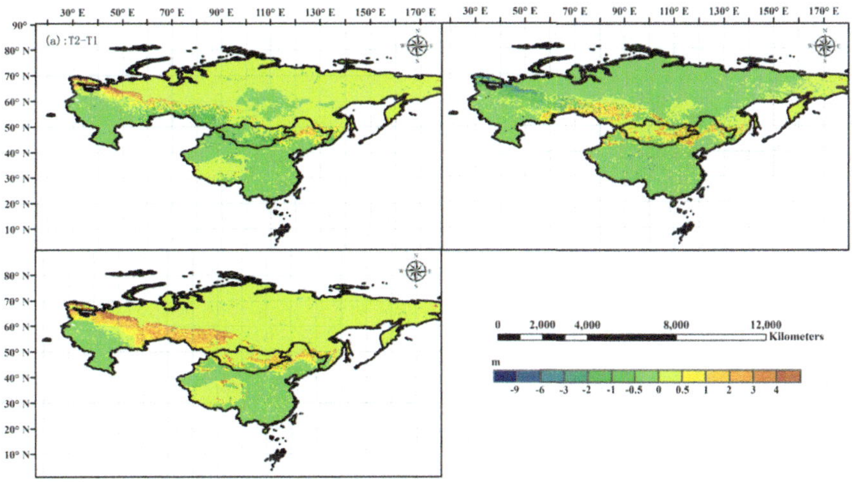


**Fig. 24** The change of the permafrost distribution: T1:1901–1930; T2:1931–1960; T3:1961–1990; T4:1991–2018

rate of the whole region is 0.67 mm/a. Areas with a change rate greater than zero accounting for 61.48% of the whole study area. The areas with a change rate less than 0 accounted for 38.52% of the total study area (Fig. 25).

In 1901, the maximum value of annual average temperature was 24.87 °C, and the high temperature was mainly distributed in southwest Russia and southeast China. The minimum value is −22.00 °C, and the low temperature is mainly located in the north of Russia. The average temperature of the whole region was −3.44 °C. The maximum value of average annual temperature in 2018 is 25.26 °C, the minimum is −20.37 °C, and the average temperature is −1.71 °C. The upper part of the temperature was shifted to the north from 1901. In the past 110 years, the temperature in this region has shown a rising trend, with the highest temperature rising by 4.67 °C and 1.72

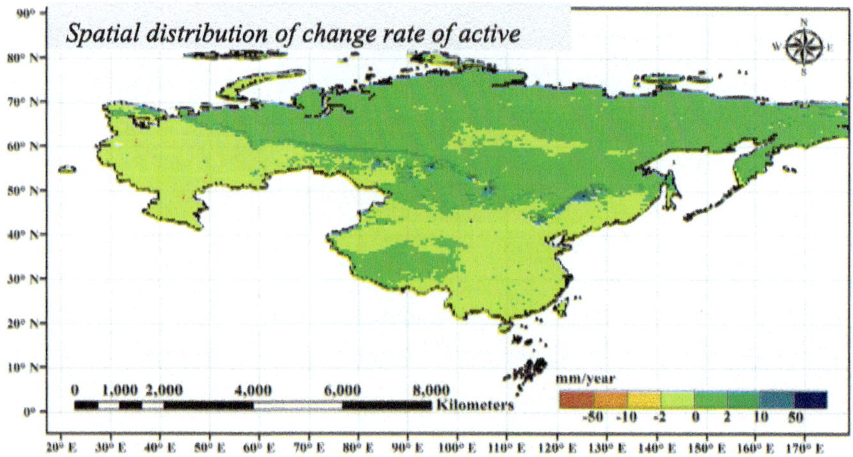


**Fig. 25** Spatial distribution of change rate of active layer

°C on average in the whole region. The most obvious warming area is located in the north and east of Russia (Fig. 26).

According to the Mann–Kendall test and Sen's estimate, the active layer thickness has shown a significant upward trend during the past 118 years, with an average annual increase of 4.62 mm. Based on Student's T test, from T3 (1961–1990) to T4 (1991–2018), the activity layer thickness changes significantly ($P < 0.01$). In recent 118 years, the relationship between the thickness and time is as follows (Fig. 27):

$$y = -3296.47 + 4.4 * t; \tag{25}$$

The annual average temperature showed a significant increasing trend, with a rising rate of 0.014 °C/a. The correlation analysis shows that the correlation coefficient between temperature and the active layer thickness is **0.76.**

The relationship between air temperature and the active layer thickness is

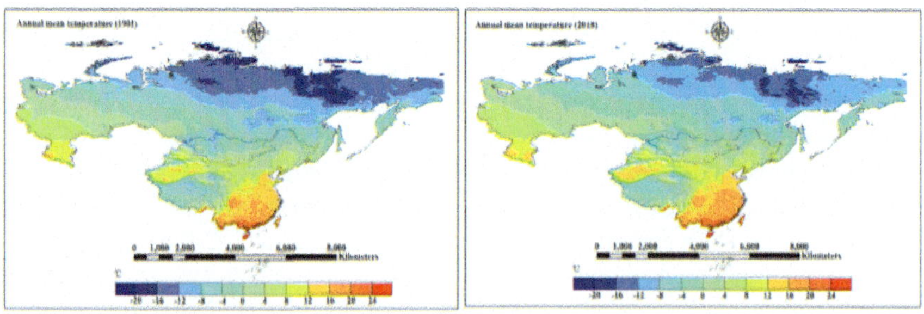

**Fig. 26** The change of temperature from 1901 to 2018

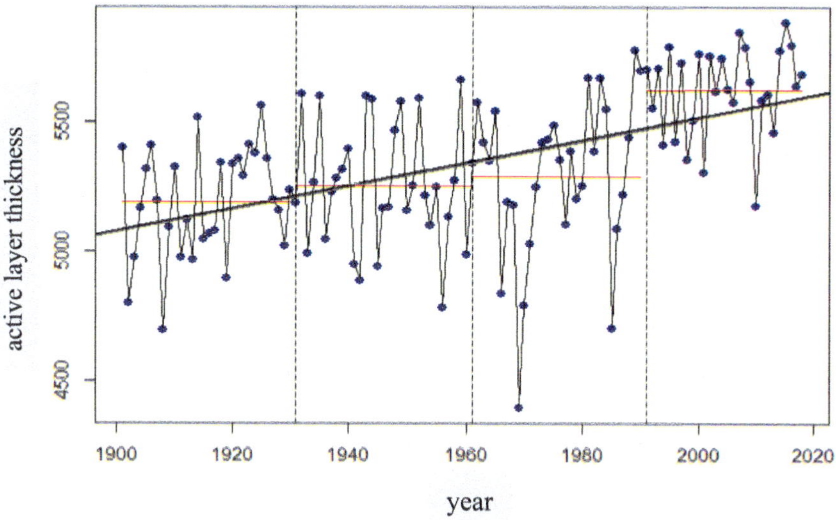


**Fig. 27** Change trend of active layer thickness

$$y = 6282.9 + 306.3 * x \tag{26}$$

$R^2 = 0.57$, indicating the annual mean temperature accounted for 57% of the variation of the permafrost.

## 7.6 Summary

We introduced the methods and models for climate variables simulation, including spatial interpolation methods, this the first stage of spatial estimation of climate variables. The second stage was remote sensing retrieval and atmospheric model calculation. Since the 1990s, spatial estimation of climate variables has developed to the third stage, that is, the stage of multi-source data merging and downscaling. Recently it has become one of the most important hotspots in meteorology and hydrology and can be used to obtain better climate variables information by integrating various kinds of information, such as gauge observations, remote sensing retrieval, and atmospheric reanalysis estimates. And then we introduce the method of HASM. A novel surface modeling method based on the fundamental theorem of surface theory. Surface modeling of digital elevation model, climate variables, soil properties, and terrestrial ecosystems indicate that HASM has increased accuracy, compared with some classical methods such as Kriging, inverse distance weighting method, and splines. HASM can be performed as an interpolation method, a downscaling method, and a data fusion method.

We gave the Trend surface simulation and pointed that a combination of interpolation methods applied through statistical transfer functions (STFs) is an efficient approach to improve the estimation error of climatic variables. For the trend surface simulation, we introduced two methods including ordinary least squares and geographical weighted regression method. Ordinary least squares method is a global method while geographical weighted regression method is a local method, it is able to better describe the spatial non-stationary or nonlinear response relationship between climate variables and related influencing variables. Then we introduced spatial stationarity analysis. Stationary analysis can be carried out from five aspects, and we gave the method of autocorrelation test of residuals, that is Moran test, which can be used to check the cluster of the data. We gave the formula of Coefficient variation to characterizate the spatial stationarity. Based on these, we gave the methods for simulating annual mean temperature and annual mean precipitation in China. Compared different methods and analyze the change trends of temperature and precipitation during 1960 to 2010 in china. We divided the landmass of china into 9 regions and found that annual temperature on average has generally shown as increasing trend in china with an average decadal increasing rate of 0.33°. On average, China became wetter during the period from 1960 to 1990s, but much drier from 1990 to 2000s. Then we give the methods for simulating climate scenarios based on the downscaling method of HASM. And show how China would become warmer and wetter under the RCP2.6, RCP4.5 and RCP8.5 scenarios. At last, we introduced the Mann–Kendall statistical test, and Sen's estimator, which are used to detect the changing trend of data series.

27 major climate extremes indices were introduced and some combined indices were presented We pointed that global warming is making some extreme events worse and many other factors contribute to individual extreme event. We investigated the variations of temperature extremes across china using Bayesian spatial temporal model and estimated the effect of urbanization on extreme climate events in beijing-tianjin-hebei region. And we introduced the K-means cluster analysis method.

And in this class, we introduced the impacts of climate change on ecosystems, we mainly focus on the impact of climate change on permafrost.

We give a brief introduction to the permafrost and pointed that global warming is melting the permafrost, and increased rainfall stimulate permafrost thaw. We presented the permafrost in china. Then we introduce the LPJml model and use this model to simulate the permafrost in china Mongolia Russia region and analyzed the impact of temperature change on permafrost in this region. And in the last we introduced the student's t test, which can be used to compare whether there is a significant difference between the two datasets.

## 7.7 Further Reading

(1) Na Zhao, Tian-Xiang Yue. 2021. High Accuracy Surface Modeling Method: The Robustness. Springer.

(2) Tian-Xiang Yue, Na Zhao, Douglas Ramsey, Chen-Liang Wang, Ze-Meng Fan, Chuan-Fa Chen, Yi-Min Lu, Bai-Lian Li. 2013. Climate change trend in China, with improved accuracy. Climatic Change 120: 137–151.
(3) Na Zhao, Yi-Meng Jiao. 2021. A new HASM-based downscaling method for high-resolution precipitation estimates. Remote Sens. 13, 2693.
(4) Na Zhao, Yi-Meng Jiao, Ting Ma, Miao-Miao Zhao, Ze-Meng Fan, Xiao-Zhe Yin, Yu Liu. 2019. Estimating the effect of urbanization on extreme climate events in the Beijing–Tianjin–Hebei region, China. Sci Total Environ. 688, 1005–1015.

### Exercises

(1) Have you ever applied a method for surface modeling of climate variables? What's the method and for what?
(2) Illustrate the methods for estimating the spatial distribution of climate variables, including their advantage and disadvantage. For the methods mentioned in this class, which one do you prefer? Why?
(3) How to establish the statistical transfer function for your studied climate variable? What factors need to be considered?
(4) How to use HASM to predict the future scenarios?
(5) What's the major climate events in your city? What's the reason?
(6) Illustrate the effects of human activities on extreme temperature/precipitation.
(7) Give an introduction of LPJml model; which fields can be used?
(8) Give an example of climate change impact on ecosystems.

## References

Alexakis, D.D., and I.K. Tsanis. 2016. Comparison of multiple linear regression and artificial neural network models for downscaling TRMM precipitation products using MODIS data. *Environmental Earth Sciences* 75 (14): 1077.

Apaydin, H., A.S. Anli, and F. Ozturk. 2011. Evaluation of topographical and geographical effects on some climatic parameters in the Central Anatolia Region of Turkey. *International Journal of Climatology* 31: 1264–1279.

Bishop, C.M. 2006. Mixture models and EM. In Pattern recognition and machine learning, 423–455. Springer-Verlag.

Bivand, R.S., V. Gomez-Rubio, and H. Rue. 2015. Spatial data analysis with R-INLA with some extensions. *Journal of Statistical Software* 63 (20): 1–31.

Brunsdon, C., A.S. Fotheringham, and M.E. Charlton. 1996. Geographically weighted regression: A method for exploring spatial nonstationary. *Geographical Analysis* 28: 281–298.

Chatterjee, S., A. Khan, H. Akbari, and Y. Wang. 2016. Monotonic trends in spatio-temporal distribution and concentration of monsoon precipitation (1901–2002), West Bengal, India. *Atmospheric Research* 182: 54–75.

Chen, T., and E. Martin. 2009. Bayesian linear regression and variable selection for spectroscopic calibration. *Analytica Chimica Acta* 631: 13–21.

Chen, Y., J. Huang, and S. Sheng. 2018. A new downscaling-integration framework for high-resolution monthly precipitation estimates: Combining rain gauge observations, satellite-derived precipitation data and geographical ancillary data. *Remote Sensing of Environment* 214: 154–172.

Chen, H., J. Sun, and H. Li. 2020. Increased population exposure to precipitation extremes under future warmer climates. *Environmental Research Letters* 15 (3): 034048.

Feng, J.M., Y.L. Wang, and Z.G. Ma. 2015. Long-term simulation of large-scale urbanization effect on the East Asian monsoon. *Climatic Change* 129: 511–523.

He, X.G., N.W. Chaney, M. Schleiss, and J. Sheffield. 2016. Spatial downscaling of precipitation using adaptable random forests. *Water Resources Research* 52: 8217–8237.

Horton, D.E., N.C. Johnson, D. Singh, D.L. Swain, B. Rajaratnam, and N.S. Diffenbaugh. 2015. Contribution of changes in atmospheric circulation patterns to extreme temperature trends. *Nature* 522: 465–469.

Hou, A.Y., R.K. Kakar, and S. Neeck. 2014. The global precipitation measurement mission. *Bulletin of the American Meteorological Society* 95: 701–722.

Huffman, G.J., R.F. Adler, D.T. Bolvin, et al. 2007. The TRMM multisatellite precipitation analysis (TMPA): Quasi-global, multiyear, combined-sensor precipitation estimates at fine scales. *Journal of Hydrometeorology* 8: 38–55.

Hutengs, C., and M. Vohland. 2016. Downscaling land surface temperatures at regional scales with random forest regression. *Remote Sensing of Environment* 178: 127–141.

Jain, A.K. 2010. Data clustering: 50 years beyond K-means. *Pattern Recognition Letters* 31: 651–666.

Joyce, R.J., J.E. Janowiak, P.A. Arkin, and P. Xie. 2004. CMORPH: A method that produces global precipitation estimates from passive microwave and infrared data at high spatial and temporal resolution. *Journal of Hydrometeorology* 5: 487–503.

Kang, L., L. Di, M. Deng, et al. 2014. Use of geographically weighted regression model for exploring spatial patterns and local factors behind NDVI-precipitation correlation. *IEEE Journal of Selected Topics in Applied Earth Observations and Remote Sensing* 7: 4520–4538.

Karl, T.R., R.W. Knight, D.R. Easterling, and R.G. Quayle. 1996. Indices of climate change for the United States. *Bulletin of the American Meteorological Society* 77: 279–292.

Lu, N., K.E. Trenberth, J. Qin, K. Yang, and L. Yao. 2015. Detecting long-term trends in precipitation water over the Tibetan Plateau by synthesis of station and MODIS observations. *Journal of Climate* 12: 1707–1721.

Ma, T., C.H. Zhou, T. Pei, S. Haynie, and J.F. Fan. 2012. Quantitative estimation of urbanization dynamics using time series of DMSP/OLS nighttime light data: A comparative case study from China's cities. *Remote Sensing of Environment* 124: 99–107.

Ninyerola, M., X. Pons, and J.M. Roure. 2007. Objective air temperature mapping for the Iberian Peninsula using spatial interpolation and GIS. *International Journal of Climatology* 27: 1231–1242.

Qi, L., and Y.Q. Wang. 2012. Changes in the observed trends in extreme temperatures over China around 1990. *Journal of Climate* 25: 5208–5222.

Qin, J., K. Yang, N. Lu, Y.Y. Chen, L. Zhao, and M.L. Han. 2013. Spatial upscaling of in-situ soil moisture measurements based on MODIS-derived apparent thermal inertia. *Remote Sensing of Environment* 138: 1–9.

Rue, H., S. Martino, and N. Chopin. 2009. Approximate Bayesian inference for latent Gaussian models by using integrated nested Laplace approximations. *Journal of the Royal Statistical Society Series B, Statistical Methodology* 71: 319–392.

Santer, B.D., T.M.L. Wigley, J.S. Boyle, et al. 2000. Statistical significance of trends and trend difference in layer average atmospheric temperature time series. *Journal of Geophysical Research* 105: 7337–7356.

Sharifi, E., B. Saghafian, and R. Steinacker. 2019. Downscaling satellite precipitation estimates with multiple linear regression, artificial neural networks, and spline interpolation techniques. *Journal of Geophysical Research* 124: 789–805.

Sorooshian, S., K. Hsu, X. Gao, H.V. Gupta, B. Imam, and D. Braithwaite. 2000. Evaluation of PERSIANN system satellite-based estimates of tropical rainfall. *Bulletin of the American Meteorological Society* 81: 2035–2046.

Trusilova, K.M. 2008. Urbanization impacts on the climate in Europe: Numerical experiments by the PSU-NCAR Meso-scale model (MM5). *Journal of Applied Meteorology and Climatology* 47: 1442–1455.

Ushio, T., et al. 2009. A Kalman filter approach to the Global Satellite Mapping of Precipitation (GSMaP) from combined passive microwave and infrared radiometric data. *Journal of the Meteorological Society of Japan* 87A: 137–151.

Wang, X.L. 2003. Comment on "detection of undocumented change points: A revision of the two-phase regression model." *Journal of Climate* 16: 3383–3385.

Wang, M.N., X.Z. Zhang, and X.D. Yan. 2013. Modeling the climatic effects of urbanization in the Beijing-Tianjin-Hebei metropolitan area. *Theoretical and Applied Climatology* 113: 377–385.

Xie, Y.H., Q.H. Weng, and P. Fu. 2019. Temporal variation of artificial nighttime lights and their implications for urbanization in the conterminous United States, 2013–2017. *Remote Sensing of Environment* 225: 160–174.

Yue, T.X. 2011. Surface modeling: High accuracy and high speed methods. New York, NY, USA: CRC Press.

Yue, T.X., Z.P. Du, D.J. Song, and Y. Gong. 2007. A new method of surface modeling and its application to DEM construction. *Geomorphology* 91: 161–172.

Yue, T.X., Na Zhao, Douglas Ramsey, Chen-Liang Wang, Ze-Meng Fan, Chuan-Fa Chen, Yi-Min Lu, and Bai-Lian-Li. 2013. Climate change trend in China, with improved accuracy. Climatic Change 120: 137–151.

Yue, T.X., N. Zhao, Z.M. Fan, et al. 2019. Methods for simulating climate scenarios with improved spatiotemporal specificity and less uncertainty. *Global and Planetary Change* 181: 102973.

Yue, T.X., N. Zhao, Y. Liu, et al. 2020. A fundamental theorem for eco-environmental surface modelling and its applications. *Science China-Earth Sciences* 63 (8): 1092–1112.

Zhang, H.H., H.A. Loaiciga, D. Ha, and Q.Y. Du. 2020. Spatial and temporal downscaling of TRMM precipitation with novel algorithms. *Journal of Hydrometeorology* 21: 1259–1278.

Zhao, N. 2021. An efficient downscaling scheme for high-resolution precipitation estimates over a high mountainous watershed. *Remote Sensing* 13 (2): 234.

Zhao, N., and T.X. Yue. 2014. Sensitivity studies of a high accuracy surface modelling method. *Science China-Earth Sciences* 57: 1–11.

Zhao, N., and Tian-Xiang Yue. 2021. High accuracy surface modeling method: The robustness. Springer.

Zhao, N., and Yi-Meng Jiao. 2021. A new HASM-based downscaling method for high-resolution precipitation estimates. Remote Sensing 13: 2693.

Zhao, N., T.X. Yue, C.F. Chen, M.W. Zhao, and Z.M. Fan. 2018. An improved statistical downscaling scheme of Tropical Rainfall Measuring Mission precipitation in the Heihe River basin, China. *International Journal of Climatology* 38: 3309–3322.

Zhao, N., Y.M. Jiao, T. Ma, et al. 2019. Estimating the effect of urbanization on extreme climate events in the Beijing–Tianjin–Hebei region, China. *Science of the Total Environment* 688: 1005–1015.

# Chapter 8
# Surface Modelling of Ecosystems and Ecological diversity

**Learning Objectives**

- Surface modelling of terrestrial ecosystems.
- Surface modelling of land cover scenarios.
- Surface modelling of ecological diversity.

## 8.1 Introduction

### *8.1.1 Concept and Mechanism of Terrestrial Ecosystem*

Terrestrial ecosystem is a land-based community of organisms and the interactions of biotic and abiotic components in a given area. Terrestrial ecosystem changes, as the major indicators of global change on Earth's surface, have an important influence on the condition of the human living environment and the sustainable development of the social economy.

Terrestrial ecosystem changes, as the major indicators of global change on Earth's surface, have an important influence on the condition of the human living environment and the sustainable development of the social economy. The key climate parameters of biotemperature, precipitation and evaporation directly determine the changes in the spatial distribution of terrestrial ecosystem. Therefore, how to identify the changes in the spatiotemporal distribution of terrestrial ecosystems based on climate data from station observations and model simulations is very important to understanding future terrestrial ecosystem changes.

The key climate parameters of biotemperature, precipitation and evaporation directly determine the changes in the spatial distribution of terrestrial ecosystems. Therefore, how to identify the changes in the spatiotemporal distribution of terrestrial ecosystems based on climate data from station observations and model simulations



T. Yue et al., *Eco-Environmental Informatics*,
https://doi.org/10.1007/978-981-96-1977-1_8

is very important to understanding future terrestrial ecosystem changes. The interaction between climate elements and terrestrial ecosystem changes could be explained in terms of long-term historical climate observation. The data on global climate scenarios could be used to effectively identify the spatial differences in climate change and further reveal the response of terrestrial ecosystems to future climate change scenarios.

Terrestrial ecosystems mainly include the Nival, Tundra, Steppe, Scrub, Thorn woodland, Dry Forest, Moist Forest, Wet Forest, Rain Forest, and deserts, which are distributed in one of Polar/Alpine, Subpolar/Alpine, Cold temperate/Subalpine, Cool temperate, Warm temperate, Subtropical, or Tropical zones. The type of terrestrial ecosystem found in a particular place is dependent on the temperature range, the average amount of precipitation received, the soil type, and amount of light it receives.

### *8.1.2 Progress of Land Cover Models*

Land cover is defined as the (bio) physical cover at the surface of the earth surface including grass, trees, bare ground and water. Land cover is a fundamental variable that impacts on and links many parts of the human and physical environments (Foody 2002). Land cover change has significant effects on biogeochemical cycling, soil erosion, ecological diversity, sustainable land use and climate change (Chapin et al. 2000). Land cover change affects the ability of biological systems to support human needs by altering ecosystem services (Vitousek et al. 1997).

Land-cover classification is closely tied to the ecological condition of the Earth's surface and have significant implications for global ecosystem health, water quality, and sustainable land management. At present, there are many kinds of land cover classification systems, among which the IGBP land cover classification system is widely used in the world.

How climatic changes might affect land cover is one of overarching questions of land use and land cover change project (LUCC), a core project of the International Geosphere-Biosphere Program (IGBP) and International Human Dimensions Programme (IHDP). Land cover dynamics and its diagnostic models are one of LUCC's focuses. Various models for simulating land cover scenarios have been developed (Yue et al. 2007; Fan et al. 2013, 2015, 2020).

SRES Model is the published SRES land cover scenarios assumes that everywhere within major world region changes occur at the same rate. The land cover changes under A1, B1 and B2 marker scenarios are highly uncertain. The A2 marker scenario did not include land cover change, so changes under the A1 scenario were assumed to apply also to A2. SRES A2 and B2 scenarios of IPCC were downscaled in a probabilistic cellular automata model (PCAM) to define the narrative scenario conditions of future urban land use change. The results of the modeling experiments illustrated the spectrum of possible land cover scenarios of New York Metropolitan Region for the years 2020 and 2050 (Solecki and Oliveri 2004). The SRES land cover scenarios do not include the effect of climate change on future land cover (Arnell et al. 2004).

CLUE Model was developed within a framework of conversion of land use and its effects under assumptions. (Verburg et al. 1999, 2002). CLUE model is a dynamic equilibrium between the total population and the agricultural production and land cover changes occur only when biophysical and human demands cannot be met any more through existing land use. The CLUE model needs to give the finally total area of each land cover type, however, which is very difficult to be obtained in a large scale (e.g., Global scale) in the future.

The input–output (I–O) model was used to develop land cover scenarios, which was not spatially explicit and did not consider possible impacts of climate changes (Hubacek and Sun 2001). Socio-economic changes are linked to different types of land via an explicit representation of land requirement coefficients associated with specific economic activities. The strong biophysical linkages are mainly manifested in the derivation of regional differences of the land requirement coefficients and the typical I–O technical coefficients by means of Agro-Ecological Zone.

SESM Model is a spatially explicit stochastic methodology (SESM) was developed for simulating land use changes at a watershed level without the need to describe the complex relationships between biophysical, economic and human factors (Luijten 2003). Its transition probabilities were based on observed frequencies of actual conversions between forest, pasture and scrub in history. Land use changes were simulated on a grid cell basis, in which each grid cell acts independently. Three explorative scenarios for the year 2025 were developed under assumptions of Business as Usual (BU), Ecological Watershed (EW), and Corporate Farming (CF). So, SESM model is difficult to use on a large scale.

LUCK Model is a land use change modeling kit (LUCK) was developed for scenario generation in a grid-based discretization-mode on catchment scale, which represents the spatial distribution of land cover types in the landscape (Niehoff et al. 2002). Thus, LUCK model also is difficult to simulate land cover change on a large scale. The potential conversion of land cover types of each grid is based on an evaluation of the characteristics of each grid as well as on its neighborhood relationships. Because land cover changes happen successively, LUCK tries to simulate the dynamics by an iterative procedure.

CA Model is a Cellular Automata (CA) model was involved to simulate the urban land use scenarios (Clarke et al. 1997; Wu 2002), and improved to downscale cultivated land and built-up land of the global land use datasets (Li and Shi 2016) and ignores the effect of climate change on future land cover. CA land cover model is mainly used to simulate the land cover scenarios at an urban or river basin scale. At a large scale, CA land cover model only has been improved to downscale the agricultural land and built-up land of global land cover data. Therefore, how to develop a new method of surface modelling in land cover on multi-scale is the main topic of this class.

### 8.1.3 Definitions of Ecological diversity

Scientists have estimated that there are around 8.7 million species of plants and animals in existence. However, only around 1.2 million species have been identified and described so far, most of which are insects. This means that millions of other organisms remain a complete mystery. As the climate change, and humans put increasing pressure on the planet, using and consuming more resources than ever before, we risk upsetting the balance of ecosystems and losing biodiversity. WWF's 2018 Living Planet Report found an average 60%decline in global populations of mammals, fish, birds, reptiles, and amphibians since 1970. The 2019 landmark Global Assessment Report by the Intergovernmental Platform on Biodiversity and Ecosystem Services (IPBES) reported one million animal and plant species are now threatened with extinction – the highest number in human history.

Biodiversity, defined by World Wildlife Fund (WWF), is all the different kinds of life you'll find in one area—the variety of animals, plants, fungi, and even microorganisms like bacteria that make up our natural world. Each of these species and organisms work together in ecosystems, like an intricate web, to maintain balance and support life. Biodiversity supports everything in nature that we need to survive: food, clean water, medicine, and shelter.

Biodiversity, defined by National Geographic Society (NGS), is a term used to describe the enormous variety of life on Earth. It can be used more specifically to refer to all of the species in one region or ecosystem. Biodiversity refers to every living thing, including plants, bacteria, animals, and humans.

Biodiversity in Millennium ecosystem assessment (MES) is defined as "the variability among living organisms from all sources including, inter alia, terrestrial, marine and other aquatic ecosystems and the ecological complexes of which they are part; this includes diversity within species, between species and of ecosystems." The importance of this definition is that it draws attention to the many dimensions of biodiversity. It explicitly recognizes that every biota can be characterized by its taxonomic, ecological, and genetic diversity and that the way these dimensions of diversity vary over space and time is a key feature of biodiversity.

Ecological diversity is measured by looking at its three levels of genetic, species and ecotope diversity.

Genetic Diversity refers to the differences in the genetic make-up of a distinct species and to the genetic variations within a single species. It concerns DNA (or RNA in some viruses) sequences. Humans, for example, have different eye and skin colors, hair textures, propensity for disease, reactions to pollutants, heights, hormone levels and so on. We are the same species but have genetic variations which make us diverse. This means that if one individual die when stung by a bee, others will not and are able to carry on to ensure the success of the species.

Species diversity relates to numbers and spread – how many different species live in an ecosystem and how are they distributed? When considering only the number of different species within an ecosystem we then talk of species richness. Species richness only considers the number of different species in an ecosystem, not their distribution.

Ecotope diversity has two components: richness based on the total number and evenness based on the relative abundance. This concept has been used to formulate ecotope diversity (Yue et al. 2001, 2005; Yue and Li 2010). An ecotope is the smallest holistic land unit, which commonly includes three characteristics: (1) smallest homogenous, mapable unit of land, (2) homogenous in general substrate conditions, potential natural vegetation, and potential ecosystem functioning, and (3) possibly being composed of patches in different successional stages of land uses. Species diversity is distinguished into individual-counting diversity and biomass-based diversity. Ecotope diversity is divided into individual ecotope-counting diversity and ecotope-area based diversity.

Functional diversity refers to those components of biodiversity that influence how an ecosystem operates or functions. The biological diversity, or biodiversity, of a habitat is much broader and includes all the species living in a site, all of the genotypic and phenotypic variation within each species, and all the spatial and temporal variability in the communities and ecosystems that these species form. Functional diversity, which is a subset of this, is measured by the values and range in the values, for the species present in an ecosystem, of those organismal traits that influence one or more aspects of the functioning of an ecosystem. Functional diversity is of ecological importance because it, by definition, is the component of diversity that influences ecosystem dynamics, stability, productivity, nutrient balance, and other aspects of ecosystem functioning.

Alpha Diversity is the biodiversity within a particular area, community or ecosystem. It is usually expressed by the number of species (i.e., species richness) in that ecosystem. This can be measured by counting the number of taxa (distinct groups of organisms) within the ecosystem (e.g., families, genera, and species).

Beta diversity (β-diversity) is a measure of biodiversity which works by comparing the species diversity between ecosystems or along environmental gradients. This involves comparing the number of taxa that are unique to each of the ecosystems. It is the rate of change in species composition across habitats or among communities. It gives a quantitative measure of diversity of communities that experience changing environments.

Gamma Diversity refers to the total species richness over a large area or region. It is a measure of the overall diversity for the different ecosystems within a region. It is the product of a diversity of component ecosystems and the P diversity between component ecosystems. Gamma diversity can be expressed in terms of the species richness of component communities as follows:

$$Y = S_1 + S_2 - c \tag{1}$$

where, $S_1$ is the total number of species recorded in the first community, $S_2$ is the total number of species recorded in the second community, c is the number of species common to both communities.

## 8.2 Change Trends and Scenarios of Terrestrial Ecosystems

### 8.2.1 The Improved HLZ Model

The HLZ (Holdridge life zone) model is a scheme that uses three bioclimatic variables derived from standard meteorological data to formulate the relationship between climate patterns and broad-scale vegetation distribution in middle-long time scale. It relates the distribution of major terrestrial ecosystems (termed "life zones") to the bioclimatic variables. The HLZ classification divides the world into over 38 terrestrial ecosystem types on the basis of mean annual biotemperature (MAB), average total annual precipitation (TAP) and potential evapotranspiration ratio (PER) (Holdridge 1947, 1967).

Biotemperature is defined as the mean of unit-period temperatures with substitution of zero for all unit-period values below 0 °C and above 30 °C. Evapotranspiration is the total amount of water that is returned directly to the atmosphere in the form of vapor through the combined processes of evaporation and transpiration. Potential evapotranspiration is the amount of water that would be transpired under constantly optimal conditions of soil moisture and plant cover. The potential evapotranspiration ratio is the ratio of mean annual potential evapotranspiration to average total annual precipitation, which provides an index of biological humidity conditions (Holdridge et al. 1971). However, the original HLZ model only computes the vegetation type in the weather observation station. Due to improve the HLZ model to simulate the vegetation distribution in a continual surface, the High Accuracy Surface Modeling (HASM) method was adopted to interpolate the station observation climatic data and downscale the climatic model scenario data to ensure high spatial resolution and high precision of parameters, and the formula of HLZ model is improved as.

$$MAB(x, y, t) = \frac{1}{365}\sum_{j=1}^{365} TEM\,(j, x, y, t) \tag{2}$$

$$TAP(x, y, t) = \sum_{j=1}^{365} P(j, x, y, t) \tag{3}$$

$$PER(x, y, t) = \frac{58.93 \times MAB(x, y, t)}{TAP(x, y, t)} \tag{4}$$

$$TEET_i(x, y, t) = \sqrt{(M\,(x, y, t) - M_{i0})^2 + (T(x, y, t) - T_{i0})^2 + (P(x, y, t) - P_{i0})^2} \tag{5}$$

where $MAB(x, y, t)$, $TAP(x, y, t)$ and $PER(x, y, t)$ are respectively the values of mean annual biotemperature (°C), average total annual precipitation (mm) and potential evapotranspiration ratio at site (x, y) at the period $t$; $TEM\,(j, x, y, t)$ is the mean accumulated temperature above 0 °C and below 30 °C on the $j$th day at the period

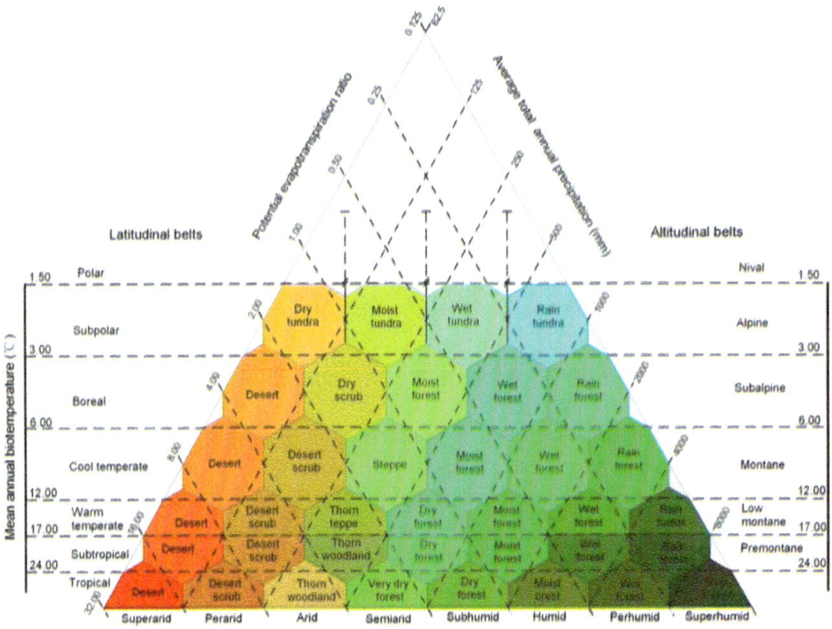

**Fig. 1** Framework of improved HLZ ecosystem model

$t$; $P(j, x, y, t)$ is the total precipitation on the $j$th day; $M(x, y, t) = lnMAB(x, y, t)$, $T(x, y, t) = lnMAP(x, y, t)$, $P(x, y, t) = lnPER(x, y, t)$; $M_{i0}$, $T_{i0}$ and $P_{i0}$ are respectively the value of *MAB* logarithm, *MAP* logarithm and *PER* logarithm at the central point of the $i$th vegetation ecosystem type in the HLZ ecosystem scheme respectively; $HLZ_i(x, y, t)$ is the $i$th ecosystem type at the grid $(x, y)$ at the period $t$ (Fig. 1).

### 8.2.2 Change Trends of Terrestrial Ecosystem in Poyang Lake Basin

Due to simulate the Change trends of terrestrial ecosystem in Poyang Lake Basin,106 meteorological stations were available over and around the Jiang-Xi province during the period from 1951 to 2010, which 85 stations are located within Jiang-Xi province and 21 stations around Jiang-Xi province. These stations can provide daily temperature and precipitation data. The mean annual temperature (MAT) and the mean annual precipitation (MAP) at every meteorological station are calculated in terms of the daily temperature and precipitation data during the period from 1951 to 2010. We downscaled a digital elevation model (DEM) of Jiang-Xi province at a spatial resolution of 90 m × 90 m from the Shuttle Radar Topography Mission (SRTM) (http://srtm.csi.cgiar.org). The DEM at a spatial resolution of 250 m × 250 m is developed by means of a cubic convolution resampling approach. The DEM was used as auxiliary data to interpolate MAT and MAP from the stations to grid cells at the spatial resolution of 250 m × 250 m.

Compared with 1960s, the main reason for the change of terrestrial ecosystem type was caused by the change of temperature in the 1970s. After the 1980s, the influence of precipitation change on the transformation of terrestrial ecosystem was intensified. In the 2000s, the precipitation exceeded the temperature becoming the main reason for the change of terrestrial ecosystem type. The results of the three-year time scale show that the change of temperature is the main reason to the change of terrestrial ecosystem type, which leads to nearly 70% of the pattern of terrestrial ecosystem changes.

The main forests with a large area are subtropical humid forests, warm temperate moist forests and warm temperate wet forests. Numerical simulations show that the small changes of climate may have a greater impact on the potential ecosystems in Poyang Lake Basin. At the time scale of 30 years, about 20% of the terrestrial ecosystem change in Poyang Lake Basin due to climate change.

### *8.2.3 Change Trends of Terrestrial Ecosystem in Loess Plateau of China*

Loess Plateau, located at 100°54′–114°43′E and 33°43′–41°16′N, covers an area of about 640,000 km$^2$ in the upper and middle reaches of China's Yellow River. It is not only the largest region in Loess but also suffers the highest rates of soil erosion in China. The mean annual temperature and mean annual precipitation respectively range from 3.6 °C to 14.3 °C and from 150 to 750 mm. Soil erosion is the principal ecological issue in Loess Plateau. Low vegetation coverage, as a result of unsuitable land use (e.g., step slope land reclamation and overgrazing), is the leading cause of severe soil erosion in the region (Fan et al. 2013a). Loess Plateau is therefore one of the key target areas for implementation of the "Grain-for-Green Programme" initiated by China's central government. Between 1998 and 2010 all cultivated lands with a slope over 25° were converted into woodland or grass land in order to effectively control the soil erosion. The project succeeded in increasing vegetation cover in Loess Plateau. This improved the soil nutrient levels, and helped recover the soil's physical properties. However, terrestrial ecosystem changes in Loess Plateau still need to be quantitatively analyzed.

Due to simulate Change trends of terrestrial ecosystem in Loess Plateau of China, the daily temperature and precipitation data from 1971 to 2005 were selected from the weather stations that are scattered around Loess Plateau of China. The bioclimatic data in Loess Plateau used of the mean annual biotemperature, average total annual precipitation and potential evapotranspiration ratio were simulated to 1 km × 1 km by operating the high accuracy and speed method of surfacing modeling (HASM), which was described in the last courses of Prof. Yue and Prof. Zhao. The DEM

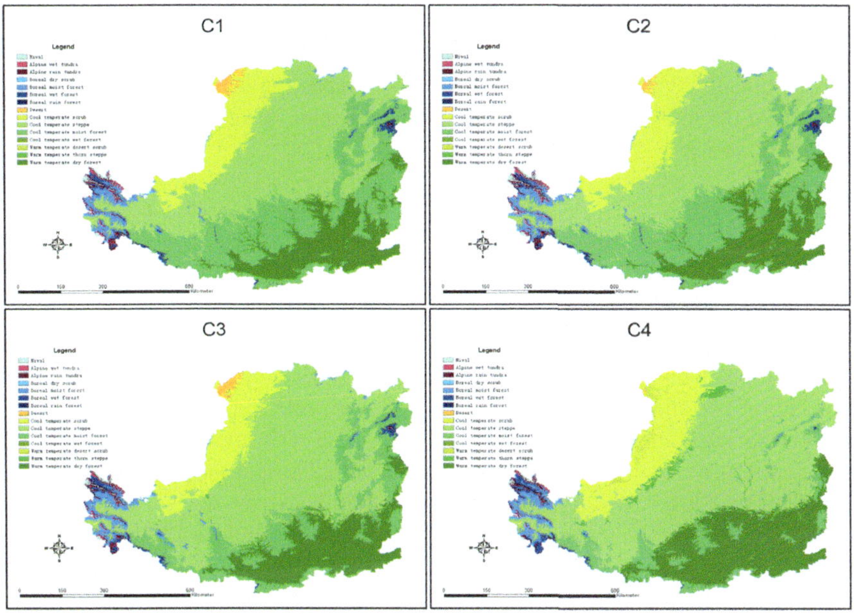


**Fig. 2** Spatial distribution of HLZ ecosystem in Loss Plateau of China from C1 to C4

data of Loess plateau were obtained from SRTM data (http://srtm.csi.cgiar.org), and resampled to a spatial resolution of 1 km × 1 km.

The simulated result shows that there are 15 types of terrestrial ecosystem in Loess Plateau and just in the Qilian Mountains, located in southwestern Loess Plateau, 9 types appear. From C1 to C4, there are significant regional differences of climate-vegetation geographic distribution pattern in Loess Plateau of China (Fig. 2).

The cool temperate steppe, cool temperate moist forest and warm temperate are the major terrestrial ecosystem types which respectively occupied 82.73%, 83.19%, 82.86% and 79.77% during the periods from C1 to C4. Alpine wet tundra and rain tundra mainly distribute in high elevation region above 3000 m; Boreal dry scrub with small area is distributed in the fringe area of boreal moist forest which is border on Inner Mongolia Plateau. Desert major distributes in the northwest of Loess Plateau of China.

Terrestrial ecosystem vertical distribution pattern can be calculated in terms of DEM data and terrestrial ecosystem spatial distribution data during the periods from C1 to C4, which shows that the average elevation of nival, alpine and boreal ecosystem distribution have continuously increased. The average elevation increasing rate (17.7 m per decade) of the boreal wet forest is the fastest, the alpine rain tundra is the second fast (6.3 m per decade), and other alpine and boreal terrestrial ecosystem types is relative slow (about 3.0 m per decade).

By integrated analyzing upper bound and lower bound of each terrestrial ecosystem type, we can find the vertical distribution changing trend and area changing trend of every type are coherent. That is to say, in most cases, if one type's area shrinks, its' vertical distribution range will also shrink. These analysis results indicate the quantitative analysis of climate change impacting on vegetation ecosystem will afford the efficient data to support the implement of reasonability ecological restoration project.

### *8.2.4 Basic Data Change Trends of Terrestrial Ecosystem in China*

Due to simulate the change trends of terrestrial ecosystem in China, daily temperature and precipitation data from 1971 to 2005 were selected from the 852 weather stations that are scattered China. The bioclimatic data of the mean annual biotemperature, average total annual precipitation and potential evapotranspiration ratio were simulated to 1 km × 1 km by operating the high accuracy and speed method of surfacing modeling (HASM). Climate scenario data include the RCP26, RCP45 and RCP85 scenario data released by CMIP5. RCP26, RCP45 and RCP85 represent the low, middle and high scenarios of greenhouse gas emissions (http://www.ipcc-data.org), which were downscaled to 1 km × 1 km by operating the high accuracy and speed method of surfacing modeling (HASM. The DEM data of China were obtained from SRTM data (http://srtm.csi.cgiar.org), and resampled to a spatial resolution of 1 km × 1 km.

The simulated results show that 28 terrestrial ecosystem types can be found in China during the period from 1960 to 2000s, among which tropical moist forest has since 1970s. The 10 dominant terrestrial ecosystem types include warm temperate moist forest (12.94%), nival area (10.66%), cool temperate moist forest (10.07%), cool temperate steppe (9.76%), warm temperate desert (7.89%), subtropical moist forest (7.89%), warm temperate dry forest (7.47%), cool temperate scrub (5.37%), boreal moist forest (4.94%) and alpine rain tundra (4.58%). Their area accounts for 81.56% of the whole of China. Although these 10 terrestrial ecosystem types are dominant ones in China, other 18 terrestrial ecosystem types play an important role in terrestrial ecosystem diversity and most of them, such as warm temperate thorn steppe, subtropical wet forest, cool temperate wet forest, are much more sensitive to climate change and human activities than the 10 dominant terrestrial ecosystem types.

Tropical moist forest has appeared in China since 1970s and its mean center has shifted northeast from Hainan province to Guangzhou province, which implies that both temperature and precipitation have an increase trend in Hainan province and its northeast neighboring area in recent decades. Area of nival zone has continuously decreased since 1960 and the mean center of nival zone has shifted towards northwest, which tallies with climate warming and precipitation increasing in southwest Tibet and northwest China. Cool temperate steppe, which accounts for 9.76% of the whole terrestrial area of China, shifted towards southwest about 500 km. Although terrestrial ecosystem types increased from 27 to 28 because tropical moist forest appeared in Hainan province in 1970s, on the average, terrestrial ecosystem diversity has a decrease trend in the four decades and its decadal decrease rate is 0.003%. Meanwhile, terrestrial ecosystem patch connectivity has continuously increased in the four decades and the decadal increase rate is 0.0439% on the average. Because the patch connectivity is defined as movement efficiency of migrants in patches of a region under consideration (Yue et al. 2005, 2006), the continuous increase of patch connectivity signifies the strengthening ability of human utilizing ecosystem services. In short, the simulated results on the spatial pattern changes reflect the great effects of climate change on terrestrial ecosystems in China.

### 8.2.5 Scenarios of Terrestrial Ecosystem in China

The simulated results of terrestrial ecosystem scenarios in China shows that the nival areas in the Qinghai-Xizang Plateau would rapidly shrink towards the northwest and the nival zones in the Tianshan and Qilian mountains would shrink back to the mountaintops in response to the projected increase in temperature and precipitation during the period from 2010 to 2100. Except for desert and desert scrub, warm temperate terrestrial ecosystem and subtropical terrestrial ecosystem types currently are distributed south of the Yangtze River, while they would cross the Yangtze River and shift towards the north in the twenty-first century (Fig. 3).

Tropical dry and tropical moist forests would expand from the southeastern coastal area towards the north. Arid terrestrial ecosystem types, such as warm temperate desert scrub and warm temperate thorn steppe, would increase continuously, which means that desertification might become a more serious problem in the twenty-first century in China. The area of arid and cold terrestrial ecosystem types would gradually decrease and be replaced by terrestrial ecosystem types characteristic of arid and warm environments, such as warm temperate desert scrub, warm temperate thorn steppe and warm temperate dry forest.

Nival areas, cool temperate moist forest, warm temperate moist forest, subtropical moist forest and boreal wet forest would be the five terrestrial ecosystem types with the biggest area. The combined area of these five terrestrial ecosystem types would be greater than 50% of China's land mass. Among these five terrestrial ecosystem types, nival areas and warm temperate moist forest and boreal wet forests would decrease continuously by 38 million ha, 43 million ha and 16 million ha, respectively, during the four periods, while subtropical moist and cool temperate forests would increase continuously by 60 million ha and 25 million ha, respectively.

In terms of the mean-center model, the shift ranges of terrestrial ecosystem types would be markedly shift in twenty-first century. Especially, boreal wet forest, subtropical moist forest, tropical dry forest, warm temperate moist forest and subtropical wet forest would have a bigger shift range, indicating that these terrestrial ecosystem types are more sensitive to the climate change. The continuous shift of both boreal dry and cool temperate scrubs towards the northwest in the twenty-first century means

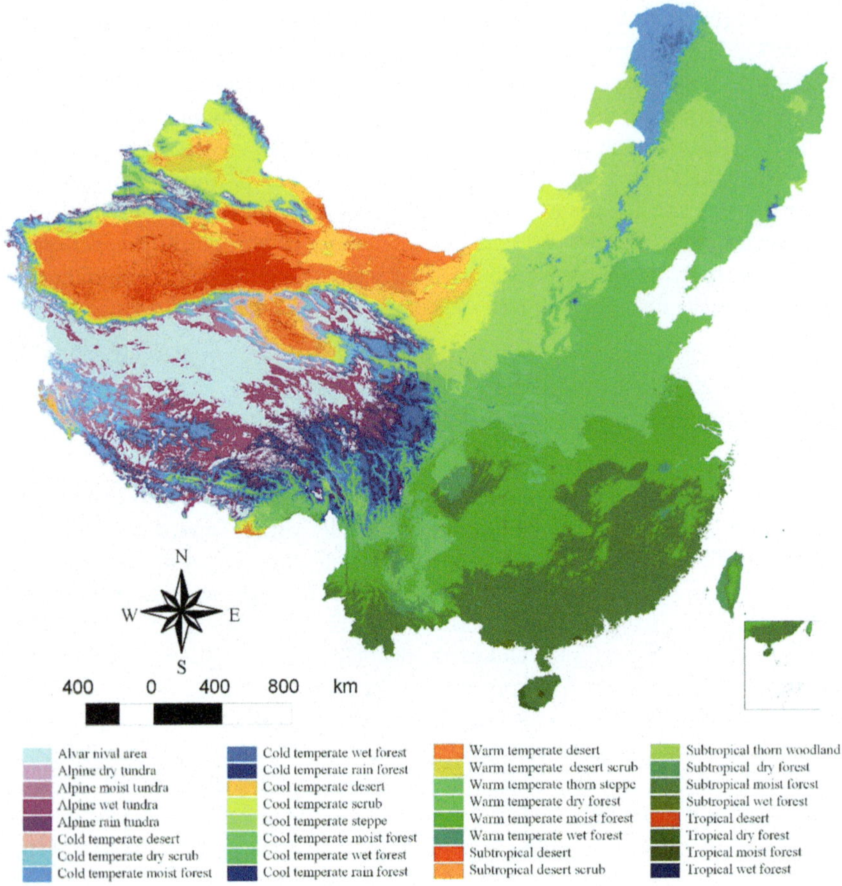


**Fig. 3** Spatial distribution of HLZ ecosystem of China

that the terrestrial ecosystem types present in arid environments would move towards the northwest in response to the projected temperature and precipitation increase in northwest China.

Mean centers of alpine moist tundra and cool temperature wet forest would shift towards the north in general, possibly because of the areal expansion of the two terrestrial ecosystem types in the high mountains of northern China and the precipitation increase in northern China. Terrestrial ecosystem types in cold and frigid environments, such as alpine moist tundra, would be replaced by other terrestrial ecosystem types, such as boreal wet forest, such that the relative ratio of boreal forest in Qinghai-Xizang Plateau to that in the Da Hinggan, Xiao Hinggan and Changbai mountains would increase in the twenty-first century because of temperature and precipitation increases in Qinghai-Xizang Plateau. This would lead to a considerable shift of the mean center of boreal forest towards the southwest.

### 8.2.6 Scenarios of Terrestrial Ecosystem in Eurasia

The distributions of complex geographic environments and numerous vegetation types exhibit obvious latitudinal and vertical zonality in Eurasia (Zheng 1996). How to quantitatively simulate and reveal the change trends and scenarios of terrestrial ecosystems in Eurasia, which is the planning area for the Belt and Road Initiative (BRI), is important for the sustainable development of the countries in the BRI region (Liu 2015).

Eurasia was classified into seven regions, China (A), Mongolia and Russia (B), West Asia (C), Central Asia (D), South Asia (E), Southeast Asia (F) and Europe (G) according to the Annual Report of Global Ecological Environmental Remote Sensing Monitoring in 2015 (Li and Shi 2016), which covers all six economic corridors of the China-Mongolia-Russia Economic Corridor (CMREC), the New Eurasian Continental Bridge Economic Corridor (NECBEC), China-Central Asia-Western Asia Economic Corridor (CCWAEC), China-Pakistan Economic Corridor (CPEC), Bangladesh-China-India-Myanmar Economic Corridor (BCIMEC) and China-Central South Peninsula Economic Corridor (CCSPEC) (Fig. 4).

Due to simulate the scenarios of terrestrial ecosystem in Eurasia, the observation monthly climatic data has been collected from 2127 weather observation stations in Eurasia during the period from 1981 to 2010. The climatic scenario data of CMIP5 RCP26, RCP45 and RCP85 scenarios has been got from IPCC website during the period from 2011 to 2100. The mean annual biotemperature, average total annual precipitation and potential evapotranspiration ratio on spatial resolution of 0.125° × 0.125° were respectively obtained by operating a high accuracy and speed method of surfacing modeling (HASM) (Yue 2010, Yue et al. 2016) during all the four periods from 1981 to 2010 (T0), 2011 to 2040 (T1), 2041 to 2070 (T2), and 2071 to 2100 (T3).

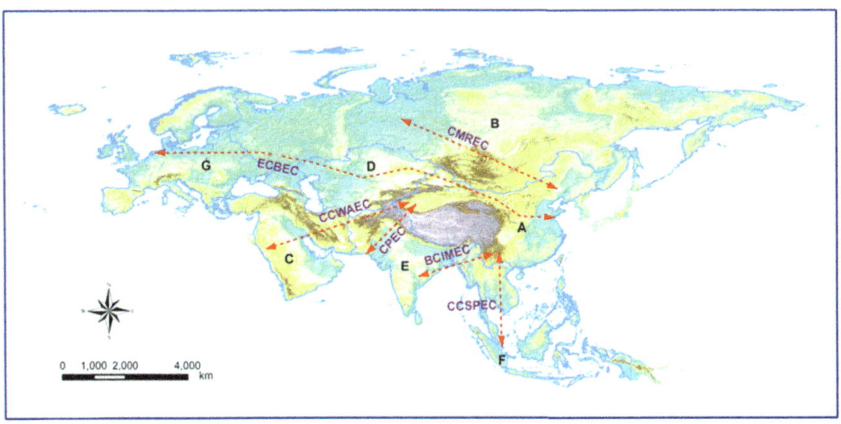


**Fig. 4** DEM and seven regions along the BRI area in Eurasia

The simulated results show that there was an average of 32 terrestrial ecosystem types in Eurasia from 1981 to 2010. Cold temperate wet forest, cool temperate moist forest and desert are the major terrestrial ecosystem types. These ecosystem types cover 14.49%, 11.61% and 10.61% of the total area of Eurasia. Cold temperate wet forest is mainly distributed in the northern part of the Eastern European Plain, the central part of the West Siberian Plain, and the mountainous region of the central and southern parts of the Central Siberian Plateau.

Cool temperate moist forest is mainly distributed in the eastern and western parts of Eurasia, especially in the southern part of the Eastern European Plain, the northwestern and southeastern parts of the Alps Mountains, the Korean Peninsula, and the Loess Plateau and Northeast Plain of China. Desert is mainly distributed in the Arabian Peninsula and its eastern area, the Turan Plain, and the western part of the Mongolian Plateau, and it is widely distributed throughout the 6 large deserts.

Under the RCP26, RCP45 and RCP85 scenarios, the subpolar/alpine moist tundra, tropical dry forest, warm temperate dry forest and tropical thorn woodland would be more sensitive to climate change than any of the other terrestrial ecosystem types during the period from 2010 to 2100. Cold temperate wet forest, cool temperate moist forest and desert are the main terrestrial ecosystem types, which account for 14.49%, 11.61% and 10.61% of the total area of Eurasia, respectively.

Climate change would lead to decreases in the ecological diversity and patch connectivity of the HLZ ecosystems in the future, and the ecological diversity and patch connectivity would decrease faster under more rapid climate change, which means that the ecological diversity and patch connectivity of the HLZ ecosystem in Eurasia would show different decreasing trends under different climate change scenarios. During the period from 2010 to 2100, the ecological diversity of the terrestrial ecosystem under the RCP26, RCP45 and RCP85 scenarios would decrease by 0.091%, 0.130% and 0.160%, respectively, per decade on average.

The terrestrial ecosystem types in Central Asia would have the highest change intensity, and those of Southeast Asia would have the lowest change intensity under the RCP26, RCP45 and RCP85 scenarios during the period from 2010 to 2100. The terrestrial ecosystem types throughout the BRI area would mainly change from the cool type to the warm type in Europe, would generally change to the warm-wet type in China, Mongolia and Russia, and Southeast Asia, and would change to the warm-dry type in West Asia, Central Asia and South Asia, which indicates that the areas of the dry and semidry terrestrial ecosystem types might expand under climate change in the future.

### *8.2.7 Scenarios of Terrestrial Ecosystem on Global Scale*

During the progress of simulating the scenarios of terrestrial ecosystem on global scale, the basic data of DEM, climatic data observed in weather station, climatic scenario data were involved. The DEM is from Shuttle Radar Topography Mission (SRTM), which can be obtained at spatial resolution of three arc seconds, about 90 m

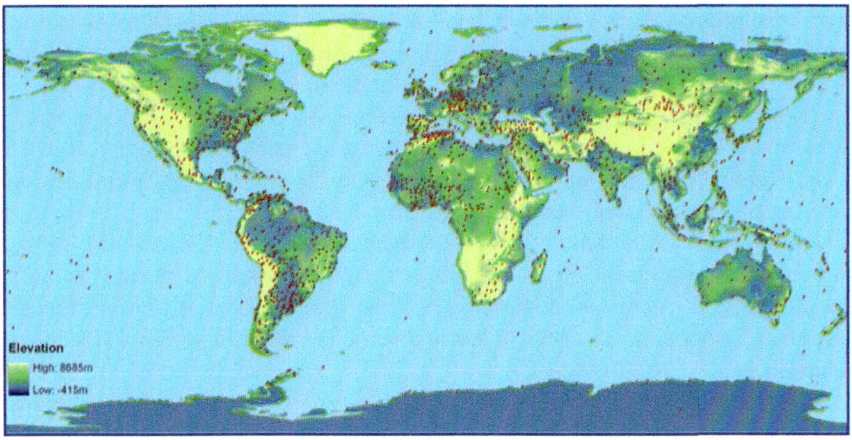


**Fig. 5** DEM on global level and 2766 weather observation stations scattered over the world

(http://srtm.csi.cgiar.org). The first source of climate change dataset on global level is from Monthly Climatic Data for the World (MCDW). MCDW is a publication of monthly summary climatic data from around the world, prepared by National Oceanic and Atmospheric Administration (NOAA) cooperating with the World Meteorological Organization (WMO) (Fig. 5).

The climatic scenario data of CMIP5 RCP26, RCP45 and RCP85 scenarios has been got from IPCC website during the period from 2011 to 2100. The RCP85 scenario describes a future world of very rapid economic growth, global population that peaks in mid-century and declines thereafter, and the rapid introduction of new and more efficient technologies. RCP45 scenario describes a very heterogeneous world with more rapid population growth but less rapid economic growth than RCP85. The RCP26 scenario describes a world, in which population increases at a lower rate than in RCP45 scenario and the emphasis is on local solutions to economic, social, and environmental sustainability. The mean annual biotemperature, average total annual precipitation and potential evapotranspiration ratio on spatial resolution of $0.125° \times 0.125°$ were respectively obtained by operating a high accuracy and speed method of surfacing modeling (HASM) (Yue 2010, Yue et al. 2016) during all the four periods from 1981 to 2010 (T0), 2011 to 2040 (T1), 2041 to 2070 (T2), and 2071 to 2100 (T3). The spatial resolution of scenario downscaled data of global MAT, MAP, and PER all are $0.125° \times 0.125°$ (Fig. 5).

The results indicate that all polar/nival, subpolar/alpine and cold ecosystem types would have a continuously decreasing trend, including nival area, moist tundra, wet tundra, rain tundra, cold temperate dry scrub, cold temperate moist forest, cold temperate wet forest, cold temperate rain forest, and cool temperate scrub. 80% of alpine moist tundra would at least disappear in the period T4, comparing with its status in T1.

Except tropical rain forest, all other tropical ecosystem types would increase, such as tropical wet forest, tropical moist forest, tropical dry forest, tropical very dry forest, tropical thorn woodland, and tropical desert scrub as well as subtropical rain forest. Tropical thorn woodland would increase by more than 97%. In general, climate change would cause a continuous decrease of ecological diversity.

Under all the three scenarios, in northern hemisphere, mean centers of subpolar/alpine moist tundra, warm temperate moist forest and subtropical dry forest would move longer distances towards west; mean center of desert would move towards west because desert area would decrease in China and central Asia; Borborema plateau in Brazil would have a rapid desertification trend. In southern hemisphere, cool temperate wet forest and cool temperate rain forest would shift longer distances towards west. Warm temperate thorn steppe and warm temperate wet forest would have their longer shift distances in stable directions in both northern and southern hemispheres.

Subpolar/alpine moist tundra would be the most sensitive ecosystem type because its area would have the rapidest decreasing rate and its mean center would shift the longest distance towards west. In other words, subpolar/alpine moist tundra could be selected as an indicator of climatic change.

Warm temperate rain forest, only 131 $km^2$, would appear in the period T4 of scenario RCP8.5 while other 33 ecosystem types would exist in all periods under all the three scenarios. Nival area would account for the biggest proportion of terrestrial surface on an average in terms of the three scenarios and the proportion would be about 12%. The first three smallest ecosystem types would be subtropical rain forest, 745 $km^2$, dry tundra, 3859 $km^2$, and tropical rain forest, 7710 $km^2$, on an average.

## 8.3 Change Trends and Scenarios of Land Cover

### *8.3.1 The Theory of Surface Modelling in Land Cover (SMLC)*

Land cover change is a complex process driven by natural factors, climate conditions, and human factors. If the different impacts of various factors all are to be considered in the process of simulated predictive analysis, the entire study will become extremely complicated, and even it is impossible to start. Climate conditions such as mean annual biotemperature, average total annual precipitation and potential evapotranspiration ratio have directly impact on the natural course of land cover (Lauenroth et al. 1993). After comparing the corresponding relationship of the spatial distribution between terrestrial ecosystem types and land cover types, the distribution of terrestrial ecosystem types and land cover types are very similar on spatial pattern (Fan et al. 2005; Yue et al. 2007, 2011; Fan et al. 2013a, b, 2015, 2020a, b) (Fig. 6).

Thus, based on the similar characteristics between the spatial distribution of terrestrial ecosystem (TE) types and land cover (LC) types, we established the corresponding probability matrix of spatial distribution between TE types and LC types. It is found that each TE type can basically correspond to a LC type in a relatively

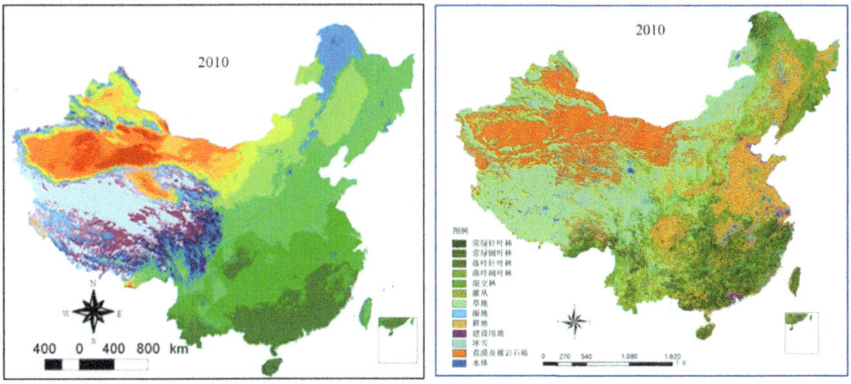


**Fig. 6.** Distribution of terrestrial ecosystem and land cover of China in 2010

large probability. Therefore, we build a prototype method of surface modelling of land cover scenarios (SLMCs) (Fan 2005; Yue et al. 2007, 2019). After more than 10 years of development, the SMLCs method has been able to simulate the land cover scenarios at regional scale, national scale and intercontinental scale. At present, the SMLCs method is being used to simulate the global land cover scenarios. The theoretical formulas of SMLCs include,

$$LP(x, y)_{k,t+1} = LP(x, y)_{k,t} \times \frac{1}{2}\left(1 + \frac{HLZP(x, y)_{k,t+1} - HLZP(x, y)_{k,t}}{HLZP(x, y)_{k,t+1} + HLZP(x, y)_{k,t}}\right) \tag{6}$$

$$PI(x, y)_{k,t} = \frac{PN(x, y)_{k,t} - Min\{PN(x, y)_t\}}{Max\{PN(x, y)_t\} - Min\{PN(x, y)_t\}} \tag{7}$$

$$TI(x, y)_{k,t} = \frac{TD(x, y)_{k,t} - Min\{TD(x, y)_t\}}{Max\{TD(x, y)_t\} - Min\{TD(x, y)_t\}} \tag{8}$$

$$EI(x, y)_{k,t} = \frac{GDP(x, y)_{k,t} - Min\{GDP(x, y)_t\}}{Max\{GDP(x, y)_t\} - Min\{GDP(x, y)_t\}} \tag{9}$$

$$\begin{aligned} LC_{P(x,y)_{k,t+1}} = {} & \alpha_1 PI(x, y)_{k,t} + \alpha_2 TI(x, y)_{k,t} + \alpha_3 EI(x, y)_{k,t} \\ & + \alpha_4 LP(x, y)_{k,t} + \alpha_5 HL\_P(x, y)_{k,t+1} \end{aligned} \tag{10}$$

$$LC\_T(x,y)_{t+1} = Value(k)_{Max\{LC\_P(x,y)_{k,t+1}|k=1,2,3,\ldots,13\}}$$
$$if\begin{cases} LC\_T(x,y)_t \in NR, LC\_T(x,y)_{t+1} \neq (Lcrop \cup Lbuilt) \\ LC\_T(x,y)_t \in ER, LC\_T(x,y)_{t+1} \neq (Lbuilt) \\ SLOPE(x,y) \geq 25, LC\_T(x,y)_{t+1} \neq (Lcrop) \\ LC\_T(x,y)_t \in BC, LC_{T(x,y)_{t+1}} = Lcrop(x,y)_t \end{cases} \quad (11)$$
$$k = 1, 2, 3, \ldots, 13; \textit{number of landcover types}$$
$$\text{t} = 2010,\ 2040,\ 2070,\ 2100;\ \text{simulation periods}$$

where $x, y$ is the coordinate of grid cell; $k$ is the type code of land cover; $t$ is the variable of time; $HLZP(x,y)_{k,t}$ represents the transition probability between land cover type $k$ and the HLZ type appeared at grid $(x, y)$ in $t$ year. $LP(x,y)_{k,t}$ is the percentage of land cover type $k$ contained in grid $(x, y)$ which should satisfy the formula $\sum_1^k LP(x,y)_{k,t} = 1$; $LP(x,y)_{k,t+1}$ is the transition probability of land cover type $k$ in $t+1$ period; $LC(x,y)_{t+1}$ is the type value of land cover at grid $(x, y)$ in $t+1$ period. $PI(x,y)_{k,t}$, $TI(x,y)_{k,t}$ and $EI(x,y)_{k,t}$ are respectively the distribution density of population, traffic density, and GDP value at the grid cell $(x, y)$ in the $t$ period; $Min\{PN(x,y)_t\}$, $Min\{TD(x,y)_t\}$ and $Min\{GDP(x,y)_t\}$ are respectively the minimum value of population, traffic density, and GDP value at the grid cell $(x, y)$ in the $t$ period; $Max\{PN(x,y)_t\}$、$Max\{TD(x,y)_t\}$ 和 $Max\{GDP(x,y)_t\}$ are respectively the maximum value of population, traffic density, and GDP value at the grid cell $(x, y)$ in the $t$ period; $LC\_T(x,y)_t$ characterized land cover type at site $(x, y)$ in the period $t$; SLOPE(x, y) characterized slope at site $(x, y)$ in the period $t$; $Lcrop(x,y)_t$ characterized cultivated land at site $(x, y)$ in the period $t$; *Lcrop* and *Lbuilt* characterized cultivated land and built-up land.

### *8.3.2 Land Cover Changes in Transition Zones of Loess Plateau in China*

Loess Plateau is not only the largest region in Loess but also suffers the highest rates of soil erosion in China, where is one of the key target areas for implementation of the "Grain-for-Green Programme" initiated by China's central government (Fan et al. 2013a, b, 2020a, b). When quantitatively identifying and classifying the different transition zones of terrestrial ecosystem (TETZs) in Loess Plateau, the location of transition zones between arid, semi-arid and semi-moist climate can be clearly recognized. Assessing the changes in land-cover between different TETZs is vital to understanding restoration efficiency and trends in land-cover change, which will assist in providing the support for the land planning and management of Loess Plateau.

Due to explicate the change trends of land cover in transition of Loess Plateau in China, mean annual biotemperature, average total annual precipitation and potential evapotranspiration ratio at a resolution of 1 km × 1 km from 1985 to 2005, which were simulated by HASM method. The land cover data was extracted from the land-cover data of the rest of China at a spatial resolution of 1 km × 1 km in 1985, 1995 and 2005, among which the classification of landcover in 1985 and 1995, with an overall accuracy of 81%, is derived from the Advanced Very High-Resolution Radiometer (AVHRR) sensors (Liu et al. 2003), and that in 2005 has an overall accuracy of 95% and is based on Landsat TM and CBERS-2 CCD data (Liu et al. 2009). Pixels are classified as cultivated land, woodland, grassland, wetland and water areas, built-up land, desert and other land.

For evaluating the transform trends of land cover in Loess Plateau, we divided them into positive transformation and negative transformation with regards to the capacity of different land cover type to provide ecosystem services. A positive and negative transformation index of land over (PNTIL) was developed to quantitatively evaluate the land cover transformation trends of TETZs; within which the positive transformation indicates that the land cover change is beneficial and reduces the eroding, overgrazing and drought ravaged in Loess Plateau, and to improve the capacity of ecosystem services. Conversely negative transformation shows the land cover change will increase the possibility of soil erosion of Loess Plateau.

$$LCTR_{ik} = \left(\frac{\Delta S_{i,k}}{S_i}\right) \times 1/t \times 100 \tag{12}$$

where $LCTR_{ik}$ is the $k$ direction transformation rate of land-cover in *ith* TETZ type; $k = 1, 2$, are respectively the positive and negative transformation rates of land-cover in the investigation area; $t$ is the variable of time; $S_i$ is the area of land-cover in *ith* TETZ types; $\Delta S_{i,k}$ is the k direction transformation area of land-cover in *ith* TETZ type. Regulations to define the transform direction of land cover in Loess Plateau: " + " and "-" respectively express the positive transformation and negative transformation of land cover (Fig. 7).

The simulated results show that the increasing ratios of built-up land were 9.61% and 8.57% per decade respectively in transition zones and adjacent biomes in Loess Plateau during the period 1985–2005. This increase in built-up land mainly occurred around urban areas, indicating that urban expansion was a major contributor to the increase of built-up land in Loess Plateau during the two decades therefore urban planning should be prioritized where it comes to Loess Plateau.

The ecological diversity and patch connectivity of land-cover in all transition zones generally shows an increasing trend during the period 1985–2005, with the same change trend in all adjacent biomes. What's more, the ecological diversity and patch connectivity of land-cover in all transition zones is higher than that of adjacent biomes, which indicates that transition zones have higher ecological diversity and patch connectivity than adjacent biomes. However, higher ecological diversity of land-cover landscape in transition zones predicates more landscape fragmentation than adjacent biomes. Furthermore, land-cover landscape with higher ecological diversity may help the survival of more species in transition zones, but it is worth noting that the ecosystem of transition zones is more unstable than other biomes

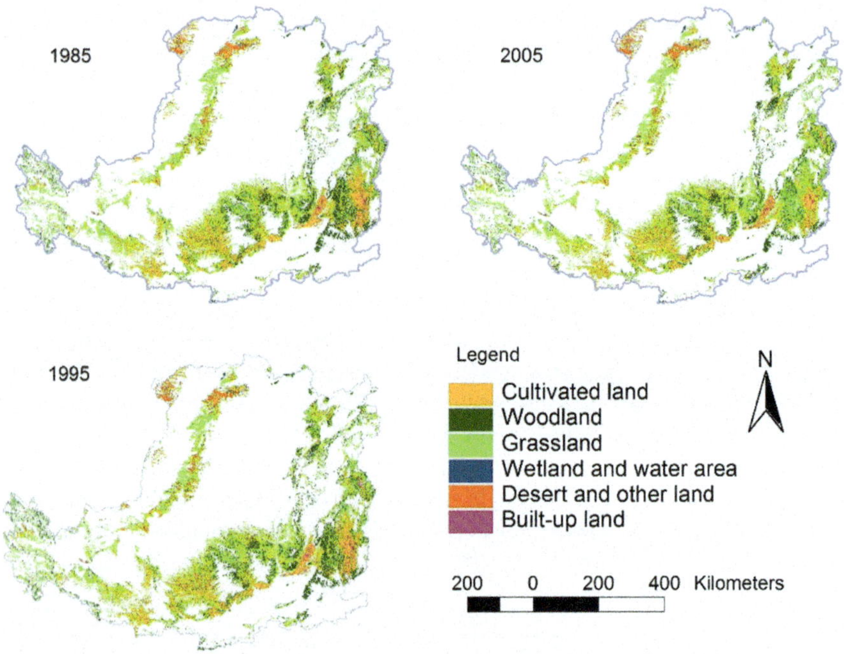


**Fig. 7** Spatial distribution of land cover in BTZ of Loess Plateau across different periods

because the land-cover landscape is more susceptible to climate change and human activities (Fan et al. 2013a).

### *8.3.3 Land Cover Changes in the NECBEC Area*

Land cover change, since 2001, has presented a clearly spatial difference in the New Eurasian Continental Bridge Economic Corridor (NECBEC) area, especially the Belt and Road Initiative (BRI) released in 2013.With the launch of Belt and Road Initiative (BRI) in 2013, more and more countries have joined the initiative. Lots of researches have been made to discuss the framework and model of cooperative development and the relationships between the involved countries in the BRI area, especially in the NECBEC area. However, the dynamic pattern and driving force of land cover change are still lack of explicitly analysis and discussion that is very beneficial to the scientific planning and efficient utilization of land resources in NECBEC area.

Due to analyze the dynamics of land cover change in the NECBEC area, land cover data used to analyze were collected from the MODIS dataset of NASA website (https://www.nasa.gov) in 2001, 2005, 2009, 2013 and 2017 at a spatial resolution of 500 m $\times$ 500 m.The land cover types were reclassified into woodland, grassland, cropland, built-up land, wetland and waterbody, and bare land, combining with the land cover classification system of University of Maryland (UMD) and a few scholars (Fan et al. 2013a, b), by operating the ArcGIS software and Python and R programming language.

Moreover, how to Quantitatively identifying the annual change intensity of land cover is important to improve the adaptation strategies used to mitigate (and in turn even benefit from) the projected impact on land use from climate change. The dynamic degree indicator was introduced to calculate the rates of land cover change

$$D_k = \frac{S_{t+1} - S_t}{S_t} \times \frac{1}{Y_{t+1} - Y_t} \times 100\% \tag{13}$$

where $D_k$ represents the annual dynamic degree of land cover type k during the periods from t to t + 1; $S_t$ and $S_{t+1}$ are the area of a certain land cover type at the periods t and t + 1, respectively; $Y_t$ and $Y_{t+1}$ are the year value at the t and t + 1 period, respectively.

Land cover change is a complex process effected by the continuous interaction of various natural elements and human activities (Fan et al. 2013a, b). If the research of driving mechanism on land cover changes only limits in some regional and administrative unit at a certain period, and ignores the grid heterogeneity and inter-annual uncertainty of land cover change, which would become more difficult to deeply understanding the driving force of land cover change and annual dynamic degree of land cover at the grid level. Thus, we developed a new spatiotemporal dynamic probability (STDP) model to compute the spatiotemporal dynamic probability of land cover at grid level in NECBEC area (Fan et al. 2021).

$$DT_k = \left(\left|\Delta S_{k,in}\right| + \left|\Delta S_{k,out}\right|\right)/S_{k,t} \tag{14}$$

$$DW_k = \begin{cases} \frac{DT_k}{\sum_{i=1}^{i=6} DT_i}, & DT_k < 1 \\ 1, & DT_k \geq 1 \end{cases} \tag{15}$$

$$DM(x, y)_k = \left\{\frac{1}{T}, 0\right\} \tag{16}$$

$$STDP(x, y)_k = \sum_{j=1}^{j=m} DW_k \times DM(x, y)_{k,j} \tag{17}$$

where x, y is the coordinate of grid cell, k is a certain type of land cover, t is the variable od time, i is the type code of land cover which value is from 1 to 6, j is the annual interval value between t and t + 1 period; $DT_k$ represent the dynamic trend and its weight index of the land cover type between t and t + 1 period; $\Delta S_{k,in}$ and $\Delta S_{k,out}$ are respectively increase and decrease area of land cover type k between t and t + 1 period; $S_{k,t}$ is the total area of land cover type k at t period; $DT_i$ is the dynamic trend of the ith land cover type between t and t + 1 period; $DM(x, y)_k$ represents the annual change index of the land cover type k at grid(x, y) between t and t + 1 period; T is the time interval value between t and t + 1 period; $STDP(x, y)_k$ represents the integrated spatio-temproal dynamic probability of land cover type k at grid(x, y) between t and t + 1 period, which value range is from 0 to 1.

The computed results show that the areas of grassland, cropland and built-up land all have showed an expansion trend and respectively increased by 114.57 million $hm^2$, 8.41 million $hm^2$ and 3.96 million $hm^2$, which the annual increased rate of Built-up land was the highest, and that of cropland was the lowest in NECBEC area between 2001 and 2017. The grassland mainly converted to woodland and accounted for 69.72% of the total increased woodland area, but on the other hand, the woodland mainly converted to grassland and accounted for 70% of the total increased grassland area that centrally distributed in the temperate continental climate zone, the Mediterranean climate zone, and the transition zone between woodland and grassland of Russia, which indicates that grassland and woodland were the easiest to convert to each other in NECBEC area (Fan et al. 2021).

The cropland mainly converted to Built-up land and accounted for 50% of the total increased Built-up land area, which mainly distributed in the eastern coastal areas and the northern slope of Tianshan Mountains, Turkey, Azerbaijan and Russia. There was 98% of the total increased cropland area transformed from grassland, which mainly distributed in the northwestern of China and the Eastern region of Europe. However, the areas of woodland, water area and wetland, and bare land all have showed a decrease trend and respectively decreased by 74.09 million $hm^2$, 6.26 million $hm^2$, and 46.59 million $hm^2$ in NECBEC area during the period from 2001 to 2017 (Fig. 8).

The results simulated by the spatiotemporal dynamic probability (STDP) model show that all land cover types, except built-up land, showed the change possibility

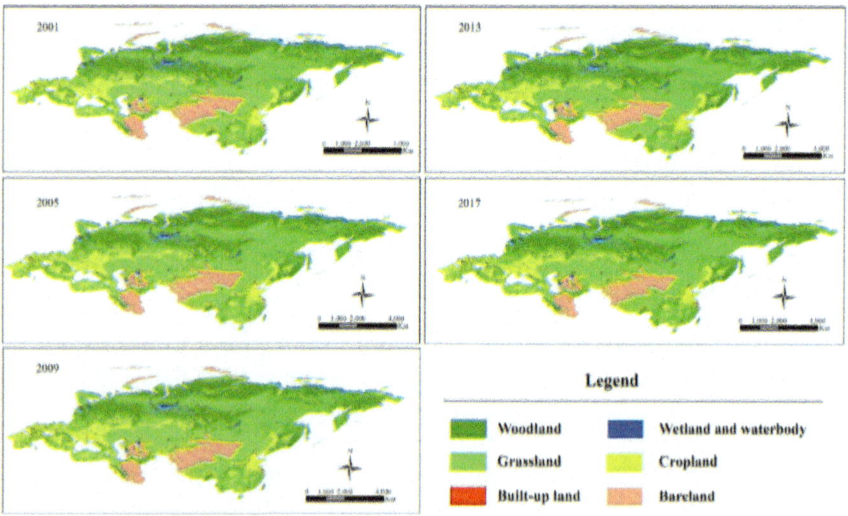


**Fig. 8** The spatial patterns of land cover in the NECBEC area region from 2001 to 2017

during the four periods of 2001–2005, 2005–2009, 2009–2013 and 2013–2017, which the STDP value of grassland was the largest, and then that of woodland and cropland on average. The annual transformed area of water area and wetland was the least, but the STDP of unchanged water area and wet land was more than that of other land cover types in 2001–2005. The STDP of unchanged bare land area was less than that of other land cover types during the four periods. Furthermore, the STDP values of cropland, water area and wetland showed a decrease trend during the three periods of 2001–2005, 2005–2009 and 2009–2013, and that only increased in 2013–2017, but their possible transformed areas showed an increase trend in all periods.

The STDP values of woodland, grassland and cropland in 2013–2017 generally showed a decrease trend, but their possible transformed areas showed a rapidly increase trend, which indicated their stability was reduced from 2013 to 2017.

### 8.3.4 Scenarios of Land Cover in the Karst Areas of Southwestern China

Karst areas, as a fragile and vulnerable ecological environment, are more sensitive to climate change and human activities. The total area of karst terrains in China is approximately one third area of Chinese territory, which type includes bare karst, covered karst and buried karst. Karst areas in southwestern China, as one of the most typical, complex and extensive karst landforms in the world record the change process of ecological environment under different geological, climatic, hydrological and biological conditions, which administratively includes Yunnan, Guizhou, Sichuan, Chongqing and Guangxi (Fig. 9).With the rapid socio-economic development, rocky desertification of karst areas in southwestern China has been deteriorated that includes serious soil erosion, extensive exposure of basement rocks, drastic decrease of soil productivity, and the appearance of a desert-like landscape (Yuan 1997; Wang et al. 2004; Ying et al. 2014), due to the high population pressure, and more and more irrational, intensive land use and resources exploitation (Ying et al. 2014).

Due to simulate the scenarios of land cover in the karst areas of southwestern China, the Bioclimatic datasets include the mean annual biotemperature, average total annual precipitation and potential evapotranspiration ratio were simulated to 1 km × 1 km by HASM method that were introduced in the last classes, which involve the observation data from 1981 to 2010, and the scenarios data of IPCC CMIP5's RCP26, RCP45 and RCP85 data from 2011 to 2100; Spatial data of terrestrial ecosystem types in 2010, 2040, 2070 and 2100 were respectively generated by operating iHLZ model; History land cover data in karst areas of western China has been separated from the land cover data with 1 km resolution of the rest of China in 1980 and 2010. The classification of land cover in 1980 with an overall accuracy of 81%, is derived from the Advanced Very High-Resolution Radiometer (AVHRR) sensors.

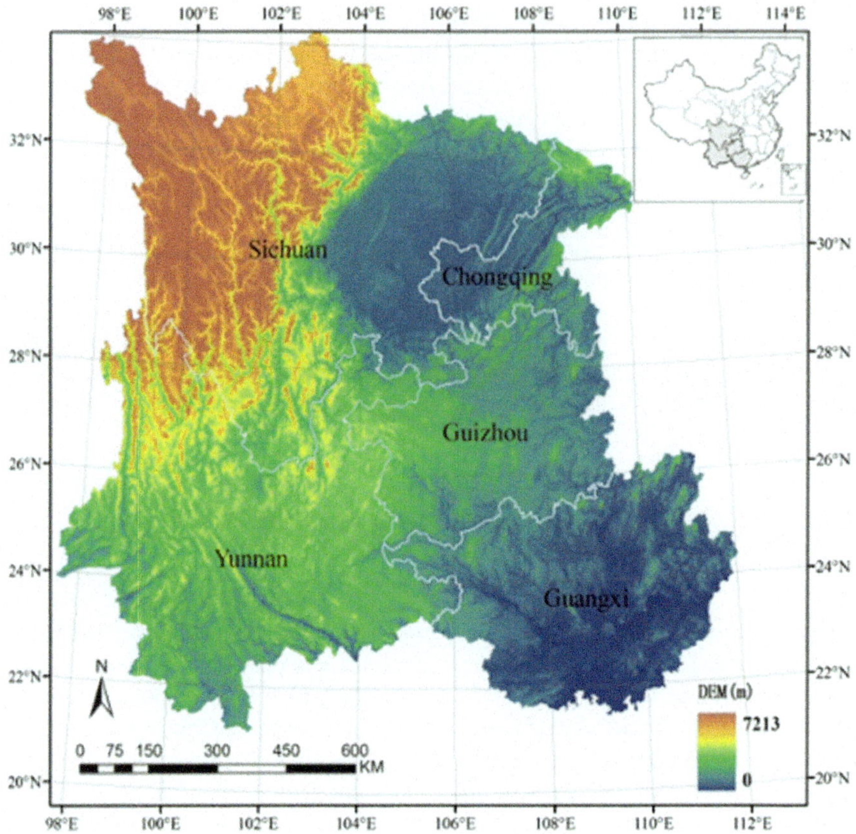


**Fig. 9** A digital elevation model of karst areas in southwestern China

The classification of land-cover in 2010 has an overall accuracy of 86% and is based on Landsat TM/ETM + images (Liu et al. 2014). The land cover types include evergreen coniferous forest, evergreen broadleaf forests, deciduous coniferous forest, deciduous broadleaf forest, mixed forest, scrublands, grassland, wetland, cropland, built land, nival area, desert and bare rock, and water area (Fan et al. 2015).

During the process of validation, the land cover data in 1980 and the terrestrial ecosystem in 1980 and 2010 were used as the initial data (Fig. 10). After operating SMLCs, we obtained the simulated land cover data in 2010. Then, we compared the simulated land cover data with actual land cover data in 2010. The assessment method of average total accuracy ($P_{ATA}$) and Kappa index (KI) were used to validate the accuracy of SMLCs, which can be respectively formulated as,

$$P_{ATA} = \sum_{i=1}^{r} p_{ii}/N \tag{18}$$

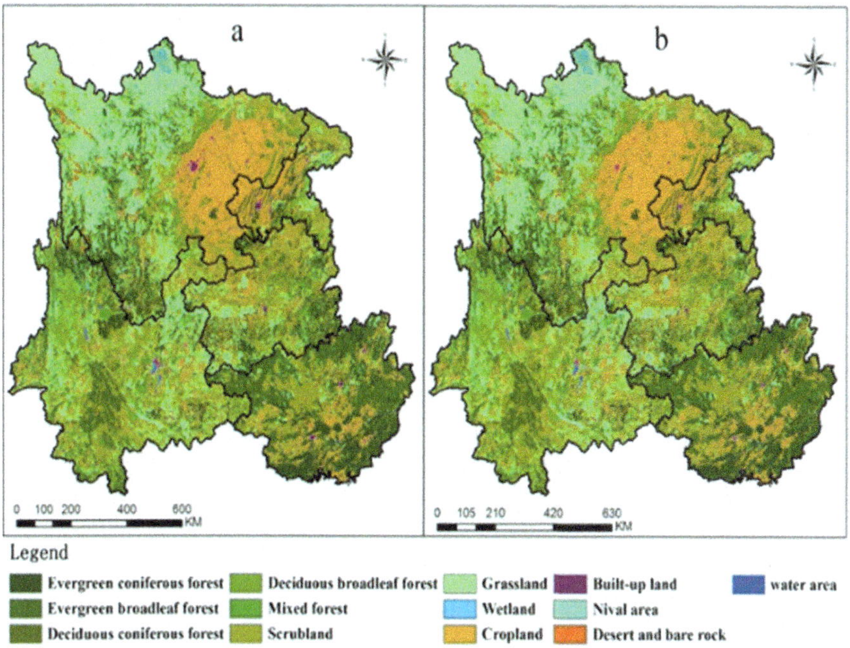


**Fig. 10** Actual land cover map (**a**) and simulated land cover map (**b**) in 2010

$$KI = \left(N\sum_{i=1}^{r} p_{ii} - \sum_{i=1}^{r}(p_i + p_{i+1})\right) \Big/ \left(N^2 - \sum_{i=1}^{r}(p_i + p_{i+1})\right) \tag{19}$$

where $N$ is the total grid cell number of research area; r is the number of land cover type; $P_{ii}$ is the grid cell number of ith land cover type simulated correctly; $P_{i+}$ is the grid cell number of ith type of actual land cover data; $P_{i+}$ is the grid cell number of ith type of simulated land cover data. The validation indicates that the $P_{ATA}$ and KI are respectively 90.25% and 87.96%.

The simulated results of land cover distribution in the karst areas of southwestern China show that the evergreen coniferous forest and evergreen broadleaf forests would be mainly distributed in the mountainous and hilly regions of the Henduan Mountains and around Sichuan Basin, Yunnan-Guizhou Plateau and the most part of hilly regions in Guangxi province. Other woodland types, including deciduous coniferous forest, deciduous broadleaf forest and mixed forest would be scattered in the south of Henduan Mountains, Dabashan Mountains, Wushan Mountains, and Yunnan-Guizhou Plateau. Scrubland would mainly occur in the hilly region with relative low altitude of Yunnan, Guizhou and Guangxi. Grassland would be continuously and intensively distributed in the northwestern Sichuan Province. Besides, there would be grassland discontinuously distributed in Yunan, Guizhou and Guangxi Province, Staggered distribution with woodland in corresponding area. Water area mainly include rivers, lakes, reservoirs and water terraces distributed in southwestern China. From 2010 to 2100, river areas would still be the mainly water body area, which include the upper reaches of Yangtze River, Mekong River and Zhujiang River within Yunnan, Guizhou, Sichuan, Chongqing and Guangxi provinces. Wetland would mainly occur in low-lying wet areas around lakes and watersheds. Nival areas include ice and snow land which would be mainly distributed in highly mountains of Western Sichuan province and Henduan Mountains in northwestern of Yunan province. Desert and bare rock land mainly refers to karst rocky desertification area in southwestern China, especially distributed in the center area on Guizhou Plateau.

The distribution of built-up land in southwestern China would be directly determined by the socioeconomic development, especially the urban extension speed, which would be mainly distribution in the areas near rivers areas or valleys where can provide abundant water resources, fertile soil, convenient traffic, plentiful food and other living products. Cropland would be continuously and centrally distributed in Sichuan Basin, and extensively and discontinuously distributed in Yunnan-Guizhou Plateau and Guangxi province. The simulated results of land cover scenarios in the karst areas of southwestern China show that the change of nival area, desert and bare rock would mainly occur in Yunnan and Sichuan under all the three scenarios, among which area would continually decrease in all the three scenarios except that nival area distributed in Sichuan would have an increase during the period from 2010 to 2040, and desert and bare rock distributed in Yunnan would increase during the period from 2070 to 2100 under RCP26 scenario. Grassland distributed in Sichuan, Guizhou and Guangxi would gradually decrease during all the three periods under all the three scenarios, but the annual decrement would be decelerating. Moreover, the change of grassland distributed in Chongqing and Yunnan would have a considerable fluctuation. The change of wetland would mainly occur in Sichuan, which would have a continuous increase trend during all the three periods under all the three scenarios. The area of evergreen broadleaf forest and evergreen coniferous forest would all continuously increase in the five provinces of southwestern China during all the three periods under the RCP45 and RCP85 scenarios, except that evergreen coniferous forest distributed in Yunnan would decrease in the period from 2010 to 2040.

In terms of all the three scenarios, the cropland area of the five provinces in southwestern China would have a continuous decrease trend during all the three periods, among which the main decrement would come from Yunnan and Sichuan. The area of built-up land in whole southwestern China would have a continuous increase trend during all the three period under the three scenarios except a decrease in Yunnan province.

### *8.3.5 Scenarios of Land Cover in China*

Due to simulate the scenarios of land cover in China, the bioclimatic datasets include the mean annual biotemperature, average total annual precipitation and potential evapotranspiration ratio were simulated to 1 km × 1 km by HASM method that were introduced in the last classes, which involve the observation data from 1971 to 2000, and the scenarios data of HadCM3 A1FI, A2 and B2 data from 2011 to 2100. Spatial data of terrestrial ecosystem types in 2010, 2040, 2070 and 2100 were respectively generated by operating iHLZ model.

Moreover, the history land cover of China has been from the land cover data with 1 km resolution of the rest of China in 2000. The classification of land cover with an overall accuracy of 81%, is derived from the Advanced Very High-Resolution Radiometer (AVHRR) sensors (Liu et al. 2003). The land cover types include evergreen coniferous forest, evergreen broadleaf forests, deciduous coniferous forest, deciduous broadleaf forest, mixed forest, scrublands, grassland, wetland, cropland, built land, nival area, desert and bare rock, and water area (Yue et al. 2007; Fan et al. 2015).

#### 8.3.5.1 Spatial Distribution of Land Cover in China

The simulated results of land cover distribution in China show that, between 2020 and 2100 years, the cultivated land in China would be in principle divided into two production areas, agricultural region and livestock farming region, which take the curve linking Da-Xiao Hinggan Mountains, Yulin, Lanzhou, the east Qinghai-Xizang Plateau and its southeast edge as their border. Cultivated land would centrally be distributed in northeast plain, north China plain, middle and lower reaches of Yangtze River, Sichuan basin, central Shaanxi plain, Hexi corridor in Gansu province, and alluvial fan areas on the north and the south of Tianshan mountains. In addition, a great quantity of cultivated land would scatter on hilly areas in the south of China.

The complicated terrain characteristics and heterogeneous climate lead to great difference of spatial woodland distribution. Woodland in northeast China would mainly be distributed in Da-Xiao Hinggan Mountains, Changbai Mountains and East Liaoning Basin; in southwestern China, woodland would mainly be distributed in areas of Himalaya Mountains and Hengduan Mountains in the east and south of Yalu Tsangpo River in Tibet, mountainous range around Sichuan basin, Yunnan-Guizhou plateau and most hilly areas in Guangxi.

In southeastern China, woodland would mainly be distributed in the low mountainous and hilly areas such as Wuyi Mountains, Nanling ridges and Taiwan Mountains. In northwestern China, the woodland would mainly be distributed in the mountainous areas of Tianshan Mountain, Altai Mountains, Qilian Mountains, Ziwu Mountain, Helan Mountain, Liupan Mountain and Yinshan Mountain. In short, woodland in China would mainly be distributed in mountainous and hilly areas.

Grassland would mainly be distributed in western China and comparatively less in eastern China. Geographically, the grassland would mainly be distributed in Qinghai-Xizang Plateau, Inner Mongolia plateau, Loess Plateau, Tianshan Mountains, Altai Mountains and areas around Tarim Basin. In the meanwhile, some grassland would scatter on hilly areas of Hunan, Hubei, Anhui, Fujian, Yunnan, Guizhou, Sichuan, Guangdong, Guangxi and Taiwan, which would mix with woodland.

Water area of inland system in China major includes the five major lake regions of the northeast lake region, the northwest lake region, the Qinghai-Xizang Lake region, the eastern lake region and the southwest lake region. They include 2800 natural lakes approximately, each with an area greater than 1 $km^2$ and many reservoirs that are artificial lakes.

Built-up land mainly be distributed in the areas near rivers or valleys, with abundant water resources, convenient traffic, fertile soil and plentiful products, which are determined by characteristics of urban development. Because of joint affection of various factors such as geographical features and social economic conditions, in eastern China, built-up land would mainly be distributed in Northeast Plateau, North China Plain, and in the middle and lower reaches of Yangtze River, Yangtze River Delta and Pearl River Delta in the future 100 years. In western China, built-up land would mainly be distributed in Sichuan Basin, central Shaanxi plain, Hetao Plain in Ningxia, Hexi Corridor in Gansu and Xinjiang Oasis. In mountainous, hilly and plateau areas, built-up land would mainly be distributed in provincial and prefectural capitals and other administrative areas.

Nival areas would mainly occur in mountainous ridges in Tianshan Mountains, Qilian Mountains, Kunlun Mountains, Himalaya Mountains and Hengduan Mountains. Deserts would mainly occur in Junggar Basin, Tarim Basin, Qaidam basin, Inner Mongolia Plateau. Bare rocks would mainly be distributed in Qinghai-Xizang Plateau, the east of Turpan Basin, and Karst area in the southwest China. Desertification land would mainly be distributed in outside areas of deserts and inside areas of meadows in northwestern China.

#### 8.3.5.2 Scenarios of Land Cover Area

Land cover types, of which area would increase in all the periods from 2010 to 2039, from 2040 to 2069, and from 2070 to 2099 under all the three scenarios, include built-up land and desertification land. Area of woodland type would increase in all the three periods under all the scenarios except a decrease in the period from 2069 to 2099 under scenario A1FI. Land cover types, of which area would decrease in all the three periods, include cultivated land, grassland and nival area. Area of water area type would decrease in all the three periods under all the three scenarios except an increase in the period from 2070 to 2099 under scenario A1FI.

During the period from 2010 to 2039, area of wetland type would decrease under all the three scenarios; from 2040 to 2069, area of wetland would increase under scenarios A1FI and A2, but decrease under scenario B2; from 2070 to 2099, the one would increase under scenarios A1FI and B2, but decrease under scenario A2. Under scenarios A1FI and A2, area of desert would increase during all the three periods; under scenario B2, area of desert would decrease during the period from 2010 to 2039, but increase during other two periods. Under scenario B2, area of bare rock would decrease during all the three periods; under scenario A2, area of bare rock would decrease during the periods from 2010 to 2039 and from 2070 to 2099, but increase during the period from 2040 to 2069.

Woodland would have the greatest increasing rate that would be 2.34% per decade; bare rock would have the biggest decreasing rate that would be 2.38% per decade. The land cover types that would have a greater transformation include cultivated land, woodland, grassland and desertification land. The cultivated land that would have been transformed would be mainly converted into woodland and the converting rate would become smaller and smaller gradually during the period from 2010 to 2099. The transformed woodland would be mainly converted into grassland. Most transformed grassland would be converted into woodland and a considerable part of the transformed grassland would be converted into desertification land. The transformed desertification land would mainly be converted into woodland and grassland.

## 8.3.6 Scenarios of Land Cover in Eurasia

Due to simulate the scenarios of land cover in Eurasia, the bioclimatic datasets include the mean annual biotemperature, average total annual precipitation and potential evapotranspiration ratio were simulated to $0.125° \times 0.125°$ by HASM method that were introduced in the last classes, which involve the observation data from 1981 to 2010, and the scenarios data of CMIP5 RCP26, RCP45 and RCP85 scenarios from 2011 to 2100. Spatial data of terrestrial ecosystem types in 2010, 2040, 2070 and 2100 were respectively generated by operating iHLZ model.

The land cover data of Eurasia have been separated from the land cover data on spatial resolution of $0.125° \times 0.125°$ of the global rest in 2010. The land cover types were classified into 13 types by combining the land cover classification systems of FAO and UNEP (Gregorio and Jansen 2001), IGBP, USGS, ESA GlobCover, and UMD, which include Evergreen broadleaf forest, Deciduous broadleaf forest, Evergreen needleleaf forest, Deciduous needleleaf forest, Mixed Forest, Shrub land, Grasslands, Wetlands, Croplands, Water bodies, Snow and Ice, Built-up land, Barren or sparsely vegetated land. (Fan et al. 2015, 2020a, b).

### 8.3.6.1 Spatial Distribution of Land Cover in Eurasia

The simulated results of land cover distribution in Eurasia show that there would be a great difference of spatial distribution of forest and shrub because of the complicated terrain characteristics and heterogeneous climate change in Eurasia. Evergreen broadleaf forest would be mostly distributed in Southeast Asia. Evergreen needleleaf forest would mainly be distributed in Nordic Europe, East European Plain, Western Siberian Plain, Japan, South Korea, and hilly areas in the south of China.

Deciduous needleleaf forest would be mainly distributed in Central Siberian Highlands and Eastern Siberia, which would cover about half of Russia's area. Shrub land would be mainly distributed in the north and east of Russia, the southwest of China, the northeast of Laos, the border zone between India and Myanmar and the border zone between India and Pakistan, the north of United Kingdom, the south of Greece, the most of Turkey, the east of Georgia, and central Azerbaijan.

Grasslands would mainly be distributed in Tianshan Mountains, Tibetan Plateau, Inner Mongolian Plateau, Loess Plateau, Altai Mountains and areas of around Tarim Basin of China, North Siberian Plain, Norway, the northwest of Iran, the north of Kazakhstan and the middle of Mongolia.

Croplands would mainly be distributed in the East European Plain and Central European Plains in Europe, and in Northeast Plain, North China Plain, middle and lower reaches of Yangtze River, Sichuan Basin and Guanzhong Basin of China. Hexi Corridor and the river alluvial sectors of Tianshan Mountains in China also would have relatively concentrated cultivated land. The middle of South Asia is the great plains, which is formed by Indus, Ganges River and Brahmaputra Rivers, with dense river networks, numerous irrigation channels, and developed agriculture, such as India. In addition, Croplands would centrally be distributed in Thailand, Myanmar, Vietnam in Indo China Peninsula, the coastal area of Malay Archipelago.

Barren or sparsely vegetated land would mainly be distributed in Uzbekistan, Turkmeni-stan and southern Kazakhstan of Central Asia, and Iran, Saudi Arabia, Yemen of Western Asia. Moreover, barren or sparsely vegetated land would be distributed in southern Mongolia and Northwest China. In China, Barren or sparsely vegetated land would mainly be distributed in the arid and desert areas in the north-west, namely Gurbantunggut Desert in Zhungeer Basin, Taklamakan Desert in Tarim Basin, the center of Qaidam Basin, and Tengger Desert, Alashan Plateau.

Water Bodies include rivers and lakes. The main water body area is river area, which include Volga River, Yenisei, Ob River, Ural River, Syr Darya, Tigris River, Euphrates River, Po, Mekong. Lakes would mainly be distributed in Finland, named Country of Thousand Lakes that major include Baikal, and Aral Sea. In China, the river system mainly originates in Qilian Mountains and Hengduan Mountains in the east of the Qinghai-Tibet Plateau, the mountainous regions around the Sichuan Basin, the Changbai Mountains in the northeast, the Big Xing'an Mountains, and the hilly regions in the south of China. The distribution of Water Bodies would be more dispersed than other land cover types. Wetlands would mainly be dis-tributed in Western Siberian Plain.

#### 8.3.6.2 Area Change of Land Cover in Eurasia

Land cover under the three scenarios RCP26, RCP45 and RCP85 indicate that deciduous needleleaf forest, mixed forest, shrub land, wetlands and snow and ice would generally decrease in Eurasia during the period from 2010 to 2100. Snow and ice would have the fastest decreasing rate that would decrease by 36.44% on average. Shrub land would have the slowest decreasing rate that would decrease by 5.19% on average. Water bodies would have the fastest increasing rate that would increase by 26.94% on average. Barren or sparsely vegetated land would have the slowest increasing rate that would increase by 0.70%.

Furthermore, the simulation results show that there would generally appear a similar change pattern of land cover driven by different levels of climate change scenarios in Eurasia. Land cover under the RCP85 scenario would generally have the fastest change rate, especially that the reducing trend of snow and ice would much faster than that of scenario RC26 and RC45. There would have the lowest change rate of land cover in the scenario RCP26. The simulated results can approve proofs that climate change would directly impact on land cover change in Eurasia. For instance, The RCP85 is a highly energy-intensive scenario as a result of high population growth, a lower rate of technology development and non-climate policy, which represent a high scenario in climate change. Under the RC85 scenario, the temperature and precipitation would have the faster increase rate than other two scenarios of RCP26 and RCP45, so that there would have the largest change rate of land cover simulated by using the climatic data of the scenario RCP85. With the rapid increase of temperature and precipitation, the snow and ice would have a fast-melting trend, the grasslands would also show a increase trend, and the succession would occur between different forest types.

## 8.4 Spatial Distribution and Scenarios of Ecological diversity

### *8.4.1 Relationship Between Species Diversity and Ecotope Diversity*

#### 8.4.1.1 Progress in Relationship Between Species Diversity and Ecotope Diversity

The relationship between species diversity and ecotope diversity has long been debated. However, diversity has been used by landscape architects, ecologists, policy analysts, politicians, artists, and so on. Each discipline has provided its own understanding and definition. For instance, the multiplicative concept expressed gamma diversity as the product of the mean alpha diversity and beta diversity, in which diversity was equated to species richness.

An understanding of relationship between species diversity and ecotope diversity, including their response to past, current and future human activities, is essential to develop models for sustainable land management. Therefore, lots of scientists have paid special attention to the relationship since early 1960s. There have appeared two camps on the relationship hypothesis. Some scientists found that an increase of ecotope diversity leads to an increase in species diversity. But others findings did not support this hypothesis. The relationship between species diversity and ecotope diversity has long been debated.

The related statistically analyzed relationship between species diversity and landscape diversity showed that both bird species diversity and plant species diversity were closely related to landscape diversity and their correlation coefficients were, respectively, 0.887 and 0.817 in United States by means of Shannon's index. Complex habitats allowed more species to co-exist in a given area after they assessed effects of surface complexity and area on species richness and abundance (Yue and Li 2010). The analyzed results of spatial components of biotic diversity in landscapes of Georgia showed that the diversity of land cover types was positively correlated with species richness.

Biodiversity was dependent on the natural richness of different parts of the continent, while it also depended on the impact of humans and the way they changed natural into cultural landscapes. Lots of researches showed that there was no positive effect of landscape complexity on species richness and no evidence supported the hypothesis that habitat heterogeneity would enhance diversity of a landscape by increasing the number of species in an area (Cramer and Willig 2005). A study area with 4 $km^2$ near village of Friedeberg of central Germany (Yue and Li 2010).

#### 8.4.1.2 Materials and Method

DongZhi Tableland is geographically in the north of JinHe River, between MaLianHe River and PuHe River. Dongzhi tableland is located in interland of Loess Plateau. Total area is 2778 $km^2$, of which flat land is 960 $km^2$. Serious soil erosion is the major ecological problem in Dongzhi tableland and is the main restriction of local economic development. The density of gullies is 2.17 km/$km^2$ on an average. Soil erosion area is 2724 $km^2$, accounting for 98.1% of total area of Dongzhi tableland. Water erosion and gravity erosion are major soil erosion types. Elevation of Dongzhi tableland is 1350 m on an average. Its elevation is lower in the south, about 882 m, and higher in the north, about 1576 m.

The quadrat for trees was 10 m $\times$ 10 m in size and for grasses 1 m $\times$ 1 m in size. The records of every quadrat included its longitude, latitude, slope, aspect, elevation, and ecological circumstances. The records of trees were species name and individual number of each species as well as crown cover, height and pectoral diameter of every species on an average. For calculating the relationship between individual-counting diversity and biomass-based diversity, 98 sample quadrats in Dongzhi tableland were selected, which include 19 sample quadrats of woodland and 79 grassland quadrats. One hundred and seventeen species can be found in the woodland quadrats, and 104 species in grassland quadrats.

Moreover, for analyzing the relationship between ecotope diversity and species diversity as well as individual ecotope counting diversity and ecotope area based diversity, 10 square regions were selected randomly in Dongzhi tableland. Every region is 8 km $\times$ 8 km in size. In each square region, six quadrats were designed in different slope aspect and landform types.

The classical diversity indexes had been widely used but in the meanwhile had been criticized by many ecologists. It has been demonstrated that all the classical diversity indexes do not satisfy the definition of diversity. However, in contrast, the scaling diversity index can formulate both evenness and richness aspects of diversity;

it is scientifically sound and could be operated at affordable cost (Yue et al. 1998, 2001, 2003, 2004, 2005).

$$D(\varepsilon, r, t) = -\frac{\ln\left(\sum_{i=1}^{m(\varepsilon,r,t)} (p_i(\varepsilon, r, t))^{\frac{1}{2}}\right)^2}{\ln \varepsilon} \quad (20)$$

where $(p_i(\varepsilon, r, t)$ is propability of the ith ecotope or of the *ith* species; $m(\varepsilon, r, t)$ is the total number of species or ecotopes under investigation; t represents time; $\varepsilon = \frac{1}{\mathrm{e}+a}$, $a$ is the whole study area, e equals 2.71828.

#### 8.4.1.3 Results Analysis

The calculation results of sampling data on the woodland indicate that correlation coefficient between individual-counting diversity and biomass-based diversity is 0.286 and the significance level is 0.235. In other words, individual-counting diversity and biomass-based diversity have no significant dependency (Yue et al. 2010).

However, the results in grassland show that individual-counting diversity and biomass-based diversity are significantly correlated and the regressive equation can be formulated as, y = 0.854x + 0.399. where y represents biomass-based diversity, x is individual-counting diversity, correlation coefficient is 0.934, and significance level is less than 0.0001.

The analyses demonstrated that correlation coefficient between individual-counting diversity and individual ecotope-counting diversity was − 0.381 and the significance level was 0.178. Individual-counting diversity and ecotope-area based diversity had a correlation coefficient of-0.253 and a significance level of 0.481. Biomass-based diversity and individual ecotope-counting diversity had a correlation coefficient of 0.19 and a significance of 0.599. The correlation coefficient between biomass-based diversity and ecotope-area based diversity was 0.188 and its significance level was 0.604. In short, species diversity and ecotope diversity have no significant correlation in Dongzhi tableland.

#### 8.4.1.4 Discussion and Conclusions

Biodiversity definitions varied within the disciplineand the relationship between species diversity and ecotope diversity was debated by speaking different languages on biodiversity (Ewers and Rodrigues 2006), which made the debates meaningless. Yue et al. (2010) adopt the definition of individual-counting diversity that combines richness and evenness. Richness is based on the total number of species present and evenness is based on the relative abundance of species and the degree of its dominance thereof. Then this concept is generalized to formulate biomass-based diversity, individual ecotope counting diversity, and ecotope-area based diversity.

Yue et al. (2010) first try to answer questions whether it is necessary to distinguish species diversity into individual-counting diversity and biomass-based diversity, and ecotope diversity into individual ecotope-counting diversity and ecotope-area based diversity. The analysis results show that individual counting diversity and biomass-based diversity have no significant correlation in woodland ecosystems. In grassland ecosystems, individual-counting diversity is significantly correlated with biomass-based diversity. Individual ecotope-counting diversity and ecotope-area based diversity have a significant correlation. Therefore, it is unnecessary to distinguish ecotope diversity into individual ecotope-counting diversity and ecotope-area based diversity and to discriminate between individual-counting diversity and biomass-based diversity for grassland ecosystems for the issues that have no special requirement for accuracy. But it is necessary for woodland ecosystems to divide species diversity into individual-counting diversity and biomass-based diversity.

Finally, Yue et al. (2010) conclude that individual-counting diversity and species-biomass have no significant correlation with ecotope diversity according to the results from analyzing the data sampled in Dongzhi Tableland.

### *8.4.2 Surface Modelling of Vascular Plant Species Abundance in Qinghai-Tibet Plateau*

#### 8.4.2.1 Definition and Classification of Vascular Plant

A vascular plant is any one of a number of plants with specialized vascular tissue. The two types of vascular tissue, xylem and phloem, are responsible for moving water, minerals, and the products of photosynthesis throughout the plant. As opposed to a non-vascular plant, a vascular plant can grow much larger. The vascular tissue within provides a means of transporting water to great heights, allowing a vascular plant to grow upward to catch the sun. Inside of a vascular plant, the structure is much different from that of a non-vascular plant. In non-vascular plants, there is little to no differentiation between the different cells. In vascular plants, the specialized vascular tissues are arranged in unique patterns, depending on the division and species the vascular plant belongs to. The xylem, made mostly of the structural protein lignin and dead cells, specializes in transporting water and minerals from the roots to the leaves.

A vascular plant does this by creating a pressure on the water on multiple fronts. In the roots, water is absorbed into the tissues. The water flows into the xylem, and creates an upward pressure. At the leaves, water is being used and evaporates out of the stoma. These small pores are said to transpire, which pulls upward on the column of water in the xylem. Through the actions of adhesion and cohesion, the water moves upward through the xylem like a drink through a straw.

Vascular plants exhibit, like all plants, an alternation of generations. This means that there are two forms of the plant, the sporophyte and the gametophyte. The sporophyte, a diploid organism, goes through meiosis to produce the haploid spore. The spore grows into a new organism, the gametophyte. The gametophyte is responsible for producing gametes, capable of fusing together during sexual reproduction. The vascular plants are embryophytes, which is a large clade or related group, consisting of both non-vascular and vascular plants. The embryophytes are further broken down into the Bryophytes including mosses, liverworts, and non-vascular plants, and Tracheophyte. As the trachea in humans is a passageway for air, the term tracheophyte refers to the vascular tissue in vascular plants.

The tracheophytes are further divided into divisions. The divisions are distinguished mostly on how their spores and gametophyte's function. In ferns and club-mosses, the gametophyte becomes a free-living generation. In gymnosperms (conifers) and angiosperms (flowering plants), the gametophyte is dependent on the sporophyte. The gametes developed within become a seed, forming the next sporophyte generation. While every vascular plant shows an alternation of generations with a dominant sporophyte, they differ on how they go about distributing spores and seeds.

#### 8.4.2.2 Basic Datasets

Long-term interaction between climatic factors, environmental factors, and human drive factors directly controls the spatial distribution of vascular plant species. These effect factors generally include biotemperature, precipitation, soil, topography, soil, land cover types, landscape pattern, human activity, and so on. Vascular plants play an important role in the interrelationship between climate and atmosphere by intercepting water and nutrient cycling to affect soil and water conservation and surface element cycling (Yue et al. 2011). As the vascular plants can provide the necessary material and energy for other species, the spatial distribution of vascular plant and its abundance directly effects other species structure distribution and diversity (Gaston 2000).

How to analyze the spatial distribution pattern of vascular plant species abundance, especially how to quantitatively explain the spatial pattern changes of vascular plant species abundance according to various influencing factors, which has gradually become the focus and difficulty issue of biodiversity research field (Fan and Bai 2019). In recent decades, there are many research on the relationship between spatial distribution of vascular plants and climate change. The existing research mainly focused on the change trend of vegetation types and land cover in the Tibetan Plateau, and the diversity studies also mainly used the sampling data to analyze the spatial statistics and analysis of the species diversity in the local pattern. While how to extend the scatter observation data, and simulate the vascular plant species abundance is rarely involved in Qinghai-Tibet Plateau.

During the progress of simulating the vascular plant species abundance in Qinghai-Tibet plateau, the land cover data, climatic data and vascular plant species data in national nature reserves of China were collected. The historical land cover data in 2010 with resolution of 1 km × 1 km is based on landsat TM and CBERS-2 CCD data and its pixels are classified according to maximum proportion of land cover type in each pixel. The percentage raster data of each land cover type in every grid in 2010 was selected to represent current spatial pattern of land cover in Qinghai–Tibet Plateau. The land cover scenarios data in 2040 (T1), 2070 (T2) and 2010 (T3) with resolution of 1 km × 1 km is obtained by operating the surface model of land cover scenarios in CMIP5 RCP2.6, RCP4.5, RCP8.5 scenarios (Fan et al. 2015, 2019; Yue et al. 2007, 2016, 2019, 2011, 2020).

The climatic data with the resolution of 1 km × 1 km is simulated by a high accuracy and speed method of surfacing modeling (HASM), including the spatial distribution data of the mean annual biotemperature an d average total annual precipitation in the Qinghai-Tibet Plateau from 1981 to 2010, the scenarios raster data of the mean annual biotemperature and average total annual precipitation in the RCP2.6, RCP4.5, RCP8.5 scenarios of CMIP5 from T1 to T3 (Yue and Li 2010; Yue et al. 2013, 2016; Fan et al. 2009, 2012).

The vascular plant species data is the vascular species survey data of 37 national nature reserves in the Qinghai-Tibet Plateau published by the Ministry of Ecology and Environment Protection of the People's Republic of China (http://www.mee.gov.cn/ywgz/zrstbh/swdyxbh/).

Moreover, the data of the Qinghai-Tibet Plateau area and scope derived from Global Change Research Data Publishing & Repository (Zhang et al. 2014).The boundary data of the protected area was based on the basic data is collected from the data sources such as the Ministry of ecology and Environment Protection of the People's Republic of China, the Forestry Bureau, the National Administration of Surveying, Mapping and Geoinformation, the National Geomatics Center of China, as well as some high-resolution remote sensing image data obtained by vectorization of ArcGIS (Fan et al. 2012). DEM's underlying data came from NASA's SRTM data and (http://srtm.csi.cgiar.org/) published with a resolu-tion of 1 km × 1 km.

#### 8.4.2.3 Theory and Key Steps of Surface Modelling of Vascular Plant Species Abundance

In the process of analysis of the number of vascular plants, land cover and other environ-mental factors in the national nature reserve, it is found that the dynamic intensity of land cover in nature reserves is lower than that in non-protected areas (Fan et al. 2012), and the spatial distribution of the vascular plant species abundance in nature reserves has a good correlation with specific environmental factors, land cover types and structural indices. Based on this analysis, the vascular plant species data of the Qinghai-Tibet Plateau National Nature Reserve are used as benchmarks and sample data, and make full use of the research, observation and statistical data of vascular plant species in existing nature reserves, the spatial simulation method of the Qinghai-Tibet Plateau vascular plant species abundance.

In the process of model construction, the spatial analysis method for most species diversity at present mainly focuses on the spatial correlation analysis of land cover type and environmental factors at a single level, in addition to considering the land cover type and environmental factors simultaneously, the landscape diversity index is introduced to analyze the spatial correlation between the spatial distribution of vascular plant species abundance and the driving mechanism.

Firstly, using the GIS spatial analysis method to achieve the area of each type of land cover and its proportion (including: arable land, woodland, grassland, wetlands, construction land, desert, snow and ice).

Secondly, mean annual biotemperature (Yue et al. 2001), average total annual precipitation (Shary et al. 2002) and topographic relief (Yue et al. 2006) are selected as the spatial correlation analysis parameters of environmental factors on the vascular plant species abundance.

Thirdly, the scaling diversity index of land cover (Fan and Yue 2005; Yue et al. 1999, 2005, 2007) and the land cover patch connectivity index (Yue 1999; Van Vuuren et al. 2011) are selected as the variables that analyze the spatial correlation between vascular plant species abundance and landscape diversity.

Fourthly, on the basis of obtaining the vascular plant species abundance (Expressed as the number of vascular plant species distributed per square kilometer), land cover type, mean annual biotemperature, average total annual precipitation and landscape diversity index in 37 national nature reserves in Qinghai-Tibet Plateau, the correlations between various factors and vascular plant species abundance are analyzed, and the optimal regression model and equation are selected to participate in the simulation. In the current species-area curve and other species abundance models, linear regression and non-linear regression models are used, but it is always controversial which model is more suitable.

Finally, according to the regression model analysis results of vascular plant species abundance and various influencing factors, in the process of constructing the spatial model of vascular plants species abundance, the relationship between land cover type factor, environmental factor and landscape diversity index and vascular plants species abundance is analyzed, the three fitting equations of logarithmic model, exponential logarithmic model and exponential model are used respectively. And then the spatial distribution model of vascular plants species abundance of the Qinghai-Tibet Plateau is characterized by land cover type factor, environmental factor and landscape diversity index factor, the theoretical formula of the model is expressed as follows:

$$VPA(x,y)_{\mathrm{t}} = Cons + \omega_1 \begin{bmatrix} \alpha_1 \ln Fa(x,y)_t \\ +\alpha_2 \ln Fo(x,y)_t \\ +\alpha_3 \ln Gr(x,y)_t \\ +\alpha_4 \ln Wa(x,y)_t \\ +\alpha_5 \ln Bu(x,y)_t \\ +\alpha_6 \ln De(x,y)_t \\ +\alpha_7 \ln Ic(x,y)_t \end{bmatrix}$$

$$+\omega_2\begin{Bmatrix}\beta_1\exp\left[MAB(x,y)_t\right]\\+\beta_2\ln MAP(x,y)_t\\+\beta_3\exp\left[Re(x,y)_t\right]\end{Bmatrix}+\omega_3\begin{Bmatrix}\gamma_1\exp\left[Cot(x,y)_t\right]\\+\gamma_2\exp\left[Dt(x,y)_t\right]\end{Bmatrix} \tag{21}$$

where $t$ is the variable of time, $(x, y)$ is the position coordinate, $Cons$ is constant, $\omega$ is the model coefficient, $VPA(x, y)_t$ is the vascular plant species abundance at site (x, y) in the $t$ period, $MAB(x, y)_t$ is the mean annual biotemperature at site (x, y) in the $t$ period, $MAP(x, y)_t$ is the average total annual precipitation at site (x, y) in the $t$ period, $Re(x, y)_t$ is the average topographic relief at site (x, y) in the $t$ period, $Cot(x, y)_t$ is the land cover patch connectivity index at site (x, y) in the $t$ period, $Dt(x, y)_t$ is the land cover ecological diversity index at site (x, y) in the $t$ period, $Fa(x, y)_t$ is the area of arable land at site (x, y) in the $t$ period, $Fo(x, y)_t$ is the area of forest at site (x, y) in the $t$ period, $Gr(x, y)_t$ is the area of grassland at site (x, y) in the $t$ period, $Wa(x, y)_t$ is the area of water at site (x, y) in the $t$ period, $Bu(x, y)_t$ is the area of construction land at site (x, y) in the $t$ period, $De(x, y)_t$ is the area of desert at site (x, y) in the $t$ period, $Ic(x, y)_t$ is the area of snow and ice at site (x, y) in the $t$ period; $\alpha_1$, $\alpha_2$, $\alpha_3$, $\alpha_4$, $\alpha_5$, $\alpha_6$, $\alpha_7$ are the model coefficients of the land cover type factors and the vascular plant species abundance, $\beta_1$, $\beta_2$, $\beta_3$ are the model coefficients of mean annual biotemperature, average total annual precipitation, topographic relief and vascular plant species abundance, $\gamma_1 and \gamma_2$ are the model coefficients of land cover ecological diversity index, land cover patch connectivity index and vascular plant species abundance, respectively. The variance of each parameter in Eq. (6) is obtained by fitting the equation of the vascular plant species in Table 1 published in 37 national nature reserves of the Qinghai-Tibet Plateau: $Cons = -11{,}582.89$, $\omega_1 = 0.38$, $\omega_2 = 0.98$, $\omega_3 = 1.02$, $\alpha_1 = 66.08$, $\alpha_2 = 0.68$, $\alpha_3 = 2.40$, $\alpha_4 = -34.22$, $\alpha_5 = 4.04$, $\alpha_6 = 31.63$, $\alpha_7 = 56.00$, $\beta_1 = 0.23$, $\beta_2 = 536.45$, $\beta_3 = 1400.28$, $\gamma_1 = -283.11$, $\gamma_2 = 6007.68$. The complex correlation coefficient of Eq. (6) is $R = 0.94$ and passed the significance test of 0.01. On the basis of constructing the model, the spatial distribution of the vascular plant species abundance in the Tibetan Plateau and the future scenarios are simulated and analyzed respectively.

Moreover, in the process of model validation, the error between the simulated value of the vascular plant species abundance per square kilometer and the statistical value of the survey is verified. The error is calculated as:

$$S_{Error} = \frac{1}{n}\sum_{i=1}^{n}\left(\frac{|Value_{is} - Value_{it}|}{S_i}\right)\pi \tag{22}$$

where $S_{Error}$ is the simulation error of the vascular plant species abundance under the model, $n$ is the number of nature reserves, $S_i$ is the area of the i-th nature reserve, $Value_{is}$ is the simulation value of the vascular plant species abundance in the i-th national nature reserve, $Value_{it}$ is the statistical value of the vascular plant species abundance in the i-th national nature reserve.

The results of the error analysis show that the overall error of the model simulation of the spatial distribution of vascular species in 37 national nature reserves of the

Qinghai–Tibet Plateau is 2.21 types per square kilometer. The above analysis shows that the simulation results based on the land cover type factor, environmental factor and landscape index factor are the same as the actual situation of the number of vascular plant species in the study area, which can be applied to the spatial distribution of regional vascular plants species abundance.

#### 8.4.2.4 Spatial Distribution of Vascular Plant Species Abundance in Qinghai-Tibet Plateau

The average abundance of vascular plant species in the Qinghai-Tibet Plateau was 496.79 species per 100 km$^2$. The overall trend of the spatial distribution gradually decreased from the southeast to the northwest, which the vascular plants species abundance on the Tibetan Plateau in the mid-west of the Qinghai-Tibet Plateau and Qaidam Basin is relatively low, and the average species abundance of vascular plants is less than 200 species per 100 km$^2$. While the vascular plants species abundance in the transition area between the western Qinling of the northeastern Qinghai-Tibet Plateau and Qilian Mountains, the eastern Minshan-Qionglai Hill-Snowy Mountains regions and the southeastern Gaoligongshan-Nushan and Yulongshan regions are relatively high, and the average species abundance of vascular plants is greater than 600 species per 100 km$^2$. Especially the vascular plants species abundance in the Bailongjiang River, Dadu River, Jinsha River, Lancang River, Nujiang and Brahmaputra of up to 1000 species per 100 km$^2$ or more, and even in its local area are more than 2000 species per 100 km$^2$ (Fig. 11).

#### 8.4.2.5 Scenario Analysis of Vascular Plants Species Abundance

The vascular plants species abundance in the Qing-hai-Tibet Plateau decreased from T0 to T4. The RCP2.6 scenario would show a decreasing trend, but the decrease rate would decrease by 0.11%, 0.03% and 0.04%, respectively, from T0 to T1, from T1 to T2 and from T2 to T3. Under RCP4.5 scenario, the vascular plant species abundance would increase by 0.05% from T0 to T1, while it would decrease, respectively, by 0.29% and 0.21%, from T1 to T2 and from T2 to T3. Under RCP8.5 scenario, the vascular plant species abundance would increase by 0.10% from T0 to T1, while it would decrease, respectively, by 0.39% and 0.84%, from T1 to T2 and from T2 to T3. In the three scenarios, the variation of vascular plant species abundance in the Qinghai-Tibet Plateau is the largest under RCP8.5 scenario, while it is the smallest under RCP2.6 scenario, and the variation of vascular plant species abundance under RCP4.5 scenario is between the first two scenarios (Figs. 12, 13, and 14).

The results show that the correlation between the species abundance and the three effect factors is 0.84, and the correlation test is carried out by $P = 0.01$. The average abundance of vascular plants in the Qinghai-Tibet Plateau are 496.79 species per 100 km$^2$. In general, the spatial distribution would gradually reduce from southeast to northwest. The vascular plant species abundance in the rest of the Qinghai-Tibet

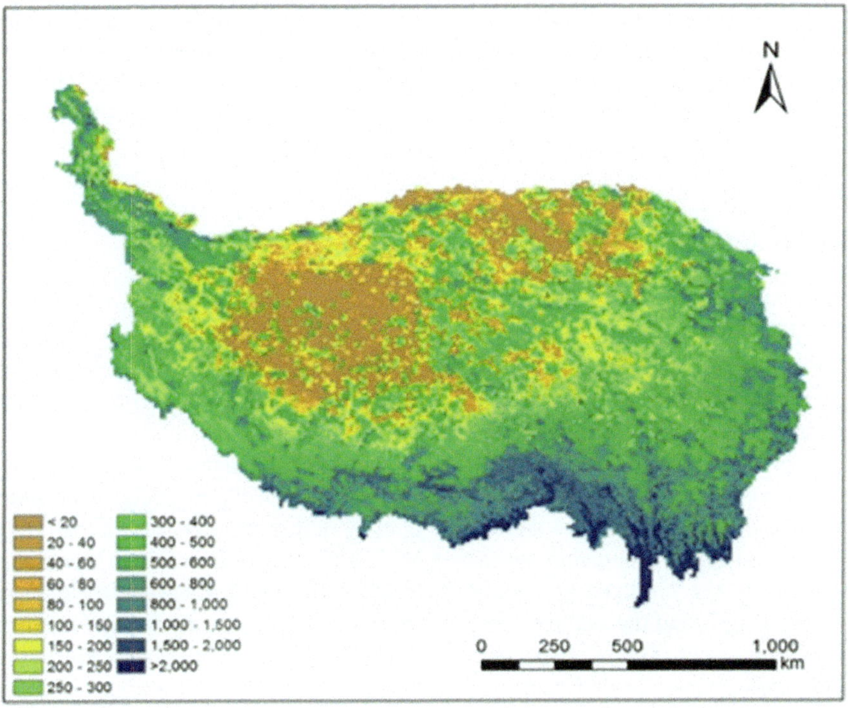


**Fig. 11** Spatial distribution of average abundance of vascular plant species in the Qinghai-Tibet Plateau

Plateau would decrease with the increasing of altitude, except in the desert area of Qaidam Basin. In addition, vascular plant species abundance in the Qinghai-Tibet Plateau would decrease from T0 to T4. The variation of vascular plant species abundance would increase in turn under RCP2.6, RCP4.5 and RCP8.5.

## 8.5 Summary

HLZ (Holdridge life zone) ecosystem model, as a spatial analysis model that on the basis of main bioclimatic factors, MAB, MAP and PER, quantitatively describe the vegetation ecosystem types and their potential distribution on mid-long-term scales (Herrick et al. 2006). It has been widely used in ecosystem mapping, biodiversity conservation, land cover scenario analysis and land use planning and management, etc. (Fan et al. 2013a, b, 2015); especially in the study of the effects of climate change on the spatiotemporal distribution of vegetation ecosystems. The improved the HLZ model to simulate the vegetation distribution in a continual surface on the scales of

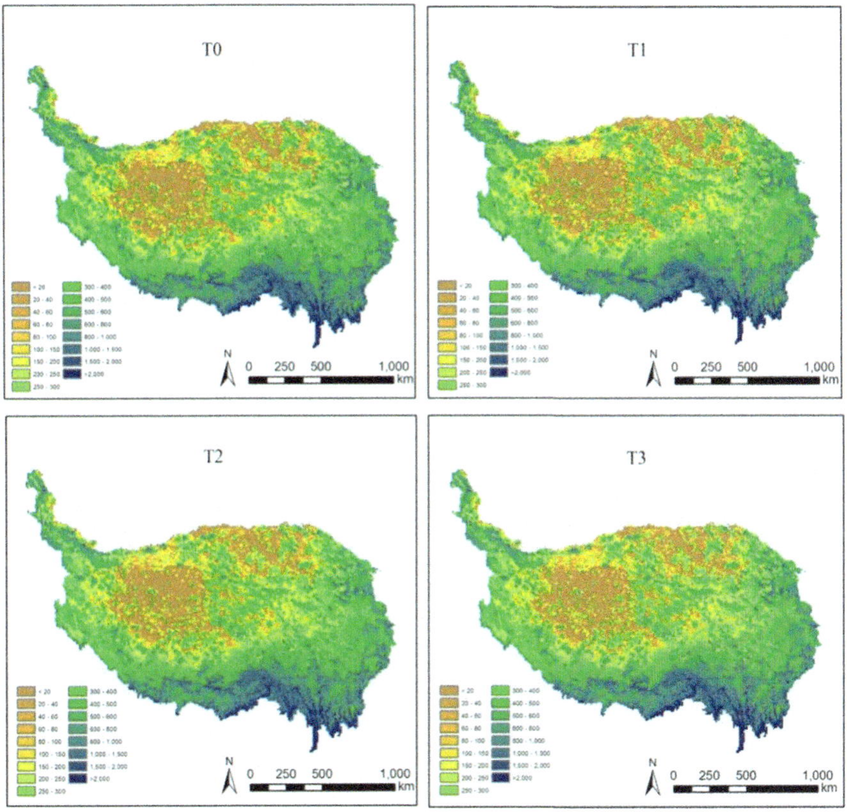


**Fig. 12** Spatial distribution of vascular plant species abundance under RCP2.6 scenario

region, national, and global level (Yue et al. 2005, 2006, 2011; Fan and Fan 2019; Fan et al. 2012, 2019a, b; Fan and Bai 2021).

The method of surface modelling in land cover scenario (SMLCs) (Yue et al. 2007; Fan et al. 2013a, b, 2015, 2020a, b) is to expand and modify the SMLC model driven by natural climate change in view of the model defects of the existing I-O, IMAGE, CLOUD, CA, SD and FLUX models for land cover simulation (Fan et al. 2015; Fan 2022). SMLCs can be used to simulate land cover types at long-term scale on the scales of region, national, and global levels (2013a, b, 2015, 2020a, b; Fan 2022).

The comparative analysis between the simulation results and biodiversity survey data show that the spatial distribution pattern is consistent with the current situation, which indicate that the method of surface modelling of vascular plant species (SMVPS) can be used to simulate the spatial distribution and the change of the vascular plant species abundance in the Qinghai-Tibet Plateau. In the process of model development, not only the land cover type and environmental factors are taken into account, but also the ecological landscape index is taken into account,

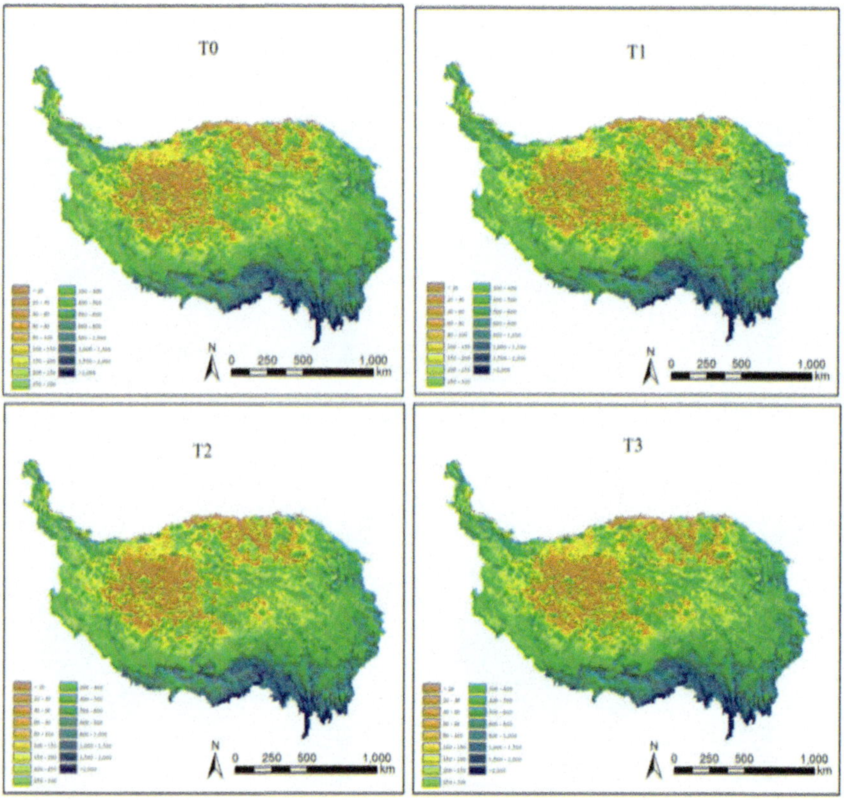


**Fig. 13** Spatial distribution of vascular plant species abundance under RCP4.5 scenario

and improved the existing models. Moreover, the SMVPS method can be used to simulate the spatial distribution of vascular plant species abundance in a large scale, also can predict and simulate the vascular plant species abundance under different climatic scenarios (Fan and Bai 2019).

## 8.6 Further Reading

(1) Tian-Xiang Yue. 2011. Surface Modelling: High Accuracy and High Speed Methods. New York: CRC Press.
(2) Fan ZM, Fan B, Yue TX, 2019. Potential distribution scenario of vegetation ecosystem and its response to climate change in Eurasia. Science China. Earth Sciences,62(10): 1607–1618
(3) Fan ZM, Bai RY, Yue TX, 2020. Scenarios of land cover in Eurasia under climate change. Journal of Geographical Sciences, 30(1): 3-17

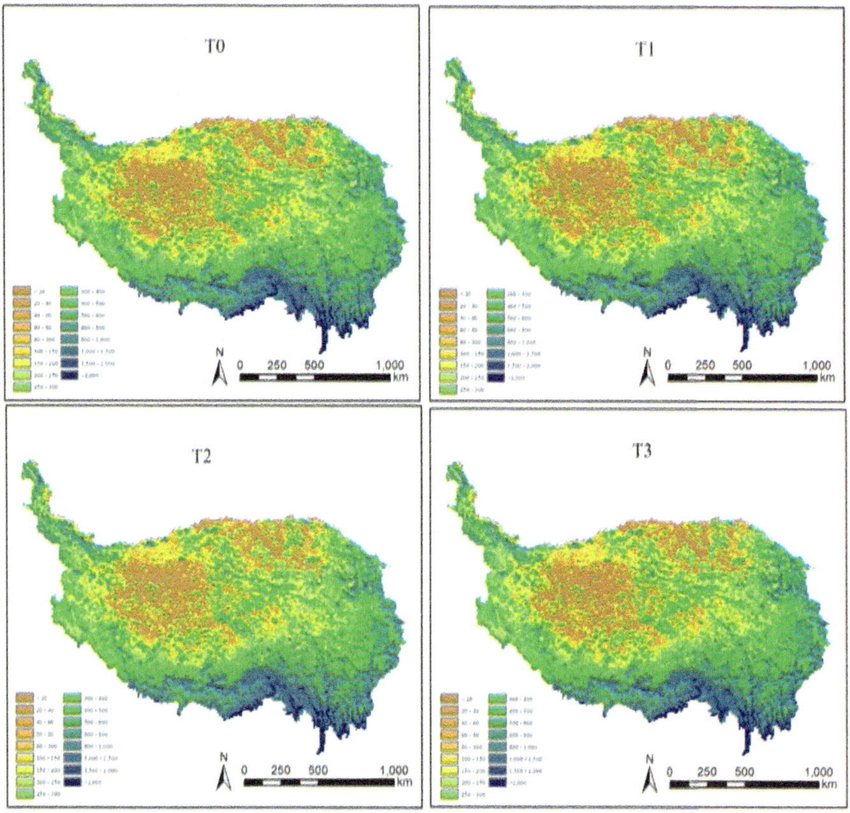


**Fig. 14** Spatial distribution of vascular plant species abundance under RCP8.5 scenario

(4) Fan ZM, Bai RY, 2019. Spatiotemporal distribution of vascular plant species abundance on Qinghai-Tibet Plateau. 2019,29(10): 1651-1662

## Exercises

(1) What is the difference between Whittaker's biome classification and HLZ classification?
(2) Design a framework of using the iHLZ model to simulate the terrestrial ecosystem of your country.
(3) What are the limitations of SRES, CLUE, I-O, SESM, LUCK, CA models?
(4) What is the theoretical scheme and advantages of SMLCs?
(5) Which basic data types are normally involved in operating the SMLCs model?
(6) What is ecotope diversity?
(7) Explicating the scaling diversity index?

## References


Chapin, F.S., III., E.S. Zavaleta, V.T. Eviner, R.L. Naylor, P.M. Vitousek, H.L. Reynolds, D.U. Hooper, S. Lavorel, O.E. Sala, S.E. Hobbie, M.C. Mack, and S. Diaz. 2000. Consequences of changing biodiversity. *Nature* 405: 234–242.

Clarke, K., S. Hoppen, and L. Gaydos. 1997. A self-modifying cellular automaton model of historical urbanization in the San Francisco Bay area. *Environment & Planning B Planning & Design* 24 (2): 247–261.

Climatic Change 109(1–2): 5.

Fan, Z.M. 2022. Simulation of land-cover change in Jing-Jin-Ji region under different scenarios of SSP-RCP. *Journal of Geographical Sciences* 22 (3): 421–440.

Fan, Z.M., and B. Fan. 2019. Shifts of the mean centers of potential vegetation ecosystems under future climate change in Eurasia. *Forests* 10 (873): 1–14. https://doi.org/10.3390/f10100873.

Fan, Z.M., and X.Y. Bai. 2021. Scenarios of potential vegetation distribution in the different gradient zones of Qinghai-Tibet Plateau under future climate change. *Science of the Total Environment* 796: 148918.

Fan, Z.M., J. Li, and T.X. Yue. 2012. Changes of climate-vegetation ecosystem in Loess Plateau of China. *Procedia Environmental Sciences* 13: 715–720.

Fan, Z.M., J. Li, and T.X. Yue. 2013a. Land-cover changes of biome transition zones in Loess Plateau of China. *Ecological Modelling* 252: 129–140.

Fan, Z.M., X. Zhang, J. Li, T.X. Yue, J.Y. Liu, B. Xiang, and W.H. Kuang. 2013b. Land-cover changes of national nature reserves in China. *Journal of Geographical Sciences*. 23 (2): 258–270.

Fan, Z.M., J. Li, T.X. Yue, X. Zhou, and A.J. Liu. 2015. Scenarios of land cover in Karst area of Southwestern China. *Environmental Earth Sciences* 74: 6407–6420.

Fan, Z.M., R.Y. Bai, and T.X. Yue. 2019a. Spatiotemporal distribution of vascular plant species abundance on Qinghai-Tibet Plateau. *Journal of Geographical Sciences* 29 (10): 1625–1636.

Fan, Z.M., B. Fan, and T.X. Yue. 2019b. Potential distribution scenario of vegetation ecosystem and its response to climate change in Eurasia. *Science China. Earth Sciences* 62 (10): 1607–1618.

Fan, Z.M., R.Y. Bai, and T.X. Yue. 2020a. Scenarios of land cover in Eurasia under climate change. *Journal of Geographical Sciences* 30 (1): 3–17.

Fan, Z.M., S.B. Li, and H.Y. Fang. 2020b. Explicitly identifying the desertification change in CMREC area based on multisource remote data. Remote Sens 12: 3170. https://doi.org/10.3390/rs12193170

Fan, Z.M., S.B. Li, and H.Y. Fang. 2021. Explicating the mechanisms of land cover change in the New Eurasian Continental Bridge Economic Corridor region in the 21st century. *Journal of Geographical Sciences* 31 (10): 1403–1418.

Foody, G.M. 2002. Status of land cover classification accuracy assessment. *Remote Sensing of Environment* 80: 185–201.

Gaston, K. 2000. Global patterns in biodiversity. *Nature* 405 (6783): 220–227.

Herrick, J.E., B.T. Bestelmeyer, S. Archerb, A.J. Tugelc, and J.R. Brown. 2006. An integrated framework for science-based arid land management. *Journal of Arid Environments* 65: 319–335.

Hubacek, K., and L.X. Sun. 2001. A scenario analysis of China's land use and land cover change: Incorporating biophysical information into input–output modeling. *Structural Change and Economic Dynamics* 12: 367–397.

Lauenroth, W.K., D.L. Urban, D.P. Coffin, W.J. Parton, H.H. Shugart, T.B. Kirchner, and T.M. Smith. 1993. Modeling vegetation structure-ecosystem process interactions across sites and ecosystems. *Ecological Modelling* 67 (1): 49–80.

Li, J.H., and J.C. Shi. 2016. *Global Ecological Environmental Remote Sensing Monitoring 2015 Annual Report (in Chinese)*. Beijing: Science Press.

Liu, J.Y., W.H. Kuang, Z.X. Zhang, X.L. Xu, Y.W. Qin, J. Ning, W.C. Zhou, S.W. Zhang, R.D. Li, C.Z. Yan, S.X. Wu, X.Z. Shi, N. Jiang, D.S. Yu, X.Z. Pan, and W.F. Chi. 2014. Spatiotemporal characteristics, patterns, and causes of land-use changes in China since the late 1980s. *Journal of Geographical Sciences* 24 (02): 195–210.

Liu, X.C., L. Yu, T. Sohl, N. Clinton, W.Y. Li, Z.L. Zhu, X.P. Liu, and P. Gong. 2016. A cellular automata downscaling based 1 km global land use datasets (2010–2100). *Science Bulletin* 61 (21): 1651–1661.

Luijten, J.C. 2003. A systematic method for generating land use patterns using stochastic rules and basic landscape characteristics: Results for a Colombian hillside watershed. *Agriculture, Ecosystems and Environment* 95: 427–441.

Niehoff, D., U. Fritsch, and A. Bronstert. 2002. Land-use impacts on storm runoff generation: Scenarios of land-use change and simulation of hydrological response in a meso-scale catchment in SW Germany. *Journal of Hydrology* 267: 80–93.

Solecki, W.D., and C. Oliveri. 2004. Downscaling climate change scenarios in an urban land use change model. *Journal of Environmental Management* 72: 105–115.

Verburg, P.H., A. Veldkamp, and L.O. Fresco. 1999. Simulation of changes in the spatial pattern of land use in China. *Applied Geography* 19 (3): 211–233.

Verburg, P.H., W. Soepboer, A. Veldkamp, A. Limpiada, V.O. Espaldon, and S. Mastura. 2002. Modeling the spatial dynamics of regional land use: The CLUE-S model. *Environmental Management* 30 (3): 391–405.

Vitousek, P.M., H.A. Mooney, J. Lubchenco, and J.M. Melillo. 1997. Human domination of earth's ecosystems. *Science* 277: 494–499.

Wu, F. 2002. Calibration of stochastic cellular automata: The application to rural-urban land conversions. *International Journal of Geographical Information Systems* 16 (8): 795–818.

Yue, T.X. 2011. *Surface modelling: High accuracy and high speed methods*. New York: CRC Press.

Yue, T.X., and Q.Q. Li. 2010. Relationship between species diversity and ecotope diversity. *Annals of the New York Academy of Sciences* 1195: E40–E51.

Yue, T.X., Z.M. Fan, and J.Y. Liu. 2005. Changes of major terrestrial ecosystems in China since 1960. *Global and Planetary Change* 48: 287–302.

Yue, T.X., Z.M. Fan, J.Y. Liu, and B.X. Wei. 2006. Scenarios of major terrestrial ecosystems in China. *Ecological Modelling* 199: 363–376.

Yue, T.X., Z.M. Fan, and J.Y. Liu. 2007. Scenarios of land cover in China. *Global and Planetary Change* 55: 317–342.

Yue, T.X., Z.M. Fan, C.F. Chen, X.J. Sun, and B.L. Li. 2011. Surface modelling of global terrestrial ecosystems under three climate change scenarios. *Ecological Modelling* 222: 2342–2361.

Yue, T.X., N. Zhao, Z.M. Fan, J. Li, C.F. Chen, Y.M. Lu, C.L. Wang, B. Xu, and J.P. Wilson. 2016. CMIP5 downscaling and its uncertainty in China. *Global and Planetary Change* 146: 30–37.

Yue, T.X., N. Zhao, Y. Liu, Y.F. Wang, B. Zhang, Z.P. Du, Z.M. Fan, W.J. Shi, C.F. Chen, M.W. Zhao, D.J. Song, Y.J. Song, C.Q. Yan, Q.Q. Li, X.F. Sun, L.L. Zhang, Y.Z. Tian, W. Wang, Y.A. Wang, S.N. Ma, H.S. Huang, Y.M. Lu, Q. Wang, C.L. Wang, Y.Z. Wang, M. Lu, W. Zhou, Y. Liu, X.Z. Yin, Z. Wang, Z.Y. Bao, M.M. Zhao, Y.P. Zhao, Y.M. Jiao, U. Naseer, B. Fan, S.B. Li, Y. Yang, and J.P. Wilson. 2020. A fundamental theorem for eco-environmental surface modelling and its applications. *Science China-Earth Sciences* 63 (8): 1092–1112.

Yue, T.X., N. Zhao, Z.M. Fan, J. Li, C.F. Chen, Y.M. Lu, C.L. Wang, J.J. Gao, B. Xu, J. Yan, and J.P. Wilson. 2019. Methods for simulating climate scenarios with improved spatiotemporal specificity and less uncertainty. Global and Planetary Change 181 https://doi.org/10.1016/j.gloplacha.2019.05.013

Zhang, Y.L., B.Y. Li, and D. Zheng. 2014. The data publish of A discussion on the boundary and area of the Tibetan Plateau in China: Geographical information system data of the boundary and area of the Tiebetan Plateau in China. *Global Change Research Data Publishing & Repository*. https://doi.org/10.3974/geodb.2014.01.12.v1.

Zheng, D. 1996. The system of physic-geographical regions of the Qinghai-Tibet Plateau (in Chinese). *Sci China Ser D-Earth Sci* 26: 336–341.

# Chapter 9
# Surface Modelling of Soil Properties

**Learning Objectives**

- To learn the progress and prospect of surface modelling of soil properties.
- To be able to map soil properties combined with environmental information.

## 9.1 Introduction

### *9.1.1 The History of Soil Maps*

#### 9.1.1.1 Ancient Soil Maps

The first one was produced based on characteristics like soil fertility, soil color, soil texture, and soil moisture in China from about 4000 years BP, which was used for taxation, soil management, medicinal usage, and construction (Fig. 1). The other example is from Mexico. The Aztecs developed a soil classification based on soil properties, topographic location, vegetation, and farmers' practices, which was used for taxation, soil management, medicinal usage, and construction (Fig. 1).

During the reign of Emperor Yao, the Chinese heartland was frequently plagued by floods that prevented further economic and social development. Yu's father, Gun, was tasked with devising a system to control the flooding. He spent more than nine years building a series of dikes and dams along the riverbanks, but all of this was ineffective, despite (or because of) the great number and size of these dikes and the use of a special self-expanding soil. As an adult, Yu continued his father's work and made a careful study of the river systems in an attempt to learn why his father's great efforts had failed.

 
T. Yue et al., *Eco-Environmental Informatics*,
https://doi.org/10.1007/978-981-96-1977-1_9

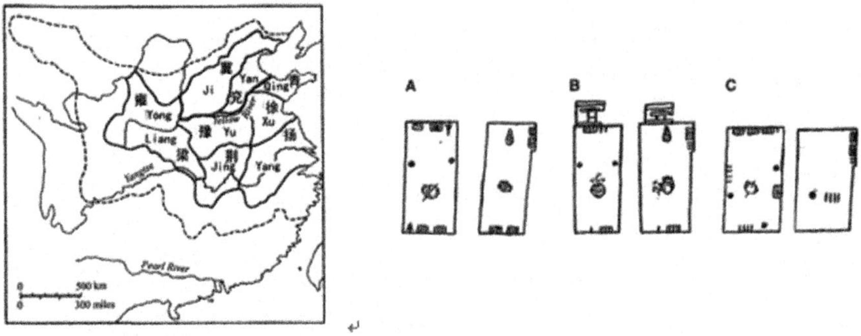


**Fig. 1** Maps by China (left) and Mexico (right)

Collaborating with Hou Ji, a semi-mythical agricultural master about whom little is concretely known, Yu successfully devised a system of flood controls that were crucial in establishing the prosperity of the Chinese heartland. Instead of directly damming the rivers' flow, Yu made a system of irrigation canals which relieved floodwater into fields, as well as spending great effort dredging the riverbeds. Yu is said to have eaten and slept with the common workers and spent most of his time personally assisting the work of dredging the silty beds of the rivers for the thirteen years the projects took to complete. The dredging and irrigation were successful, and allowed ancient Chinese culture to flourish along the Yellow River, Wei River, and other waterways of the Chinese heartland. The project earned Yu renown throughout Chinese history, and is referred to in Chinese history as "Great Yu Controls the Waters". In particular, Mount Longmen along the Yellow River had a very narrow channel which blocked water from flowing freely east toward the ocean. Yu is said to have brought a large number of workers to open up this channel, which has been known ever since as "Yu's Gateway" (Fig. 2).

#### 9.1.1.2 Early Scientific Regional Soil Maps in Nineteenth Century

Some early scientific soil maps in regions or countries emerged in nineteenth century, including Eastern Europe, Germany, Ireland, Netherlands, America, Russia and so on. The legends of these soil maps were based on geology, agrogeology, soil quality, zonal soil types and so on.

#### 9.1.1.3 The Origin World Maps by Dokuchaev

World soil maps followed the first climatic and geological maps (Fig. 3). As there was little global soil information, a global geographic theory was needed similar to that developed for the geological map. An important step in the development of this theory was when V.V. Dokuchaev related the distribution of soils to the factors of

**Fig. 2** Yu (left) and The scene of water control (right)

their formation. Although he was a geologist by training, Dokuchaev linked soils to the climate and corresponding vegetation. His ideas were possibly inspired by the biogeographical zoning of Alexander von Humboldt and earlier studies of soil-horizon division proposed by Fallou and Orth (Asio 2005). The bioclimatic aspects of soil formation were emphasized for two reasons. First, the distribution of soils on the Russian Plain, with its relatively uniform periglacial deposits, was mainly related to climate and to a lesser extent to the parent material. Secondly, Dokuchaev found that the climate was a much better predictor for the soil distribution than the complex geology.

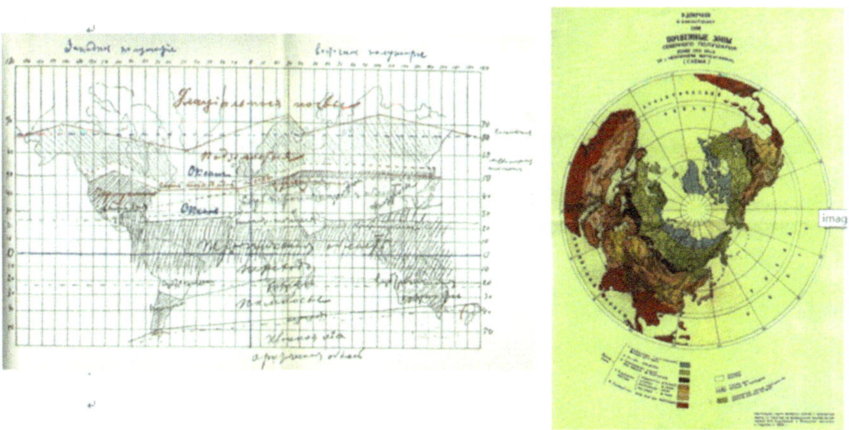

**Fig. 3** Sketch soil map of the world (1883) and soil map of the Hemishere (1899) by Dokuchaev

Dokuchaev considered making a global soil map. His archives contain an undated sketch drawing of a soil map of the world. He drew the soil zones: glacial soils, podzolic soils, grey transitional soils, and chernozems. Further to the south the soil names were mixed with the climatic names: kastanozems (with a question mark) in Mexico, subtropical soils across the Eurasian continent, krasnozems (red soils) and zheltozems (yellow soils) in China. The southern hemisphere of the Earth was less well marked; it had mostly the names of the climatic vegetative zones such as "tropical areas", "transition", "pampas", and "coniferous forest". Dokuchaev was not sure if the distribution of soils was symmetrical on the Earth, and the amount of data from the Southern Hemisphere was insufficient for good extrapolation. For that reason he probably decided to make a map for the Northern Hemisphere alone that was published in 1899 for the World Exhibition in Paris in 1900. The map had a polar projection and reflected the views of Dokuchaev on the global latitudinal zonality of soils. The legend included 11 soil units, namely the boreal, forest, chernozem steppes, wind-affected (with subdivision into stony, sandy, solonchak and loess deserts), and laterite zones plus forested stony territories, mountainous territories, and alluvial soils.

#### 9.1.1.4 The World Soil Maps of in the Twentieth Century

**Glinka's world soil maps**. K.D. Glinka, one of Dokuchaev's students, continued the work on the development of the soil-geographic concepts. Glinka contributed to the development of soil mapping and prepared the first schematic soil map of the world at a scale of 1:80 million in 1908 for the first edition of his textbook on soil science. In the Russian school of soil science, Glinka made the next step after Dokuchaev from soil-climatic zones to particular soils. However, this map showed that the spatial distribution of soils still closely followed the climatic distribution. The legend of the first version of the map published in 1908 included 18 units (soil types) plus a unit for water bodies: 1—podzolic (and soddy) soils, 2—forest soils and degraded chernozem, 3—chernozem (and regoor), 4—chestnut soils, 5—stratified columnar soils of temperate deserts, 6—esert crusts of temperate deserts, 7—krasnozem (terra rossa), 8—laterite, 9—zheltozem (Gehänmgelehm of Rihtgofen), 10—soils of dry tundra, 11—meadow (and meadow-steppe) soils, 12—krasnozem of tropical and subtropical deserts, 13—desert crusts of subtropical deserts, 14—vertical zones of mountainous regions, 15—large lakes, 16—bog soils, 17—sands of the deserts, 18—solonetz, and 19—dark soils of tropical savannas (Fig. 4).

**The USDA soil maps**. Some soil maps were produced in the USA since the mid-1800s, but systematic soil mapping begun with the start of the national soil survey program in 1899. Soil mapping was mainly done on the basis of soil series (Simonson 1989). In 1909 Milton Whitney published the first small-scale map of soil regions of the United States (Whitney 1909), followed by the maps by Coffey (1912) and Marbut et al. The legends of these maps included the landscapes, sediments, or geographical entities rather than soils (Simonson 1989). The development of fine-scale soil

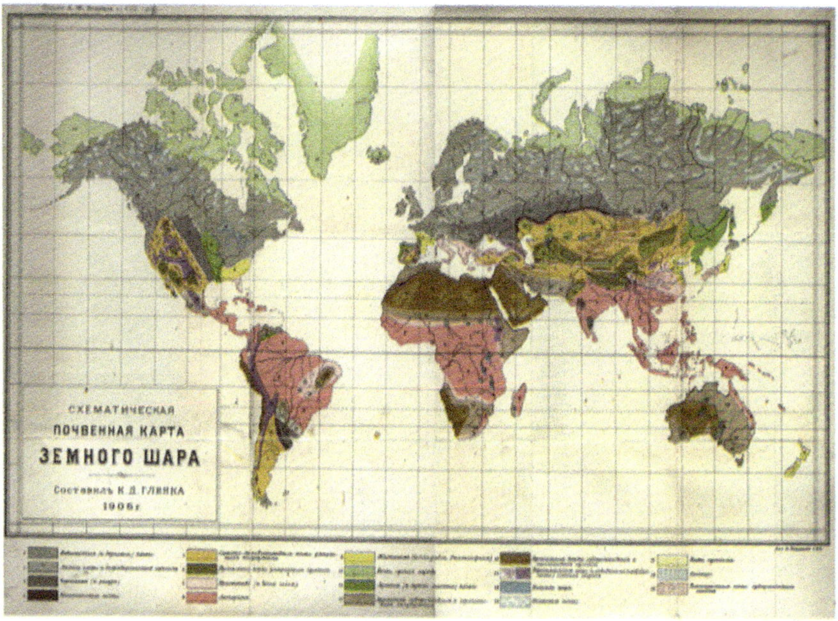


**Fig. 4** Glinka's schematic soil map of the world at a scale of 1:80 million (published in Glinka 1908)

cartography followed the introduction of soil classification by Curtis Marbut. The first soil map of the entire country appeared 16 years later, together with the US soil classification by Baldwin, Kellogg and Thorp. This map served as an illustration to the proposed classification and was published at a scale of 1:80 million; only 9 units were listed in the legend (Fig. 5). The classification included only zonal soil sequence, and intrazonal and azonal soils were omitted from the world map. The world soil map based on the same classification system was published in the 1957 Yearbook of Agriculture by Simonson (1957). The map has six broad zones: tundra soils, podzolic soils, chernozemic soils, desertic (arid) soils, latosolic soils, and soils of mountains. Each zone generally has similar processes of horizon differentiation prevailing over it and many kinds of soils are present in every zone (Simonson 1957).

**Other world soil maps**. Hollsteins (1930) published a schematic world soil map of the scale 1:125 million for the extensive Handbuche der Bodenlehre edited by E. Blank. Ten years after the last version of Glinka's map, L. I. Prasolov, the Director of the Dokuchaev Soil Institute in Moscow, prepared a 1:50 million soil map of the world for the Big Soviet Atlas of the World. It was the first world map completely based on the map of continents and regions of lrger scales. The map had a more extensive legend than the previous ones (Fig. 6). A characteristic of this map was a broad usage of landscape and land use names in the legend, such as "forest", "meadow", and "desert" soils (Rozanov 1977); this legend influenced the development of soil classification in Russia for several decades.

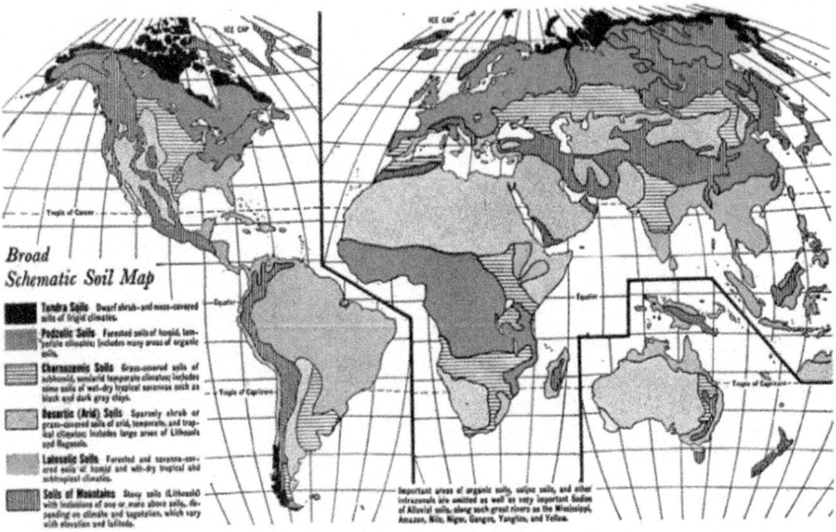


**Fig. 5** Map of the world showing six broad soil zones (Simonson 1957)

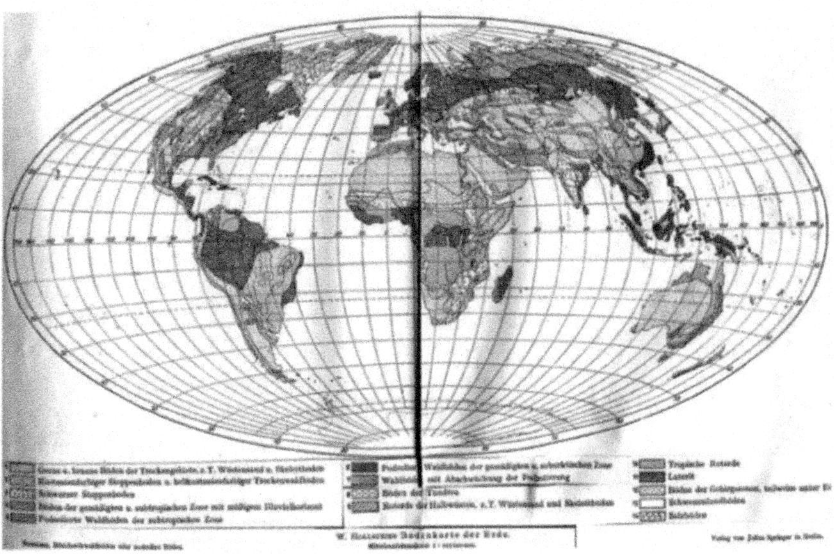

**Fig. 6** Hollsteins (1930), schematic world soil map (scale 1:125 million) from the Handbuche der Bodenlehre edited by E. Blank

#### 9.1.1.5 World Soil Maps Published Since 1906

There are several world soil maps published since 1906 listed as below (Table 1).

**Table 1** World soil maps published between 1906 and 2006

| Year of publication | Author(s) | Scale | Map name | Number of major units in the legend |
|---|---|---|---|---|
| 1906 | Glinka | 1:80 million | Schematic soil map of the globe | 18 |
| 1915 | Glinka | 1:80 million | Schematic soil map of the globe | 15 |
| 1927 | Glinka | 1:82 million | Schematic soil map of the globe | 15 |
| 1930 | Hollsteins | 1:125 million | Soil map of the world | 14 |
| 1937 | Prasolov | 1:50 million | Soil map of the world | 30 |
| 1957 | Simonson | No scale | Schematic world soil map | 6 |
| 1960 | Gerasimov | 1:60 million | Soil map of the world | 93 |
| 1968 | Ackerson et al | 1:50 million | Soils of the world | 117 |
| 1964 | Papadakis | 1:4 to1:20 million | Major soil regions of the world | 238 |
| 1970 | Papadakis | 1:100 million | Soil map of the world | 11 |
| 1971–1975 | FAO–UNESCO | 1:5 million | Soil map of the world | 106 |
| 1972 | USDA Soil Survey | 1:50 million | Soils of the world: the distribution of orders and major suborders | 43 |
| 1974 | Kovda et al | 1:10 million | Soil map of the world | 293 |
| 1982 | Glazovskaya and Fridland | 1:15 million | Soil map of the world for higher school | 79 |
| 2006 | USDA | 1:130 million | Global soil regions | 12 |

Modified after Rozanov (1977)

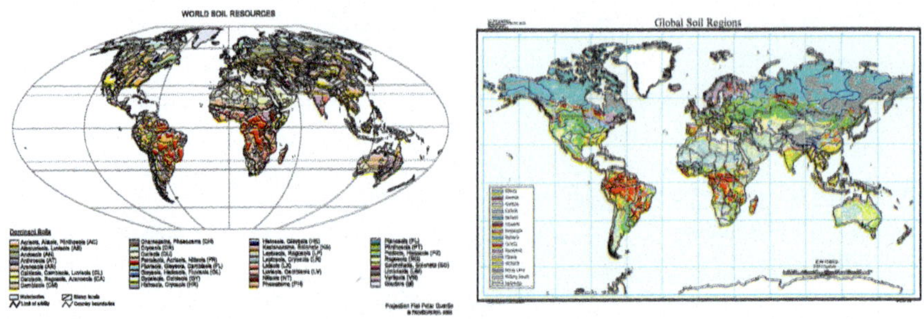


**Fig. 7** World soil maps based on FAO-EC-ISRIC, (2003) and Eswaran and Reich (2005)

#### 9.1.1.6 A Global Map Using the 12 Orders of Soil Taxonomy in 2006

With increasing information following extensive soil surveys in all parts of the world, world soil maps have become more precise, less schematic. Over time, world soil maps show an increasing complexity of the depicted soil pattern (Fig. 7).

Soil Taxonomy was developed through a series of approximations resulting in the seventh approximation in 1960, followed by publication of the first edition in 1975 (Soil Survey Staff 1975) and a second edition in 1999 (Soil Survey Staff 1999). Soil maps were produced with each new version of the classification system. The maps based on the seventh approximation and the first edition of Soil Taxonomy displayed 10 soil orders. The map published in the second edition of Soil Taxonomy is at a scale of 1:7.5 million and shows two additional soil orders (andisols and gelisols) that were added between 1975 and 1999. In 2006 a global map using the 12 orders of Soil Taxonomy was presented at the 18th World Congress of Soil Science in 2006. The map was derived by overlapping the reclassified FAO world soil map and the map of soil climates. Also a 1:5 million map of soil Suborders is available at the USDA web site.

#### 9.1.1.7 Limitations

Unfortunately, the existing soil maps, produced by conventional soil survey and mapping techniques, are often coarse in scales and lack details. There is a need to design more sophisticated technologies for modeling soil properties and processes at fine resolution and with high accuracy. Over time, world soil maps show an increasing complexity of the depicted soil pattern, but most of the maps were concept-dependent rather than data-derived and are influenced by the underlying systems of soil classification.

#### 9.1.1.8 The Success of Digital Soil Mapping

The success of digital soil mapping is from a confluence of several factors in the beginning of 2000 including: the increased availability of spatial data, like digital elevation model and satellite imagery; the availability of computing power for processing data; the development of data-mining tools and GIS; numerous applications beyond geostatistics; an increased global demand for spatial data, including uncertainty assessments, a rejuvenation of many soil survey, university centres which helped in the spreading of digital soil mapping technologies and knowledge.

#### 9.1.1.9 Information Technology in Pedometrics

Digital soil mapping is dependent on computer power and information technology, because the ability to model and dealing with spatial data needs higher computing power. We can found that the pixel number we should deal with has an exponential growth with time. Here is the history of information technology use in digital soil mapping. The success of digital soil mapping cannot live without information technology use.

Before 1960s, the earliest days of quantitative soil science, all computed by hand. Going back to the earliest days of quantitative soil science, in 1910 Mercer and Hall (1911) studied the variability of wheat yield in an apparently uniform field, one of a series of uniformity trials popular at the time (Cochran 1937). Although they did not measure the soil, they inferred that the substantial local variability in crop yield and grain/ straw ratio could in part be attributed to local differences in soil nutrients. They computed numerical summaries, linear regression, and probable errors, all computed by hand. To simplify hand computations for linear regression they grouped observations, rather than use all 500 to fit the model. Fifteen years later Haines and Keen (1925) carried out a series of dynometer experiments and produced a contour map of the drawbar pull required over a field, again by hand; this may be the first two-dimensional continuous soil property map. In a subsequent article (Haines and Keen 1925) they expanded this to a 3D solid model. In 1937 Youden and Mehlich (1937) computed what were effectively variograms of soil properties at spacings of a nested sampling design, along with a nested analysis of variance, again all by hand.

During 1960s, the first moves beyond hand computation were made possible by the development of the digital computer. Also during the 1960s, models of individual soil processes, such as heat and water transport.

During the early development of soil information systems (1975–1983) computers came into wide use. Typical is the information system of STIBOKA (Dutch Soil Survey Institute) in 1977. This was also the period where geostatistical approaches, primarily various forms of kriging, were first applied to pedometrics. Towards the end of this period the so-called "personal" computer, on the desktop of the individual pedometrician, became available. Computer programs specifically for these were soon developed, for example the IDRISI GIS grid-based GIS, developed from 1987

(Eastman and Fulk 1993). For more computationally-demanding applications, this era also saw the advent of desktop Unix workstations, although these do not appear to have been used by pedometricians until about ten years later.

Towards the end of the 1980s, the development of more powerful computers and improved numerical methods (Press 1986) integrated into subroutine packages allowed the development of soil models that integrated different types of processes (physical, chemical, biological) and their interaction; these were applied at various scales from pore to global at different levels of process generalization (Vereecken et al. 2010).

By 1992 Burrough et al. (1994) could state "Cheap, powerful personal computers have made geostatistics available to many people and have stimulated the teaching of the methods in universities", and list 11 programs, including some that have continued in use and others that were once popular but now have been superseded. Among the methods which were computationally feasible by 1992 were multivariate geostatistics, cross validation, jackknifing, principal components, factor analysis, varimax rotation, spectral analysis by Fast Fourier Transform, sets of partial differential equations, and simulation by process models. By 1995 large numbers of Monte Carlo simulations became feasible on desktop computers. Advances in GIS allowed terrain parameters to be computed over moderately-sized areas, as covariates for digital soil mapping (DSM). 1997 was a breakthrough year for electronic communication. Journals publishing pedometrics papers started to print authors' e-mail addresses, and institutions reserved e-mail address blocks for their staff. Desktop computer power continued to increase.

### 9.1.2 The Theoretical Framework for Digital Soil Mapping

Here, we generalise and formalise the approach that has begun to emerge in papers published lately. We use a Jenny-like formulation not for explanation but for empirical quantitative descriptions of relationships between soil and other spatially referenced factors with a view to using these as soil spatial prediction functions. We consider seven factors:

s: soil, other properties of the soil at a point;
c: climate, climatic properties of the environment at a point;
o: organisms, vegetation or fauna or human activity;
r: topography, landscape attributes;
p: parent material, lithology;
a: age, the time factor;
n: space, spatial position.

We have included soil as a factor because soil can be predicted from its properties, or soil properties from its class or other properties. We shall call this the scorpan model, which can be written as: Sc = f (s, c, o, r, p, a, n) or Sa = f (s, c, o, r, p, a, n). Where Sc is soil classes and Sa is soil attributes. The s refers to soil information either from a prior map, or from remote or proximal sensing or expert knowledge.

The key predictor factors are r (80% of studies) followed by s (35%), o and p (both 25%), n (20%) and c (5%) whereas a does not seem to have been used as a factor. A single factor is used in 40% of studies, two factors are used in 40% of studies, around 10% of studies considered three of the seven factors and 2% considered four factors. No studies considered five or more factors. The most common combination was r and s. Most studies used a DEM as the main source of ancillary data, followed by remotely sensed imagery and preexisting soil coverages.

#### 9.1.2.1 Sources of Data—The Seven SCORPAN Factors

There are seven factors or sets of variables in the SCORPAN model which makes it different from Jenny's model. The aim is to obtain information on all of these. It will be a matter of convenience (access to data sources) and scientific contention which variables are used to represent the factors. Indeed, this is an area that has not been well enough studied. The creation of these digital maps of the input environmental variables representing the six factors in the SCORPAN model is seen as an integral part of the digital soil resource assessment approach and a very valuable, environmentally useful, by-product of the new approach. The layers can be used for other modelling purposes. Much of the earth science and ecological research of the last twenty years has been contributing towards the creation of these layers.

**S**. Remote and proximal active and passive sensing gives detailed information on the soil themselves— these reflections or emissions or transmissions are intrinsic properties of the soil material and profile they may indicate other soil attributes like texture or mineralogy. This factor is likely to becoming increasingly important as technology advances.

**C**. Climate could be represented by mean annual temperature (T) and mean annual rainfall (P) and perhaps some measure of potential evapotranspiration (E). In the old literature P/E was used to separate pedalfers from pedocals. A P/E < < 1 would imply semiarid conditions and precipitation of carbonates, etc. at depth.

Attempt to use soil moisture and temperature regimes for classification of soil has also been made (Donatelli et al. 2002). They made a classification using dynamic simulation of physical processes which describe the system soil-grass as affected by weather. The daily step simulation of a reference grass offered an insight into the soil moisture and temperature regimes of different soils.

Climate surfaces can be produced from meteorological stations interpolated by Laplacian smoothing splines. This has been implemented in a program called ANUCLIM. The climate variables used are monthly mean values for minimum temperature, maximum temperature, precipitation, solar radiation, evaporation and others. The climate surfaces can be used to generate secondary information, e.g., bioclimatic parameters such as mean temperature of warmest period, precipitation of driest quarter, etc., which are useful in determining the climatic envelope for plant and animal species.

Published work suggests that remote-sensing analysis can be used for estimating and mapping air temperature, soil moisture and atmospheric humidity at regional to global scales. Air temperature can be inferred from normalised difference vegetation index (NDVI) data from the NOAA Advanced Very High Resolution Radiometer (AVHRR). Several studies have shown that the surface albedo can be estimated using remote sensing data, and that net radiation can be calculated with sufficient accuracy.

The main soil forming or altering organisms are vegetation or humans, although other organisms can have an appreciable soil-modifying effect locally (Hole 1981). In pristine or newly developed environments, the 'natural' vegetation class should represent some kind of equilibrium relation with soil type. Human effects can be seen through land use changes whereupon humans choose different soil types for various purposes. Therefore, land use and land cover are useful indicators of soil properties and class. More recently, estimates of biomass for both crops and more natural vegetation have been obtained; once again these can reflect soil differences. Estimates of vegetation type, land use and land cover and biomass have all been obtained from visible and infrared reflectance by remote sensing and have been enhanced more recently by microwave imagery.

**R.** This is now mainly derived from digital elevation models. Sources of elevation data can be from digitising contour and streamline data, point measurements of elevation from traditional land surveys or from vehicle-mounted high-resolution GPS receivers, or remotely sensed elevation data. The first step to use this information is parametisation of the surface model or the numerical description of continuous surface form (Pike 1988; Wood 1996). Parametisation is to quantitatively measure properties of a landscape that can be used to describe form (Wood 1996). These parameters are aimed to characterise the geomorphometry of the surface or for landform classification.

Different attributes can be parameterised from a DEM, such as altitude, slope, aspect, different curvatures, upslope area, compound topographic index, etc. Topography has been recognised as one of the soilforming factors (Jenny 1941). Aandahl (1948) is perhaps the first scientist to quantitatively relate landscape attributes to soil properties. He derived the distribution of N based on slope length. Troeh (1964) fit a cylindrical parabola to contour lines to derive slope and curvatures. The landform parameters were derived to correlate to soil drainage classes. Many would argue that digital terrain modelling is the most useful and quantitatively developed factor for predicting soil attributes and soil classes (McKenzie et al. 2000).

**P**. Parent material information can be obtained from digitised geological maps—maps that focus on lithology and not so much on stratigraphy will probably be more useful for soil prediction. Some kind of quantitative information about surface mineralogy and texture (related to parent material) can be obtained by gamma radiometrics. Geomorphological and weathering models (Dickson et al. 1996) have been used to identify the distribution of soil-forming materials.

Additionally, the natural fields of the earth, gravitational, electrical (Andriani et al. 2001), magnetic and electromagnetic (Beard 2000) can be used to provide information on underlying geological structure. Taylor and Eggleton (2001) discuss regolith mapping. Regolith should be predictive of soil, as soil is the upper part of it. The regolith can be thought of as representing either s or p. Indeed, regolith itself may more useful than the soil profile for hydrological environmental modelling. Regolith maps are produced in Australia from a combination of multi- and hyperspectral and airborne geophysical data and expert knowledge: reflecting the scorpan approach.

A represents age or elapsed time. This will give limits on how long pedogenesis has been occurring and should differentiate soil classes and properties. One useful estimate of a is the age of the ground surface, which may be very old indeed (Twidale 1985). Alternatively, a can be represented by the age of the material in which soil has developed, suggesting that the scorpan approach will not deal well with polycyclic soils. It is theoretically likely that soil development will follow some logarithmic or square-root time relationship, suggesting more need to differentiate between younger materials than older ones.

N. Soil can be predicted from spatial coordinates alone. Obtaining these is now much easier due to the advent of GPS with 5-m accuracy receivers costing less than US$1000. This may indeed reflect some other environmental variable such as climate, and because of this it can be argued that n is not really a factor, but simply putting the coordinates is a simple way to ensure that spatial trends not included in the other environmental variables are not missed. Therefore, n could also be described by some linear or nonlinear (nonaffine) transformation of the original spatial coordinates, e.g., a new coordinate could be the closest distance of each location to the coast (Webster 1977, p. 201), or distance uphill from the nearest discharge area (Bui and Moran 2000). This factor is potentially a valuable yet cheap source of environmental information, and should never be disregarded.

#### 9.1.2.2 The General Approach

If we write the equation as $S = f(Q) + e$, then the general approach we shall use is to take some observations of S in the field at known locations [x, y] and fit some kind of function a set of pedologically meaningful predictor variables Q which will be generally raster data layers of size M in a GIS. Once the model is fitted at the m observation points, the prediction can be extended to the M points or cells in the raster thereby giving a digital map. The efficiency of the method relies on the fact that hopefully $m < < M$, and because S is much more difficult and expensive to measure than the Q. S is soil classes and soil attributes as a function of soil, climate, organisms, relief, parent material, age and spatial position, Soil attributes have been estimated more often (70% of studies) than soil classes (30% of studies). f() is some form of empirical quantitative function f linking S to the scorpan factors. The most common used methods include Kriging, IDW, Splines, Machine learning models, Regression models, HASM and so on; 'e' is spatially correlated residuals.

### 9.1.3 Global Workshops on Digital Soil Mapping

Since the digital soil mapping scorpan concept was introduced, and following a series of global workshops, there has been huge interest in this topic of research. The first global digital soil mapping workshop was held in Montpellier in September 2004. The IUSS working group on Digital Soil Mapping was formed following the first workshop. Successive global workshops were held in Rio De Janeiro, Brazil in 2006, Logan, USA in 2008, Rome, Italy in 2010, Sydney, Australia in 2012 and in Nanjing, China in 2014. It has resulted in a series of books. Following the second global workshop in 2006, the Global Soil Map project was initiated.

There are three milestones for the development of digital soil mapping. The first milestone is the international workshop on digital soil mapping which was held in Montpellier, France, 2004. It provided an opportunity to put in common a wide range of skills and tools that have a role to play in DSM: soil surveying, soil information systems, expert systems, GIS, pedometric techniques, data mining techniques and remote sensing procedures. The second milestone is the setting up of the working group on digital soil mapping under International Union of Soil Science (IUSS) in 2005. This working group organized such global workshops regularly (every two years). The workshops, combined with the Pedometrics meetings which holds every two years since 1997, ensure that every year can have an opportunity to bring together soil mapping scientists to communicate the progresses of soil mapping researches. The third milestone is the official launch of the GlobalSoilMap.net Project at University of Columbia, USA in 2009. Eight nodes across the world were set up for the project. The establishment of the GlobalSoilMap.net Consortium (GSC 2015), putting together the major players in digital soil mapping in the world, has initiated a process that will deliver a new digital soil map of the world at fine resolution (Sanchez et al. 2009). This project aims to predict and map most of the ice-free land surface of the globe over the next decade at a grid resolution of 90 by 90 m using state-of-the-art and emerging technologies.

### 9.1.4 Soil Database

**Global and national soil datasets of soil-type map and soil profiles**. Based on these workshops, Global and national soil datasets of soil-type map and Soil profiles were applied in the digital soil mapping. There are some famous databases at the global level, including FAO-UNESCO Soil Map of the World (SMW), The World Inventory of Soil Emission Potentials (WISE), and The World Soil Information Service (WoSIS). There are also some databases at the regional and national levels, including SOTER, and the databases from Europe, China, US, Canada, Australia, Africa, Mexico, Brazil and so on (Table 2).

The soil databases are produced mainly by two kinds of methods—the linkage method and digital soil mapping. The linkage method, involves linking soil profiles

**Table 2** Global and national soil datasets of soil-type map and soil profiles

| Soil data | Databases at the global level | Databases at the regional and national levels |
|---|---|---|
| Soil-type map | FAO-UNESCO Soil Map of the World (SMW) | The Soil and Terrain Database (SOTER) databases for different regions<br>The European Soil Database (ESB)<br>The 1:1 million Soil Map of China<br>The US General Soil Map (GSM)<br>The 1:1 million Soil Map of Canada<br>The Australian Soil Resource Information System (ASRIS) |
| Soil profiles | The World Inventory of Soil Emission Potentials (WISE),<br>The World Soil Information Service (WoSIS) | The USA National Cooperative Soil Survey Soil Characterization database<br>The USA National Soil Information System<br>The Africa Soil Profiles database<br>The ASRIS<br>The Chinese National Soil Profile database<br>The Canadian Soil Information System<br>Soil profiles from SOTER<br>The soil profile analytical database for Europe<br>The Mexico soil profile database<br>The Brazilian national soil profile database |

and soil mapping units on soil-type maps, sometimes with ancillary maps such as topography and land use. Digital soil mapping is the creation and population of a geographically referenced soil database, generated at a given resolution by using field and laboratory observation methods coupled with environmental data through quantitative relationships (Dai et al. 2019) (Table 3).

**Table 3** Global and national soil datasets by linkage method and DSM

| Method | Database | |
|---|---|---|
| | The global level | The regional and national levels |
| The linkage method | The FAO SMW with derived soil properties<br>Data and Information System of International Geosphere-Biosphere Programme (IGBP-DIS) database<br>The Soil and Terrain Database for multiple regions and countries<br>The HWSD<br>The Global Soil Dataset for Earth System Model (GSDE)<br>WISE30sec | The USA State Soil Geographic Database (STATSGO2)<br>Soil Landscapes of Canada;<br>The ASRIS<br>The Soil-Geographic Database of Russia<br>The European Soil Database<br>The China dataset of soil properties |
| Digital soil mapping | GlobalSoilMap<br>The SoilGrids system<br>The Global SOC (soil organic carbon) Map of the Global Soil Partnership | |

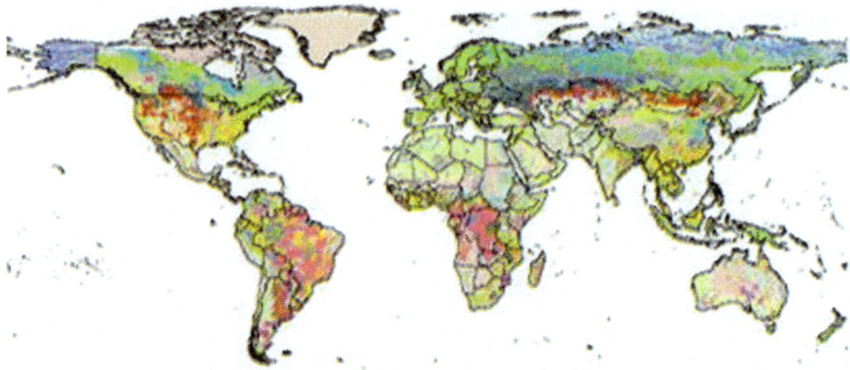

**Fig. 8** HWSD

#### 9.1.4.1 HWSD

The HWSD attribute database provides information on the soil unit composition for each of the 15,773 soil mapping units. The database shows the composition of each soil mapping unit, and standardized soil parameters for top- and subsoil. A soil mapping unit can have up to 9 soil unit/topsoil texture combination records in the database. Coordinated by FAO, several universities and institutes integrated the regional soil maps to a new comprehensive Harmonized World Soil Database (HWSD). Physical and chemical characteristics of topsoil (0–30 cm) and subsoil (30–100 cm) (Fig. 8).

#### 9.1.4.2 SoilGrids 250 m

SoilGrids™ (hereafter SoilGrids) is a system for global digital soil mapping that uses state-of-the-art machine learning methods to map the spatial distribution of soil properties across the globe. SoilGrids prediction models are fitted using over 230,000 soil profile observations from the WoSIS database and a series of environmental covariates. Covariates were selected from a pool of over 400 environmental layers from Earth observation derived products and other environmental information including climate, land cover and terrain morphology. The outputs of SoilGrids are global soil property maps at six standard depth intervals (according to the GlobalSoilMap IUSS working group and its specifications) at a spatial resolution of 250 m. Prediction uncertainty is quantified by the lower and upper limits of a 90% prediction interval. The additional uncertainty layer displayed at soilgrids.org is the ratio between the inter-quantile range and the median. The SoilGrids maps are publicly available under the CC-BY 4.0 License.

Maps of the following soil properties are available: pH, soil organic carbon content, bulk density, coarse fragments content, sand content, silt content, clay

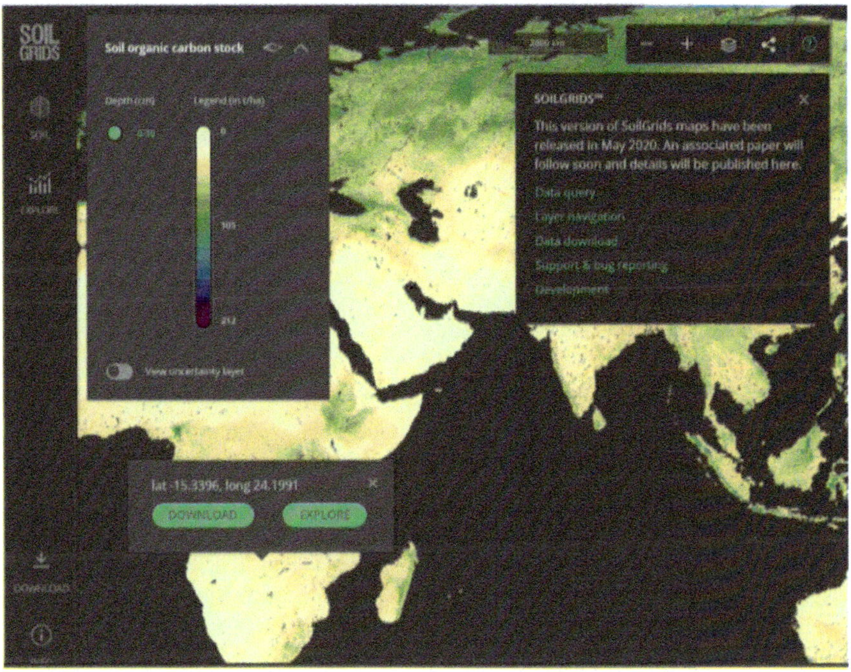


**Fig. 9** SoilGrids 250 m

content, cation exchange capacity (CEC), total nitrogen as well as soil organic carbon density and soil organic carbon stock (Fig. 9).

#### 9.1.4.3 Comparison of Available Global Soil Datasets

Researchers compared several available soil datasets and soil profiles for several soil properties, including IGBP, HWSD, GSDE, WISE30sec, and SoilGrids, which are freely available. Four statistics were used in the evaluation, including mean error (ME), root mean squared error (RMSE), coefficient of variation (CV), and coefficient of determination ($R^2$). Based on these evaluation, SoilGrids performs better than the others. Distributions of soil properties in the soil dataset produced by the digital soil mapping method are different from those using the linkage method.

**GlobalSoilMap project**. The official launch of the GlobalSoilMap.net Project at University of Columbia, USA in 2009 has initiated a process that will deliver a new digital soil map at a global scale on a 3″support grid (approximately 90 × 90 m). The GlobalSoilMap project provides a collaborative scientific framework to process this legacy soil data into consistent, spatially explicit and continuous global soil information, freely accessible and in a gridded format at a high resolution. The targeted information includes predicted values of selected key soil properties at 6 standard depth intervals (0–5; 5–15; 15–30; 30–60; 60–100; and 100–200 cm), at a global scale on a 3″ support grid (approximately 90 × 90 m) along with their uncertainties. The key primary soil properties include clay, silt and sand content, coarse elements, pH, soil organic carbon (SOC), effective cation exchange capacity (ECEC) and soil depth to bedrock and effective root zone depth. Additional key properties include bulk density, plant-available water holding capacity and electrical conductivity.

### *9.1.5 The Application of Surface Modelling of Soil Properties*

#### 9.1.5.1 Agricultural Production

The soil is the most crucial component for maintaining ecosystem services in terrestrial ecosystems (Bishop and McBratney 2001). The surface modelling of soil properties play a crucial role in agricultural development, providing a valuable foundation for subsequent and future measurement assessments (Wang et al. 2008). For example, Soil sodicity is a particularly significant issue in the subsoil, where reduced porosity and limited oxygen diffusion can constrain root growth once a constraint is reached. This causes an agronomic problem, and reduces the exploitable depth for roots, meaning that there is less moisture and nutrients that is accessible at depth by plants. This can lead to an overestimation of water and nutrient availability by farmers if the depth to a sodicity constraint is not known, possibly leading to incorrect fertilizer rates being applied and the waste of input resources. Given the high cost and time-consuming nature of soil sampling, research on methods for reconstructing soil maps from sparse data becomes increasingly important.

#### 9.1.5.2 Environmental Management

Surface modelling of soil properties is also useful for environmental management. For example, large numbers of livestock and poultry feces were applied into the soils of the intensive vegetable cultivation areas (Bound and Voulvoulis 2004), some of the antibiotics via the manure application were imported in soils and been strongly adsorbed (Sarmah et al. 2006; Martínez-Carballo et al. 2007; Aust et al. 2008). The spatial distributions of soil antibiotics are complex, because they are all from the external sources, and have irregular transportation into water and the irregular absorption by different plants (Shi et al. 2016). Because an inaccurate spatial pattern estimation of the soil antibiotics will result in considerable bias in risk assessment, the spatial interpolation of soil antibiotics is vital to evaluate the potential risks to human and environment (Xie et al. 2012).

#### 9.1.5.3 Mitigation and Adaptation of Climate Change

Soils represent the largest terrestrial carbon sink on Earth, showing a high spatial variability (Yang et al. 2020). It plays an important role in the global carbon cycle (Zhang et al. 2014). Climate change has greatly affected soil organic carbon (SOC) (Zhang et al. 2019), and small changes in soil carbon pools may lead to large fluctuations in atmospheric $CO_2$ concentrations, which will further affect the global climate. Consequently, the stability of SOC is an essential factor in the global carbon cycle and climate change (Belay-Tedla et al. 2009). The use of soil property surface modeling approach to assess SOC stocks is necessary and at the same time one of the feasible methods. Detailed, accurate, and up-to-date soil carbon information is urgently needed to aid in developing solutions for these issues and to concerning mitigation and adaptation of climate change (Zhang and Wu 2018).

#### 9.1.5.4 Assessment of Ecosystem Services

Soil is an essential component of the terrestrial ecosystem. Many physical and chemical processes in soil science, including water and heat fluxes, nutrient dynamics, water and nutrient holding capacity, soil structure, stability, and erosion, play a crucial role in supporting ecosystem services (Delbari et al. 2011; Morgan et al. 1998; Odeh et al. 2003; Rawls et al. 1982). For instance, soil PSF interpolation and soil texture classification provide crucial information for understanding soil water-holding and nutrient capacities, soil structure morphology and stability, as well as soil erosion, making it of significant importance in the context of ecosystem services. Accurate spatial prediction of soil properties helps understand the simulation processes of the models of assessment of ecosystem services (Shi et al. 2023).

## 9.2 Methods and Models

### *9.2.1 Classical Interpolators*

#### 9.2.1.1 Kriging

The presence of a spatial structure where observations close to each other are more alike than those that are far apart (spatial autocorrelation) is a prerequisite to the application of geostatistics (Goovaerts 1992). The experimental variogram measures the average degree of dissimilarity between unsampled values and a nearby data value (Deutsch and Journel 1998), and thus can depict autocorrelation at various distances. The value of the experimental variogram for a separation distance of h (referred to as the lag) is half the average squared difference between the value at z(xi) and the value at z(xi + h) (Lark 2000):

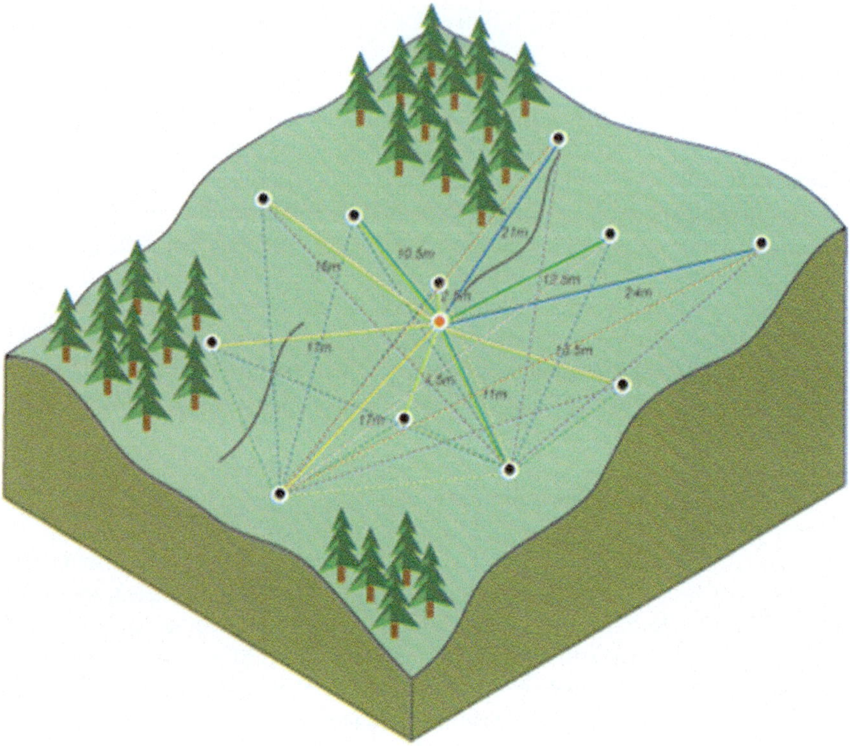

**Fig. 10** Kriging

$$\hat{\gamma}(h) = \frac{1}{2N(h)}\sum_{i=1}^{N(h)}[z(x_i) - z(x_i + h)]^2 \tag{1}$$

where N(h) is the number of data pairs within a given class of distance and direction. If the values at $z(x_i)$ and $z(x_i + h)$ are autocorrelated the result of Eq. (1) will be small, relative to an uncorrelated pair of points. From analysis of the experimental variogram, a suitable model (e.g. spherical, exponential) is then fitted, usually by weighted least squares, and the parameters (e.g. range, nugget and sill) are then used in the kriging procedure (Fig. 10).

#### 9.2.1.2 IDW

Similar to kriging, inverse distance weighting directly implements the assumption that a value of an attribute at an unsampled location is a weighted average of known data points within a local neighborhood surrounding the unsampled location. The formula of this exact interpolator is:

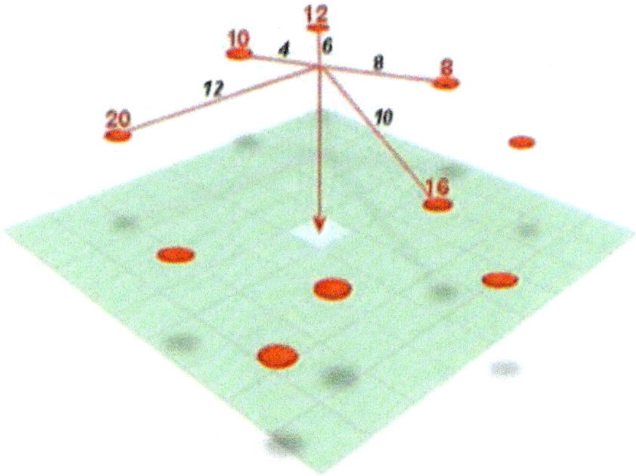


**Fig. 11** IDW

$$\hat{z}(x_0) = \frac{\sum_{i=1}^{n} z(x_i) d_{ij}^{-r}}{\sum_{j=1}^{n} d_{ij}^{-r}} \tag{2}$$

where $x_0$ is the estimation point and xiare the data points within a chosen neighborhood. The weights (r) are related to distance by dij, which is the distance between the estimation point and the data points. The IDW formula has the effect of giving data points close to the interpolation point relatively large weights whilst those far away exert little influence. The higher the weight used the more influence points close to $x_0$ are given (Fig. 11).

#### 9.2.1.3 Splines

Splines produce good results with gently varying surfaces, and thus are often not appropriate when there are large changes in the surface values within a short horizontal distance. A surface does not usually fit the formulation of the spline surface theoretically, which leads to a considerable interpolation error (Fig. 12).

### *9.2.2 A New Method—HASM*

Surface modelling is a useful tool for soil property interpolation in agricultural and environmental fields. The application of precision agriculture and management of soil contamination need the development of this technique. However, the effectiveness of the application relies on the accuracy of spatial interpolation which is used to describe the spatial variability of soil properties. It is necessary to study the interpolating method with high accuracy and improve interpolation quality for soil properties.

High accuracy surface modelling (HASM), constructed by Prof. Yue Tian-Xiang, is a spatial interpolation technique based on the fundamental theorem of surfaces.

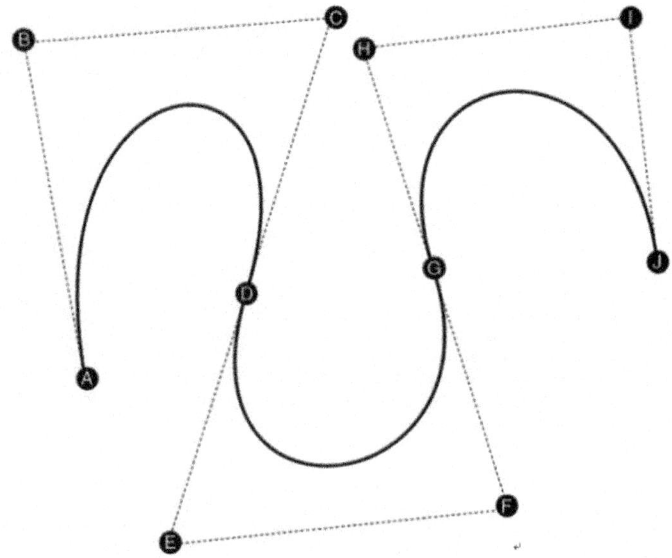


**Fig. 12** A cubic Bezier spline

It has good performance in spatial interpolation, spatial Up-scaling, spatial Down-scaling, Data fusion and Model-data assimilation. For example: HASM was much more accurate than the classical methods in numerical tests (Yue et al. 2007, 2008a, b, in press), and theoretically gave a solution to the error problem that had long troubled DEM construction (Yue et al. 2007); Interpolation of climate and ecosystem change showed that HASM had highly accelerated computational speed, especially for interpolation with huge computational work, and it also had greatly increased interpolation accuracy (Yue et al. 2008a, b).

According to the fundamental theorem of surfaces, a surface is uniquely defined by the first and the second fundamental coefficients (Henderson and Taimina 1997). The first fundamental coefficients are used to express information about some geometric properties of the surface, by which we can calculate the lengths of the curves, angles of tangent vectors, areas of regions and geodesics on the surface (Yue et al. 2010). The second fundamental coefficients reflect the local warping of the surface, namely its deviation from tangent plane at the point under consideration.

If a surface is a graph of a function $z = u(x, y)$, the basic theoretical equations of HASM could be formulated as:

$$u_{xx} = \Gamma_{11}^{1} u_x + \Gamma_{11}^{2} u_y + \frac{L}{\sqrt{EG - F^2}} \tag{3}$$

$$u_{yy} = \Gamma_{22}^{1} u_x + \Gamma_{22}^{2} u_y + \frac{N}{\sqrt{EG - F^2}} \tag{4}$$

where $E = 1 + u_x^2$; $F = u_x u_y$; $G = 1 + u_y^2$; $L = \frac{u_{xx}}{\sqrt{1+u_x^2+u_y^2}}$; $N = \frac{u_{yy}}{\sqrt{1+u_x^2+u_y^2}}$;

$$\Gamma_{11}^1 = \frac{1}{2}\left(GE_x - 2FF_x + FE_y\right)\left(EG - F^2\right)^{-1};$$

$$\Gamma_{22}^1 = \frac{1}{2}\left(2GF_y - GG_x - FG_y\right)\left(EG - F^2\right)^{-1}$$

$$\Gamma_{11}^2 = \frac{1}{2}\left(2EF_x - EE_y - FE_x\right)\left(EG - F^2\right)^{-1}$$

$$\Gamma_{22}^2 = \frac{1}{2}\left(EG_y - 2FF_y + FG_x\right)\left(EG - F^2\right)^{-1}$$

If the maximum lengths of the computational domain in the $x$ and $y$ directions are respectively $L_x$ and $L_y$, the computational domain can be included in the rectangular domain $[0, L_x] \times \left[0, L_y\right]$. If $h$ represents interpolation step length and $I + 2$ and $J + 2$ represent the lattice numbers in directio $x$ and in direction $y$, the central point of lattice $(0.5h + (i - 1)h, 0.5h + (j - 1)h)$ could be expressed as $(x_i, y_j)$, in which $i = 0, 1, \cdots, I, I + 1$ and $j = 0, 1, \cdots, J, J + 1$. $u(x + h, y)$ and $u(x - h, y)$ could be formulated as the following Taylor expansion in series,

$$u(x + h, y) = u(x, y) + h\frac{\partial u(x, y)}{\partial x} + \frac{h^2}{2!}\frac{\partial^2 u(x, y)}{\partial x^2} + \frac{h^3}{3!}\frac{\partial^3 u(x, y)}{\partial x^3} + \mathrm{O}(h^4) \quad (5)$$

$$u(x - h, y) = u(x, y) - h\frac{\partial u(x, y)}{\partial x} + \frac{h^2}{2!}\frac{\partial^2 u(x, y)}{\partial x^2} - \frac{h^3}{3!}\frac{\partial^3 u(x, y)}{\partial x^3} + \mathrm{O}(h^4) \quad (6)$$

Formulation (5) minus formulation (6) gives that,

$$u(x + h, y) - u(x - h, y) = 2h\frac{\partial u(x, y)}{\partial x} + \frac{2h^3}{3!}\frac{\partial^3 u(x, y)}{\partial x^3} + \mathrm{O}(h^5) \quad (7)$$

Therefore,

$$u_x(x, y) = \frac{\partial u(x, y)}{\partial x} = \frac{u(x + h, y) - u(x - h, y)}{2h} + \frac{h^2}{3!}\frac{\partial^3 u(x, y)}{\partial x^3} + \mathrm{O}(h^4) \quad (8)$$

For sufficiently small $h$, the finite difference approximation of $u_x(x, y)$ and $u_y(x, y)$ could be expressed as,

$$u_x(x, y) \approx \frac{u(x + h, y) - u(x - h, y)}{2h} \quad (9)$$

$$u_y(x, y) \approx \frac{u(x, y + h) - u(x, y - h)}{2h} \quad (10)$$

Formulation (9) plus formulation (10) gives that,

$$u(x+h,y)+u(x-h,y)=2u(x,y)+\frac{2h^2}{2!}\frac{\partial^2 u(x,y)}{\partial x^2}+\mathrm{O}(h^4) \quad (11)$$

Therefore,

$$u_{xx}(x,y)=\frac{\partial^2 u(x,y)}{\partial x^2}=\frac{u(x+h,y)-2u(x,y)+u(x-h,y)}{h^2}+\mathrm{O}(h^2) \quad (12)$$

For sufficiently small h, the finite difference approximation of and could be expressed as,

$$u_{xx}(x,y)\approx\frac{u(x+h,y)-2u(x,y)+u(x-h,y)}{h^2} \quad (13)$$

$$u_{yy}(x,y)\approx\frac{u(x,y+h)-2u(x,y)+u(x,y-h)}{h^2} \quad (14)$$

If $\{\overline{u}_{i,j}\}$ are the sampled values of $u$ at sampling points $\{(x_i,y_j)\}$, $u^n_{i,j}$ ($n\geq 0$; $0\leq i\leq I+1$ and $0\leq j\leq J+1$) are the $n$th iteration values of lattices whose centers are points of $(x_i,y_j)$, in which $u^0_{i,j}=\overline{u}_{i,j}$ and $\{\overline{u}_{i,j}\}$ are the interpolated values based on the sampled values $\{\overline{u}_{i,j}\}$. In terms of numerical mathematics, the ($n$ + $1$) th iterative formulation of finite difference of basic equations of HASM given by (3) and (4) could be formulated as (Yue et al. 2007, 2008a, b, c, 2009),

$$\begin{aligned}\frac{u^{n+1}_{i+1,j}-2u^{n+1}_{i,j}+u^{n+1}_{i-1,j}}{h^2}&=\left(\Gamma^1_{11}\right)^n_{i,j}\frac{u^n_{i+1,j}-u^n_{i-1,j}}{2h}\\&\quad+\left(\Gamma^2_{11}\right)^n_{i,j}\frac{u^n_{i,j+1}-u^n_{i,j-1}}{2h}+\frac{L^n_{i,j}}{\sqrt{E^n_{i,j}+G^n_{i,j}-1}}\end{aligned} \quad (15)$$

$$\begin{aligned}\frac{u^{n+1}_{i,j+1}-2u^{n+1}_{i,j}+u^{n+1}_{i,j-1}}{h^2}&=\left(\Gamma^1_{22}\right)^n_{i,j}\frac{u^n_{i+1,j}-u^n_{i-1,j}}{2h}\\&\quad+\left(\Gamma^2_{22}\right)^n_{i,j}\frac{u^n_{i,j+1}-u^n_{i,j-1}}{2h}+\frac{N^n_{i,j}}{\sqrt{E^n_{i,j}+G^n_{i,j}-1}}\end{aligned} \quad (16)$$

where $n\geq 0;0<i<I+1;\ 0<j<j+1;\ E^n_{i,j}=1+\left(\frac{u^n_{i+1,j}-u^n_{i-1,j}}{2h}\right)^2 F^n_{i,j}=\left(\frac{u^n_{i+1,j}-u^n_{i-1,j}}{2h}\right)\left(\frac{u^n_{i,j+1}-u^n_{i,j-1}}{2h}\right);\ G^n_{i,j}=1+\left(\frac{u^n_{i,j+1}-u^n_{i,j-1}}{2h}\right)^2$

$$L^n_{i,j}=\frac{u^n_{i+1,j}-2u^n_{i,j}+u^n_{i-1,j}}{h^2\sqrt{1+\left(\frac{u^n_{i+1,j}-u^n_{i-1,j}}{2h}\right)^2+\left(\frac{u^n_{i,j+1}-u^n_{i,j-1}}{2h}\right)^2}}$$

$$N_{i,j}^n = \frac{u_{i,j+1}^n - 2u_{i,j}^n + u_{i,j-1}^n}{h^2\sqrt{1 + \left(\frac{u_{i+1,j}^n - u_{i-1,j}^n}{2h}\right)^2 + \left(\frac{u_{i,j+1}^n - u_{i,j-1}^n}{2h}\right)^2}}$$

$$\left(\Gamma_{11}^1\right)_{i,j}^n = \frac{G_{i,j}^n\left(E_{i+1,j}^n - E_{i,j}^n\right) - 2F_{i,j}^n\left(F_{i+1,j}^n - F_{i-1,j}^n\right) + F_{i,j}^n\left(E_{i,j+1}^n - E_{i,j-1}^n\right)}{4\left(E_{i,j}^n G_{i,j}^n - \left(F_{i,j}^n\right)^2\right)h}$$

$$\left(\Gamma_{22}^1\right)_{i,j}^n = \frac{2G_{i,j}^n\left(F_{i,j+1}^n - F_{i,j-1}^n\right) - G_{i,j}^n\left(G_{i+1,j}^n - G_{i-1,j}^n\right) - F_{i,j}^n\left(G_{i,j+1}^n - G_{i,j-1}^n\right)}{4\left(E_{i,j}^n G_{i,j}^n - \left(F_{i,j}^n\right)^2\right)h}$$

$$\left(\Gamma_{11}^2\right)_{i,j}^n = \frac{2E_{i,j}^n\left(F_{i+1,j}^n - F_{i-1,j}^n\right) - E_{i,j}^n\left(E_{i,j+1}^n - E_{i,j-1}^n\right) - F_{i,j}^n\left(E_{i+1,j}^n - E_{i-1,j}^n\right)}{4\left(E_{i,j}^n G_{i,j} - \left(F_{i,j}^n\right)^2\right)h}$$

$$\left(\Gamma_{22}^2\right)_{i,j}^n = \frac{E_{i,j}^n\left(G_{i,j+1}^n - G_{i,j-1}^n\right) - 2F_{i,j}^n\left(F_{i,j+1}^n - F_{i,j-1}^n\right) + F_{i,j}^n\left(G_{i+1,j}^n - G_{i-1,j}^n\right)}{4\left(E_{i,j}^n G_{i,j}^n - \left(F_{i,j}^n\right)^2\right)h}$$

$u_{0,j}^{n+1} = u_{0,j}^0(0 \le j \le J+1)$; $u_{i,0}^{n+1} = u_{i,0}^0(0 \le i \le I+1)$; $u_{I+1,j}^{n+1} = u_{I+1,j}^0(0 < j < J+1)$; $u_{i,J+1}^{n+1} = u_{i,J+1}^0(0 < i < I+1)$; $u_{0,j}^{n+1}$, $u_{i,0}^{n+1}$, $u_{I+1,j}^{n+1}$ and $u_{i,J+1}^{n+1}$ are boundary conditions of HASM The matrix formulation of equation set (5) and (6) can be expressed as,

$$A_1 U^{n+1} = b_1^n \tag{17}$$

$$A_2 U^{n+1} = b_2^n \tag{18}$$

where $U^{n+1} = (u_{1,1}^{n+1}, \cdots, u_{1,J}^{n+1}, u_{2,1}^{n+1}, \cdots, u_{2,J}^{n+1}, \cdots, u_{I-1,1}^{n+1}, \cdots, u_{I-1,J}^{n+1}, u_{I,1}^{n+1}, \cdots, u_{I,J}^{n+1})^T$; $A_1$ and $b_1^n$ are respectively coefficient matrix and right-hand item vector of Eq. (15); $A_2$ and $b_2^n$ are respectively coefficient matrix and right-hand item vector of Eq. (16).

If $Z = \begin{bmatrix} A_1 \\ A_2 \end{bmatrix}$ and $q^n = \begin{bmatrix} b_1^n \\ b_2^n \end{bmatrix}$, the following equality-constrained least squares problem can be developed to make the interpolated values equal to or approximate to the sampled values at the sampling points,

$$\begin{cases} \min\left\| ZU^{n+1} - q^n \right\|_2 \\ \text{s.t.}CU^{n+1} = d \end{cases} \tag{19}$$

where $C(k, (i-1) \cdot J + j) = 1$ and $d(k) = \overline{u}_{i,j}$, which means that the sampled value is $\overline{u}_{i,j}$ at the $k$th sampling point $(x_i, y_j)$.

To solve the algorithm (19) which is the least squares problem, a positive weight $\lambda$ is introduced on the basis of the well-known method of weights (van Loan 1985). The parameter $\lambda$ is the weight of the sampling points and determines the contribution of the sampling points to the simulated surface (Yue et al. 2009). For sufficiently large $\lambda$, the algorithm (19) can be transferred into unconstrained least squares approximation (Van Loan 1985),

$$\min_f \left\| \begin{bmatrix} Z \\ \lambda C \end{bmatrix} U^{n+1} - \begin{bmatrix} q^n \\ \lambda d \end{bmatrix} \right\|_2 \tag{20}$$

or

$$\begin{bmatrix} Z^T & \lambda C^T \end{bmatrix} \begin{bmatrix} Z \\ \lambda C \end{bmatrix} U^{n+1} = \begin{bmatrix} Z^T & \lambda C^T \end{bmatrix} \begin{bmatrix} q^n \\ \lambda d \end{bmatrix} \tag{21}$$

Equation (20) can be solved by the preconditioned conjugate gradients method (PCG) (Golub and van Loan 1996). The parameters of PCG include the tolerance of the method, the maximum number of iterations and preconditioner. In this study, the tolerance of PCG is 1e-9; the maximum number of iterations is 1000; preconditioner is the main diagonal of coefficient matrix.

Data of soil sampling points are used to globally fit a surface through several iterative simulation steps in HASM. This surface is then used to interpolate soil pH values at unknown points. The iterative simulation steps are summarized as follows (in press): (1) conducting interpolation on the computational domain in terms of sampling data $(X_i, Y_j, \overline{U}_{i,j})$, from which we can get interpolated approximate values $\{\tilde{U}_{i,j}\}$ at point $(X_i, Y_j)$; (2) letting $U^0_{i,j}=\tilde{U}_{i,j}$,jand calculating the first fundamental coefficients $E^n_{i,j}$, $F^n_{i,j}$ and $G^n_{i,j}$ nand the second fundamental coefficients $L^n_{i,j}$and $N^n_{i,j}$ as well as coefficients of HASM equations in terms of $\{U^n_{i,j}\}$; (3) for n $\geq$ 0, we can get $\{U^{n+1}_{i,j}\}$ by solving the HASM equations; 4) the iterative process is repeated until simulation accuracy is satisfied. The parameters of HASM include$\lambda$in Eq. (19), the iteration number (n) and the tolerance of the method. In this study, $\lambda$ is 2; the iteration number (n) is 10; the tolerance of HASM is 1e−12.

### 9.2.3 Validation

The usual methods of validation for comparing the interpolation methods are known as cross-validation and validation with an independent data set. Cross-validation involves eliminating each observation in turn, estimating the value at its site from the remaining observations and comparing the predicted value with the measured value. This procedure is a rapid, inexpensive one for comparing predicted and measured values. Unfortunately, it has limitations in many cases. For kriging estimators, it retains the same variogram, and to be true cross-validation the variogram should be recomputed andfitted afresh when each observation is removed. These shortcomings can be avoided by using an independent data set for validation. Validation with an independent data set which is a superior and more dependable method directly estimates the spatial uncertainty, as validation points are located randomly throughout the field. However, when the sample size is small, not enough and separate data are available for modelling and validation. Consequently, a modified jackknifing technique was adopted to resample the base sample data 20 times for the purpose of assessing the performance of the four methods. The resampling size each time was 30 samples for validation and 120 samples for interpolation in this study.

Three indices were calculated from the measured and interpolated values at each validation sample site for each test data set. The mean error (ME), the mean absolute error (MAE) and the root mean square error (RMSE) are calculated by the following equations:

$$ME = \frac{1}{n}\sum_{i=1}^{n}\{z*(x_i) - z(x_i)\} \tag{22}$$

$$MAE = \frac{1}{n}\sum_{i=1}^{n}\{|z*(x_i) - z(x_i)|\} \tag{23}$$

$$RMSE = \sqrt{\frac{1}{n}\sum_{i=1}^{n}\{z*(x_i) - z(x_i)\}^2} \tag{24}$$

where n are the amounts of validation data set; $z(x_i)$ is the measured values, and $z*(x_i)$ are the predicted value.

## 9.3 A Case—Surface Modelling of Soil pH

### 9.3.1 The Study Area

The study area (26°26′55″–27°37′35″N, 114°17′49″–115°30′13″E), covering 6156.92 km$^2$, is located in the middle part of Jiangxi Province, China (Fig. 13). It includes the Ji'an municipal district, Ji'an and Taihe Counties. Respectively their precipitation is 1458, 1438 and 1381 mm per annum and the mean annual temperature is 18.1, 18.4 and 19.0 °C which are typical of a subtropical monsoon climate. The elevation decreases from the periphery towards the center with altitude ranging from 1204.5 to 42.0 m. There are 6 land use types including woodlands, croplands, grasslands, water bodies, built-up areas and unused lands. According to the 1/1,000,000 scale soil maps reported by National Soil Census Office, the soils are classified into 7 groups: red soils, paddy soils, purple soils, fluvo-aquic soils, yellow soils, alluvial

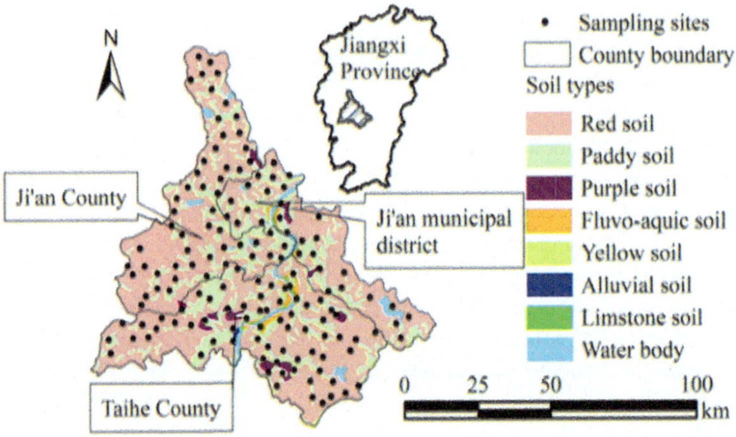


**Fig. 13** Soil types of the study area and the distribution of soil samples

soils and limestone soils. Red soils are the main types in the study area, predominantly derived from Quaternary red earth, Tertiary red sandstone and granite, whose areas are 3945.22 km$^2$.

The study area was covered by 150 sampling sites. The average distance between soil sampling locations is approximately 6 km. The sampling sites were designed to cover evenly the whole area and to include different soil types, different land use types, different vegetation types and different altitudes and aspects, etc. We collected sample points over areas with higher variability, so the limited samples could represent information of study area as much as possible. Soil samples, taken in 2007 from 150 sites, were air-dried and passed through a 2 mm sieve. Soil pH was measured in a 1:2.5 soil/water suspension with the pH meter method. Topsoil (0–20 cm) pH of the 150 samples was used to compare the performance of different interpolators (Fig. 13).

### *9.3.2 The Interpolated Maps of Soil pH by HASM and Three Classical Methods*

The patterns of soil pH distribution produced by different interpolators using 150 points are shown in Fig. 14. In the maps from the four techniques, kriging map (Fig. 14) shows rather gradual transitions with low level of details. Owing to its smoothing effect, the maximum value (6.00) is the lowest and the minimum value (4.53) is the highest in the four techniques. The range of soil pH in the prediction map by kriging is somewhat narrower because the interpolator creates smoothed surfaces that have a smaller range than the observations to obtain smaller RMSE. On the contrary, splines (Fig. 14d) have a serious oscillation problem which makes

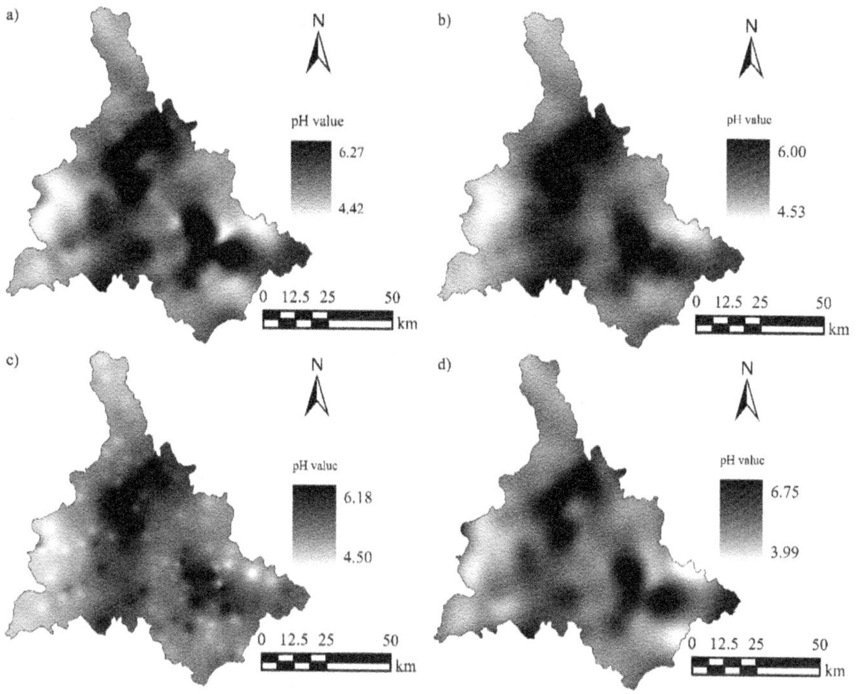


**Fig. 14** Interpolated maps of soil pH by four different methods, **a** HASM, **b** kriging, **c** IDW and **d** splines

the minimum output (3.99) lower and the maximum output (6.75) higher than those of other techniques. IDW interpolation produces many "Bull's eyes" on the soil pH surface (Fig. 14c). HASM obtains soil pH surface without narrower ranges, oscillation problem and "Bull's eyes" (Fig. 14).

This is an initial attempt to apply HASM to interpolate the spatial pattern of a soil property. We can conclude that HASM is more accurate, and the map from HASM is also better than the three classical interpolation methods. The generally superior performance of HASM is due to the less error incorporated into HASM when using the development of least squares problem and enough iteration numbers in the procedure. Therefore, HASM can be considered as an alternative and accurate method for interpolating soil properties.

We also have some limitations in this study. The first one is speed, the computing time of HASM was proportional to the third power of the total number of grid cells. And the second one is accuracy. The soil properties may vary severely within a short horizontal distance, so it is difficult to interpolate accurately using HASM when we ignore environmental features.

### 9.3.3 The Results from HASM Based on Muti-grid

We combined multi-grid in HASM to solve the problem of long computing time, so we can call it as HASM-MG. The pre-smoothing, coarse grid correction and post-smoothing parts form a two-grid cycle, which is repeated until expected accuracy is reached. The simulation result indicates that computing time of HASM-MG is proportional to the first power of the total number of grid (Fig. 15).

We also compared the HASM-MG method with kriging, IDW and splines. We can find that the HASM-MG is also the most accurate method. We found that HASM-MG can be considered as an accurate method with high speed, and it advances the practicability of HASM applied in soil property interpolation (Table 4).

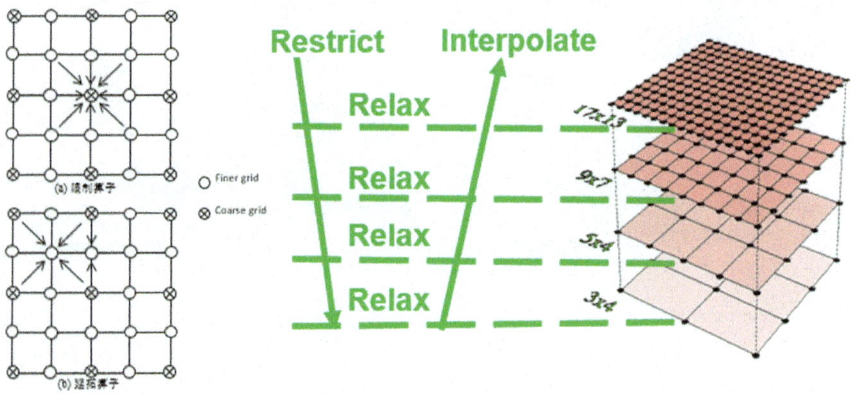


**Fig. 15** Information fusion on multigrids **a** information on the fine grid ○ is transferred to the coarse grid ⊗, and **b** information on the coarse grid ⊗; is transferred to the fine grid ○

**Table 4** MEs, MAEs, RMSEs and correlation coefficients of HASM-MG and the classical methods

| | | HASM-MG | Kriging | IDW | Spline |
|---|---|---|---|---|---|
| $V_{20\%}$ | ME | −0.02 | 0.06 | 0.03 | 0.01 |
| | MAE | 0.14 | 0.19 | 0.16 | 0.22 |
| | RMSE | 0.20 | 0.25 | 0.21 | 0.29 |
| $V_{40\%}$ | ME | 0.01 | 0.05 | 0.03 | –0.03 |
| | MAE | 0.19 | 0.24 | 0.21 | 0.23 |
| | RMSE | 0.27 | 0.31 | 0.29 | 0.32 |
| $V_{60\%}$ | ME | –0.04 | 0.01 | 0.02 | –0.05 |
| | MAE | 0.20 | 0.26 | 0.22 | 0.30 |
| | RMSE | 0.28 | 0.34 | 0.30 | 0.40 |

## 9.4 A Case—Surface Modelling of Soil Properties Based on Land Use Information

High accuracy surface modelling (HASM) is a spatial interpolation technique based on the fundamental theorem of surfaces. We proposed a modified HASM method based on the incorporation of ancillary land use information (HASM_LU) for improving the interpolation of soil properties. To assess the accuracy of the developed method, we collected 150 samples in a typical red soil region which was in the middle part of Jiangxi Province, China. In addition, the methods including HASM, ordinary kriging with land use information (OK_LU), stratified kriging (SK) and regression-kriging using a generalized linear model (RK_GLM) were also applied. We used several indices to compare the performace of different methods, including the mean error (ME), mean absolute error (MAE), root mean square error (RMSE) and prediction efficiency (PE).

### *9.4.1 Methods and Models*

#### 9.4.1.1 Secondary Information

The interpolator cannot improve accuracy much only through limited soil samplings without soil environmental information. When measurements are sparse or poorly correlated in space, the estimation of the primary attribute of interest is generally improved by accounting for secondary information originating from other related continuous or categorical attributes. We used the following environmental variables to interpolate (Table 5).

Combining auxiliary variables not only improves the interpolation accuracy of soil attributes, but also provides more detailed information and more accurate spatial structure information and not only reduce prediction errors but also can be accordant with the distribution of geographic information, which make the spatial simulation of soil property more reasonable. When the spatial correlation of soil attribute values is poor or changes violently, using other relevant continuous variables or category information can improve the quality of the prediction map and better understand the spatial distribution of soil attributes. Combined with auxiliary information, the overall trend and local details of interpolation are closer to the actual situation compared with the traditional method.

In this study, we used the data of land use, which was interpreted from Landsat TM images for 2004/2005 with 30 m × 30 m resolution. Six groups including woodlands, croplands, grasslands, built-up areas, water bodies and unused lands were classified in this land use data, and the accuracy was 95.3% for the classification (Fig. 16).

**Table 5** Environmental covariates used in this case

| Environmental covariates | Nature of the soil variable derived | Name of covariate | Type |
|---|---|---|---|
| Topographic attributes | DEM | Elevation (EI) | Quantitative |
| | | Topographic wetness index (TWI) | |
| | | Wetness index (WI) | |
| | | Curvature (profile and plan) | |
| | | Flow accumulation | |
| | | Stream power index | |
| | | Aspect | |
| | | Slope | |
| Remote sensing attributes | Images | NDVI | Quantitative |
| | | Ratio vegetation index (RVI) | |
| | | Enhanced vegetation index (EVI) | |
| | | Perpendicular vegetation index (PVI) | |
| | | Land-use type | Categorical |
| Maps | Existing maps | Geomorphology map | Categorical |
| | | Geology map | |
| | | Legacy soil survey maps | |
| | | Parent materials maps | |
| | | Lithology type map | |
| | | Vegetation type map | |
| Climatic indices | Meteorological stations or satellite | Precipitation | Quantitative |
| | | Temperature | |
| | | Surface moisture | |
| | | Evapotranspiration | |
| | | Land surface temperature (LST) | |

#### 9.4.1.2 Interpolators

**HASM**. According to the fundamental theorem of surfaces, a surface is uniquely defined by the first and the second fundamental coefficients. The first fundamental coefficients are used to express information about some geometric properties of the surface, by which we can calculate the lengths of the curves, angles of tangent vectors, areas of regions and geodesics on the surface (Yue et al. 2010). The second

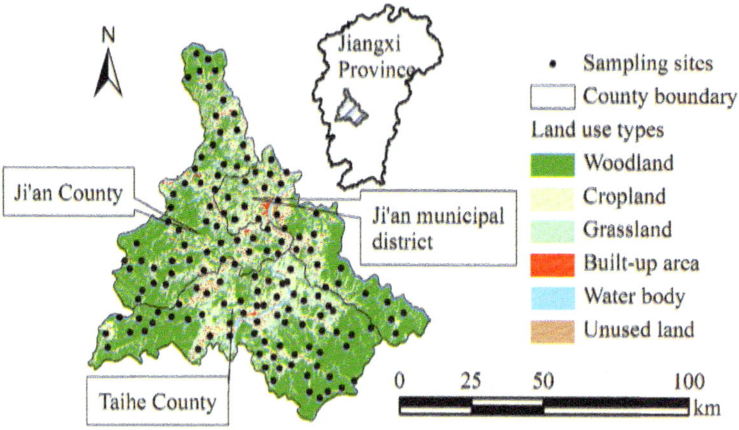


**Fig. 16** Land use types of the study area and the distribution of soil samples

fundamental coefficients reflect the local warping of the surface, namely its deviation from tangent plane at the point under consideration.

**HASM_LU and OK_LU**. Each observation $z(x_{kj})$ on one specific soil property can be expressed as,

$$z(x_{kj}) = m(L_j) + r(x_{kj}) \tag{25}$$

where $x_{kj}$ is the sampling location of $z(x_{kj})$, which is the kth sampling points on the j th land use type $L_j$, $m(L_j)$ is the mean value of $z(x_{kj})$ for each land use type $L_j$, and $r(x_{kj})$ is the residual which is computed by subtracting the average soil property of the relative land use type from the original value of each soil property. We assume that $m(L_j)$ and $r(x_{kj})$ are independent mutually and the variation of $r(x_{kj})$ is homogeneous over the whole study area. The residuals are then used to interpolate the surface of residuals in the whole study area by HASM and OK, respectively. The interpolated values of residuals are finally summed to the soil property averages of the relevant land use types as the final interpolated values of HASM with land use information and OK with land use information for the soil property, namely that it is a modification of the mean with surface modelling of residuals.

**SK**. We assume that soil variables are isotropic and obey the intrinsic hypothesis within each soil stratum (Stein et al. 1988). Stratification of the region was based upon land use types. In the SK analysis, a separate variogram was computed for each land use stratum and only the data for that land use type was used to interpolate within that type (Stein et al. 1988). In this study, the total study area was partitioned into woodlands, croplands and grasslands. In general, this method needs enough data to compute separate variograms for each land use type (Voltz and Webster 1990). Because only 5 sample points were collected in grasslands, and grasslands were generally near to woodlands in this study area, we stratified the whole area into two strata. Woodlands and grasslands were one stratum, and croplands were the other one. Kriging was then performed within each stratum using a semivariogram model and predictions in the entire region had to be carried out with respect to the stratified area (Stein et al. 1988).

**RK_GLM**. The trend was modeled by means of a generalized linear model (GLM). Next, we computed the residuals by subtracting the trend from the observations, then ordinary kriging was applied for interpolating the residuals over the whole study area, and the predicted residuals were added to the GLM-predictions. More details of a generalized linear model can be found in McCullagh and Nelder.

#### 9.4.1.3 Validation

Jackknife validation proposed by Deutsch and Journel (1998) was used for the validation of the interpolators in this study. The procedure involves randomly splitting the data into the interpolation and validation subsets, estimating the value using interpolation subset and comparing the interpolated value at every validation point with its measured value. A total of 120 training points were randomly created as interpolation data sets, and the remaining 30 samples were used as validation data sets. We used the most common indices, the mean error (ME), the mean absolute error (MAE) and the root mean square error (RMSE) as measures of interpolation quality.

#### 9.4.1.4 ANOVA Analysis

The variation $\sigma_z^2$ of $z(x_{kj})$ can be shown as the sum of $\sigma_m^2$ between land use types and $\sigma_r^2$ within land use types:

$$\sigma_z^2 = \sigma_m^2 + \sigma_r^2 \tag{26}$$

where $\sigma_m^2$ represents the effect of land use type on the spatial variation of soil properties, and $\sigma_r^2$ shows the overall variation within land use types. To compare the difference of soil properties among land use types, the soil property data were grouped into three classes based on the main land use types. There were 61, 84 and 5 samples in woodlands, croplands and grasslands, respectively. The variances of each soil property between and within land use types were determined by using ANOVA with the software package SPSS 14.0 for Windows. In ANOVA, Scheffee's test was conducted for post hoc multiple comparisons because of the different sample numbers in each land use type.

Table 6 showed analysis of variances to compare the effects of different land use types on the tested soil properties. We found that the variances $\sigma_m^2$ of most soil properties (except for Zn) among the land use categories were significant at the 0.01 or 0.05 level, but that of Zn was significant at the 0.1 level. ANOVA analysis revealed that variances of the tested soil properties among land use types might play an important role in their spatial prediction in the study area.

## 9.4.2 Results

### 9.4.2.1 Geostatistical Analysis

After removing local means in the same land use types, semivariograms of original values and the residuals for soil properties showed good performance (Fig. 17). Variograms of residuals for pH, AN, C, N, K, Al, Ca, and Mg showed a shorter range and a smaller sill, indicating that the drift had been removed (Hengl et al. 2014). For Zn, the variograms of original values and residuals were quite similar, indicating that the spatial distribution of this variable might not be significantly impacted by the land use types (Fig. 17).

### 9.4.2.2 Comparison of the Performance of HASM_LU, HASM, OK_LU, SK and RK_GLM

We compared the five methods including HASM_LU, HASM, OK_LU, SK and RK_GLM in interpolating soil properties, and used three indices including MEs, MAEs and RMSEs to compare the predicted values with the measured values (Table 7). Most of the MEs of soil properties (except for C and Zn) from OK_LU, HASM_LU and RK_GLM which interpolated with land use information, were much closer to 0 than HASM (without the information of land use), indicating that land use information could make interpolators more unbiased. Although SK also included land use information, MEs in most variables (except for Al) were not closer to 0. For MAEs, most of them from HASM_LU were much lower than both OK_LU (except for N and Mg) and RK_GLM (except for Mg and Zn). In addition, MAEs from HASM_LU (except for pH and Zn) were much lower than HASM. The lowest RMSEs of HASM_LU were achieved in the five interpolators. For example, the RMSEs of AN from HASM_LU were lower than HASM, OK_LU, SK and RK_GLM by 33.4%, 1.6%, 41.5% and 67.59%, respectively. We found higher MAEs and RMSEs in most soil properties interpolated by SK than OK_LU and HASM_LU due to the limited sample points to estimate the variogram (Stein et al. 1988). HASM_LU generally achieves the lowest RMSEs (except for Zn) in different methods. The higher RMSE for Zn is due to the insignificant differences among land use types (Table 1 in Chap. 10).

**Table 6** ANOVA analysis for testing the significance of land use type effects on the variances of soil properties

| Soil property | Source of variance | df | Sum of squares | Mean square | F value | Sig. |
|---|---|---|---|---|---|---|
| pH | Between land use types | 2 | 2.35 | 1.18 | 10.03 | <0.01** |
| | Within land use types | 147 | 17.26 | 0.12 | | |
| | Total | 149 | 19.61 | | | |
| AN, mg $kg^{-1}$ | Between land use types | 2 | 66,648.95 | 33,324.47 | 14.16 | <0.01** |
| | Within land use types | 147 | 346,016.39 | 2353.85 | | |
| | Total | 149 | 412,665.34 | | | |
| C, g $kg^{-1}$ | Between land use types | 2 | 386.79 | 193.39 | 3.84 | 0.02* |
| | Within land use types | 147 | 7410.55 | 50.41 | | |
| | Total | 149 | 7797.33 | | | |
| N, g $kg^{-1}$ | Between land use types | 2 | 3.79 | 1.9 | 13.9 | <0.01** |
| | Within land use types | 147 | 20.06 | 0.14 | | |
| | Total | 149 | 23.85 | | | |
| K, g $kg^{-1}$ | Between land use types | 2 | 453.74 | 226.87 | 4.24 | 0.02* |
| | Within land use types | 147 | 7859.17 | 53.46 | | |
| | Total | 149 | 8312.91 | | | |
| Al, g $kg^{-1}$ | Between land use types | 2 | 8928.5 | 4464.25 | 20.16 | <0.01** |
| | Within land use types | 147 | 32,553.08 | 221.45 | | |
| | Total | 149 | 41,481.58 | | | |
| Ca, g $kg^{-1}$ | Between land use types | 2 | 466.23 | 233.12 | 64.01 | <0.01** |
| | Within land use types | 147 | 535.39 | 3.64 | | |
| | Total | 149 | 1001.62 | | | |
| Mg, g $kg^{-1}$ | Between land use types | 2 | 140.5 | 70.25 | 16.72 | <0.01** |
| | Within land use types | 147 | 617.75 | 4.2 | | |

(continued)

**Table 6** (continued)

| Soil property | Source of variance | df | Sum of squares | Mean square | F value | Sig. |
|---|---|---|---|---|---|---|
| | Total | 149 | 758.24 | | | |
| Zn, mg $kg^{-1}$ | Between land use types | 2 | 6605.64 | 3302.82 | 2.59 | 0.08 |
| | Within land use types | 147 | 187,322.39 | 1274.3 | | |
| | Total | 149 | 193,928.03 | | | |

** Significant at the 0.01 level
* Significant at the 0.05 level

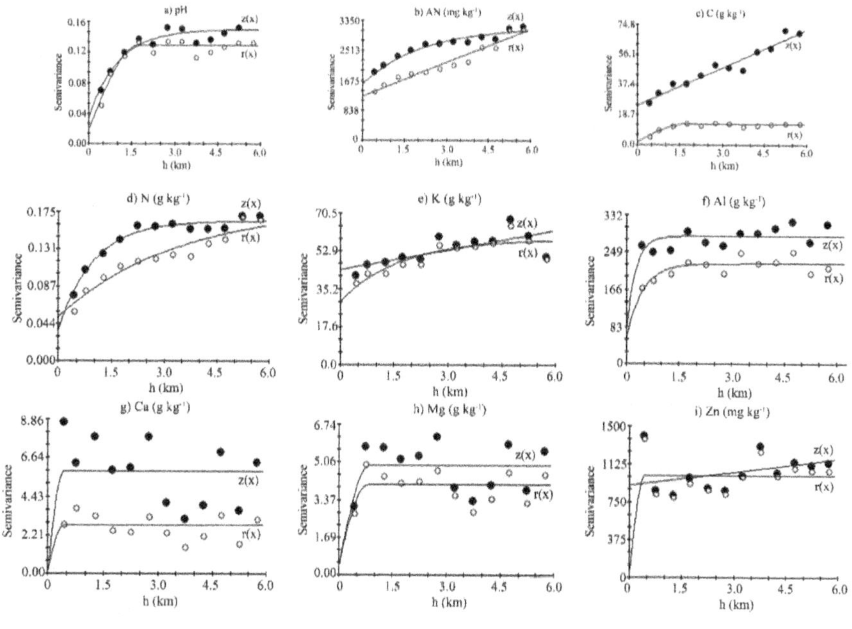


**Fig. 17** Semivariograms of original values and the residuals for soil properties: **a** pH, **b** AN, **c** C, **d** N, **e** K, **f** Al, **g** Ca, **h** Mg and **i** Zn

#### 9.4.2.3 Prediction Efficiency (PE)

A reduction-in-error index, referred to as goodness-of-prediction, was also determined to assess the quality of soil property maps. It is calculated as,

$$PE(\%) = \left(MSE_{average} - MSE_{interpolation}\right) \times 100/MSE_{average} \tag{27}$$

**Table 7** The MEs, MAEs and RMSEs of HASM_LU, HASM, OK_LU, SK and RK_GLM

| Method | Index | pH | AN | C | N | K | Al | Ca | Mg | Zn |
|---|---|---|---|---|---|---|---|---|---|---|
| | | / | mg kg$^{-1}$ | g kg$^{-1}$ | | | | | | mg kg$^{-1}$ |
| HASM_LU | ME | −0.01 | 0.65 | −0.2 | 0 | −1.45 | −5.38 | −0.09 | −0.58 | −12.58 |
| HASM | | −0.01 | 4.73 | 0.1 | 0.01 | −2.15 | −6.1 | −0.54 | −0.87 | −12.53 |
| OK_LU | | 0.01 | 1.47 | −0.15 | 0 | −1.35 | −5.32 | −0.09 | −0.59 | −12.31 |
| SK | | −0.01 | −4.72 | −0.92 | −0.04 | −2.3 | −3.49 | −0.45 | −0.69 | 51.96 |
| RK_GLM | | −0.01 | 1.68 | 2.23 | −0.01 | −1.53 | −4.51 | −0.11 | −0.52 | −13.31 |
| HASM_LU | MAE | 0.14 | 29.84 | 4.13 | 0.22 | 4.68 | 10.72 | 1 | 1.49 | 27.98 |
| HASM | | 0.14 | 41.58 | 4.83 | 0.34 | 5.5 | 11.75 | 1.29 | 1.56 | 26.63 |
| OK_LU | | 0.15 | 30.37 | 4.15 | 0.22 | 4.99 | 10.79 | 1.01 | 1.49 | 28.69 |
| SK | | 0.19 | 40.7 | 5.28 | 0.29 | 6.23 | 11.59 | 1.34 | 1.72 | 53.36 |
| RK_GLM | | 0.21 | 52.43 | 5.29 | 0.35 | 6.05 | 14.24 | 1.75 | 1.48 | 25.99 |
| HASM_LU | RMSE | 0.19 | 36.69 | 5.07 | 0.31 | 5.94 | 13.71 | 2.6 | 2.23 | 36.87 |
| HASM | | 0.19 | 48.95 | 5.88 | 0.43 | 6.61 | 14.93 | 3.6 | 2.44 | 34.62 |
| OK_LU | | 0.2 | 37.28 | 5.1 | 0.32 | 6.31 | 13.77 | 2.61 | 2.23 | 37.14 |
| SK | | 0.25 | 51.92 | 6.67 | 0.39 | 7.35 | 14.19 | 3.73 | 2.58 | 55.25 |
| RK_GLM | | 0.26 | 61.49 | 6.34 | 0.42 | 7.40 | 17.06 | 4.08 | 2.42 | 35.15 |

where $MSE_{average}$ represents the mean square error (MSE) by using the average of the training data as estimate for all validation data points; $MSE_{interpolation}$ is the square of the RMSE from comparing the interpolated values with measured values. Positive prediction efficiencies can be interpreted as the percent reduction in MSE compared to the field average approach. This index can be used to compare the map quality for the interpolations of different soil properties from the two estimators because this one is a relative index which is quite different from ME, MAE and RMSE.

The HASM_LU, HASM, OK_LU, SK and RK_GLM reduced 68.89%, 69.23%, 62.05%, 46.14% and 41.75% of MSE for soil pH compared to the field average approach. The smallest PE of HASM was in the case of soil N, but for HASM_LU, OK_LU and SK, they were in Zn. The accuracy of HASM is generally better than OK. However, it was not appropriate for Zn, because the land use types might not be the major factor of its spatial distribution (Table 7). Superior performance in the two methods of OK_LU and HASM_LU was due to the consideration of land use types (Fig. 18).

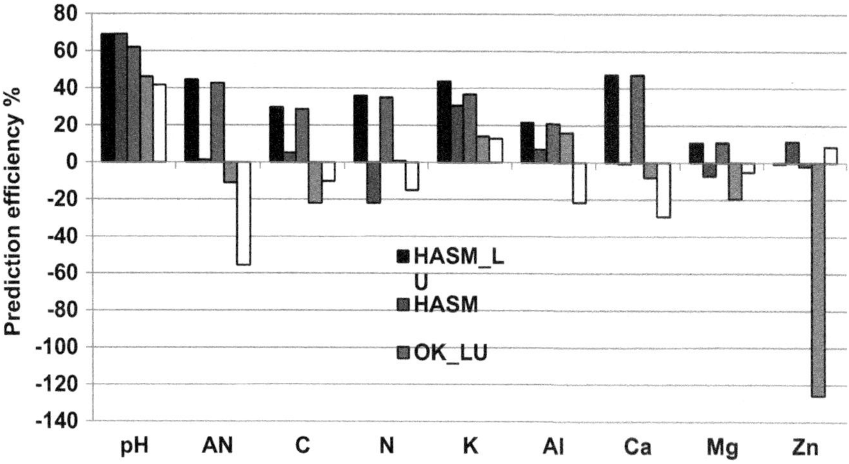


**Fig. 18** Comparison of the PEs for interpolations of 9 soil properties by HASM_LU, HASM, OK_LU, SK and RK_GLM

#### 9.4.2.4 Comparison of Soil Property Maps Interpolated by HASM_LU, HASM, OK_LU, SK and RK_GLM

To compare the maps generated by HASM_LU, HASM, OK_LU, SK and RK_GLM, we take soil AN as an example (Fig. 19a–e). The methods of HASM_LU (Fig. 19a), OK_LU (Fig. 19c), SK (Fig. 19d) and RK_GLM (Fig. 19e) provided many details, but the methods without land use consideration (Fig. 19b: HASM) exhibited worse interpolation performance especially in the presence of sharp boundaries among different land use types (Fig. 19). In summary, methods combined with land use types describe spatial variability more clearly.

Differences of between the four methods of HASM, OK_LU, SK and RK_GLM, with HASM_LU for soil AN were presented in Fig. 20a–d, respectively. The difference between HASM and HASM_LU (Fig. 20a) had similar spatial pattern with land use types, indicating that the difference between with and without land use types using HASM. Map pattern from OK_LU is similar with HASM_LU (Fig. 20b), but larger differences existed at the locations of sample points. Soil AN from SK were lower in woodland of mid-eastern part of study area than HASM_LU (Fig. 20c), while soil AN values interpolated by RK_GLM were lower in most of cropland than HASM_LU (Fig. 20d). HASM_LU was an relatively accurate method, which could improve the performance of soil property interpolation.

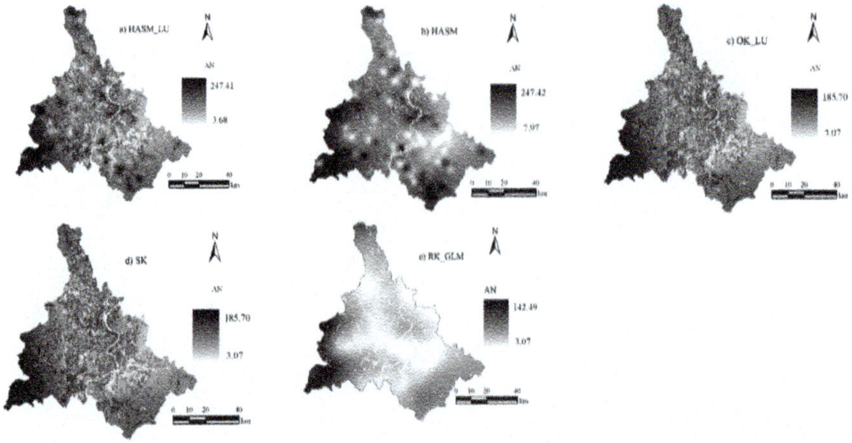


**Fig. 19** Soil AN (mg kg$^{-1}$) maps obtained by **a** HASM_LU, **b** HASM, **c** OK_LU, **d** SK and **e** RK_GLM

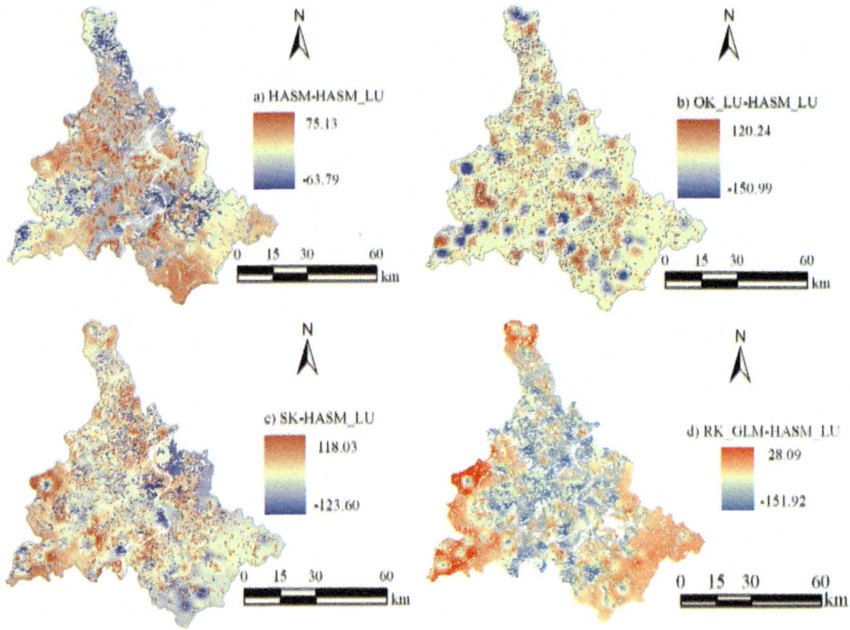


**Fig. 20** The differences of the performance of **a** HASM, **b** OK_LU, **c** SK, and **d** RK_GLM with HASM_LU in interpolating soil AN

### 9.4.3 Conclusions

There are several reasons for better performance of HASM_LU. First, the fundamental theorem for eco-environmental surface modelling (FTEEM) stated that an eco-environmental surface is uniquely defined by both extrinsic information and intrinsic information of the surface, which can be simulated with an appropriate method for integrating the global and local information, such as HASM, when the spatial resolution of the surface is fine enough to capture the attribute(s) of interest. In other words, when ground observation data of soil properties at observation points are used only, the surface modelling of soil properties is not so accurate because of its lack of extrinsic information even though the appropriate method, HASM, is employed. Second, HASM_LU takes into account land use information, a kind of extrinsic information, which makes the surface modelling follow FTEEM. Therefore, HASM combined with land use types (HASM_LU) has better performance in surface modelling of soil properties.

## 9.5 A Case—Surface Modeling of Soil Antibiotics

Due to the restricted use of chemical fertilizers for organic vegetable productions, large numbers of livestock and poultry feces were used in the intensive vegetable cultivation areas. Because of widely use of antibiotics for protecting animal health and improving the feeding efficiency of animals, some antibiotics via the manure application were imported in soils (Kuchta and Cessna 2009). However, the soil antibiotic patterns are complex, because that they are from external sources, and have irregular transportation into water and irregular absorption by plants (Sarmah et al. 2006). Because soil antibiotics is vital to evaluate the potential risks to human and environment, it is needed to develop an effective surface modeling method to interpolate the spatial patterns of soil antibiotics (Xie et al. 2012).

The interpolation of soil antibiotics is more difficult than other normal soil properties due to the special sources of soil antibiotics. The aims of this case are to develop an interpolator of HASM combined with vegetable types which are suitable for interpolation of soil antibiotics, and to evaluate the performance of HASM in soil antibiotic interpolation compared with other methods including kriging with External Drift (KED), stratified kriging (StK), and also ordinary kriging (OK) and inverse distance weighting (IDW).

## 9.5.1 Data and Methods

### 9.5.1.1 The Study Area

The study area is located in the east of China (Fig. 21), which is a typical vegetable planting area covering 160 km$^2$. The main soil types are cinnamon soil and fluvo-aquic soil. The vegetable types include cucumber–cucumber (16%), tomato–tomato (39%), pepper (11%), tomatoleaf (13%), eggplant (6%), leaf (5%) and tomato–melon (10%). The main manure applied into the soils was chicken dung, ranging from 3 to 9 kg/m$^2$ (Li et al. 2013). The elevation are from 5 to 34 m (Fig. 21c).

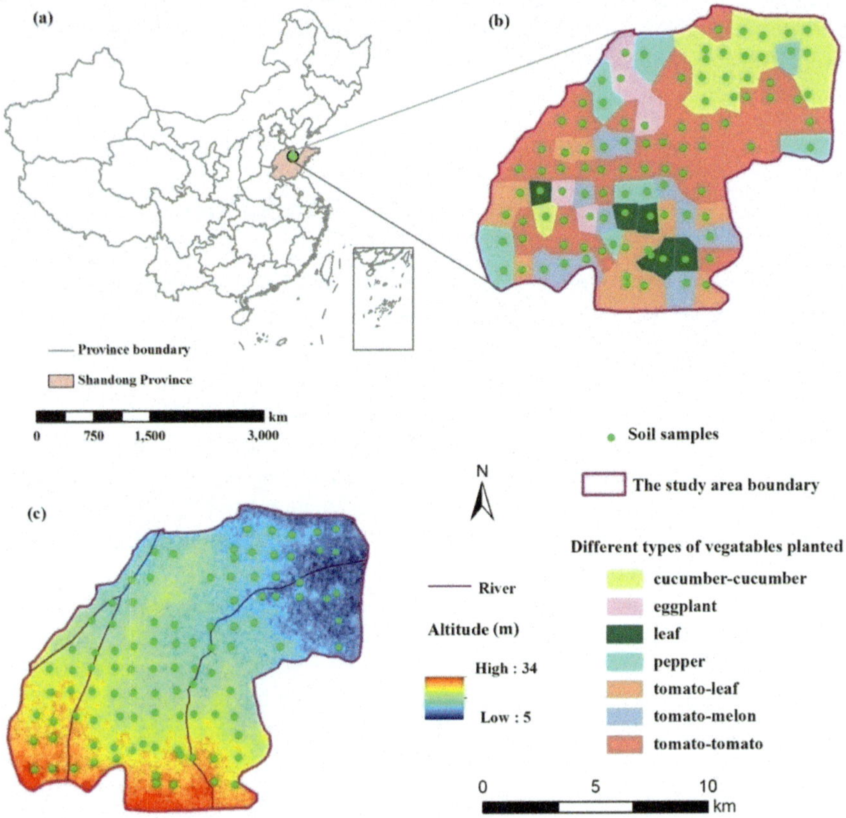


**Fig. 21** The location of the study area in China (**a**), the soil sample distributions in different vegetable planted types in the study area (**b**) and the spatial distribution of altitude along with river in the study area (**c**)

#### 9.5.1.2 Soil Samples and Analysis

A total of 100 greenhouses were selected and in each of them the surface soil samples were collected following an S-shape in November 2010. The sampling sites were designed to cover evenly in the whole area and different vegetable planted types, manure types, planting years, etc., and the intervals of the soil samples were approximately 1 km. Three types of FQs including ciprofloxacin (CFX), enrofloxacin (EFX), and norfloxacin (NFX), were analyzed using high-performance liquid chromatography and fluorescence detection (HPLC–FD). The details of the measurement can be found in Li et al.

#### 9.5.1.3 Methods

**The improvement of HASM.** For the surface modeling of soil antibiotics, we propose a modified HASM combined with the information of vegetable type which is more suitable for the prediction of the spatial patterns of soil antibiotics. Because the detailed equations and the explanations of HASM have been published in the previous references (Shi et al. 2009, 2011, 2012), we only listed the main equations and the modified part in this study for HASM.

The matrix formulation of HASM master equations can be respectively expressed as (Yue 2011; Yue et al. 2013, 2014, 2015),

$$\begin{cases} AF^{n+1} = c^n \\ BF^{n+1} = d^n \end{cases} \tag{28}$$

where A and B represent coefficient matrixes of the first equation and the second equation; $c^n$ and $d^n$ are right-hand side vectors of the two equations respectively; $F^{n+1} = \left(f_{1,1}^{n+1}, \cdots, f_{1,J}^{n+1}, \cdots, f_{I,1}^{n+1}, \cdots, f_{I,J}^{n+1}\right)^{\mathrm{T}}$ $(1 \le i \le I, 1 \le j \le J)$, and $f_{i,j}^{n}$ is the value of the $n$th iteration of $f(x, y)$ at grid cell $(x_i, y_i)$.

The solution procedure of HASM, taking the soil sampled values as its constraints, can be transformed into solving the following linear equation set in terms of least squares principle,

$$\begin{cases} \min \left\| \begin{bmatrix} A \\ B \end{bmatrix} F^{n+1} - \begin{bmatrix} c \\ d \end{bmatrix}^n \right\|_2 \\ S \times F^{n+1} = t \end{cases} \tag{29}$$

where $n \ge 0$, $S \in R^{K \times (I \times J)}$ *and* $t \in R^{K \times 1}$ were sampling matrix and sampling vector, and $K$ is the sampling numbers.

Equation (5) is to solve the unconstrained least squares problem for a Lagrange parameter $\lambda$, and it is equivalent to the symmetric positive definite linear system,

$$F^{n+1} = \left(\mathbf{A}^T \cdot \mathbf{A} + \mathbf{B}^T \cdot \mathbf{B} + \lambda^2 \cdot \mathbf{S}^T \cdot \mathbf{S}\right)^{-1}\left(\mathbf{A}^T \cdot \mathbf{c}^n + \mathbf{B}^T \cdot \mathbf{d}^n + \lambda^2 \cdot \mathbf{S}^T \cdot \boldsymbol{k}\right) \quad (30)$$

where $\lambda$ is the weight of the sampling points and determines the contribution of the sampling points to the simulated surface.

Due to the external sources of soil antibiotics, the spatial variation of soil antibiotic residues is different from other soil properties. The algorithm of HASM should be modified according the characters of the spatial distributions of soil antibiotic. There are two aspects needed to be modified on HASM.

First, because the measurements of soil antibiotics are sparse and poorly correlated in space, combination of environment information with ground measurements can considerably improve simulation accuracy. To combine the vegetable types as secondary information, we take the mean value of the soil antibiotic residues in each vegetable type as global approximate information which is the driving field of HASM, and the soil samplings for antibiotic residues as its optimum control constraints. Thus the iteration initial-values $f_{i,j}^0$ of Eq. (6) at the point $(x_i, y_j)$ can be achieved according to the types of vegetable planted,

$$f_{i,j}^0 = F_m \quad (31)$$

where $F_m$ is the mean value of the soil antibiotic residues in the vegetable type m.

Second, for the large variance of the spatial pattern of soil antibiotics, each soil sampling point should have its own weight in dependent of the variance degree. For the soil sample k ($k = 1, 2, 3, \cdots, K$), we compute the variance $V_k$ of soil antibiotic residues between the soil sample and its R neighbor samples, so we can get the variance sequence for each sample points $\{V_1, V_2, V_3, \cdots, V_K\}$. Supposed the maximum and minimum values are $V_{max}$ and $V_{min}$, the weight of *kth* point is,

$$\lambda_k = \lambda \frac{100 \times V_{max} - 99 \times V_k - V_{min}}{V_{max} - V_{min}} \quad (32)$$

where $\lambda$ can be supposed as from 1 to 100, and R can be supposed as from 10 to 20 according to the experience. For the larger variance of the spatial patterns of soil antibiotics, each sampling point affects smaller region, so smaller $\lambda$ and $R$ should be chosen; conversely, larger $\lambda$ and $R$ should be used.

**Kriging with external drift (KED).** Each observation $z(x_{kj})$ on one specific soil property can be expressed as,

$$z(x_{kj}) = m(V_j) + r(x_{kj}) \quad (33)$$

where $x_{kj}$ is the sampling location of $z(x_{kj})$, which is the *kth* sampling points on the *jth* vegetable type $V_j$, $m(V_j)$ is the mean value of $z(x_{kj})$ for each vegetable type $V_j$, and $r(x_{kj})$ is the residual computed by subtracting the average soil antibiotic of the relative vegetation type from the original value of each soil antibiotic. We assume that $m(V_j)$ and $r(x_{kj})$ are independent mutually and the variation of $r(x_{kj})$ is homogeneous over the whole study area. The residuals are then used to interpolate the surface of residuals in the whole study area by OK, respectively. The interpolated values of residuals are finally summed to the soil antibiotic averages of the relevant vegetable types as the final interpolated values of KED.

**Stratified kriging (StK)**. StK is another kriging method which can be combined with categorical variable (Shi et al. 2011). The stratification was based on vegetable types in this study. Each separate variogram was computed for each stratum, and only the data for the same types were used to interpolate within those vegetable types. Because the method needs enough soil samplings to compute the variograms for each stratum, we stratified the whole area into two strata according to the local mean values for each vegetable type for the interpolation of FQs, CFX, EFX and NFX.

**OK and IDW**. OK and IDW were also used for the interpolation of soil FQs. We employed them to compare their performance with that of HASM method. The best parameters for both OK and IDW were decided by choosing the smallest root mean squared error (RMSE) statistic. For OK, different models were fitted using weighted least squares, and the model with the smallest residual sum of squares (RSS) was chosen. For IDW, we also choose the powers of 1, 2, 3, and 4 for the best accuracy. For the number of the closest samples of OK and IDW, the best one from 5 to 30 with 5 step interval was chosen.

## *9.5.2 Results*

### 9.5.2.1 The Spatial Variations of Soil FQ Concentrations

The soil FQ residues show large variations in different types of planted vegetable (Fig. 22). In the cucumber–cucumber planted soil, FQ residues show the highest average concentration with 246.3 ± 181.5 μg/kg, and followed by those in tomato–tomato (196.6 ± 168.3 μg/kg), eggplant (187.9 ± 117.1 μg/kg), leaf (176.1 ± 102.9 μg/kg), pepper (160.5 ± 73.0 μg/kg) and tomato–melon (149.3 ± 140.7 μg/kg) planted areas. The highest FQ residues are found in the samples collected from cucumber–cucumber and tomato–tomato planted areas, which are more than 650 μg/kg. This is because the maximums of the soil CFX in these two vegetable types are 493.4 μg/kg and 651.6 μg/kg, respectively. The averages of soil CFX in the two types are also the highest, which are 147.5 μg/kg and 132.4 μg/kg, respectively. The average of soil EFX is less compared with the other two soil FQs, but the coefficients of variation are larger than 100% only except for pepper and leaf planted areas. The maximum of soil NFX is found in tomato–melon planted areas (288.3 μg/kg), and cucumber–cucumber and tomato–tomato followed with 206.3 μg/kg and 206.0 μg/kg.

Table 8 showed the analysis of variances to compare the effects of different vegetable types on the tested soil antibiotics. We found that the variances of soil FQs and CFX among the vegetable types were significant at 0.05 level, but that of

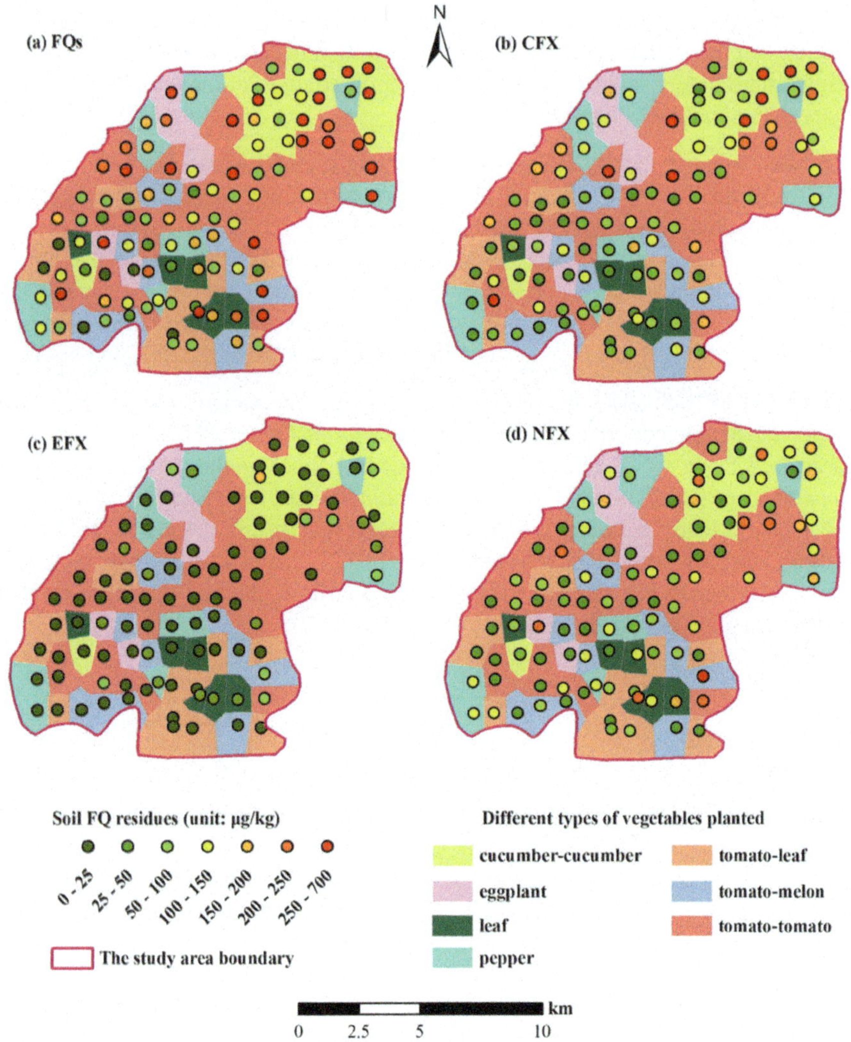


**Fig. 22** Spatial variations of soil FQ residues in different vegetable types: total FQs (**a**), CFX (**b**), EFX (**c**) and NFX (**d**)

NFX and EFX were not significant. ANOVA analysis revealed that variances of the tested soil antibiotics among vegetable types especially in soil FQs and CFX might play an important role in their spatial prediction in the study area. Soil FQ residues in tomato-leaf are the lowest with average of 47.4 μg/kg, which may be due to the high absorption by leaf especially by spinach for soil NFX and EFX with 411 μg/kg and 24.1 μg/kg (Li et al. 2014). The variation of soil FQs in different vegetable planted areas may be due to the different sources of manure with antibiotics from different animal breeding.

**Table 8** ANOVA analysis for testing the significance of vegetable type effects on the variances of soil antibiotics

| Soil property | Source of variance | df | Sum of squares | Mean square | F value | Sig. |
|---|---|---|---|---|---|---|
| FQ, μg/kg | Between vegetation types | 6 | 325,986.08 | 54,331.01 | 2.61 | < 0.05* |
| | Within vegetation types | 93 | 1,933,382.29 | 20,789.06 | | |
| | Total | 99 | 2,259,368.36 | | | |
| NFX, μg/kg | Between vegetation types | 6 | 21,075.47 | 3512.58 | 1.11 | 0.36 |
| | Within vegetation types | 93 | 293,487.92 | 3155.78 | | |
| | Total | 99 | 314,563.39 | | | |
| CFX, μg/kg | Between vegetation types | 6 | 168,871.08 | 28,145.18 | 2.19 | < 0.05* |
| | Within vegetation types | 93 | 1,197,971.04 | 12,881.49 | | |
| | Total | 99 | 1,366,842.12 | | | |
| EFX, μg/kg | Between vegetation types | 6 | 4658.58 | 776.43 | 1.21 | 0.31 |
| | Within vegetation types | 93 | 59,477.27 | 639.54 | | |
| | Total | 99 | 64,135.85 | | | |

* Significant at the 0.05 level

#### 9.5.2.2 Comparisons of the Interpolated Maps by HASM, KED, StK, OK and IDW

The methods including HASM, KED and StK combined with vegetable types had better performance with more details. HASM provided the spatial patterns and pollution assessment of soil antibiotics clearly (Fig. 23). Approximately half area of soil FQ concentrations are greater than 165 μg/kg, higher than the technical guidance from European registration procedure for human and veterinary medicines. The main polluted source in the study area is from CFX, resulting in 44% area greater than 100 μg/kg, especially in the northeast part planted cucumbers and tomatoes. Soil EFX residues in most of the study area are lower than 25 μg/kg. Soil NFX residues in half study areas are lower than 50 μg/kg. Higher CFX and lower EFX are always together shown in the study area, because metabolic conversion from EFX to CFX happens in many species' metabolism. Other factors may also affect the soil antibiotic patterns like water movement. For example, higher FQ residues are distributed in the southeast where is in the lower river region in the study area (Figs. 23, 24, 25, and 26).

#### 9.5.2.3 Comparisons of the Performance of HASM, KED, StK, OK and IDW

Cross-validation method was used to test the performance of HASM, KED, StK, OK and IDW (Table 9), indicating that HASM performs better than KED, StK, OK, and IDW. The methods combined with vegetable types, including HASM, KED and StK, achieve higher accuracies especially for FQs and CFX, indicating that vegetable type is a key factor for soil antibiotic patterns.

The MAEs of HASM for FQs, CFX, EFX and NFX are almost half of other four methods, and KED and StK which are also combined with the vegetable type information perform better than OK and IDW especially for FQs and CFX.

#### 9.5.2.4 Conclusions

Antibiotics are substances from extraneous sources into soils, which results in large variance of soil antibiotics related to the manure application and the absorption of the vegetable planted. We developed a modified method of HASM, taking the information of vegetable types as a driving factor. In addition, different weight of sampling points were used as its optimum control constraints. We concluded that HASM has better accuracy combined with the vegetable information. Furthermore, HASM can provide reasonable patterns and details of the spatial distributions of soil antibiotics.

There are three reasons for better performance of the modified HASM when predicts the spatial pattern of soil antibiotics. First, HASM takes the vegetable type information as global approximate information, and takes local sampling data as its optimum control constraints. Second, spatial heterogeneity of the neighbor samplings was considered in the prediction. Third, the output of HASM satisfies the iteration stopping criterion which is determined by application requirement for accuracy.

## 9.6 Further Reading

(1) Yue TX, Zhao N, Liu Y, et al. 2020. A fundamental theorem for eco-environmental surface modelling and its applications. Science China-Earth Sciences 63: 1092–1112.

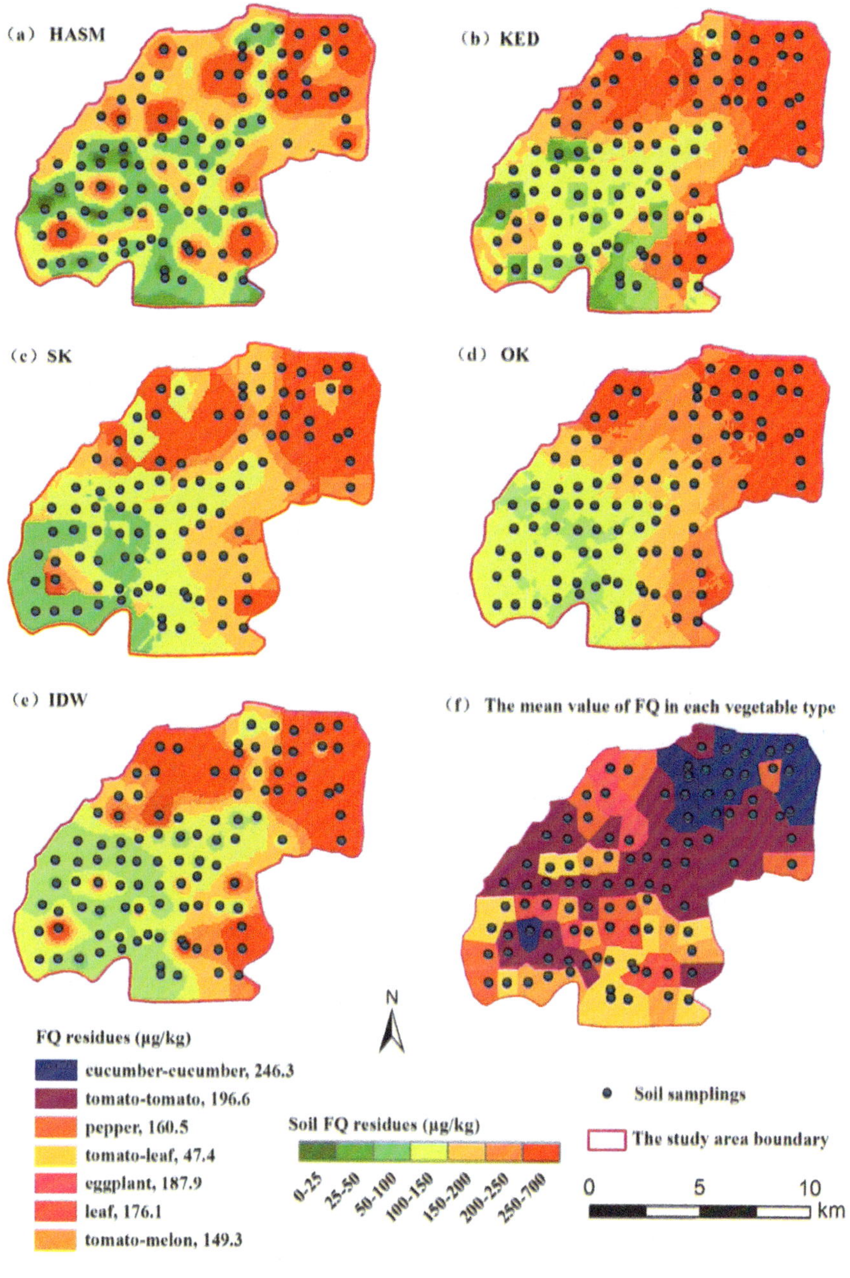


**Fig. 23** The spatial patterns of soil FQ residues obtained by HASM (**a**), KED (**b**), StK (**c**) OK (**d**) and IDW (**e**), and the mean values of FQ residues in each vegetable type (**f**)

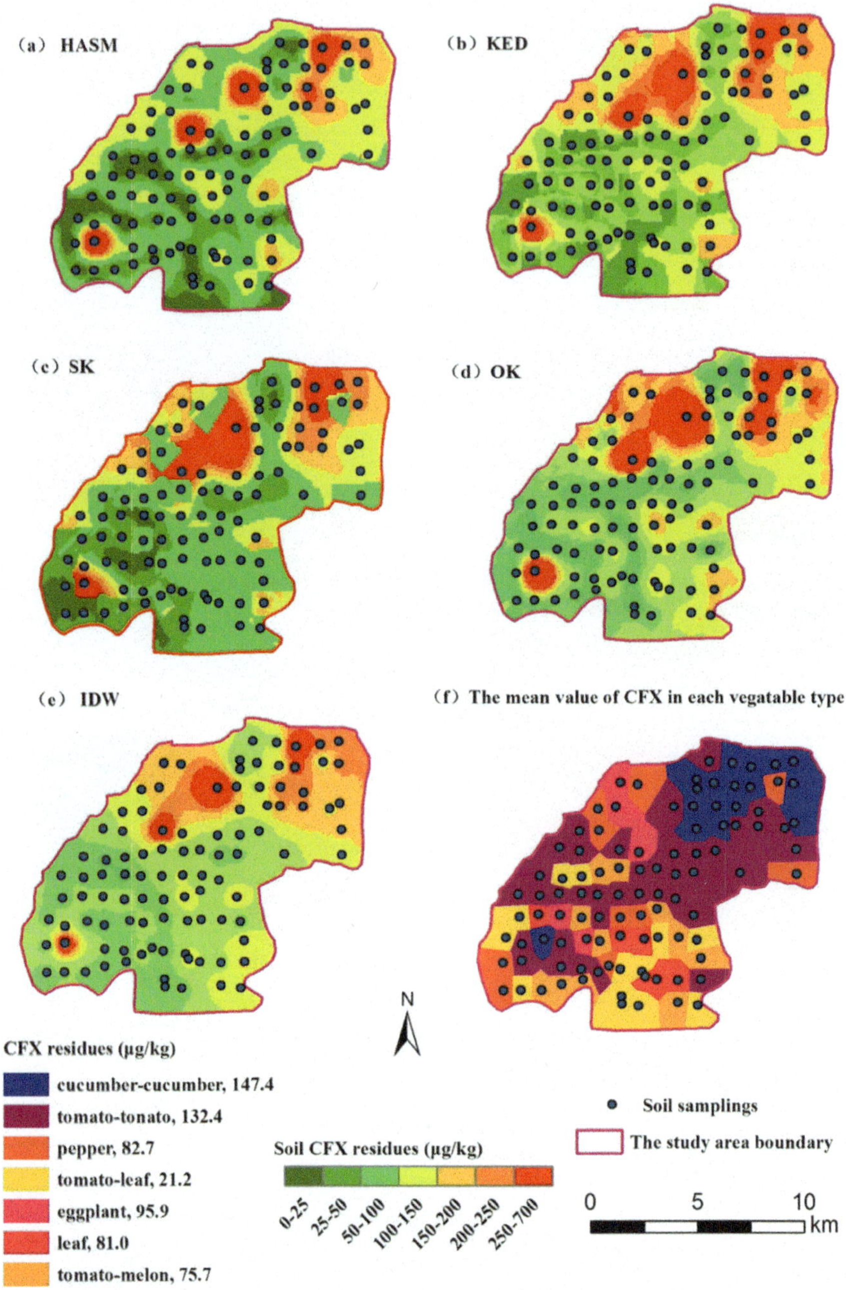


**Fig. 24** The spatial patterns of soil CFX residues obtained by HASM (**a**), KED (**b**), StK (**c**) OK (**d**) and IDW (**e**), and the mean values of CFX residues in each vegetable type (**f**)

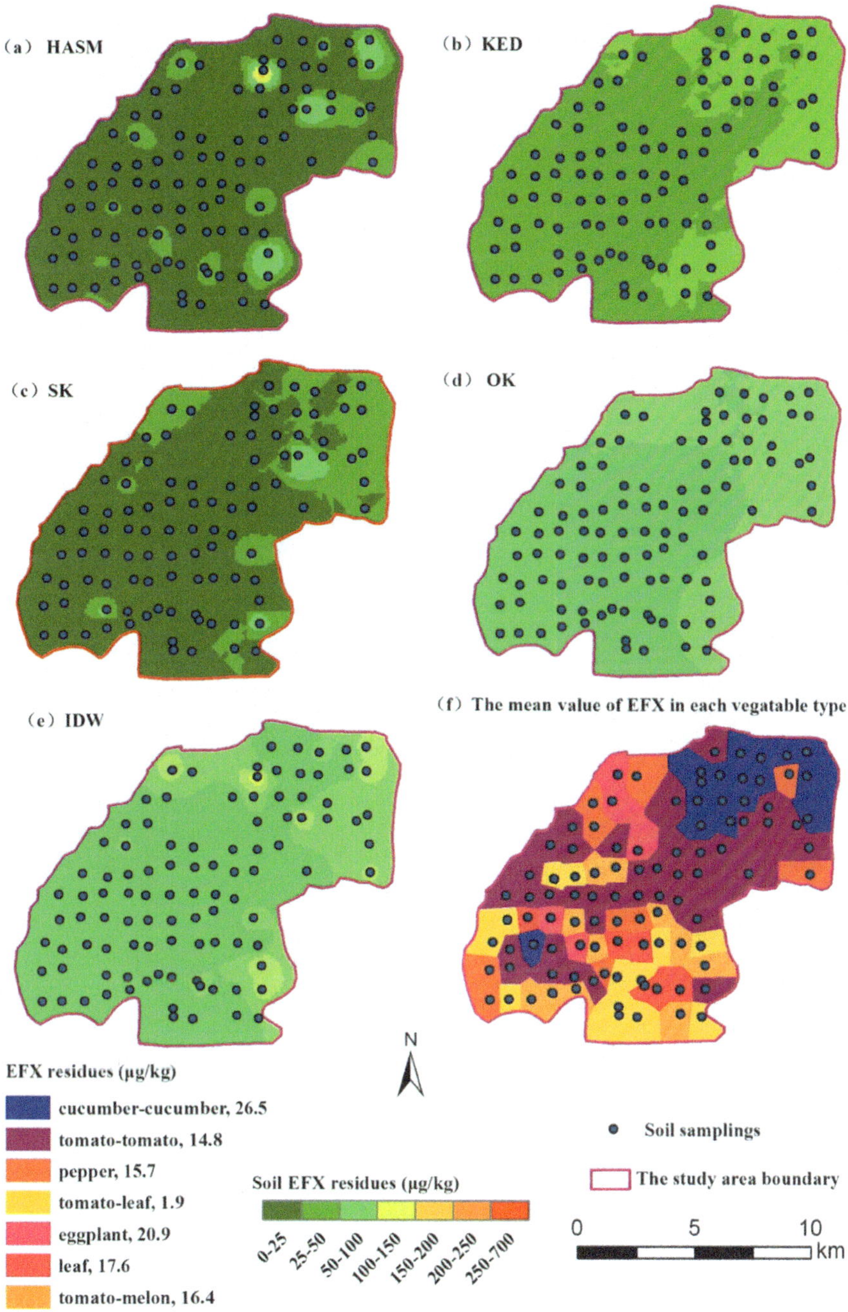


**Fig. 25** The spatial patterns of soil EFX residues obtained by HASM (**a**), KED (**b**), StK (**c**) OK (**d**) and IDW (**e**), and the mean values of EFX residues in each vegetable type (**f**)

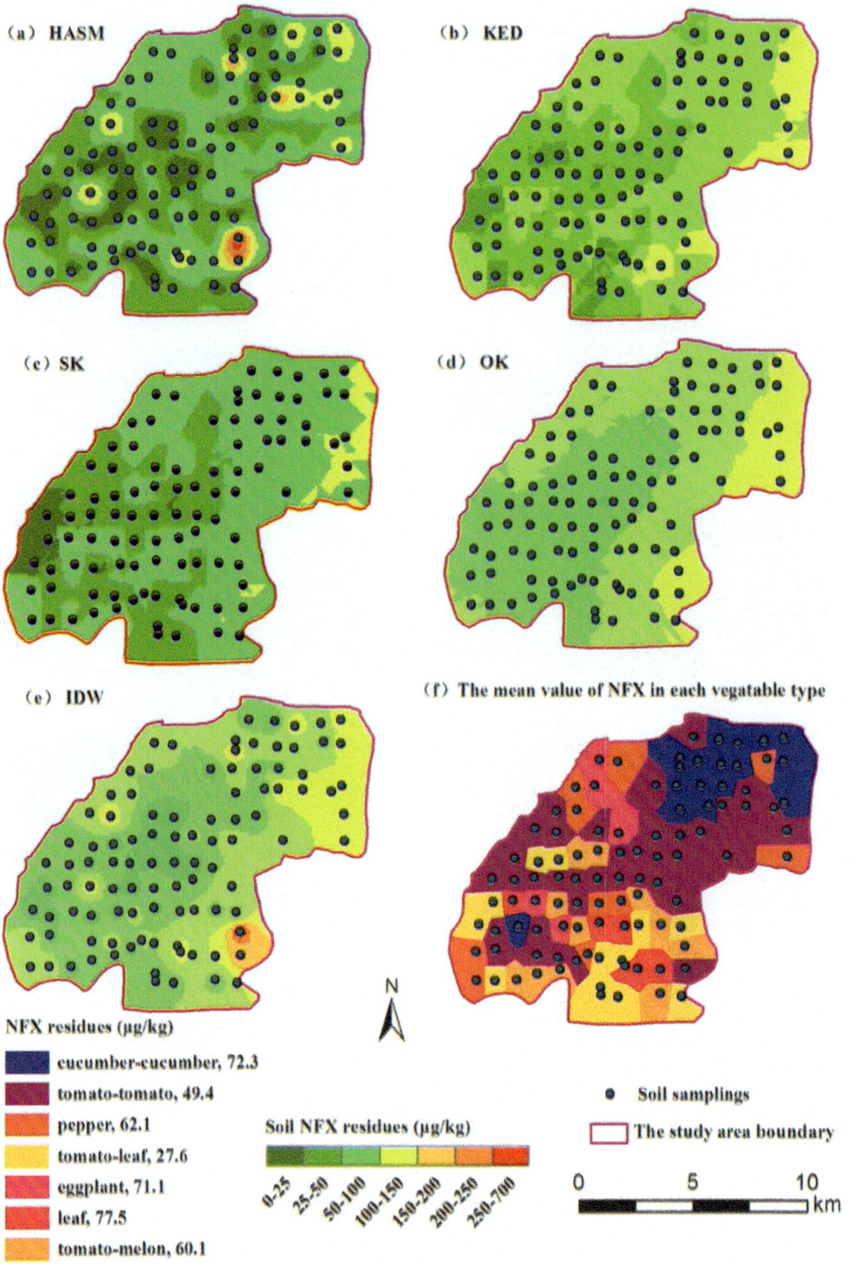


**Fig. 26** The spatial patterns of soil NFX residues obtained by HASM (**a**), KED (**b**), StK (**c**) OK (**d**) and IDW (**e**), and the mean values of NFX residues in each vegetable type (**f**)

**Table 9** The MAEs and RMSEs of cross-validation for HASM, KED, StK, OK and IDW (unit: μg/kg)

| Soil antibiotics | Indices | HASM | KED | StK | OK | IDW |
|---|---|---|---|---|---|---|
| FQs | MAE | 55.1 | 99.0 | 102.8 | 106.3 | 108.7 |
| | RMSE | 106.2 | 138.3 | 141.4 | 146.1 | 150.9 |
| CFX | MAE | 34.8 | 70.8 | 67.4 | 75.1 | 72.7 |
| | RMSE | 88.6 | 112.3 | 112.6 | 118.3 | 116.9 |
| EFX | MAE | 8.9 | 15.4 | 16.4 | 15.6 | 16.8 |
| | RMSE | 20.4 | 24.8 | 26.0 | 25.1 | 27.6 |
| NFX | MAE | 23.4 | 40.0 | 41.6 | 42.1 | 45.1 |
| | RMSE | 39.2 | 53.5 | 56.2 | 55.2 | 59.1 |

(2) Yue TX, Liu Y, Zhao MW, Du ZP, Zhao N. 2016. A fundamental theorem of Earth's surface modelling. Environmental Earth Sciences 75: 1–12.
(3) Yue TX. 2011. Surface Modelling: High Accuracy and High Speed Methods. New York: CRC Press.
(4) Yue TX, Du ZP, Song DJ, Gong Y. 2007. A new method of surface modeling and its application to DEM construction. Geomorphology 91: 161–172.
(5) Shi WJ, Liu JY, Du ZP, Stein Alfred, Yue TX. 2011. Surface modeling of soil properties based on land use information. Geoderma 162: 347–357.
(6) Shi WJ, Liu JY, Du ZP, Song YJ, Chen CF, Yue TX. 2009. Surface modelling of soil pH. Geoderma 150: 113–119.
(7) Shi WJ, Liu JY, Du ZP, Yue TX. 2012. Development of a surface modeling method for mapping soil properties. Journal of Geographical Sciences 22: 752–760.
(8) Shi WJ, Yue TX, Du ZP, Wang Z, Li XW. 2016. Surface Modeling of Soil Antibiotics. Science of the Total Environment 543: 609–619.

## Exercises

(1) Please introduce the spatial distributions of soil types and soil properties of your hometown, or your country, or your study area? You can show the spatial location of your hometown, or your country, or your study area, and tell us the sources of the soil sampling data, or which surface modelling method is used? Then tell us the spatial patterns of soil types and soil properties.
(2) Do your think surface modelling of soil properties will be used in your research? and Why? Please introduce your field to us, and tell us why you need surface modelling of soil properties.
(3) Have you ever applied a method in surface modelling of soil properties? What's that and for what?
(4) Which methods or which soil properties are you most interested in? and why? Could you please introduce one research about surface modelling of soil properties you are interested in?

(5) Could you please give us a review for methods of surface modelling of soil properties? Give us the advantages and disadvantages for different methods.

## References


Andriani, T., R. Balia, M. Loddo, G. Pecorini, and A. Tramacere. 2001. Structural features of the Middle Tirso Valley (Central Sardinia-Italy) from geoelectrical and gravity data. *Annali di Geofisica* 44: 739–753.

Asio, V. 2005. Comments on "historical development of soil and weathering profile concepts from Europe to the United States of America." *Soil Science Society of America Journal* 69: 571–572.

Aust, M.O., F. Godlinski, G.R. Travis, X. Hao, T.A. McAllister, P. Leinweber, and S. Thiele-Bruhn. 2008. Distribution of sulfamethazine, chlortetracycline and tylosin in manure and soil of Canadian feedlots after subtherapeutic use in cattle. *Environmental Pollution* 156: 1243–1251.

Beard, L.P. 2000. Comparison of methods for estimating earth resistivity from airborne electromagnetic measurements. *Journal of Applied Geophysics* 45: 239–259.

Belay-Tedla, A., X. Zhou, B. Su, S. Wan, and Y. Luo. 2009. Labile, recalcitrant, and microbial carbon and nitrogen pools of a tallgrass prairie soil in the US Great Plains subjected to experimental warming and clipping. *Soil Biology & Biochemistry* 41: 110–116.

Bishop, T.F.A., and A.B. McBratney. 2001. A comparison of prediction method for the creation of field-extent soil property maps. *Geoderma* 103: 149–160.

Bound, J.P., and N. Voulvoulis. 2004. Pharmaceuticals in the aquatic environment: A comparison of risk assessment strategies. *Chemosphere* 56: 1143–1155.

Bui, E.N., and C.J. Moran. 2000. Regional-scale investigation of the spatial distribution and of soluble salts in central north Queensland. *Hydrological Processes* 14: 237–250.

Burrough, P.A., J. Bouma, and S.R. Yates. 1994. The state of the art in pedometrics. *Geoderma* 62 (1): 311–326.

Clevers, J.G.P.W., and H.J.C. Van Leeuwen. 1996. Combined use of optical and microwave remote sensing data for crop growth monitoring. *Remote Sensing of the Environment* 56: 42–51.

Cochran, W.G. 1937. A catalogue of uniformity trial data. *Supplement to the Journal of the Royal Statistical Society* 4: 233–253.

Coffey, G.N. 1912. A study of the soils of the United States. USDA Bureau of Soils Bull.,85. U.S. Gov. Print. Office, Washington, DC, USA.

Dai, Yongjiu, Wei Shangguan, Nan Wei, Qinchuan Xin, Hua Yuan, Shupeng Zhang, Shaofeng Liu, et al. 2019. A review of the global soil property maps for earth system models. *The Soil* 5 (2): 137–158. https://doi.org/10.5194/soil-5-137-2019.

Delbari, M., P. Afrasiab, and W. Loiskandl. 2011. Geostatistical analysis of soil texture fractions on the field scale. *Soil Water Res.* 6: 173–189.

Deutsch C.V., and A.G. Journel. 1998 GSLIB: geostatistical software library and user's guide, 2nd edn. Oxford University Press, New York.

Dickson, B.L., S.J. Fraser, and A. Kinsey-Henderson. 1996. Interpreting aerial gamma-ray surveys utilising geomorphological and weathering models. *Journal of Geochemical Exploration* 57: 75–88.

Donatelli, M., C. Stockle, E.A. Constantini, and R. Nelson. 2002. SOILR: A model to estimate soil moisture and temperature regimes.

Eastman, J.R., and M. Fulk. 1993. Long sequence time-series evaluation using standardized principal components. *Photogrammetric Engineering & Remote Sensing* 59: 991–996.

Eswaran, H., and P. Reich, 2005. Global Soil Regions. 1: 130,000,000. Global Soil Suborder Map 1, 5,000,000 http://soils.usda.gov/use/worldsoils/mapindex/order.html.

Golub, G.H., and C.F. van Loan. 1996. Matrix Computations, 3rd Edition. Johns Hopkins University Press, Baltimore.

Goovaerts, P. 1992. Factorial kriging analysis: A useful tool for exploring the structure of multivariate spatial soil information. *Journal of Soil Science* 43: 597–619.
Glinka, K.D. 1927. The great soil groups of the world and their development (Translated by C.F. Marbut). Edwards Bros., Ann Arbor, Michigan.
GSC (GlobalSoilMap Science Committee). 2015. GlobalSoilMap specifications-Tiered1 Global SoilMap products. Release 2.4. [2015-07-12]. http://globalsoilmap.net/specifications.
Haines, W.B., and B.A. Keen. 1925. Studies in soil cultivation. II. A test of soil uniformity by means of dynamometer and plough. *Journal of Agricultural Science* 15: 387–394.
Henderson, D.W., and D. Taimina. 1997. Differential Geometry: a Geometric Introduction. Prentice Hall, London.
Hengl, T., J.M. De Jesus, R.A. MacMillan, N.H. Batjes, G.B. Heuvelink, E. Ribeiro, A. Samuel-Rosa, B. Kempen, J.G. Leenaars, M. Walsh, and G. Gonzalez. 2014. SoilGrids1km–global soil information based on automated mapping. *PLoS ONE* 9: e105992.
Hole, F.D. 1981. Effects of animals on soil. *Geoderma* 25: 75–112.
Hollsteins, W., 1930. Bodenkarte de Erde (scale 1:125,000,000). In: Blanck, E. (Ed.). Handbuch der Bodenlehre, vol. 3. Verlag von Julius Springer, Berlin, 550 pp.
Johnston, R.M., S.J. Barry, E. Bleys, E.N. Bui, C.J. Moran, D.A.P. Simon, P. Carlile, N.J. McKenzie, B.L. Henderson, G. Chapman, M. Imhoff, D. Maschmedt, D. Howe, C. Grose, and N. Schoknecht. 2003. ASRIS: The database. *Australian Journal of Soil Research* 41: 1021–1036.
Jenny, H. 1941. Factors of soil formation, a system of quantitative pedology. McGraw-Hill, New York.
Kuchta, S.L., and A.J. Cessna. 2009. Lincomycin and spectinomycin concentrations in liquid swine manure and their persistence during simulated manure storage. *Archives of Environmental Contamination and Toxicology* 57: 1–10.
Lark, R.M. 2000. Regression analysis with spatially autocorrelated error: simulation studies and application to mapping of soil organic matter. *International Journal of Geographical Information Science* 14: 247–264.
Lark, R.M., and T.F.A. Bishop. 2007. Cokriging particle size fractions of the soil. *European Journal of Soil Science* 58: 763–774.
Martínez-Carballo, E., C. González-Barreiro, S. Scharf, and O. Gans. 2007. Environmental monitoring study of selected veterinary antibiotics in animal manure and soils in Austria. *Environmental Pollution* 148: 570–579.
McKenzie, N.J., P.E. Gessler, P.J. Ryan, and D. O'Connel. 2000. The role of terrain analysis in soil mapping. In *Terrain analysis—principles and applications*, ed. J.P. Wilson and J.C. Gallant, 245–265. New York: Wiley.
Mercer, W.B. and A.D. Hall, 1911. The experimental error of field trails. *J. Agric. Sci.* 4: 107–132.
Morgan, R.P.C., J.N. Quinton, R.E. Smith, G. Govers, J.W.A. Poesen, K. Auerswald, G. Chisci, D. Torri, and M.E. Styczen. 1998. The European Soil Erosion Model (EUROSEM): A dynamic approach for predicting sediment transport from fields and small catchments. *Earth Surface Processes and Landforms* 23: 527–544.
Odeh, I.O.A., A.J. Todd, and J. Triantafilis. 2003. Spatial prediction of soil particle-size fractions as compositional data. *Soil Science* 168: 501–515.
Pike, R.J. 1988. The geometric signature: quantifying landslide terrain types from digital elevation models. *Mathematical Geology* 20: 491–511
Rawls, W.J., D.L. Brakensiek, and K.E. Saxton. 1982. Estimation of soil-water properties. *Transactions of ASAE* 25: 1316–1320.
Rozanov, B.G. 1977. Pochvenniy pokrov Zemnogo shara [Soil Cover of the Globe]. Moscow University Publishers, Moscow (248 pp. (In Russian)).
Sarmah, A.K., M.T. Meyer, and A. Boxall. 2006. A global perspective on the use, sales, exposure pathways, occurrence, fate and effects of veterinary antibiotics (VAs) in the environment. *Chemosphere* 65: 725–759.
Sanchez, P.A., S. Ahamed, F. Carré, A.E. Hartemink, J. Hempel, J. Huising, P. Lagacherie, A.B. McBratney, N.J. McKenzie, M. Mendonça-Santos, L.M. Budiman Minasny, P. Okoth, C.A.

Palm, J.D. Sachs, K.D. Shepherd, T.G. Vågen, B. Vanlauwe, M.G. Walsh, L.A. Winowiecki, G.L. Zhang. 2009. Digital soil map of the world. Science 325: 680–681.

Shi, W., J. Liu, Z. Du, Y. Song, C. Chen, and T. Yue. 2009. Surface modelling of soil pH. *Geoderma* 150: 113–119.

Shi, W., J. Liu, Z. Du, A. Stein, and T. Yue. 2011a. Surface modelling of soil properties based on land use information. *Geoderma* 162: 347–357.

Shi, W.J., T.X. Yue, Z.P. Du, Z. Wang, and X.W. Li. 2016. Surface modeling of soil antibiotics. *Science of the Total Environment* 543: 609–619.

Shi, W.J., and M. Zhang. 2023. Progress on spatial prediction methods for soil particle-size fractions. *Journal of Geographical Sciences* 33, 1553–1566.

Soil Survey Staff (NRCS). United States Department of Agriculture: Web Soil Survey, available at: http://websoilsurvey.nrcs.usda.gov/, last access: 1 January 2017.

Simonson 1957 Simonson, R.Q. 1957. What soils are. Soil the 1957 Yearbook of agriculture, 17–31. U.S. Department of Agriculture Washington, D.C.

Simonson, R.W. 1989. Historical highlights of soil survey and soil classification with emphasis on the United States, 1899–1970. Technical Paper No. 18.

Sarmah, A.K., M.T. Meyer, and A. Boxall. 2006. A global perspective on the use, sales, exposure pathways, occurrence, fate and effects of veterinary antibiotics (VAs) in the environment. *Chemosphere* 65: 725–759

Stein, A., M. Hoogerwerf, and J. Bouma. 1988. Use of soil-map delineations to improve (co-) kriging of point data on moisture deficits. *Geoderma* 43: 163–177.

Taylor, R., and R.A. Eggleton. 2001. Regolith Geology and Geomorphology, Wiley, Chichester, UK.

Troeh, F.R. 1964. Landform parameters correlated to soil drainage. *Soil Science Society of America Proceedings* 28: 808–812

Van Loan, C.F. 1985. On the method of weighting for equality-constrained least-squares problems. *SIAM Journal on Numerical Analysis* 22 (5): 851–864.

Vereecken, H., M. Weynants, M. Javaux, Y. Pachepsky, M.G. Schaap, and M.T.V. Genuchten. 2010. Using pedotransfer functions to estimate the van Genuchten-Mualem soil hydraulic properties: A review. *Vadose Zone Journal.* 9: 795–820.

Voltz, M., and R. Webster. 1990. A comparison of kriging, cubic splines and classification for predicting soil properties from sample information. *Journal of Soil Science* 41: 473–490.

Twidale, C.R. 1985. Old land surfaces and their implications for models of landscape evolution. *Revue De Ge Ómorphologie Dynamique* 34: 131–147.

Wang, Z.M., B. Zhang, K.S. Song, and D.W. Liu. 2008. Soil organic carbon under different landscape attributes in croplands of Northeast China. *Plant, Soil and Environment* 54: 420–427.

Webster, R. 1977. Canonical correlation in pedology: How useful? *Journal of Soil Science* 28: 196–221.

Webster, R., and M.A. Oliver. 2001. *Geostatics for Environmental Scientists.* West Sussex, UK: John Wiley and Sons.

Whitney, M. 1909. Soils of the United States. USDA Bureau of Soils Bulletin No. 55 (243 pp. + map).

Wood, J. 1966. The geomorphological characterisation of digital elevation models. PhD thesis, University of Leicester, UK.

Xie, Y.F., X.W. Li, J.F. Wang, G. Christakos, M.G. Hu, L.H. An, and F.S. Li. 2012. Spatial estimation of antibiotic residues in surface soils in a typical intensive vegetable cultivation area in China. *Science of the Total Environment* 430: 126–131.

Yang, Y.Y., A. Goldsmith, I. Herold, S. Lecha, and G.S. Toor. 2020. Assessing soil organic carbon in soils to enhance and track future carbon stocks. *Agronomy* 10: 1139.

Youden, W.J. and A. Mehlich, 1937. Selection of efficient methods for soil sampling. *Contributions of the Boyce Thompson Institute for Plant Research* 9: 59–70

Yue, T.X. 2011. *Surface modelling: High accuracy and high speed methods.* New York: CRC Press.

Yue, T.X., D.J. Song, Z.P. Du, and W. Wang. 2010. High-accuracy surface modelling and its application to DEM generation. *International Journal of Remote Sensing* 31: 2205–2226.

Yue, T.X., L. Zhang, N. Zhao, M. Zhao, C. Chen, and Z. Du. 2015. A review of recent developments in HASM. *Environ. Earth Sci.* 74 (8): 6541–6549.

Yue, T.X., M.W. Zhao, and X.Y. Zhang. 2014. A high-accuracy method for filling voids on remotely sensed XCO2 surfaces and its verification. *J. Clean. Prod.* 106: 819–827.

Yue, T.X., N. Zhao, R.D. Ramsey, C.L. Wang, Z.M. Fan, C.F. Chen, Y.M. Lu, and B.L. Li. 2013. Climate change trend in China, with improved accuracy. *Clim. Chang.* 120: 137–151.

Yue, T.X., N. Zhao, Y. Liu, Y. Liu, Y.F. Wang, B. Zhang, Z.P. Du, Z.M. Fan, W.J. Shi, C.F. Chen, and M.W. Zhao. 2020. A fundamental theorem for eco-environmental surface modelling and its applications. *Science China-Earth Sciences* 63: 1092–1112.

Yue, T.X., Y. Liu, M.W. Zhao, Z.P. Du, and N. Zhao. 2016. A fundamental theorem of Earth's surface modelling. *Environmental Earth Sciences* 75: 1–12.

Yue, T.X., Z.P. Du, D.J. Song, and Y. Gong. 2007. A new method of surface modeling and its application to DEM construction. *Geomorphology* 91: 161–172.

Zhang, G.L., and H.Y. Wu. 2018. From "problems" to "solutions": Soil functions for realization of Sustainable Development Goals. *Bulletin of Chinese Academy of Sciences* 33: 124–134.

Zhang, J., X. Wang, and J. Wang. 2014. Impact of land use change on profile distributions of soil organic carbon fractions in the Yanqi Basin. *CATENA* 115: 79–84.

Zhang, L., Q. Zheng, Y. Liu, S. Liu, D. Yu, X. Shi, S. Xing, H. Chen, and X. Fan. 2019. Combined effects of temperature and precipitation on soil organic carbon changes in the uplands of eastern China. *Geoderma* 337: 1105–1115.

# Chapter 10
# Mapping Soil Compositional Data

**Learning Objectives**

- To understand the special characteristic in mapping soil compositional data.
- To be able to map soil compositional data combined with environmental information.

## 10.1 Introduction

### *10.1.1 Background of Soil Texture and Soil Particle-Size Fractions*

Soil texture (such as loam, sandy loam or clay) refers to the proportion of sand, silt and clay sized particles that make up the mineral fraction of the soil. The USDA soil taxonomy and WRB soil classification systems use 12 textural classes.

For example, light soil refers to a soil high in sand relative to clay, while heavy soils are made up largely of clay.

Texture is important because it influences:

- the amount of water the soil can hold
- the rate of water movement through the soil
- how workable and fertile the soil is

For example,

- Sand is well aerated but does not hold much water and is low in nutrients.
- Clay soils generally hold more water, and are better at supplying nutrients.



T. Yue et al., *Eco-Environmental Informatics*,
https://doi.org/10.1007/978-981-96-1977-1_10

### *10.1.2 Determine Soil Texture by Feel*

There are three steps to determine soil texture:

- Take about 2 tablespoons of soil in one hand and add water, drop by drop, while working the soil until it reaches a sticky consistency.
- Squeeze the wetted soil between thumb and forefinger to form a flat ribbon.
- Determine the texture based on the length of the ribbon that can be formed without breaking—see following table.

  Let's see one by one. Look at soil in your hand. Can you form a ball?

- If no, the soil texture type is sand. The length of ribbon is less than 15 mm. Sandy soil has little resistance to root growth, high infiltration rate, and Low plant available water.
- If yes, Roll on your hand width-wise. Can you form a cylinder 6–7 cm long? If no, the soil texture type is sandy loam. The length of ribbon is between 15 and 25 mm. In this type, root growth is not restricted, but it is highly susceptible to mechanical compaction. It may be hard setting, and has moderate infiltration rate and moderate plant available water.
- If yes, continue rolling. Can you form a cylinder 14 cm long? If no, the soil texture type is silty loam. The length of ribbon is about 25 mm. In this type, root growth is not restricted either, but it is moderately susceptible to mechanical compaction. It may have Moderate plant available water and have Low to moderate infiltration rate.
- If yes, can you fold the cylinder into a u-shape? If no, the soil texture type is loam. It is like silty loam but with Moderate infiltration rate.
- If yes, can you fold the u-shape into a ring? If no, the soil texture type is clay loam. It has Moderate to high plant available water.
- If yes, are there cracks in the surface of the ring? If no, the soil texture type is heavy clay. In this type, Root growth is frequently restricted. It has Moderately to highly susceptible to mechanical compaction, some restriction on water movement leading to periodic waterlogging, and Moderate to high plant available water.
- If yes, the soil texture type is fine clay. In this type, root growth is moderately to severely restricted, and it has High susceptibility to mechanical compaction, and Water drains are very slowly except in self-mulching soils.

### *10.1.3 The Importance of Soil PSF*

- Most of physical and chemical processes in soil science, such as water, heat, nutrient fluxes, water and nutrient holding capacity, soil structural form and stability, and soil erosion, are partially affected by distributions of soil particle-size fractions (PSF).

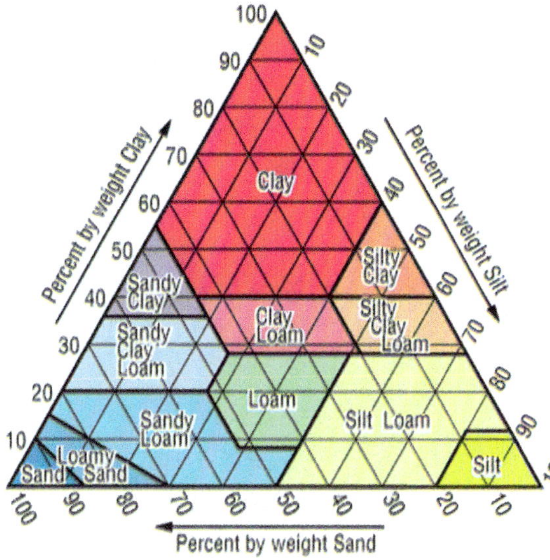


Fig. 1 Soil texture classification and soil PSF interpolation

- Digital mapping of soil particle-size fractions is helpful to understand most of physical and chemical processes in soil science.
- The distribution of PSF, including sand, silt and clay, is a key parameter in many process models and pedotransfer functions, which are important for modeling most soil physical and chemical processes and different hydrology processes (Fig. 1).

#### 10.1.3.1 The Importance of the Distribution of Soil Particle-Size Fractions (PSF)

Most of physical and chemical processes in soil science, such as water, heat, nutrient fluxes, water and nutrient holding capacity, soil structural form and stability, and soil erosion, are partially affected by distributions of soil particle-size fractions (PSF). Digital mapping of soil particle-size fractions is helpful to understand the above processes (Fig. 2).

#### 10.1.3.2 Problems and Requirements

Most of the traditional methods may not satisfy the special requirements of soil compositional data interpolation. Traditional methods such as ordinary kriging (OK) and cokriging (COK) may not satisfy these requirements and cannot guarantee that the sum of components is 1 (or 100%).The previous studies are lack of the comparative analysis of the interpolation methods for soil compositional data under different environmental factors.

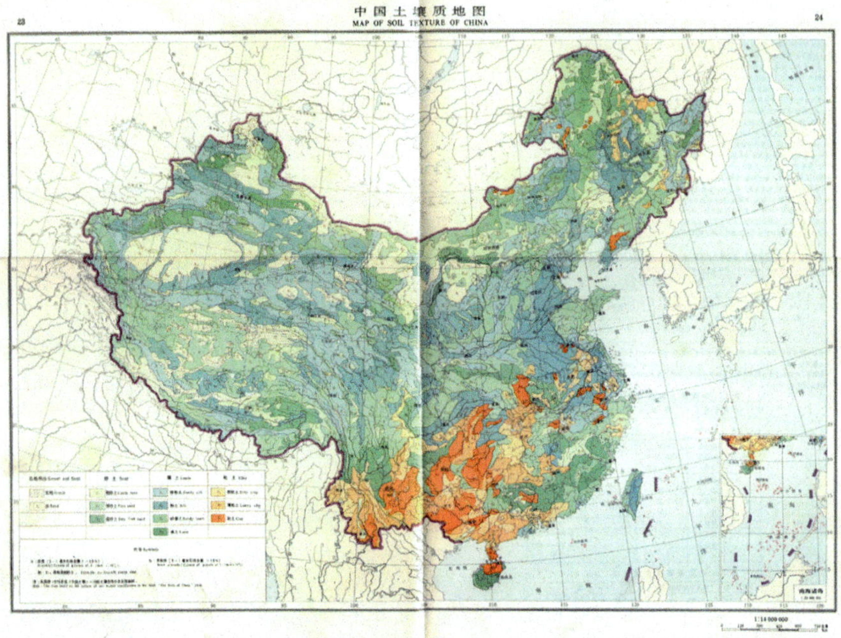


**Fig. 2** The soil quality map of China

#### 10.1.3.3 The Other Characters of Soil PSF

In log-ratio methods, the ILR method stands out for its ability to eliminate model collinearity while preserving desirable properties, such as isometry and scale invariance. It has been shown to offer superior performance compared to the ALR and CLR methods (Filzmoser and Hron 2009; Wang and Shi 2018; Zhang et al. 2020). However, from the mathematical principles of the transformation method, the ILR method is constructed within orthonormal coordinate systems (so-called balances), employing a sequential binary partition (SBP) (Egozcue and Pawlowsky-Glahn 2005). The selection of balances is not unique, allowing for the generation of multiple sets of ILR transformed data through permutations of components (different SBPs) in the compositional data. The choice of an SBP is typically guided by various factors, including prior expert knowledge, compositional biplots (Lloyd et al. 2012), and statistical analysis such as variograms and cross-variograms (Molayemat et al. 2018). In statistical science, it has been established that the choice of a specific SBP for compositions profoundly influences the interpretation of coordinates (Fiserova and Hron 2011). Further more, the decision regarding SBP is intertwined with hypotheses, research questions, or the broader context of data analysis (Coenders et al. 2017; Facevicova et al. 2018).

### 10.1.4 The Objectives of This Case

To comprehensively compare four log-ratio transformation methods, two estimators for variograms, and two CK methods,

To provide a reference for choosing a suitable interpolator according to the data characteristics by comparing the performances of different interpolators.

## 10.2 A Case of Mapping Soil Particle-Size Fractions: A Comparison of Compositional Kriging and Log-Ratio Kriging

### 10.2.1 Data and Methods

#### 10.2.1.1 The Tianlaochi

The Tianlaochi Basin (38°23′56″–38°26′47″N, 99°53′57″–99°57′10″E), covering 12.6 km$^2$, is located in Sidalong River Basin which is in the upper reaches of Heihe River (Fig. 3). The altitude is from 2600 to 4550 m. The annual mean temperature is 0.6 °C, and the annual rainfall is 437.2 mm. The annual average relative humidity is 59%. The soil types include gray-cinnamon soil, castanozems, alpine meadow soil, frigid desert soil, etc.

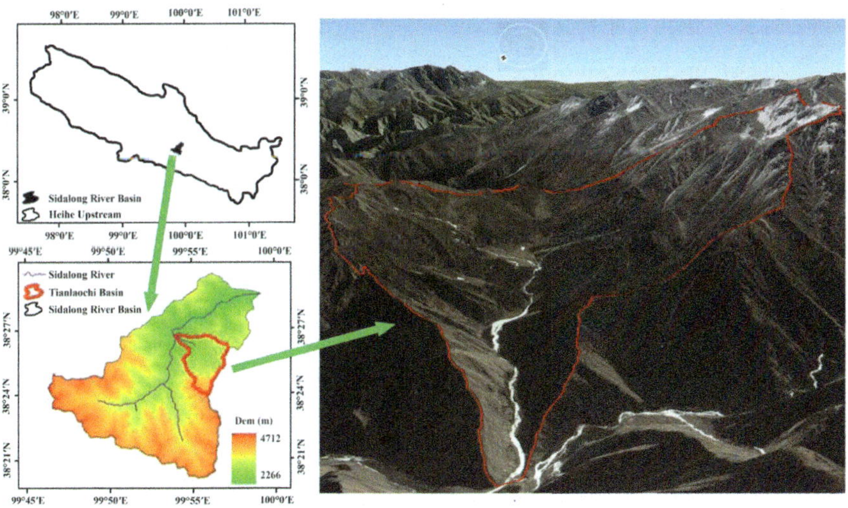


**Fig. 3** The geographical location of Tianlaochi in the Heihe River Basin, China

#### 10.2.1.2 Sampling Points

A total of 133 soil sampling points in the study area were collected, and the position information was recorded using the global position system (GPS) (Fig. 4). The average distance among sampling sites was approximately 0.35 km. We desinged the sampling sites to cover evenly the whole study area, including different environmental variables of land use types, vegetation types, soil types, elevations, slopes and aspects. The samples represent the spatial characters of soil PSF in the study area with higher variability. There were no samples collected in the southern part due to glacier. We mixed five samples of top soil (0–20 cm) to one soil sample, which can reduce the effects of sampling noise and nugget variance.

#### 10.2.1.3 The Objectives of This Case

- compare the differences of log-ratio kriging methods;
- compare the performance between CK and log ratio kriging;
- compare the accuracy of different interpolators using environmental variables.

#### 10.2.1.4 Methods

Log-ratio transformations for compositional data include additive log-ratio (alr) transformation, centered log-ratio (clr) transformation, isometric log-ratio (ilr) transformation, and symmetry log-ratio (slr) transformation. All of the log-ratio kriging methods including OK_alr, OK_clr, OK_ilr and OK_slr are satisfied with the interpolation requirements of component data.

(i) **Log-ratio kriging methods**

An appropriate sample space is the basic requirement for modeling any data (Odeh et al. 2003). The restricted part of real space, termed positive simplex (Rd), is properly proposed for compositions (z), and the D-dimensional real space (Sd) has gained the advantage of symmetry (Aitchison 1986, 1990).

The purpose of log-ratio transformations for soil PSF is to change the sample space from nature space (Sd) to the real space (Rd).

The transformation equation of ilr is defined as (Egozcue et al. 2003),

$$ilr_{ij}(x) = \sqrt{\frac{j}{j+1}} ln \frac{\sqrt[j]{\prod_{j=1}^{D-1} \mu_{ij}(x)}}{\mu_{i,j+1}(x)} \tag{1}$$

and the back-transformed equation is defined as (Filzmoser and Hron 2008).

$$Y_{ij}(x) = \sum_{j=1}^{D} \frac{ilr_{ij}(x)}{\sqrt{j(j+1)}} - \sqrt{\frac{j-1}{j}} ilr_{i,j-1}(x) \tag{2}$$

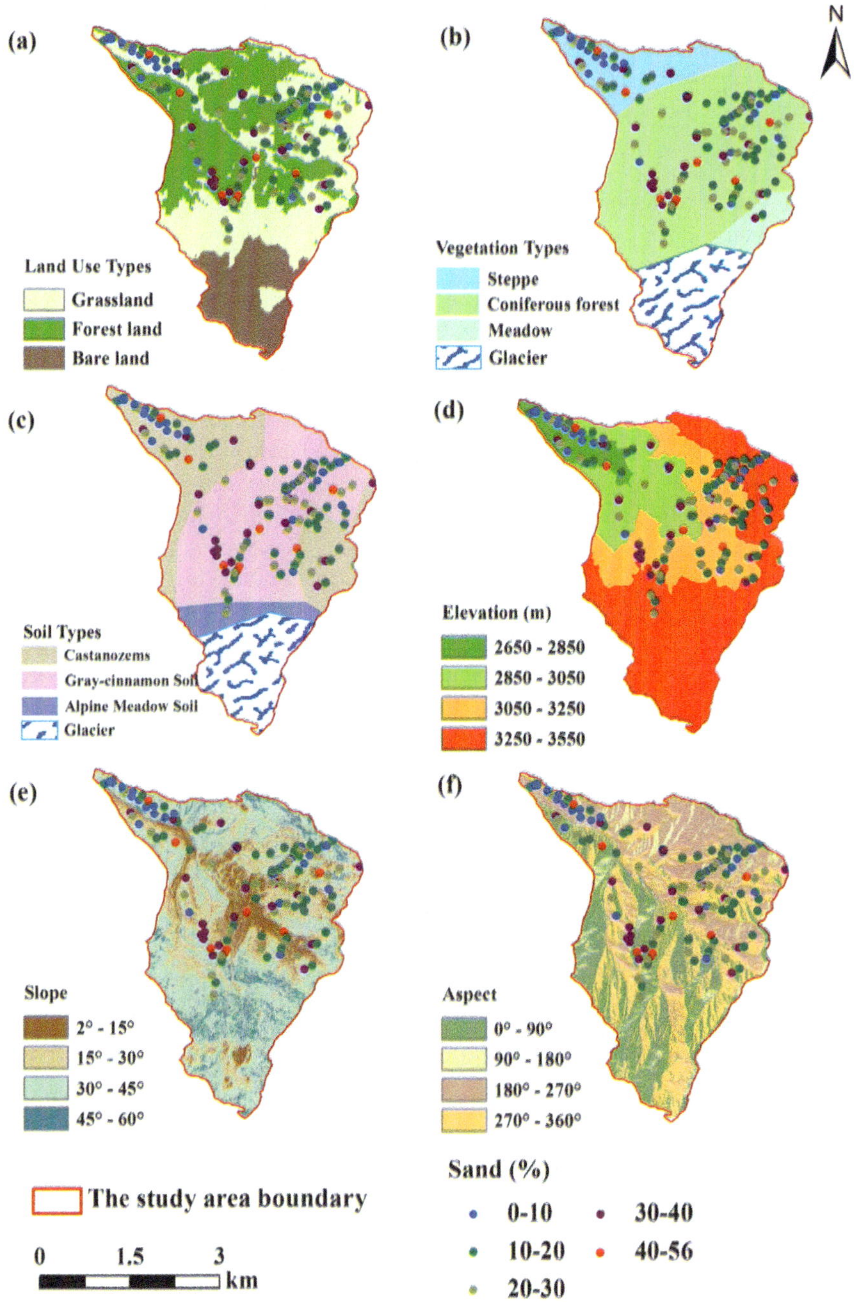


**Fig. 4** The spatial distributions of **a** land use types, **b** vegetation types, **c** soil types, **d** elevation, **e** slope and **f** aspect in the Tianlaochi Basin

The transformation equation of slr can be defined as (Odeh et al. 2003),

$$\mathrm{slr}_{ij}(x) = \ln \frac{\mu_{ij}(x) + \delta_j}{\left(\prod_{j=1}^{D}(\mu_{ij}(x) + \delta_j)\right)^{1/D}} \tag{3}$$

and the back-transformed equation can be defined by the antisymmetric log-rati transformation,

$$\overline{slr}_{ij}(x) = \left[\frac{expslr_{ij}(x)}{\sum_{j=1}^{D} expslr_{ij}(x)} - \frac{\delta_j}{1 + \sum_{j=1}^{D}\delta_j}\right]\left(1 + \sum_{j=1}^{D}\delta_j\right) \tag{4}$$

After transformations of original data using the above equations, the semivariograms for ordinary kriging were computed and modeled using the geostatistical package GS + 7.0, and ArcGIS 10.1 was used for ordinary kriging mapping of the transformation soil PSF data. The parameters for the semivariogram, such as nugget, sill and range, coefficient of determination and residual sum of squares (RSS), were determined using weighted least squares. The models with the highest R2 and RSS were chosen for the data set. We analyzed the spatial structure of the models by the nugget to sill ratio (Cambardella et al. 1994) and chose the best one from 2000 to 3000 with 100 step interval for the kriging model. Then, the back-transformed equations were used to transform the final results of the interpolation maps. The log-ratio transformations including alr, clr, ilr and slr combined with ordinary kriging (OK_alr, OK_clr, OK_ilr and OK_slr) were used to interpolate the soil PSF data in our study.

(ii) **Compositional kriging (CK)**

Tolosana-Delgado (2006) has also demonstrated CK is an optimal predictor for minimizing. The expected Aitchison distance between the true composition and its prediction. In these two procedures for CK, we used the latter procedure and implemented it through the R package compositions in our research (Tolosana-Delgado et al. 2008). The procedure is briefly summarized as follows (Taboada et al. 2013; Saavedra et al. 2010).

First, the original space of composite data is transformed into a new space dimension by changing the coordinates, and obtain optimal predictors by the random functions.

For the second step, (p−1) auto-variograms and (p−1) × (p−1)/2 cross-variograms need to be calculated and verify the variance of any linear combination of these variables is not negative.

#### 10.2.1.5 Validation

Cross-validation, which is a usual method of validation verification, is selected to compare interpolation methods. It removes each sample point in turn and estimates the soil PSF value at its site from the remaining observations, and then evaluates the prediction accuracy by comparing predicted values with measured values. We selected the mean error (ME), the root mean square error (RMSE) (Shi et al. 2009, 2011, 2012), Aitchison distance (AD), the standardized residual sum of squares (STRESS) (Lark and Bishop 2007; Martin-Fernandez et al. 2001; Sun et al. 2014) and right ratio of soil texture type (RR) as the indicators to verify the accuracy of different interpolation methods. The equations of the above indexes are as follows,

$$\mathrm{ME} = \frac{1}{n}\sum_{i=1}^{n}\left(z_i - z_i^*\right) \tag{5}$$

$$\mathrm{RMSE} = \sqrt{\frac{1}{n}\sum_{i=1}^{n}\left(z_i - z_i^*\right)^2} \tag{6}$$

$$\mathrm{AD} = \sqrt{\sum_{k=1}^{D}\left(\ln\frac{z_k}{\left(\prod_{k=1}^{D} z_k\right)^{1/D}} - \ln\frac{z_k^*}{\left(\prod_{k=1}^{D} z_k^*\right)^{1/D}}\right)^2} \tag{7}$$

$$\mathrm{STRESS} = \sqrt{\left(\frac{\sum_{i<j}\left(AD_{ij} - AD_{ij}^*\right)^2}{\sum_{i<j} AD_{ij}^2}\right)} \tag{8}$$

$$\mathrm{RR} = \mathrm{RN}/\mathrm{TN} \tag{9}$$

where, $Z$ is the observed value, $Z^*$ is the predicted value, n is the number of validation samples, $D$ is the number of dimensions, and $AD_{ij}$ and $AD_{ij}{}^*$ are the ADs between the observed soil PSF and the predicted soil PSF at sites $i$ and $j$, respectively. *RN*: the number of right points which mean that the predicted types of soil texture for the points are in the same types with the original data; *TN*: the number of total points of soil texture which generated by the five interpolators.

ME measures total prediction bias of the interpolation which should be close to zero for unbiased methods. RMSE is often used as the accuracy of prediction, which measures prediction precision and bias. AD is a measure of similarity for a prediction to observation. STRESS measures the overall similarity of the observations and the estimates (Lark and Bishop 2007). In addition, according to the observed data of sand, silt, clay for each sampling point, soil texture types can be identified according to the classification system of United States Department of Agriculture (USDA).

The predicted values of soil PSF in different maps by the five interpolators were also extracted and calculated to compare with the soil texture types of the sampling points. Thus the right ratios of soil texture type for predicted points (RRs) can be calculated as a measurement to compare the performance of different method.

#### 10.2.1.6 The Statistical Analysis of Transformed Data

We selected the unbiased total variance estimator (Hron and Kubácek 2011; Pawlowsky-Glahn et al. 2007), defined for a random composition $x = (x_1,\ldots,x_D)$ as,

$$\text{totvar}(x) = \frac{1}{D}\sum_{i=1}^{D-1}\sum_{j=i+1}^{D}\text{var}\left(\ln\frac{xi}{xj}\right) \tag{10}$$

where $D$ is the number of compositional parts and *var* is the sample variance.

For the compositions, the median absolute deviation (MAD) which has no unit and represents stability or homogeneity of the data is calculated to replace the variance of compositions.

$$\text{MAD}(x) = \text{median}(\{|x1 - \overline{x}|, \ldots, |xm - \overline{x}|\}) \tag{11}$$

### 10.2.2 Results

#### 10.2.2.1 The Description Statistics for the Original and Log-Ratio Transformation Data of the Soil PSF

Estimations of the center of the three compositions are shown in Table 1. The means of original data are different from the medians and the centers of transformation data. The results of sand contents generated by log-ratio transformation methods (17.21% for alr, clr and ilr; 17.25% for slr) are lower than mean of original data (20.33%) and closer to the median (17.54%).

**Table 1** The center of the raw and log-ratio transformation data of soil PSF

| Compositions | Mean | Median | Mean-alr | Mean-clr | Mean-ilr | Mean-slr |
|---|---|---|---|---|---|---|
| Sand (%) | 20.33 | 17.54 | 17.21 | 17.21 | 17.21 | 17.25 |
| Silt (%) | 54.40 | 55.90 | 56.73 | 56.73 | 56.73 | 56.59 |
| Clay (%) | 25.27 | 24.94 | 26.06 | 26.06 | 26.06 | 26.16 |

The total variance and median absolute deviations (MADs) were computed to evaluate the distributions of the compositional values (Table 2). The log-ratio transformation data have higher stability than the raw data with the small values of the MADs. Slr transformation data have the most optimal stability among the four log-ratio transformation methods with the MADs of 0.32, 0.23, 0.19 for sand, silt and clay, respectively.

The comparisons of histogram plots between the original and log-ratio transformation data for the 133 soil samplings of sand, silt and clay fractions are shown in Fig. 5, which provide knowledge about the statistical distributions of the soil PSF. The first column of Fig. 5 presents histogram plots of the original data of sand (Fig. 5a), silt (Fig. 5f) and clay fractions (Fig. 5k), which are right-skewed (sand) or left-skewed (silt). In contrast, the log-ratio transformation results, especially ilr sand and ilr silt, are in more symmetric images with low skewness values −0.60 and 0.18, respectively. Clay (Fig. 5k) and all of its transformations (Fig. 5l–o) are relatively symmetric. Filzmoser et al. (2009) also indicated that the log-ratio transformation, e.g. ilr transformation allowed for a correct geometrical presentation of compositional data in Aitchison geometry.

The results of the ANOVA analysis indicate that sand has significant correlations with different classes of land use types, vegetation types, elevations and aspects. Higher values of sand contents are mainly distributed in forest land with mean-clr of 19.80% (especially in the coniferous forest with the maximum value of 55.46%), also in the area with elevation between 3050 and 3250 m (mean-clr, 22.27%) and shady slope area (mean-clr with 23.75% in the aspects of 3–90° and with 22.26% in the aspects of 270–360°). Similarly, land use types, elevations and aspects also have significant effects on the distributions of silt fractions. However, higher values of silt fractions mostly distribute in the grassland (mean-clr, 59.68%), higher altitude area with elevation above 3050 m (mean-clr, 63.22%) and sunny slope area (mean-clr with 62.38% in the aspects of 90–180° and with 59.97% in the aspects of 180–270°). In contrast, vegetation types, soil types and elevation are significant factors for the distribution of clay fractions. Higher values of clay fractions are located in steppe area (mean-clr, 31.74%), castanozems area (mean-clr, 28.51%) and lower altitude area below 2850 m (mean-clr, 33.45%).

**Table 2** The total variance and median absolute deviations of original and log-ratio transformation data

| Compositions | Totvar | MAD | MAD-alr | MAD-clr | MAD-ilr | MAD-slr |
|---|---|---|---|---|---|---|
| Sand | 0.66 | 7.64 | 0.52 | 0.34 | 0.43 | 0.32 |
| Silt | | 8.16 | 0.25 | 0.23 | 0.29 | 0.23 |
| Clay | | 4.85 | 0.60 | 0.21 | 0.26 | 0.19 |

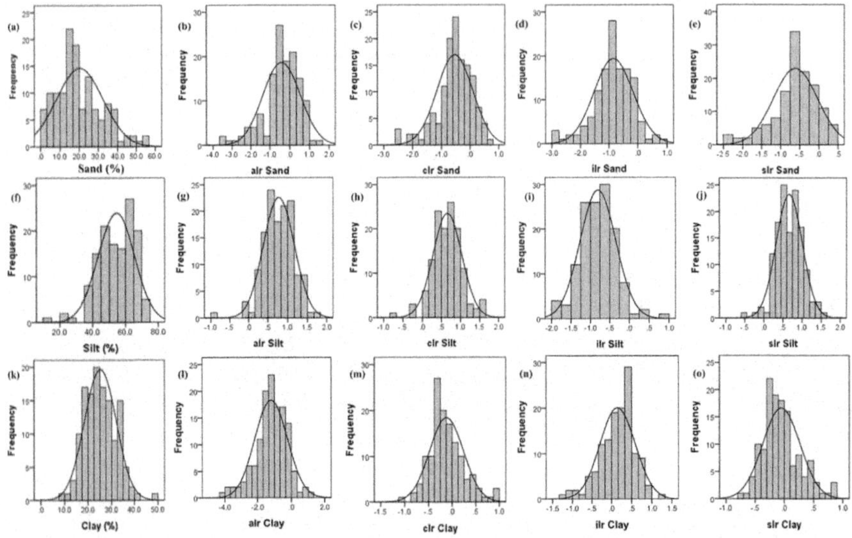


**Fig. 5** The comparisons of the histograms of the raw and alr, clr, ilr and slr transformed data for soil sand (**a–e**), silt (**f–j**) and clay (**k–o**) particle-size fractions of the 133 samplings

#### 10.2.2.2 The Geostatistical Analysis of the Original and Log-Ratio Transformation Data

Geostatistical analysis of the raw and log-ratio transformation data for soil PSF was conducted to compare the differences of the spatial structure among the raw and the four log-ratio transformation data (Fig. 6).

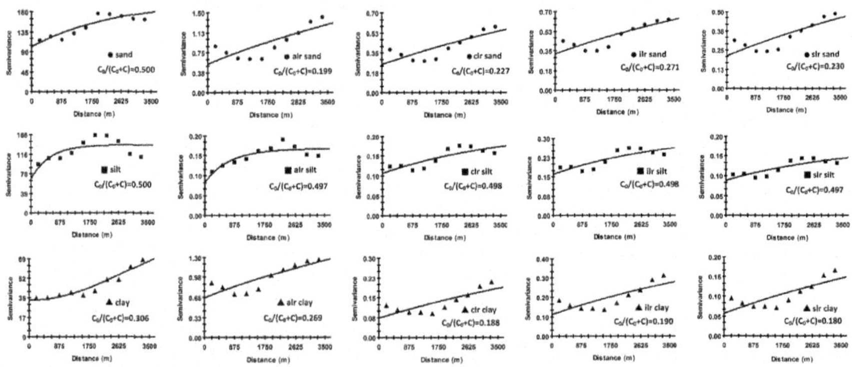


**Fig. 6** The semivariograms of the raw data and alr, clr, ilr and slr transformed data

#### 10.2.2.3 The Comparisons of the Accuracy for Maps by CK and Log-Ratio Kriging Methods

The ME, RMSE, AD, STRESS and RR were used to evaluate the performance of the five methods (Table 3). MEs of the five methods are close to 0 for all cases, indicating that they are all relatively unbiased in interpolating the soil PSF.

Figure 7 shows the distributions of soil texture types presented in USDA texture triangle and we have counted the correct number for soil texture types of predicted soil samples from different interpolators.

**Table 3** The comparisons of the accuracy for maps by CK, OK_alr, OK_clr, OK_ilr and OK_slr

| Methods | ME (%) | | | RMSE (%) | | | AD | STRESS | RR (%) |
|---|---|---|---|---|---|---|---|---|---|
| | Sand | Silt | Clay | Sand | Silt | Clay | | | |
| CK | 2.13 | −1.44 | 0.69 | 9.27 | 7.67 | 4.17 | 0.45 | 0.60 | 58.65 |
| OK_alr | 2.17 | −1.53 | −0.64 | 10.59 | 8.68 | 4.86 | 0.53 | 0.63 | 50.38 |
| OK_clr | 2.07 | −1.47 | −0.60 | 9.35 | 8.21 | 5.31 | 0.49 | 0.64 | 52.63 |
| OK_ilr | 2.09 | −1.50 | −0.60 | 9.51 | 8.19 | 5.69 | 0.50 | 0.67 | 53.38 |
| OK_slr | 2.13 | −1.43 | −0.70 | 10.05 | 8.54 | 5.47 | 0.52 | 0.68 | 51.88 |

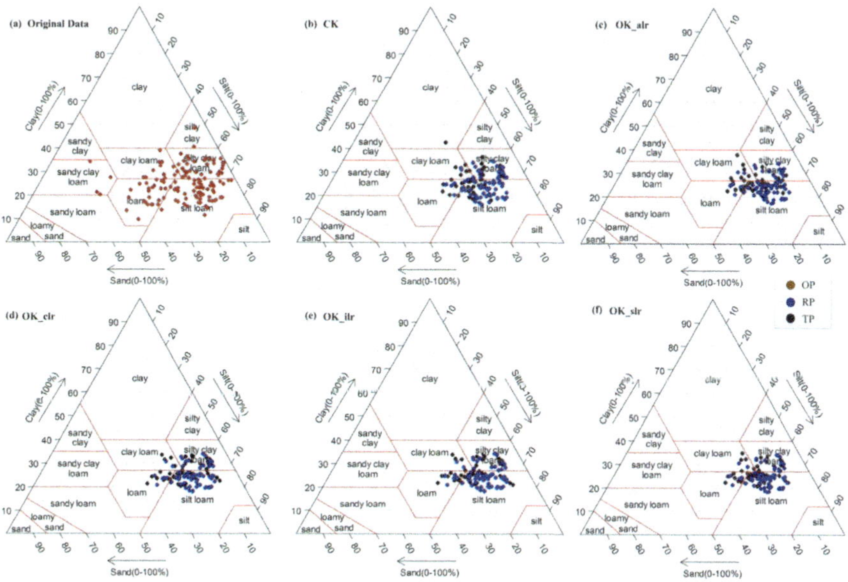


**Fig. 7** The distributions of soil texture types presented in USDA texture triangle

#### 10.2.2.4 The Comparisons of the Maps from Different Interpolators

The interpolation maps for soil PSF by CK are significantly different from maps by log-ratio kriging methods in detail, while the four maps by log-ratio kriging methods display similar patterns (Figs. 8, 9, and 10). For sand, higher values (29–35%) mainly distribute in the central region, and lower values (7–13%) are located in the northwest of the study area. For silt, higher values (64–70%) locate in northeast of the study area, and lower values (42–48%) appear in the central region. For clay, higher values (32–38%) are mainly in the northwest and lower values (14–20%) are in the eastern regions. CK has a wider range of medium values than the log-ratio kriging methods, and the maps produced by log-ratio kriging methods have seriously fluctuated because they cannot guarantee the transformed values meeting optimally unbiased results.

#### 10.2.2.5 The Differences of Maps by CK and the Log-Ratio Kriging Methods

The results of different methods by making differences between log-ratio kriging methods and CK were compared in Figs. 11, 12, and 13. The log-ratio kriging methods have higher values compared with CK for sand and clay in most areas.

The overestimated values of sand generated by log-ratio kriging are mainly distributed in the central regions at higher altitude where coniferous forests located. For silt, the overestimated values mainly distribute in the northwest and some central areas where are sunny slope areas with lower altitude (less than 2850 m) and less slope (less than 30). For clay, the differences show the values of log-ratio kriging methods are slightly less than the values generated by CK (only in partial eastern area), indicating that CK and the log-ratio kriging methods have similar performance for the components with higher values.

#### 10.2.2.6 The Analysis for the Performance of Different Interpolators Combined with Environmental Factors

Close relationship between soil PSF and environment features during the long-term development of geochemical process (Zhang 2013). In general, the soil particle size is smaller with the decrease of elevation, affecting the distributions of sand, silt, and clay fractions (Meul and Van Meirvenne 2003).

Lower values of sand were mainly distributed in steppe area of grassland with sunny slopes, but higher values of sand were mainly appeared in the forestland. The soil PSF in the steppe areas were generally overestimated, which were opposite in most forest land areas. Meanwhile, OK_clr and OK_ilr methods had lower predicted error than the methods of OK_alr and OK_slr in most shady slopes areas, where were the main distribution area of the middle value range of sand.

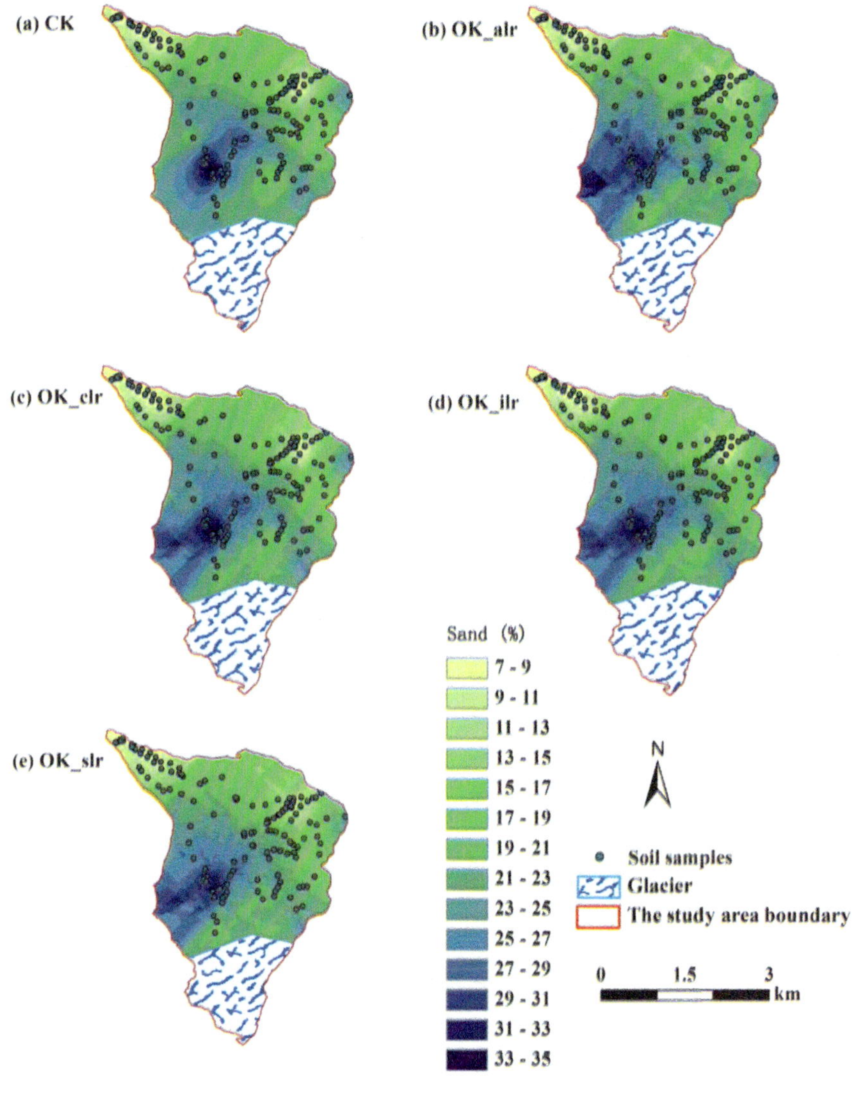


**Fig. 8** The interpolation maps of sand fractions by **a** CK, **b** OK_alr, **c** OK_clr, **d** OK_ilr and **e** OK_slr

All log-ratio kriging methods significantly overestimated lower silt in shady areas and underestimated higher values in the steppe area. Because of lower values of clay, the regularity was not so significant. Log-ratio kriging methods underestimated the values especially in lowlands, where clay fractions were relatively high and overestimated the values in the coniferous forest with lower values of clay.

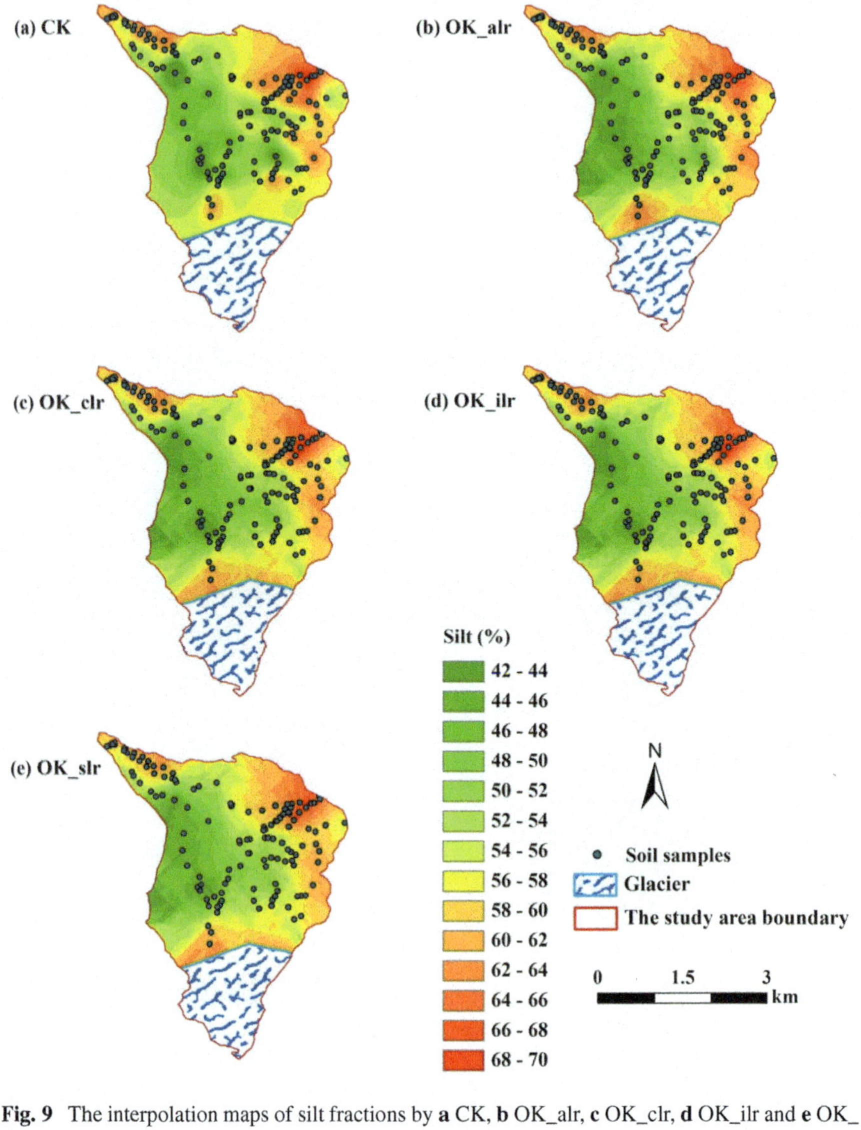


**Fig. 9** The interpolation maps of silt fractions by **a** CK, **b** OK_alr, **c** OK_clr, **d** OK_ilr and **e** OK_slr

## 10.2.3 Summary

CK and log-ratio kriging methods were powerful and effective methods for mapping compositional data like soil PSF. The performance of CK was better than log-ratio kriging methods.

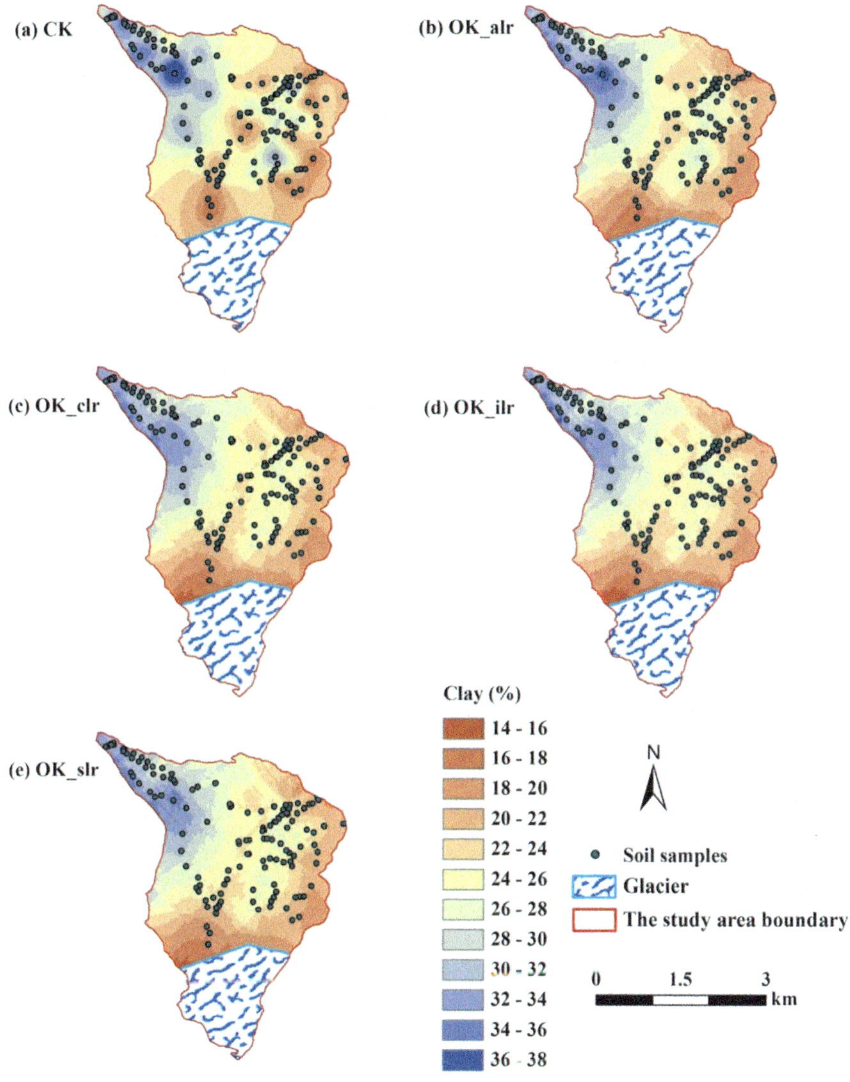


**Fig. 10** The interpolation maps of clay fractions by **a** CK, **b** OK_alr, **c** OK_clr, **d** OK_ilr and **e** OK_slr

Alr and ilr conversions are subject to the effects of composition arrangement, and the changes directly affected the data transformation and the normality, influencing the interpolation accuracy directly.

The spatial patterns of soil PSF are related to environmental variables including terrain, land use types, vegetation types, so ancillary data should be combined with the interpolators for compositional data to improve interpolation performance in the future.

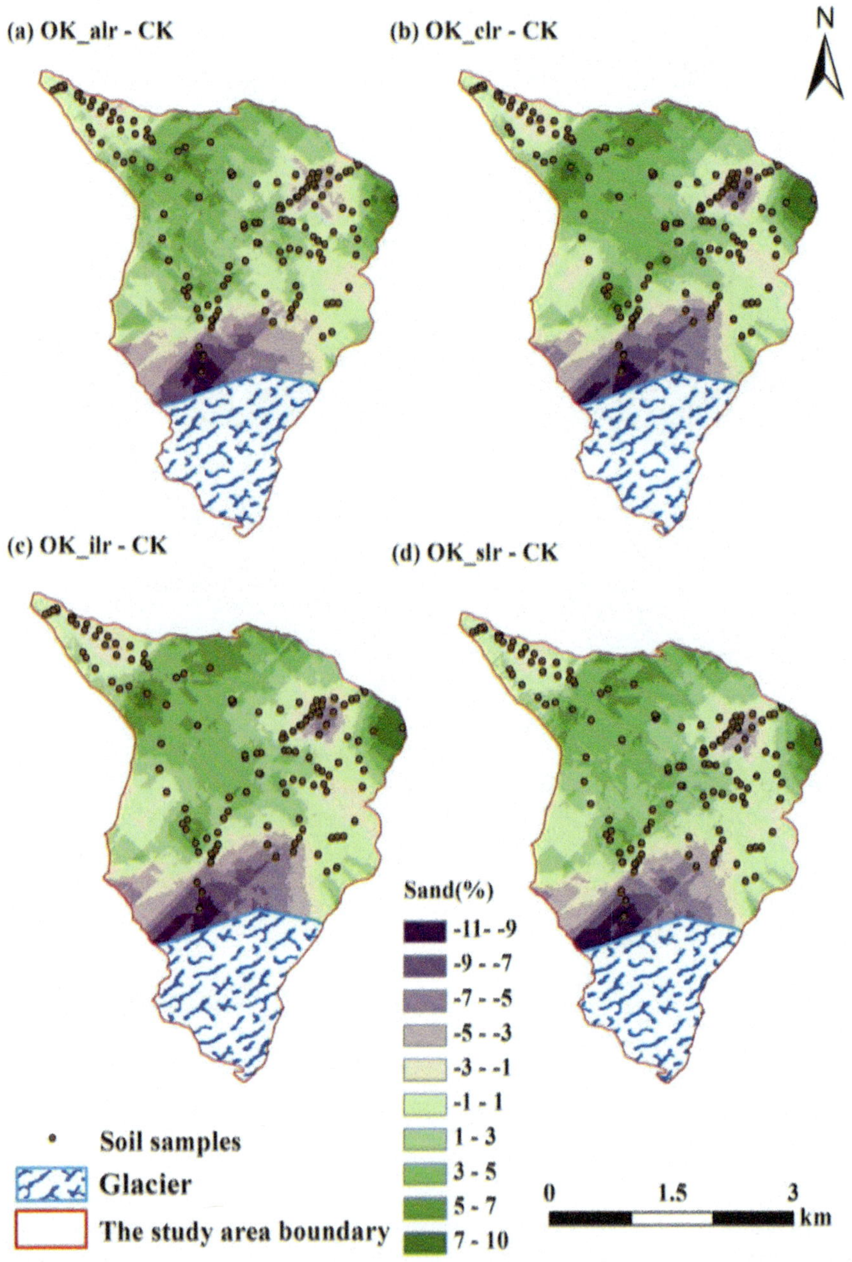


**Fig. 11** The difference of predicted sand fractions between log-ratio kriging methods and CK. **a** OK_alr—CK, **b** OK_clr—CK, **c** OK_ilr—CK and **d** OK_slr—CK

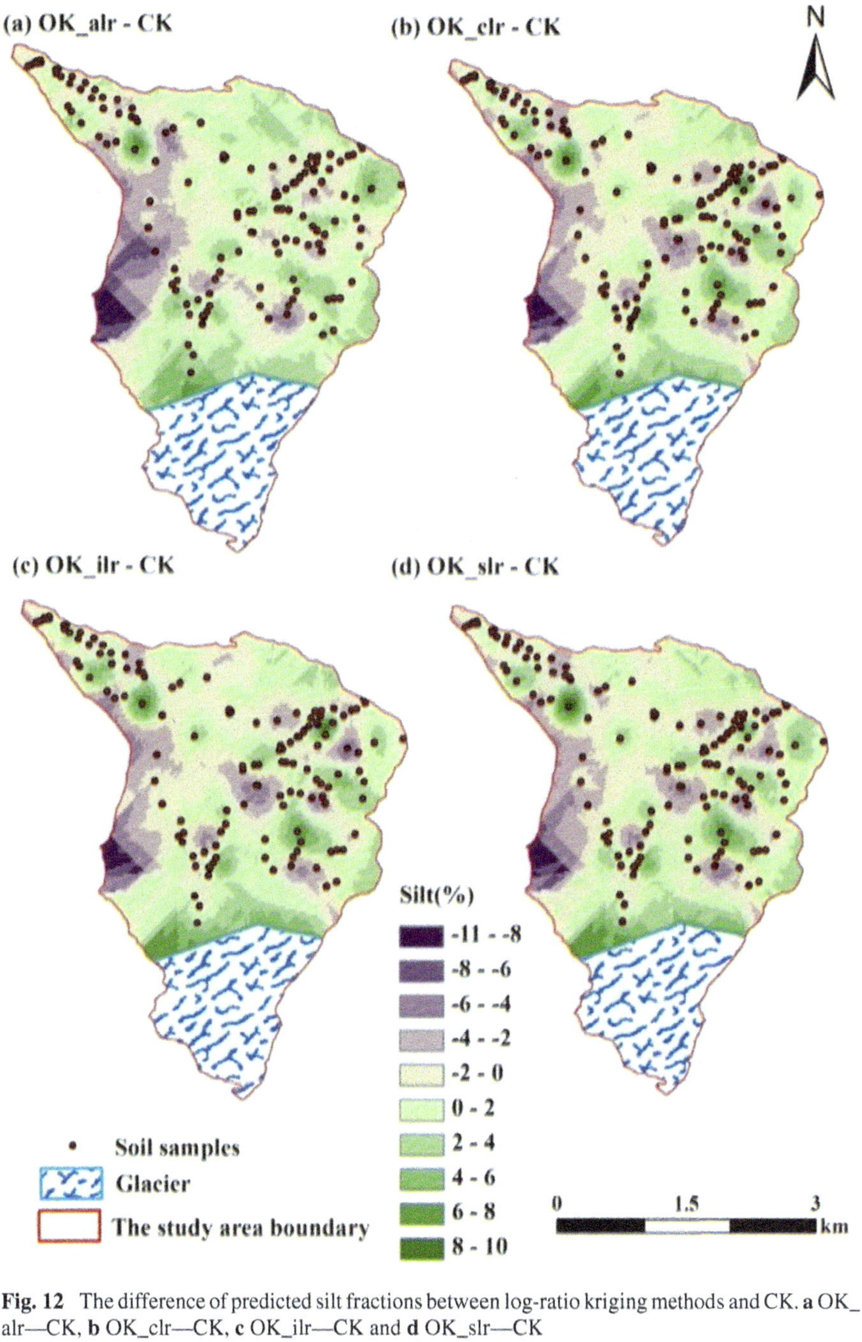


**Fig. 12** The difference of predicted silt fractions between log-ratio kriging methods and CK. **a** OK_alr—CK, **b** OK_clr—CK, **c** OK_ilr—CK and **d** OK_slr—CK

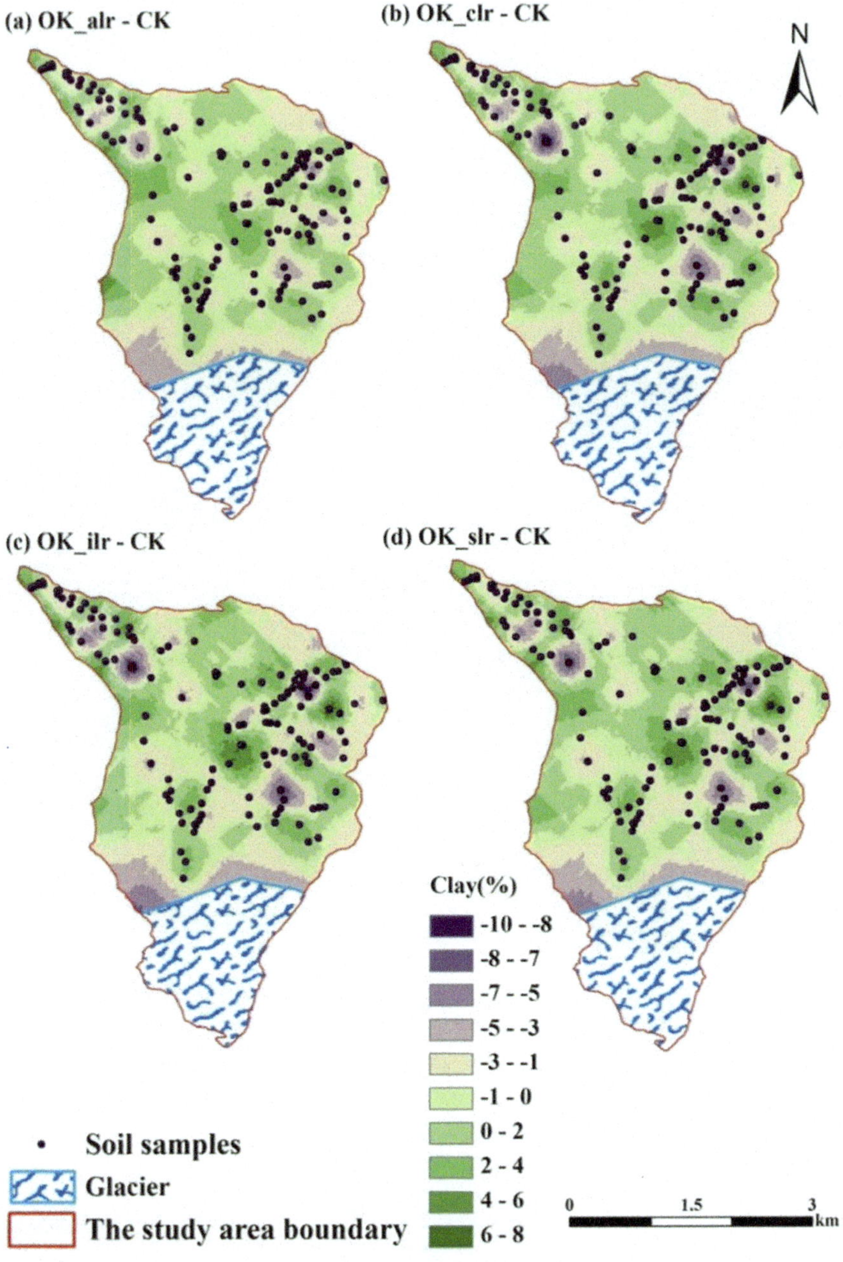


**Fig. 13** The difference of predicted clay fractions between log-ratio kriging methods and CK. **a** OK_alr—CK, **b** OK_clr—CK, **c** OK_ilr—CK and **d** OK_slr—CK

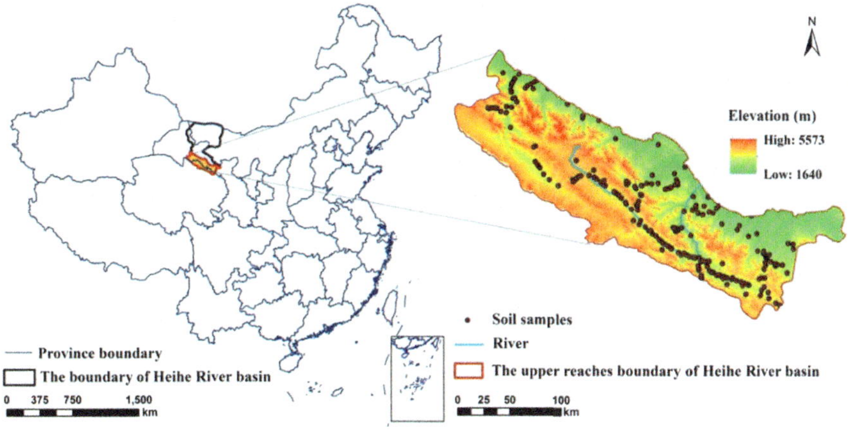


**Fig. 14** The geographical location of the upper researches of Heihe River basin, China

## 10.3 A Case of Improved Accuracy of Soil Particle-Size Fraction Mapping: Robust Variogram Estimation Combined with Isometric Log-Ratio Transformation

### *10.3.1 Data and Methods*

#### 10.3.1.1 The Upper Researches of Heihe River

The study area is mainly distributed in the Qilian mountain region, which is in the northern area of the Tibetan Plateau covering an area of 32,000 km$^2$ (Fig. 14). The elevation varies from 1640 to 5573 m (Fig. 14), and the climate is cold and damp, with an annual mean temperature of < 4 °C and an average annual rainfall of approximately 350 mm. The main vegetation types are meadow and steppe (Fig. 14b), and the main forest types are Qinghai spruce and Qilian juniper. The main land use types include grassland and forest land (Fig. 15b). The aspect and slope are shown in Fig. 15c and Fig. 15d. The main soil types in the soil genetic classification of China (Shi et al. 2004) are frigid felty soil, cold desert soil, gray-brown desert soil and castanozems (Fig. 15e).

#### 10.3.1.2 Soil Samples

We collected a total of 262 soil samples in different vegetation types, soil types, land use types, in the upper reaches of Heihe River basin (Fig. 15), and recorded the position information using the global position system (GPS). We mixed five topsoil (0–20 cm) samples for each soil sample to represent a 5 m × 5 m square, and also collected a parallel sample at each sampling site. To reduce the sampling

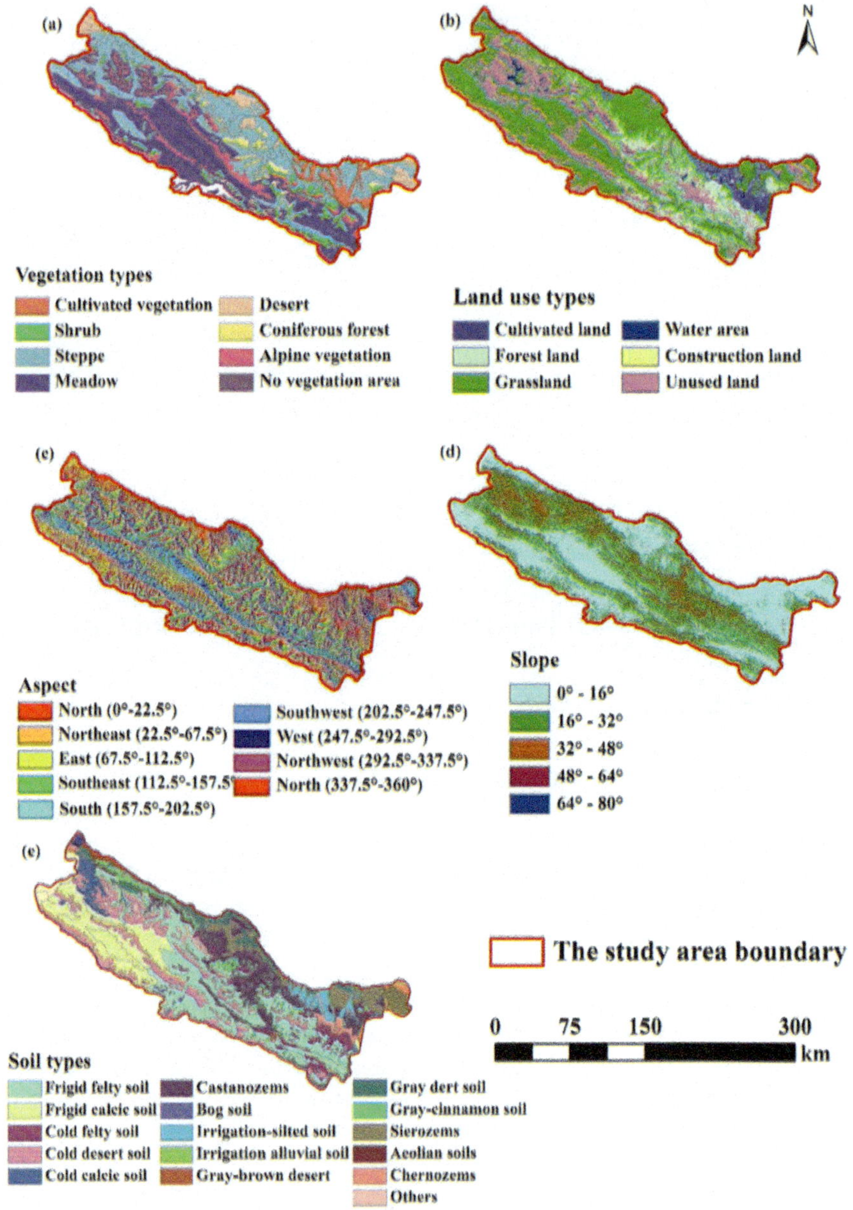


**Fig. 15** The spatial distributions of **a** vegetation types, **b** land use types, **c** aspect, **d** slope and **e** soil types in the upper reaches of Heihe River basin

noise and nugget variance effects, we used the average of two samples as the final measurement. In addition, we removed the plant debris and stones at the sampling points. After sieving and air drying, we used approximately 30 g of each sample for soil PSF analysis using Malvern Mastersizer 2000 laser diffraction particle size analyzer. Additionally, some of the soil PSF data were from the Science Data Center of Cold and Arid Regions (West DC) in China.

#### 10.3.1.3 Interpolators

There are some methods for mapping soil compositional data have been proposed.

(i) **Log-ratio kriging and log-ratio CK**

The log-ratio kriging methods, which are log-ratio transformations combined with ordinary kriging (OK), include ALR_OK (additive log-ratio transformation + kriging), CLR_OK (centered log-ratio transformation + kriging), ILR_OK (isometric log-ratio transformation + kriging) and SLR_OK (symmetry log-ratio transformation + kriging).

(ii) **Log-ratio COK**

Cokriging, which is widely used in geostatistics (Akpa et al. 2014), has been a predominant technique in mapping soil properties (Odeh et al. 1997). The log-ratio transformations combined with COK form the log-ratio cokriging methods with a standard variogram estimator (ALR_COK, CLR_COK, ILR_COK and SLR_COK). For the log-ratio cokriging methods, the semi-variograms are calculated with the same lag classes, and the models are then fitted by weighted least-squares. A lmc function of gstat was fitted to estimate the variograms for COK.

(iii) **Log-ratio RCOK**

In addition, the standard estimator of the variogram in log-ratio cokriging is sensitive to the outliers, which may result in the overestimation of the variogram and eventually lead to errors in the interpolators. In order to solve this problem, robust variogram estimators were combined with Log-ratio cokriging (log-ratio RCOK). So, log-ratio RCOK has been proposed. The log-ratio transformations combined with a robust variogram estimator (ALR_RCOK, CLR_RCOK, ILR_RCOK, and SLR_RCOK).

The problems in the previous study are that the performances of these methods were relatively uncertain, and only a few methods were used to compare their performance in previous studies. Moreover, previous studies did not consider the specific data characteristics when mapping soil PSF.

The methods used in our study included log-ratio kriging, log-ratio cokriging and CK combined with the standard variogram estimator. In addition, for the interpolators of CK and the log-ratio cokriging methods, we introduced the robust variogram estimator to weaken the effect of the outliers. There are 14 methods in the system (Table 4).

**Table 4** The method system of the interpolators

| Variogram estimators | Interpolators | | |
|---|---|---|---|
| | Log-ratio transformations combined with ordinary kriging (OK) (log-ratio kriging methods) | Log-ratio transformations combined with cokriging (COK) (log-ratio cokriging methods) | Compositional kriging (CK) |
| $\hat{\gamma}_{u,v}(h)$ | ALR_OK, CLR_OK, ILR_OK, SLR_OK | ALR_COK, CLR_COK, ILR_COK, SLR_COK | Log-ratio CK |
| $\hat{\gamma}^{M}_{u,v}(h)$ | – | ALR_RCOK, CLR_RCOK, ILR_RCOK, SLR_RCOK | Log-ratio RCK |

#### 10.3.1.4 Validation

A total of 196 points were selected randomly from the dataset as interpolation data, and the remaining 66 points were used as validation data. This process was repeated 10 times. The mean error (ME), root mean square error (RMSE), Aitchison distance (AD), and STRESS were used to verify the accuracies of the different interpolation methods. $AD_{ij}$ is the distance measured between two observed components. The equations for the validation index AD and STRESS are as follows:

$$AD = \left(\sum_{k=1}^{D}\left(\ln\frac{Z_k}{(\prod_{k=1}^{D} Z_k)^{1/D}} - \ln\frac{Z_k^*}{(\prod_{k=1}^{D} Z_k^*)^{1/D}}\right)^2\right)^{\frac{1}{2}} \tag{12}$$

$$STRESS = \left(\frac{\sum_{i<j}\left(AD_{ij} - AD_{ij}^*\right)^2}{\sum_{i<j} AD_{ij}^2}\right)^{\frac{1}{2}}, \tag{13}$$

where, $Z$ is the observed value, $Z^*$is the predicted value, $n$ is the number of validation samples, $D$ is the number of dimensions, and $AD_{ij}$ and $AD^*{}_{ij}$ are the ADs between the observed soil PSF and the predicted soil PSF at sites $i$ and $j$, respectively.

## 10.3.2 Results

### 10.3.2.1 The Descriptive Statistics for the Original and Log-Ratio-Transformed Soil PSF Data

The estimations for the center of soil PSF show that the means of the original soil PSF are different from the medians (Table 5). For example, sand has a lower median (17.06%) than mean (19.61%), while silt and clay have higher medians (silt, 67.44%; clay, 14.51%) than means (silt, 66.84%; clay, 13.55%). The mean values of sand from different log-ratio-transformed methods (18.09% for ALR, CLR and ILR; 18.12% for SLR) are similar and closer to the median (17.06%). However, the mean silt and clay contents of the log-ratio transformations are not closer to the median contents than the means of the raw data.

The descriptive statistics indicate that the log-ratio-transformed data have relatively small MAD values, and are more more stable than the raw data (Table 6). For sand and silt, SLR has the lowest MAD values (sand 0.33, silt: 0.18). For clay, the MAD of ILR (clay: 0.23) is the most stable among different transformation methods. The MADs of the silt from the transformed data are the smallest among the three components, indicating that the log-ratio transformations (except for ILR_silt) improve the silt distributions better than the sand and clay distributions. The original soil PSF values do not follow a normal distribution according to the statistic of *P* (significant at the level of 0.05). The transformed values of sand have lower skewness values (ALR_sand: 1.2, CLR_sand: 0.79, ILR_sand: – 0.37, SLR_sand: 0.82) than the original sand (1.82). In addition, all transformed values of sand (except ALR_sand) follow the normal distributions and have significant *P* values (CLR_sand: 0.12, ILR_sand: 0.7, SLR_sand: 0.11). For silt and clay, the skewness values of the transformed data are higher than those of the original data and the values of p are not satisfactory. These results are consistent with the performance of the means in Table 6. For octile skew, most values (except the values for original silt, CLR_sand, ILR_sand, and SLR_sand) are out of the range of [−0.2, 0.2], indicating that the outliers affected the distribution of most of the log-ratio-transformed data. However, the outliers may not be unambiguously wrong and represent an important phenomenon to the mapping. Thus, the robust variogram estimator should be considered in the interpolators to weaken the effects of outliers.

**Table 5** The center of the raw and log-ratio-transformed data of soil PSF

| Components | Mean | Median | Mean-ALR | Mean-CLR | Mean-ILR | Mean-SLR |
|---|---|---|---|---|---|---|
| Sand (%) | 19.61 | 17.06 | 18.09 | 18.09 | 18.09 | 18.12 |
| Silt (%) | 66.84 | 67.44 | 71.52 | 71.52 | 71.52 | 71.55 |
| Clay (%) | 13.55 | 14.51 | 10.39 | 10.39 | 10.39 | 10.33 |

Table 6 Description statistics for the original and log-ratio-transformed data

| Components | MAD | Min | Max | Skew | Kurt | Octile skew | k-s, p |
|---|---|---|---|---|---|---|---|
| Sand | 6.96 | 2.45 | 84.33 | 1.82 | 5.67 | 0.22 | 0.02 |
| Silt | 5.44 | 10.57 | 95.87 | −0.50 | 2.74 | −0.06 | 0.01 |
| Clay | 3.81 | 0.12 | 37.84 | −0.22 | −0.24 | −0.34 | 0.00 |
| ALR_sand | 0.62 | −2.54 | 6.37 | 1.20 | 2.28 | 0.27 | 0.01 |
| ALR_silt | 0.23 | 0.45 | 6.30 | 1.65 | 1.52 | 0.73 | 0.00 |
| ALR_clay | 0.47 | −3.67 | 2.08 | 0.37 | 1.74 | −0.02 | 0.70 |
| CLR_sand | 0.34 | −1.84 | 2.42 | 0.79 | 1.78 | 0.06 | 0.12 |
| CLR_silt | 0.19 | −0.45 | 2.70 | 1.08 | 1.14 | 0.54 | 0.00 |
| CLR_clay | 0.28 | −3.94 | 0.70 | −1.51 | 2.14 | −0.51 | 0.00 |
| ILR_sand | 0.34 | −1.47 | 2.59 | −0.37 | 1.74 | 0.02 | 0.70 |
| ILR_silt | 0.35 | −4.83 | 0.86 | −1.51 | 2.14 | −0.51 | 0.00 |
| ILR_clay | 0.23 | −0.55 | 3.30 | 1.08 | 1.14 | 0.54 | 0.00 |
| SLR_sand | 0.33 | −1.74 | 2.38 | 0.82 | 1.79 | 0.05 | 0.11 |
| SLR_silt | 0.18 | −0.45 | 2.65 | 1.07 | 1.11 | 0.55 | 0.00 |
| SLR_clay | 0.28 | −3.86 | 0.65 | −1.50 | 2.01 | −0.51 | 0.00 |

#### 10.3.2.2 Geostatistical Analysis of the Original and Log-Ratio-Transformed Data

The variogram models of the log-ratio-transformed data described better spatial structure than those of the original soil PSF data (Table 7). The $R^2$ values were significantly improved after log-ratio transformation, and the RSS values of the transformed data were much smaller than those of the original data. The ranges generated from the log-ratio-transformed data significantly increased compared to those of the original data. The nugget to sill ratios ($C_0/(C_0 + C)$) generated from the semi-variograms of the original sand, silt and clay contents varied from 44 to 50%, indicating that the raw data have medium spatial dependence. Similarly, the ratios from the transformed data tended to show higher spatial dependence (ALR: 13–49%; CLR: 12–49%; ILR: 13–50%; SLR sand: 11–16%) than those from the original data. The ratios from the transformed data, especially for sand component, are less than 13%. The SLR transformation has lower nugget to sill ratios for the three components than the other three transformation methods, and the ratios of the three components all achieved high spatial dependence among the log-ratio transformation methods.

The differences in spatial structure of the original and transformed data through analyzing the semi-variograms estimated by the different estimators were compared. Here, we compared the semi-variograms generated from the original data with the representative semi-variograms used in SLR_COK and SLR_RCOK to display the difference intuitively (Fig. 16). This result indicates that the variances are overestimated by the standard estimator because of the effects of outliers in the variograms.

**Table 7** The fitted variogram parameters for the raw and log-ratio-transformed data

| Components | Model | Nugget | Sill | $C_0/(C_0 + C)$ | Range | R2 | RSS |
|---|---|---|---|---|---|---|---|
| Sand | Exp | 16.40 | 143.30 | 0.44 | 34,200 | 0.68 | 1511 |
| Silt | Gau | 23.50 | 149.60 | 0.50 | 14,415 | 0.57 | 1716 |
| Clay | Exp | 6.30 | 54.10 | 0.50 | 19,800 | 0.53 | 119 |
| ALR_sand | Exp | 1.05 | 2.38 | 0.13 | 503,100 | 0.93 | 0.045 |
| ALR_silt | Exp | 1.06 | 2.13 | 0.49 | 810,900 | 0.75 | 0.065 |
| CLR_sand | Exp | 0.19 | 0.38 | 0.12 | 170,700 | 0.87 | 0.003 |
| CLR_silt | Exp | 0.03 | 0.26 | 0.48 | 20,700 | 0.74 | 0.001 |
| CLR_clay | Exp | 0.41 | 0.83 | 0.49 | 456,000 | 0.88 | 0.009 |
| ILR_sand | Exp | 0.04 | 0.30 | 0.13 | 26,100 | 0.74 | 0.003 |
| ILR_silt | Exp | 0.61 | 1.28 | 0.50 | 496,200 | 0.89 | 0.019 |
| SLR_sand | Exp | 0.18 | 0.37 | 0.11 | 169,200 | 0.88 | 0.003 |
| SLR_silt | Exp | 0.03 | 0.26 | 0.16 | 21,000 | 0.74 | 0.001 |
| SLR_clay | Exp | 0.40 | 0.80 | 0.12 | 435,000 | 0.88 | 0.009 |

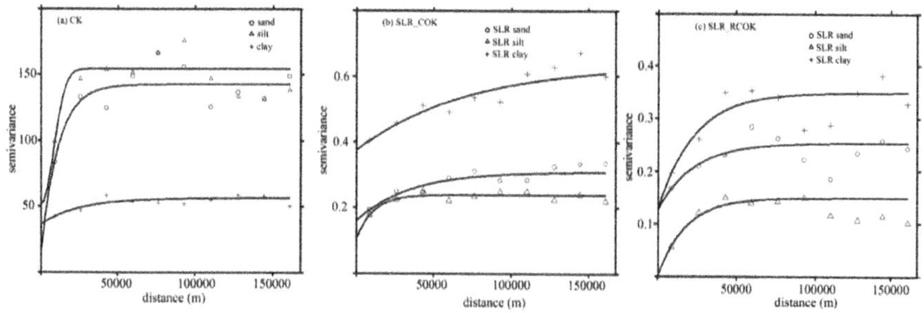


**Fig. 16** The variograms used in **a** original data, **b** SLR_COK and **c** SLR_RCOK

#### 10.3.2.3 Comparisons of the Accuracies of Maps Generated Using Different Interpolators

##### (i) Log-ratio kriging methods

The accuracy indices (ME, RMSE, AD and STRESS) for different interpolators show that. ILR_OK has the most excellent performance, followed by SLR_OK and CLR_OK (Table 7). For sand and silt, ILR_OK has the lowest RMSE (12.00% and 12.43%, respectively). For clay, SLR_OK has the lowest RMSE (7.69%). CLR_OK and SLR_OK have similar AD and STRESS values due to the similar transformed equations of the two transformations. For silt and clay, the accuracy performances of SLR_OK are slightly better than those of CLR_OK.

(ii) **Log-ratio cokriging methods**

For the log-ratio cokriging, ILR_COK is more accurate than the other log-ratio cokriging methods with the lowest RMSE (sand, 10.84%; silt, 11.59%; clay, 7.35%), AD (0.78), and STRESS (0.73). CLR_COK has the lowest accuracies for sand and silt, and ALR_COK has a weak performance for clay, with a RMSE of 7.50%. For different log-ratio cokriging methods combined with the robust variogram estimator, the accuracy performances are better than those of log-ratio cokriging methods. ILR_RCOK also has the best performance with the lowest RMSE (sand, 10.50%; silt, 11.24%; clay, 7.32%), AD (0.76), and STRESS (0.70). CLR_RCOK and SLR_RCOK have similar performances for AD (CLR_RCOK: 0.80; SLR_RCOK: 0.79) and the same STRESS (0.73).

(iii) **Compositional kriging methods**

The log-ratio RCK has better performance than the log-ratio CK method based on RMSE (sand: 10.98%, silt: 11.82% and clay: 7.40%), AD (0.79) and STRESS (0.73) values. The performance of log-ratio CK combined with the robust variogram estimator is similar to the performance of log-ratio cokriging combined with a robust estimator for mapping soil PSF. For clay, the accuracy was improved more than for sand and silt according to the RMSE values (Table 8).

**Table 8** The comparisons of the accuracy for maps by different methods

| Methods | ME (%) | | | RMSE (%) | | | Mean AD | STRESS |
|---|---|---|---|---|---|---|---|---|
| | Sand | Silt | Clay | Sand | Silt | **Clay** | | |
| ALR_OK | 1.93 | −5.22 | 3.29 | 12.77 | 13.39 | 7.87 | 0.89 | 0.88 |
| CLR_OK | 1.13 | −4.03 | 2.37 | 12.22 | 12.69 | 7.80 | 0.87 | 0.88 |
| ILR_OK | 0.96 | −3.32 | 2.36 | 12.00 | 12.43 | 7.72 | 0.84 | 0.86 |
| SLR_OK | 0.54 | −3.12 | 2.58 | 12.25 | 12.49 | 7.69 | 0.86 | 0.87 |
| ALR_COK | 1.53 | −4.46 | 2.94 | 11.08 | 11.90 | 7.50 | 0.81 | 0.77 |
| CLR_COK | 1.57 | −4.49 | 2.92 | 11.13 | 11.91 | 7.48 | 0.81 | 0.77 |
| ILR_COK | 1.45 | −3.73 | 2.27 | 10.84 | 11.59 | 7.35 | 0.78 | 0.73 |
| SLR_COK | 0.99 | −3.52 | 2.47 | 11.09 | 11.67 | 7.41 | 0.80 | 0.75 |
| ALR_RCOK | 1.47 | −3.76 | 2.35 | 10.85 | 11.58 | 7.49 | 0.78 | 0.71 |
| CLR_RCOK | 1.49 | −3.99 | 2.51 | 10.99 | 11.85 | 7.47 | 0.80 | 0.73 |
| ILR_RCOK | 1.18 | −3.67 | 2.46 | 10.50 | 11.24 | 7.32 | 0.76 | 0.70 |
| SLR_RCOK | 1.01 | −2.95 | 1.85 | 10.94 | 11.51 | 7.34 | 0.79 | 0.73 |
| Log-ratio CK | 1.48 | −4.38 | 2.91 | 11.09 | 11.97 | 7.66 | 0.83 | 0.78 |
| Log-ratio RCK | 1.41 | −4.20 | 2.72 | 10.98 | 11.82 | 7.40 | 0.79 | 0.73 |

#### 10.3.2.4 The Prediction Performance of Soil Texture Types from Different Interpolators

We used the right ratio (RR) of predicting soil texture types to estimate the prediction performances for different soil texture types (Table 9). ILR combined with COK (91.57%) or RCOK (90.04%) has a higher RR, followed by ALR_RCOK (90.04%). In contrast, ALR_COK and ALR_OK, which are both combined with ALR transformation have the lowest accuracies in the log-ratio COK and log-ratio OK methods, respectively. Log-ratio RCK (87.36%) have a relatively better performance than log-ratio CK (85.06%).

Furthermore, we compared the prediction performances of each soil texture type among different interpolators shown in the United States Department of Agriculture (USDA) texture triangle (Fig. 17). Here, the soil texture types from the original soil PSF and the predicted soil PSF generated from ILR_OK, ILR_COK and ILR_RCOK were selected to represent different performances of the methods for predicting soil texture types. The results of the ILR-transformed method combined with the robust variogram estimator are much closer to the original soil texture types than the results of the other methods. Log-ratio CK, ALR_OK and ALR_COK provided accurate predictions for almost all silt loam (SiLo) samples (Table 9). However, the prediction results from these three methods overestimated the clay content and underestimated the sand content, resulting that the predicted types of soil texture were concentrated in the SiLo. For loam (Lo) soils, ILR_COK and ILR_RCOK had relatively more accurate results, indicating that these two methods are optimal for predicting sand in

**Table 9** The prediction of soil texture types generated from different interpolators

| Methods | Sum | SiLo | Lo | SiClLo | LoSa | SaLo | Si | RR (%) |
|---|---|---|---|---|---|---|---|---|
| Original data | 262 | 214 | 10 | 3 | 2 | 4 | 29 | – |
| ALR_OK | 213 | 212 | | | | | 1 | 82.00 |
| CLR_OK | 232 | 209 | | | | | 23 | 89.66 |
| ILR_OK | 228 | 206 | 2 | | 1 | 1 | 18 | 88.12 |
| SLR_OK | 232 | 209 | | | | | 23 | 89.66 |
| ALR_COK | 211 | 211 | | | | | | 80.46 |
| CLR_COK | 215 | 203 | | | | | 12 | 83.14 |
| ILR_COK | 237 | 208 | 4 | | 1 | 1 | 23 | 91.57 |
| SLR_COK | 216 | 201 | 1 | | | | 14 | 82.38 |
| ALR_RCOK | 232 | 204 | 3 | | | | 25 | 90.04 |
| CLR_RCOK | 218 | 199 | 1 | | | | 18 | 84.67 |
| ILR_RCOK | 232 | 200 | 5 | | | | 27 | 90.04 |
| SLR_RCOK | 220 | 201 | 1 | | | | 18 | 85.44 |
| Log-ratio CK | 220 | 211 | | | | | 9 | 85.06 |
| Log-ratio RCK | 225 | 197 | 1 | | | | 27 | 87.36 |

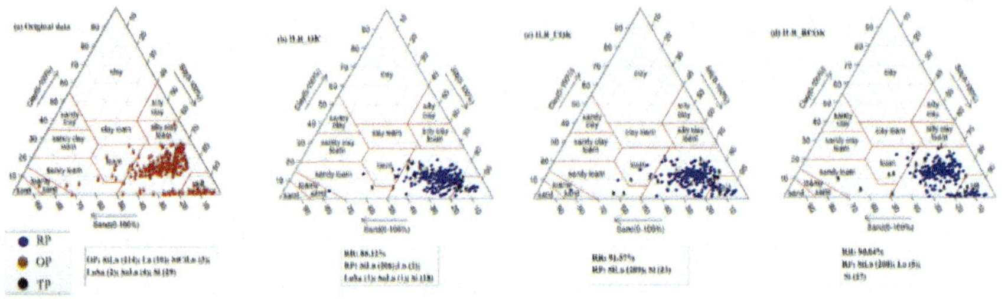


**Fig. 17** The soil texture types for the 262 points presented in USDA texture triangle. The soil components of sand, silt and clay are from **a** original data, **b** ILR_OK, **c** ILR_COK, and **d** ILR_RCOK. SiLo: Silt Loam, SiClLo: Silty Clay Loam, ClLo: Clay Loam, SaClLo: Sandy Clay Loam, SiCl: Silty Clay. OP: observed points; RP: right points mean that the predicted types of soil texture for the points are in the same types with the original data; TP: the total points of soil texture types generated by interpolators

the range of 24–52%, silt in the range of 28–50% and clay in the range of 7–27%. For loamy sand (LoSa) and sandy loam (SaLo) soils, ILR_OK and ILR_COK has satisfactory results with relatively better accuracy when the sand content was higher and the silt content was lower. For silt (Si) soils, the logratio RCK and ILR_RCOK methods have the best performance compared to the other methods, with a relatively better accuracy performance for low sand (< 18%) and clay (< 12%) contents. In addition, the results of log-ratio CK and log-ratio kriging show relatively concentrated distribution characteristics due to the overestimation and underestimation effects.

#### 10.3.2.5 Comparisons of the Maps from Different Interpolators

The maps generated by log-ratio kriging methods display similar patterns (Figs. 18, 19, and 20). The higher values for sand (> 50%) and lower values for silt (< 50%) are mainly distributed in the steppe area with frigid calcic soil and at elevations above 3300 m near the Tuolai River in the southwestern part of the study area. Lower values for sand (0–10%) and higher values for silt (80–95%) are mainly observed in the steppe area with sierozems and elevations below 3000 m in the mid-northern and southeastern regions of the study area. For clay fractions, higher values (22–30%) are mainly distributed in meadow and grassland regions with elevations above 3500 m in the mid-eastern region. Lower values (0–6%) are concentrated mainly in the steppe area with castanozems and gray-cinnamon soil and at elevations below 3500 m in the mid-northern and northwestern regions. Maps generated from CLR_OK and SLR_OK are more detailed than those generated from ILR_OK and ALR_OK. The maps generated from ILR_OK theme closely correspond with the original data regarding the maximum value, minimum value, and range of values.

For the distributions of soil PSF, the higher and lower values generated from the log-ratio cokriging methods are similar to the maps of log-ratio kriging methods. The maps of the log-ratio cokriging methods, especially for the clay components,

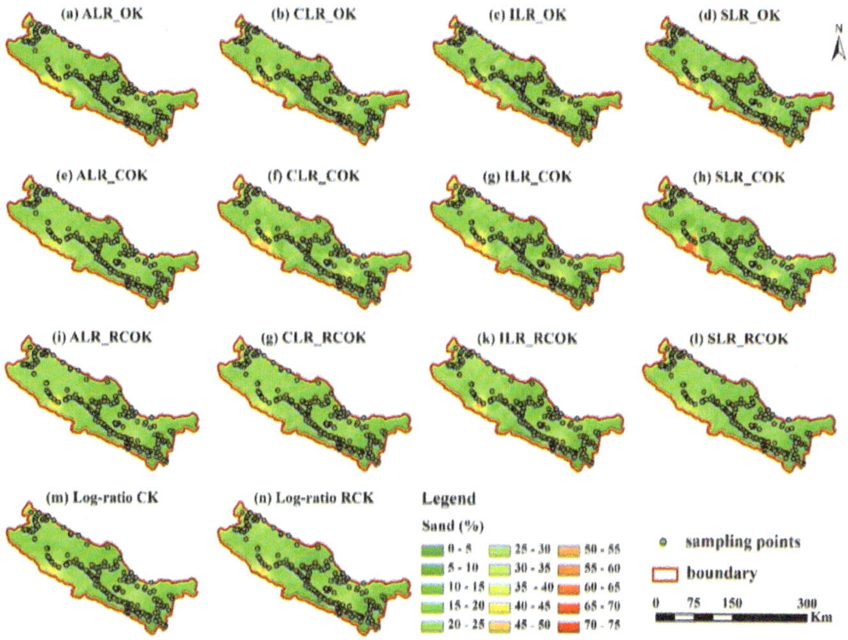


**Fig. 18** The interpolation maps of sand fraction

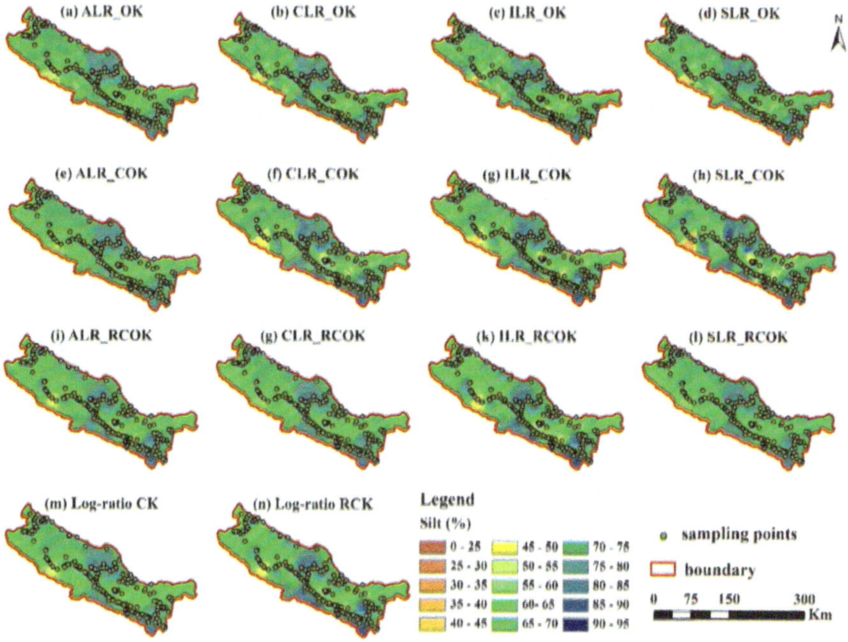


**Fig. 19** The interpolation maps of silt fraction

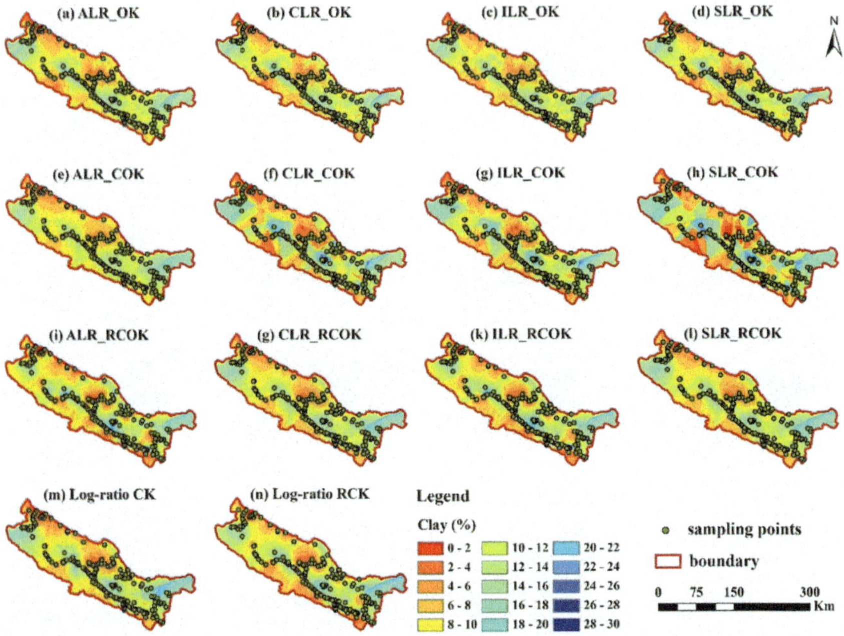


**Fig. 20** The interpolation maps of clay fraction

are relatively smooth. Meanwhile, the ranges of clay values (1.09–26.56%) generated by ILR_COK are closer to those of the original data distribution. The interpolation maps of the log-ratio cokriging methods combined with the robust variogram estimator are smoother than those with the standard variogram estimator. All the maps have significant graphic gradient features and are more consistent with the spatial distribution characteristics of the raw data. The ranges of the mapping values generated by ALR_RCOK (sand: 5.28–62.98%, clay: 0.44–28.12%) and ILR_RCOK (silt: 18.91–95.17%, clay: 0.82–25.93%) are much closer to the original data distribution.

The interpolation maps of log-ratio RCK are smoother and more gradual than those from log-ratio CK. In addition, the mapping results of log-ratio RCK are closer to the original data regarding the ranges of sand (4.8–59.56%) and silt (28.35–92%).

### 10.3.3 Summary

This is the first time that different CK methods, and different log-ratio transformation methods combined with different interpolators and different variogram estimators have been comprehensively compared and systematically analyzed relative to the data characteristics, which could provide guidance for choosing a suitable interpolator for mapping soil PSF.

ILR_RCOK would significantly improve the mapping performance especially for transformed data with many outliers.

The interpolators combined with ancillary data including continuous variables (climate data, electrical conductivity, elevation, soil bulk density, other indicators from remote sensing, etc.) and categorical variables (soil type, land use type, vegetation type, parent material, etc.) should be developed for the interpolation of soil PSF to achieve more accurate maps based on environmental information especially in larger study areas.

## 10.4 A Case of Mapping Soil Compositional Data Combined with Environmental Information

### *10.4.1 Data*

#### 10.4.1.1 The Heihe River

The Heihe River basin (HRB; 97°60′–102°30′E, 37°43′–42°40′N) is situated in the northwest of China, covering the Inner Mongolia Autonomous Region, including Gansu and Qinghai provinces, and is the second largest inland river basin in China with an area of 146,700 km$^2$ (Fig. 21a). The elevation ranges from 669 to 5573 m (Fig. 21b). For the upper reaches of the HRB, the mean annual precipitation is 350 mm, the annual mean temperature ranges from − 5 to 4°C, and the annual average evaporation is 1000 mm. For the middle reaches of the HRB, the mean annual precipitation is between 50 and 250 mm, the annual average evaporation increases from 2000 (east) to 4000 mm (west), and the mean annual temperature ranges from 2.8 to 7.6 °C. The lower reaches of the HRB are situated in Ejin Banner on the Alxa Plateau, which has an arid desert climate with an annual precipitation of under 50 mm and an annual average evaporation of above 3500 mm; the mean annual temperature ranges from 8 to 10 °C.

#### 10.4.1.2 The Vegetation

The vegetation in the upper reaches of the HRB (Fig. 21c) is influenced by hydrothermal conditions from the southeast to the northwest. The main vegetation types are alpine vegetation (4000–5000 m), the alpine meadow vegetation belt (3000–4000 m), alpine shrub meadow (3200–3800 m), the mountain forest meadow belt (2400–3200 m), the mountain grassland belt (1800–2400 m), and the desert base belt (less than 1800 m). The main vegetation types in the middle and lower reaches of the HRB are relatively fewer, and the shrub and steppe are mainly located in the area near the lower reaches of the Heihe River.

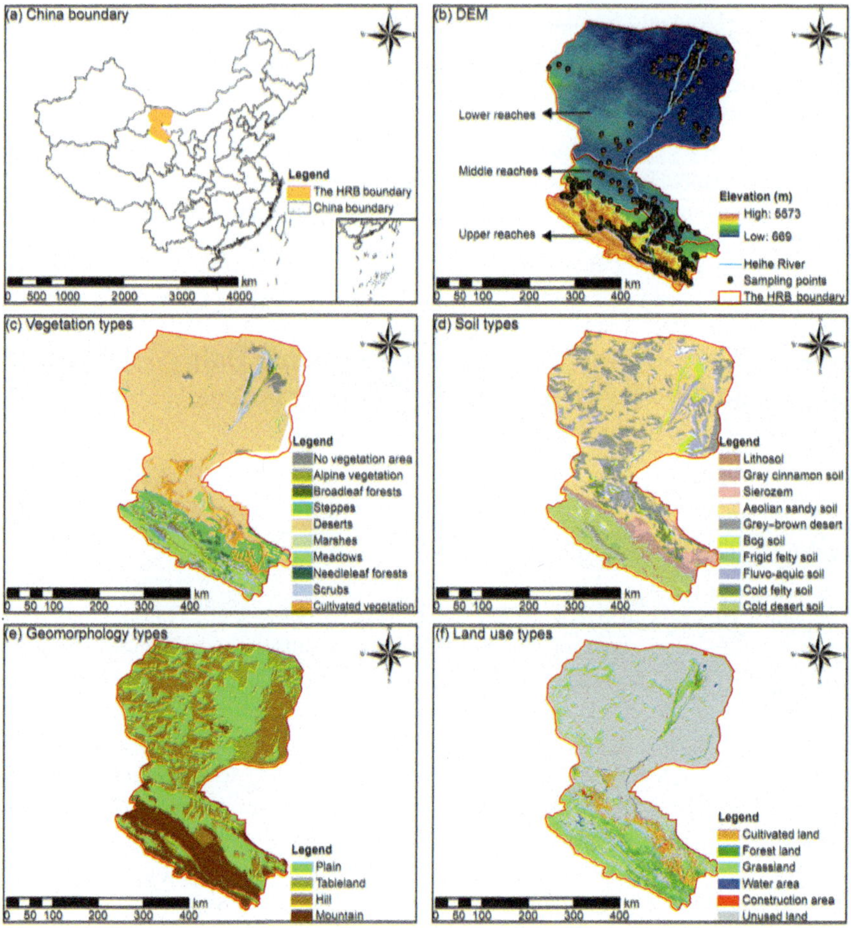


**Fig. 21** The HRB boundary and DEM

#### 10.4.1.3 The Main Soil Types

The main soil types (Fig. 21d) are frigid desert soils (higher than 4000 m), alpine meadow soil and alpine steppe soil (3600–4000 m), grey cinnamon soil and Chernozem (3200–3600 m), Sierozem and grey cinnamon soil (2600–3200 m), grey cinnamon soil (2300–2600 m), and Sierozem (1900–2300 m) in the upper reaches of the HRB. The main soil types in the middle reaches of the HRB are aeolian sandy soil, frigid frozen soil, and grey brown desert soil. The main soil types in the lower reaches of the HRB are aeolian sandy soil, grey–brown desert soil (northwest), and Lithosol (northeast).

#### 10.4.1.4 The Main Geomorphology Types

The main geomorphology types in the upper reaches of the HRB are modern glaciers, alpine, hilly, and intermountain basin (Fig. 21e). Narrow plains are distributed in the middle reaches of the HRB. In the lower reaches, the main types of geomorphology are hilly (northwest), plain, sandy land, and platform (east), as well as a flood plain located in the area near the Heihe River.

#### 10.4.1.5 The Main Land Use Type and Soil Sampling

The main land use type in the upper reaches, middle reaches, and lower reaches of the Heihe River were forest land and grassland, cultivated land, and unused land respectively (Fig. 21f). The water area and construction area were mainly distributed near the river in the middle reaches of the HRB.

A total of 640 soil sampling points was collected in the HRB, including 392 samples from the upper reaches and 248 samples from the middle and lower reaches of the Heihe River basin. The distribution of the soil types, vegetation types, the digital elevation model (DEM), and the geomorphology types of the HRB were considered in the soil sample collection, in terms of their location and proportion, in order to obtain samples that were more representative of soil PSFs using limited soil samples. Purposive sampling was used as the sampling strategy to collect soil samples and to characterize the spatial variability of soil PSFs. In this strategy, sample sites were chosen based on the variability of soil formation factors, which represented the heterogeneity of the soil PSFs in the HRB, such as the distribution of climate, categorical maps, and so on. Due to the complicated soil types and vegetation types in the middle and upper reaches of the HRB, there were more soil sampling points in these areas. In contrast, fewer samples were collected in the lower reaches because of the relatively similar vegetation types. To reduce the noise effect of soil samples, the average of three to five mixed topsoil (0–20 cm) samples for each soil sample and its parallel sample was used as the final measurement. The global position system (GPS) information and related environmental covariates were recorded. Subsequently, the samples were dried, analysed, and measured for soil PSFs (approximately 30 g of each sample). Soil PSFs were analysed using a Malvern Panalytical Mastersizer 2000 laser diffraction particle size analyser (the average measurement error is less than 3%).

#### 10.4.1.6 The Environmental Covariates

The environmental covariates, such as topographic variables, remote sensing variables, climate and position variables, soil physicochemical variables, and categorical maps, are related to the distributions of the soil PSFs. System for Automated Geoscientific Analysis (SAGA) GIS (Conrad et al. 2015) was used to compute the topographic variables from the DEM, including the slope, aspect, convergence

**Table 10** The environmental covariates

| The environmental covariates | 26 variables |
|---|---|
| Topographic variables | Elevation, slope, aspect, convergence index, general curvature, plane curvature, profile curvature, and valley depth |
| Remote sensing variables | Normalized difference vegetation index (NDVI), the brightness index (BI), and the soil adjusted vegetation index (SAVI) |
| Climate and position variables | The mean annual precipitation and the mean annual temperature, latitude and longitude |
| Soil physicochemical variables | Soil organic carbon, saturated water content, field water holding capacity, wilt water content, saturated hydraulic conductivity, and soil thickness |
| Categorical maps | Geomorphology types, soil types, land use types, and vegetation types |

index, general curvature, plane curvature, profile curvature, and valley depth. Remote sensing variables, including the normalized difference vegetation index (NDVI; Huete et al. 2002), the brightness index (BI; Metternicht and Zinck 2003), and the soil adjusted vegetation index (SAVI; Huete 1988), were derived from the Landsat 7 based on band operation. We also collected climate variables such as the mean annual precipitation and the mean annual temperature from the National Meteorological Information Center. Furthermore, latitude and longitude were also considered because of the large area of the HRB. Mean annual surface evapotranspiration data (Wu et al. 2012) were gathered from the NTPDC (http://data.tpdc.ac.cn/zh-hans/, last access: 5 May 2020) as well as soil physicochemical variables, including soil organic carbon, saturated water content, field water holding capacity, wilt water content, saturated hydraulic conductivity, and soil thickness (Yi et al. 2015; Song et al. 2016; Yang et al. 2016). Additionally, categorical maps were also used, such as geomorphology types, soil types, land use types, and vegetation types (Table 10).

## *10.4.2 Methods and Models*

### 10.4.2.1 The Choosing Process of Mapping Methods of Soil Compositional Data

Log-ratio transformations play significant roles in compositional data analysis, and several methods exist, including the additive log-ratio (ALR), centered log-ratio (CLR) (Aitchison 1986), and isometric logratio (ILR) (Egozcue et al. 2003) transformations. These log-ratio methods are widely used for transforming soil PSF data. However, the choice of the transformation method should take into account factors such as the study area scales and model selections during the modeling process. For local-scale study areas, geostatistical models combined with log-ratio transformed data can be sufficient for mapping spatial patterns, as demonstrated in previous study (Wang and Shi 2017). Alternatively, functional compositions combined with the kriging method have also been employed to generate soil particle-size curves (PSCs) (Menafoglio et al. 2014), furnishing an abundance of information. This approach involves the utilization of complete and continuous data rather than discrete data, enabling the extraction of soil PSFs from the predicted soil PSCs (Menafoglio et al. 2016a; b). In middle scale study areas, the presence of outliers may lead to an over estimation of the variogram, resulting in prediction errors (Lark 2000). Hence, it is imperative to consider the use of robust variogram estimators in spatial interpolation to enhance model performance (Lark 2003). Previous research has shown that the application of robust variogram estimators in log-ratio co-kriging significantly improved mapping ac curacy (Wang and Shi 2018). For large-scale study areas, geostatistical models may encounter limitations due to the factors such as the scarcity of soil samples and heightened spatial variation. Consequently, studies have focused on mapping soil PSFs using diverse machine-learning models in conjunction with ancillary data (i.e., environmental covariates). Log-ratio transformed data have been employed successfully on various scales, ranging from a broad basin scale (Zhang et al. 2020) to a national scale (Akpa et al. 2014) and even on a global scale (Hengl et al. 2017).

#### 10.4.2.2 Interpolators

There are four requirements for the spatial prediction of soil PSF, including unbiasedness, non-negativity, constant sum, and a minimum sum of prediction error variances. Due to these requirement of the interpolation of compositional, numerous different interpolators of compositional data have been suggested in soil science, and the most extensively used method has been a combination of log-ratio transformation methods, including the additive log ratio (ALR; Aitchison 1986), the centred log ratio (CLR; Aitchison 1986), the isometric log ratio (ILR; Egozcue et al. 2003), and symmetry log ratio. Soil PSFs have been predicted using kriging (Wang and Shi 2018; Zhang 2013) combined with log-ratio transformation methods.

In addition, the standard estimator of the variogram in log-ratio cokriging is sensitive to the outliers, which may result in the overestimation of the variogram and eventually lead to errors in the interpolators. In order to solve this problem, robust variogram estimators were combined with Log-ratio cokriging (log-ratio RCOK) or compositional krging (RCK). So, log-ratio RCOK and RCK has been proposed.

Here, five machine-learning models–K-nearest neighbour (KNN), multilayer perceptron neural network (MLP), random forest (RF), support vector machines (SVM), and extreme gradient boosting (XGB)—combined with the original data and three log-ratio transformation methods—additive log ratio (ALR), centred log ratio (CLR), and isometric log ratio (ILR)—are used in the study areas.

(i) **K-nearest neighbour**

K-nearest neighbour (KNN) is a simple and nonparametric classifier that is based on using the known instance to label the unknown instance (Cover and Hart 1967). For the test set, K-nearest training set vectors (k) were found based on distance, and the maximum summed kernel densities were computed for classification. Moreover, continuous variables can also be predicted for regression with the average values of K-nearest neighbours. The parameters of KNN contain the maximum value of k (kmax), the distances of the nearest neighbours (distance), and different kernel functions (kernel).

(ii) **Multilayer perceptron neural network**

Multilayer perceptron neural network (MLP), which is one of the most common multilayer feed-forward back-propagation networks (Zhang et al. 2018), was selected to train the artificial neural network (ANN) models due to the rapid operation, the small set of training requirements, and the ease of implementation. MLP neurons can perform classification or regression depending on whether the response variable is categorical or continuous. The MLP has three sequential layers: the input layer, the hidden layer, and the output layer. The resilient back-propagation algorithm was chosen because the learning rate of this algorithm was adaptive, avoiding oscillations and accelerating the learning process (Behrens and Scholten 2006). The range of the data set should be standardized because MLPs operate in terms of a scale from zero to one.

(iii) **Random forest**

Random forest (RF) was developed by Breiman (2001), combining the bagging method (Breiman 1996) with random variable selection, and the principle was to merge a group of "weak learners" together to form a "strong learner". Bootstrap sampling is used for each tree of RF, and the rules to binary split data are different for regression and classification problems. For classification, the Gini index is used to split the data; for regression, minimizing the sum of the squares of the mean deviations can be selected to train each tree model. The benefits of using RFs are that the ensembles of trees are used without pruning. In addition, RF is relatively robust to overfitting. Standardization or normalization is not necessary because it is insensitive to the range of input values. Two parameters should be adjusted for the RF model: the number of trees (ntree) and the number of features randomly sampled at each split (mtry).

(iv) **Support vector machine**

Support vector machine (SVM), proposed by Cortes and Vapnik (1995), is a type of generalized linear classifier that is widely applied to classification and regression problems in soil science (Burges 1998). The main principle of SVM is to classify different classes by constructing an optimal separating hyperplane in the feature space (so-called "structural risk minimization"). Regression problems can also be solved

**Table 11** The method system of soil texture classification and soil PSF interpolation

| Transformation methods | Soil texture classification (direct) | Soil texture classification (indirect) and soil PSF interpolation |
|---|---|---|
| Original data (ORI) | KNN, MLP, RF, SVM, XGB | KNN_ORI, MLP_ORI, RF_ORI, SVM_ORI, XGB_ORI |
| Log ratio transformed data (ALR, CLR, ILR) | – | KNN_ALR, KNN_CLR, KNN_ILR, MLP_ALR, MLP_CLR, MLP_ILR, RF_ALR, RF_CLR, RF_ILR, SVM_ALR, SVM_CLR, SVM_ILR, XGB_ALR, XGB_CLR, XGB_ILR |

by minimization of the structural risk using loss functions (Vapnik 1998) in SVM, which is known as support vector regression. The SVM is more effective in high dimensional spaces. A linear function was selected for SVM as the kernel function in our study. Additionally, cost and gamma are two other parameters that needed to be tuned, as these parameters control the trade-off between the classification accuracy and complexity and the ranges of the radial effect respectively.

(v) **Extreme gradient boosting**

Extreme gradient boosting, put forward by Chen and Guestrin (2016), is an efficient method of implementation for gradient boosting frames, tree learning algorithms, and efficient linear model solvers to solve both classification and regression problems (Chen et al. 2018). Like the boosted regression trees (Elith et al. 2008), it follows the principle of gradient enhancement; however, more regularized model formalization is applied to XGB to control over-fitting, making it perform better in terms of accuracy assessment. The residuals of the first tree can be fitted by the second tree to enhance the model accuracy, and the sum of the prediction of each tree generates the ultimate prediction. There are seven parameters in XGB—the learning rate (eta), the maximum depth of a tree (max_depth), the max number of boosting iterations (nrounds), the subsample ratio of columns (colsample_bytree), the subsample ratio of the training instance (subsample), the minimum loss reduction (gamma), and the minimum sum of instance weight (min_child_weight).

(vi) **The method system of soil texture classification and soil PSF interpolation**

The method system of soil texture classification and soil PSF interpolation was shown as Table 11.

#### 10.4.2.3 Validation

(i) **Validation indicators for soil texture classification**

We used the overall accuracy, kappa coefficients, area under the precision recall curve (AUPRC), and abundance index to validate the performance of different models. The first two indicators were selected to evaluate the overall prediction performance of soil texture types, and the last two were applied to evaluate the performance of each soil texture type.

The overall accuracy represents all samples of all soil texture types correctly classified by machine-learning models, divided by the total number of samples of soil texture types used in the validation. The overall accuracy is defined as follows (Brus et al. 2011):

$$Overall\ Accuracy = \frac{TP + TN}{TP + TN + FP + FN}, \tag{14}$$

where TP, TN, FP, and FN are true positive, true negative, false positive, and false negative respectively. The kappa coefficient demonstrates the agreement between the observed classes and the measured classes, which is calculated based on the confusion matrix; the equation is defined as follows:

$$kappa = \frac{p_o - p_e}{1 - p_e} \tag{15}$$

where $p_o$ is the probability of observed agreement (overall accuracy), and $p_e$ is the probability of agreement when two classes are unconditionally independent. The strength of the kappa coefficients is interpreted in the following manner: 0.01–0.20 represents slight, 0.21–0.40 represents fair, 0.41–0.60 represents moderate, 0.61–0.80 represents substantial, and 0.81–1.00 represents almost perfect (Landis and Koch 1977). The probabilities of different soil texture types (sum to 1) obtained during the training and predicting processes of machine-learning models were selected to calculate the precision and recall, which indicate the extent of identifying positive cases:

$$Recall = \frac{TP}{TP + FN} \tag{16}$$

$$Precison = \frac{TP}{TP + FP} \tag{17}$$

The confusion index (COI) based on prediction probability was calculated to evaluate the uncertainties of machine-learning models of classification (Burrough et al. 1997). The equation was as follows:

$$COI = \frac{\sum_{i=1}^{n}\left[1 - \left(P_{\max,i} - P_{secmax,i}\right)\right]}{n} \tag{18}$$

where $P_{\max,i}$ refers to the maximum value of the probability of soil sampling point i, and Psecmax, i represents the second highest value of the probability of soil sampling point i. A lower COI indicates better model performance. The abundance index was applied to describe the proportion of all soil texture types and well-classified soil texture types in prediction maps and was defined as follows:

$$Abundanceindex = p/t \tag{19}$$

where $p$ is all soil texture types in prediction maps, and $t$ is all soil texture type(s) of soil samples.

(ii) **Validation indicators for soil PSF interpolation**

Five statistical indicators, including the Spearman rank correlation coefficient (RCC), root-mean-square error (RMSE), mean absolute error (MAE), Aitchison distance (AD; Aitchison 1990), and standardized residual sum of squares (STRESS; Martin-Fernandez et al. 2001), were used to validate the methods of soil PSF interpolation. The equations for the validation indicators RCC, RMSE, MAE, AD, and STRESS are as follows:

$$RCC = \rho_{xy}(rank) = \frac{\sigma_{xy(rank)}}{\sigma_{x(rank)}\sigma_{y(rank)}} \tag{20}$$

$$RMSE = \sqrt{\frac{1}{n}\sum_{i=1}^{n}(Y_{i,m} - Y_{i,e})^2} \tag{21}$$

$$MAE = \frac{1}{n}\sum_{i=1}^{n}\left|Y_{i,m} - Y_{i,e}\right| \tag{22}$$

where $Y_{i,m}$, $Y_{i,e}$, and $Y_{i,m}$ are the measured value, the estimated value, and the mean of measured soil PSFs respectively, and n is the number of observations (soil sampling points for validation). $\sigma_{x(rank)}$ and $\sigma_{y(rank)}$ are variances for measured and estimated data respectively. $\sigma_{xy(rank)}$ is covariance. Rank refers to assigning a rank of 1 to the smallest value, a rank of 2 to the next highest value, and so on (Mishra and Datta-Gupta 2018). A higher RCC and lower RMSE and MAE show better model performance.

$$AD = \left[\sum_{i=1}^{D}\left[\log\frac{x_i}{g(x)} - \log\frac{X_i}{g(X)}\right]^2\right]^{1/2} \tag{23}$$

$$STRESS = \left[\frac{\sum_{i<j}(AD_{x,ij} - AD_{X,ij})^2}{\sum_{i<j}(AD_{x,ij})^2}\right]^{1/2} \tag{24}$$

where x is the observed value, X is the predicted value, D is the number of dimensions (for soil PSFs are three), g(x) denotes the geometric mean (x1 ... xD)1/D, and $AD_{x,ij}$ and $AD_{X,ij}$ are the ADs between the observed soil PSFs and the predicted soil PSFs at sites i and j. Both of these parameters show that model performance is better when the values are lower. The standard deviation (SD) of prediction values and the ranges of the 95% confidence interval (CI; Streiner 1996) of the indicators were derived from 30 model runs to assess the model uncertainty.

#### 10.4.2.4 Mapping Soil Particle-Size Fractions: A Comparison of Compositional Kriging and Log-Ratio Kriging Statistical Analysis for the Original and Log-Ratio-Transformed Data

The standard deviation (SD), coefficient of variation (CV), mean value, minimum value (Min), maximum value (Max), median absolute deviation (MAD), skewness (Skew), kurtosis, and the Kolmogorov–Smirnov (k–s) test ($p > 0.05$) were employed for descriptive statistical analysis of the original and log-ratio-transformed data. The means of the log-ratio transformed data are calculated as follows:

(i) transform the data using a log-ratio method,
(ii) calculate the mean values of transformed values (ALRs, CLRs, or ILRs), back-transform the calculated mean values to the initial closed space.

Furthermore, the multivariate median values based on depth measures (Bedall and Zimmermann 1979; Small 1990) were used because of the sum-constraint of compositional soil PSF data. The arithmetic mean of log-ratio transformed data should be back-transformed to the original space. For $X = [X_1, \ldots, X_n]$, the MAD can be calculated:

$$\mathrm{MAD}(X) = \mathrm{median}(|X_i - \mathrm{median}(X)|) \quad (25)$$

### 10.4.3 Results

#### 10.4.3.1 The Descriptive Statistics for the Original and Log-Ratio-Transformed Data of Soil PSFs

For the original data of sand content, the mean (30.64%) was much higher than that of the median centre (26.06%), while silt and clay contents were the opposite, with lower means (silt, 55.79%; clay, 13.57%) than median centers (silt, 59.51%; clay, 14.43%) (Fig. 22). For the log-ratio-transformed data, different log-ratio methods had the same means for sand, silt, and clay. In addition, the means of sand (28.69%) and silt (60.54%) were closer to the median centers of the original data, except for clay (10.78%). With respect to the SD and CV, soil PSF data in the log-ratio geometry had more stability and less variability than the original data. ILR and CLR had the lowest MAD for the first component (0.66) and the second component (0.43) respectively. Although the *P* values of the original and the various log-ratio-transformed data were not significant, log ratios made the data more symmetric according to the skews. All log-ratio-transformed data had lower skews (ALR: 0.77; CLR: 0.88; ILR: − 1.20)

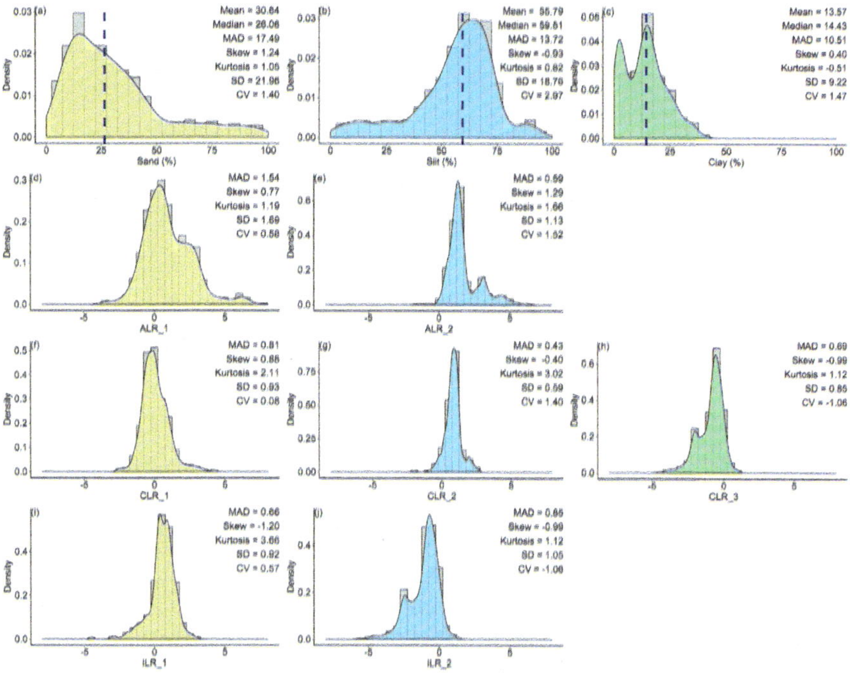


**Fig. 22** The HRB boundary and DEM Descriptive statistical analysis for the original and log-ratio-transformed data for **a** sand, **b** silt, **c** clay, **d** ALR_1, **e** ALR_2, **f** CLR_1, **g** CLR_2, **h** CLR_3, **i** ILR_1, and **j** ILR_2. "MAD" refers to median absolute deviation, "SD" refers to standard deviation, "CV" refers to the coefficient of variation, and "Median" is the multivariate median based on depth measures. ALR and ILR transformed S3 (the simplex) to R2 (the real space), and CLR transformed S3to R3. Blue dashed lines showed the multivariate medians of the original data

than those of the original data (1.24) of the first component. All of the kurtosis values for log-ratio-transformed data were much higher than for the original data.

#### 10.4.3.2 Comparison of the Machine-Learning Models in the Classification of Soil Texture Types

(i) **Comparison of the validation indicators for soil texture classification**

The overall accuracy of all models ranged from 0.613 to 0.636. (Fig. 23a). RF had the highest overall accuracy (0.636) among the five models, followed closely by KNN (0.630) and MLP (0.627). In addition, SVM (0.618) and XGB (0.613) had relatively lower accuracy than the other models. The highest kappa coefficient was generated from MLP (0.242), followed by RF (0.238), XGB (0.229), KNN (0.213), and SVM (0.213) (Fig. 23b). With respect to the confusion indices (COIs), XGB (0.278) delivered the best performance, and RF (0.501) demonstrated the highest confusion of models (Fig. 23c).

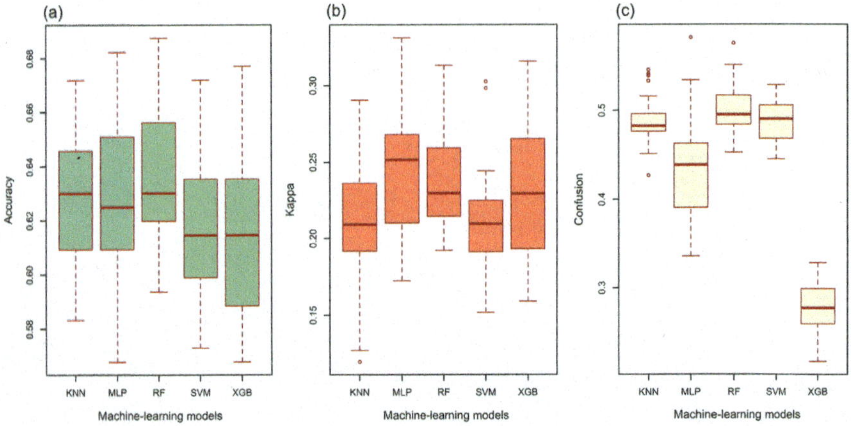


**Fig. 23** The HRB boundary and DEM **a** the overall accuracy, **b** the kappa coefficients, and **c** the confusion indices (COIs) for KNN, MLP, RF, SVM, and XGB

We combined the PRCs of the five machine-learning models to evaluate the performance of predicting each soil texture type using imbalanced data with different samples of each type (Fig. 23). The AUPRCs of the types with fewer positive examples were typically small, especially for SaClLo (only four samples), and showed unsatisfying results, because the lack of soil sampling points made models learn poorly during the training process. In contrast, the soil texture types (Lo, SaLo, SiLo, and SiClLo) with more positive examples delivered superior results to those with fewer positive examples. Moreover, these soil texture types had significant differences in AUPRCs. For example, SiLo, which had the largest number of samples, was the most effective among the nine types. For soil texture types with more samples, RF and XGB performed better. For soil texture classes with less samples according to the AUPRCs, RF and SVM showed better performance (Fig. 24).

(ii) **Comparison of the prediction maps for soil texture classification**

Prediction maps of soil texture types showed quite different spatial distributions in the overall performance of different models (Fig. 25). The abundance indices pointed out that SVM could predict all nine types, KNN and XGB predicted eight of nine types, followed closely by RF (seven of nine types) and MLP (six of nine types). The maps predicted by RF, SVM, and XGB illustrated that the main soil texture types in the northwest of the lower reaches of the HRB were mostly LoSa, while other prediction models produced SaLo. In the upper reaches of the HRB, soil texture types generated from RF were more abundant and more in accordance with the real environment.

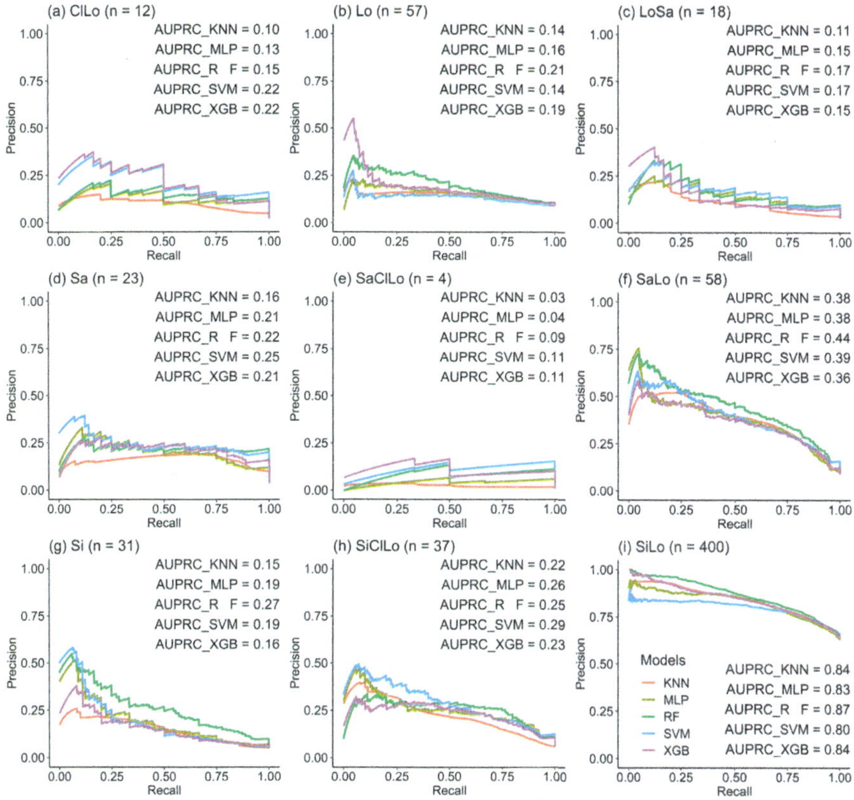


**Fig. 24** The HRB boundary and DEM. The AUPRCs for different machine-learning models in the prediction of each soil texture type: **a** ClLo, **b** Lo, **c** LoSa, **d** Sa, **e** SaClLo, **f** SaLo, **g** Si, **h** SiClLo, and **i** SiLo. "n" denotes the number of sampling points for different soil texture types

#### 10.4.3.3 Comparison of the Machine-Learning Methods Combined with Log-Ratio-Transformed Data in Soil PSF Interpolation

**(i) Comparison of the validation indicators for the interpolation of soil PSFs**

We compared the performance of each machine-learning model using both the original and log-ratio-transformed data. The results showed that the STRESS of the methods using log-ratio-transformed data were superior to the methods using original data (Table 12). The RMSE, MAE, RCC, and AD generated from KNN, MLP, RF, and XGB using original data outperformed the results using log-ratio-transformed data.

By comparison, among different log-ratio-transformed data of the same machine-learning model, ILR and CLR outperformed ALR. KNN_CLR demonstrated the most remarkable performance with the highest RCC and the lowest RMSE and MAE for KNN using the three log-ratio transformation methods. Furthermore, RF and SVM

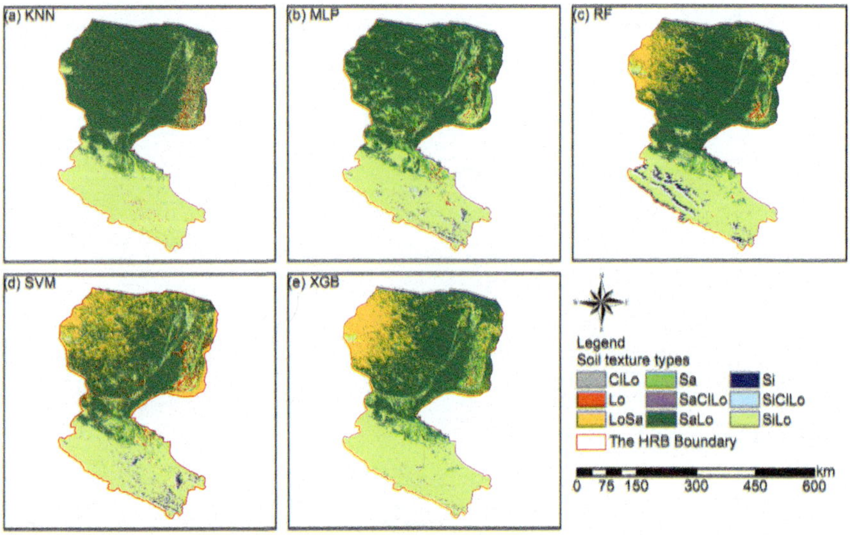


**Fig. 25** The HRB boundary and DEM Soil texture classification prediction maps of different soil texture types for **a** KNN, **b** MLP, **c** RF, **d** SVM, and **e** XGB

generated relatively similar results using CLR- and ILR-transformed data. XGB_ILR showed the best performance with most of the indicators except for RMSE (6.75%) and MAE (5.36%) of clay, and STRESS (0.63). RF had the lowest RMSE and MAE, the highest RCC, and the lowest AD and STRESS for ALR, CLR, and ILR. For original data, RF also outperformed other models.

(ii) **Comparison of the interpolation prediction maps of soil PSFs**

The Prediction maps of soil PSFs using log-ratio-transformed data (ILR) and original data are shown in Fig. 26. The maps generated from ILR-transformed data showed closer ranges to the original soil sampling data in terms of the ranges of sand (0.98–99.66%), silt (0.17–95.87%), and clay (0.03–39.77%), and the texture features were more consistent with the distributions of the real environment. With respect to different machine-learning models, RF and XGB delivered prediction maps that were closer to the range of the distribution of the original data than the methods of KNN, SVM, and MLP.

#### 10.4.3.4 Comparison of Direct and Indirect Soil Texture Classification

(i) **Comparison of the validation indicators for direct and indirect soil texture classification**

The overall accuracy and kappa coefficients of the indirect classification were improved by using log-ratio-transformed data, especially for RF and XGB. ILR

**Table 12** Comparisons of the accuracy of different machine-learning models combined with original and transformed data

| | RMSE (%) | | | MAE (%) | | | RCC | | | AD | STRESS |
|---|---|---|---|---|---|---|---|---|---|---|---|
| | Sand | Silt | Clay | Sand | Silt | Clay | Sand | Silt | Clay | | |
| KNN_ALR | 16.05 | 15.04 | 7.12 | 11.35 | 10.93 | 5.59 | 0.65 | 0.60 | 0.63 | 0.90 | 0.62 |
| KNN_CLR | 15.82 | 14.77 | 7.09 | 11.21 | 10.74 | 5.58 | 0.66 | 0.61 | 0.63 | 0.88 | 0.62 |
| KNN_ILR | 15.82 | 14.82 | 7.14 | 11.22 | 10.84 | 5.60 | 0.66 | 0.61 | 0.63 | 0.88 | 0.64 |
| KNN_ORI | 15.51 | 14.47 | 7.05 | 11.12 | 10.51 | 5.49 | 0.67 | 0.62 | 0.63 | **0.84** | 0.66 |
| MLP_ALR | 15.83 | 15.07 | 7.43 | 11.42 | 11.06 | 5.97 | 0.64 | 0.57 | 0.64 | 0.92 | 0.66 |
| MLP_CLR | 15.84 | 15.07 | 7.41 | 11.45 | 11.05 | 5.96 | 0.64 | 0.57 | 0.64 | 0.92 | 0.66 |
| MLP_ILR | 15.84 | 15.07 | 7.40 | 11.46 | 11.04 | 5.95 | 0.64 | 0.57 | 0.64 | 0.92 | 0.66 |
| MLP_ORI | 15.80 | 14.72 | 6.96 | 11.50 | 10.85 | 5.52 | 0.65 | 0.58 | 0.65 | 0.90 | 0.68 |
| RF_ALR | 15.50 | 14.43 | 6.62 | 10.90 | 10.52 | 5.24 | **0.69** | 0.65 | 0.68 | 0.86 | **0.61** |
| RF_CLR | 15.28 | 14.22 | 6.61 | 10.70 | 10.25 | 5.21 | **0.69** | 0.66 | 0.68 | 0.86 | **0.61** |
| RF_ILR | 15.27 | 14.25 | 6.66 | 10.66 | 10.26 | 5.26 | **0.69** | 0.66 | 0.68 | 0.86 | **0.61** |
| RF_ORI | **15.09** | **13.86** | **6.31** | **10.65** | **9.99** | **5.00** | **0.69** | **0.67** | **0.69** | **0.84** | 0.66 |
| SVM_ALR | 15.66 | 14.59 | 6.76 | 11.66 | 10.88 | 5.34 | 0.66 | 0.57 | 0.66 | 0.88 | 0.66 |
| SVM_CLR | 15.27 | 14.36 | 6.87 | 11.01 | 10.41 | 5.41 | 0.66 | 0.60 | 0.65 | 0.87 | 0.65 |
| SVM_ILR | 15.29 | 14.37 | 6.84 | 10.92 | 10.43 | 5.42 | 0.67 | 0.61 | 0.65 | 0.87 | 0.65 |
| SVM_ORI | 15.30 | 14.38 | 6.92 | 10.94 | 10.32 | 5.43 | 0.67 | 0.61 | 0.66 | 0.87 | 0.67 |
| XGB_ALR | 15.82 | 14.92 | 6.72 | 11.32 | 11.01 | 5.35 | 0.67 | 0.62 | 0.67 | 0.88 | 0.64 |
| XGB_CLR | 15.70 | 14.80 | 6.75 | 10.96 | 10.67 | 5.39 | 0.67 | 0.63 | 0.67 | 0.88 | 0.62 |
| XGB_ILR | 15.45 | 14.57 | 6.75 | 10.91 | 10.52 | 5.36 | 0.67 | 0.62 | 0.66 | 0.88 | 0.63 |
| XGB_ORI | 15.15 | 14.05 | 6.47 | 10.88 | 10.15 | 5.15 | 0.67 | 0.66 | 0.67 | 0.86 | 0.68 |

Bold values denote the best model performance for different indicators

showed the highest overall accuracy among the three log-ratio transformations and also demonstrated the best performance in terms of the kappa coefficients, except for MLP. We compared direct classification with indirect classification and found that the differences in the overall accuracy of direct and indirect classification methods can not be negligible. However, the kappa coefficients were greatly modified using indirect classification compared with direct classification, except for MLP; peculiarly, RF_ILR increased the kappa coefficient to 0.291 (21.3% improvement) and the accuracy remained stable.

(ii) **The prediction performance of soil texture types from different methods**

The distributions of soil texture types using original and ILR-transformed data are illustrated in Fig. 27 using the United States Department of Agriculture (USDA) soil texture triangle. The triangle of the original data of soil PSFs has wider ranges of spatial dispersion than the interpolated data using machine-learning methods. These predictions reveal the properties of aggregating from the sides to the centre of

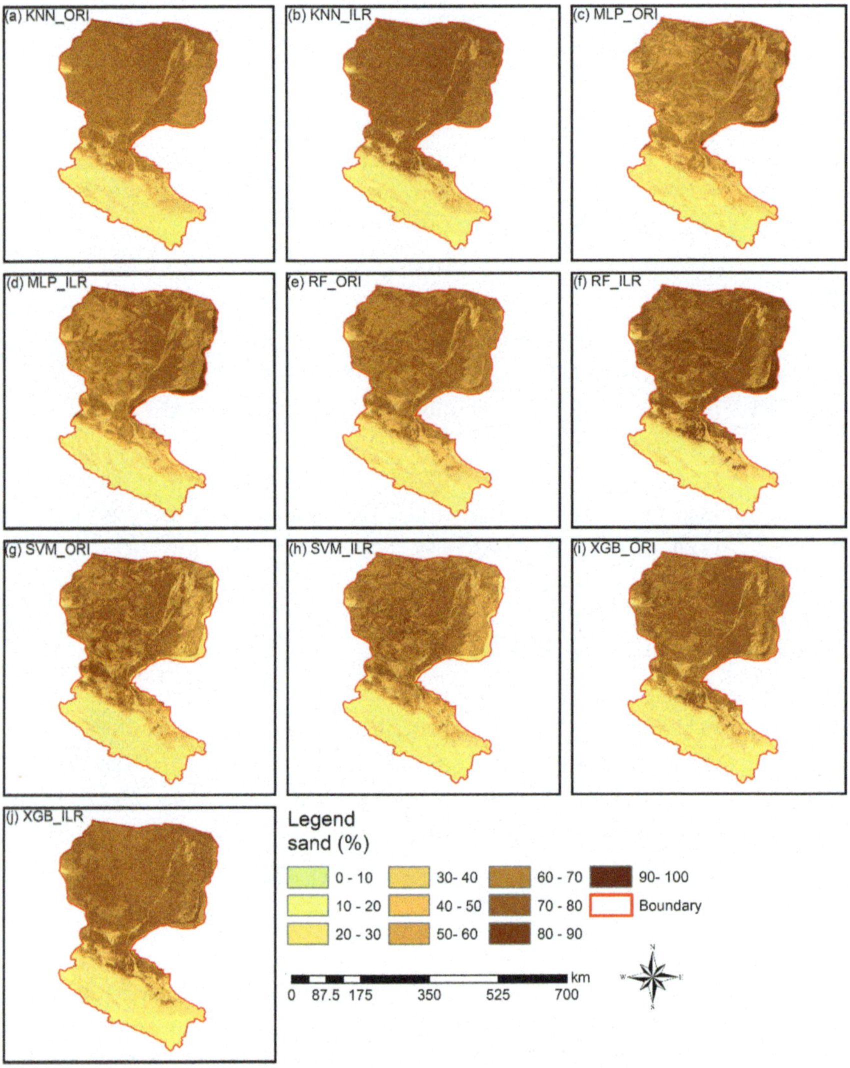


**Fig. 26** The HRB boundary and DEM Prediction maps of the sand fraction. All of the ranges of the prediction maps of sand (approximately 9.0–90.0%) were within the range of the original data (0.98–99.66%). RF_ILR (7.9–94.7%) and XGB_ORI (1.8–92.4%) generated wider output distributions and were relatively closer to the range of the distribution of the original data than other prediction maps, such as KNN_ILR (7.3–88.6%), KNN_ORI (7.8–80.8%), MLP_ILR (8.8–90.8%), MLP_ORI (9.0–90.3%), RF_ORI (9.0–81.0%), SVM_ILR (6.5–85.6%),SVM_ORI (7.3–90.0%), and XGB_ILR (5.0–88.5%)

triangles. For the machine-learning models, RF shows the most dispersed feature in accordance with the original soil PSF data. However, the predictions from models combined with ILR-transformed data are more discrete and more associated with the original soil PSF data than those resulting from ORI methods. The prediction results represent significant differences in the error ratio (yellow symbols, Fig. 27) of the soil sampling points with respect to soil types between the left part (LoSa, SaLo, and Lo) and right part of the triangles (SiLo and Si) for most of the models, especially for KNN and MLP. The log-ratio methods over-calculate the mean value of silt in the process of transformation, so these points are biased to the right of the USDA soil texture triangle based on overall contraction (regression smoothing effects), crossing the classification boundary and turning to other soil texture types. RF_ILR (Fig. 27f) delivers the highest right ratio (RR) among these models, and the classification accuracy is enhanced using the ILR method (83.9%) compared with ORI (81.7%). In the case of other models, the differences between ORI and ILR can be negligible. We also compared the RRs of indirect classification models with those of direct classification, indicating that all RRs of direct classification were higher (KNN, 67.97%; MLP, 75.16%; RF, 100%; SVM, 66.09%; XGB, 81.09%), especially for RF and XGB. However, we removed this evaluation indicator because the same data sets were employed in the processes of training and predicting.

(iii) **Comparison of prediction maps of direct and indirect soil texture classification**

There is some differences between the soil texture maps predicted using original data and log-ratio-transformed data. The classification maps from the machine-learning models combined with the log-ratio-transformed data had more detailed information. The results of machine-learning models using three log-ratio-transformed datasets were similar to the number of predicted types; however, there were significant differences between the results using original data and log-ratio-transformed data. All machine-learning models combined with original data predicted more Lo and SaLo soil texture types but fewer LoSa and Si types (Fig. 28). We also compared the prediction of soil texture types by direct classification with those generated from indirect classification using the same machine-learning models, revealing that different distributions of LoSa existed among them in the lower reaches of Heihe River basin. For the upper reaches, prediction maps of the ILR methods generated more Si and less Lo than the ORI method. Si soil texture types were mainly distributed in the middle and southeast of the upper reaches of the HRB in the predictions combined with ILR methods. For the middle reaches, because more SaLo and less Lo soil texture types were predicted in the middle reaches of the HRB, ILR prediction maps were recommended and were more in line with the real environment than the ORI methods. Furthermore, the predicted soil texture using indirect methods was more abundant than the directly predicted soil texture in the middle reaches (Fig. 28).

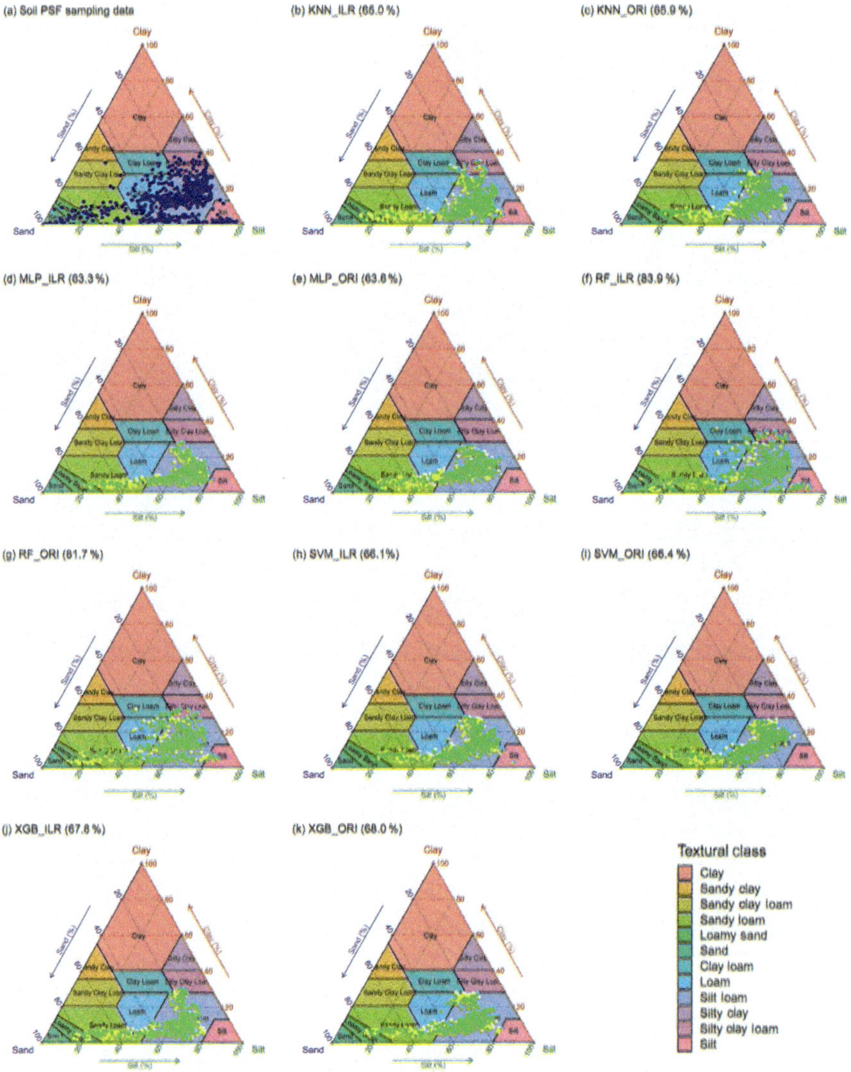


**Fig. 27** The HRB boundary and DEM Soil texture types of 640 soil samples shown using the USDA texture triangle. The results of soil PSFs were generated from **a** soil PSF samples, **b** KNN_ILR, **c** KNN_ORI, **d** MLP_ILR, **e** MLP_ORI, **f** RF_ILR, **g** RF_ORI, **h** SVM_ILR, **i** SVM_ORI, **j** XGB_ILR, and **k** XGB_ORI. Note that the green symbols represent that the predicted and original soil texture classes were the same, whereas the yellow symbols represent the misclassification of the soil texture classes. The predicted right ratios (RRs) of the soil texture classes are given in parentheses after the interpolators above the plots

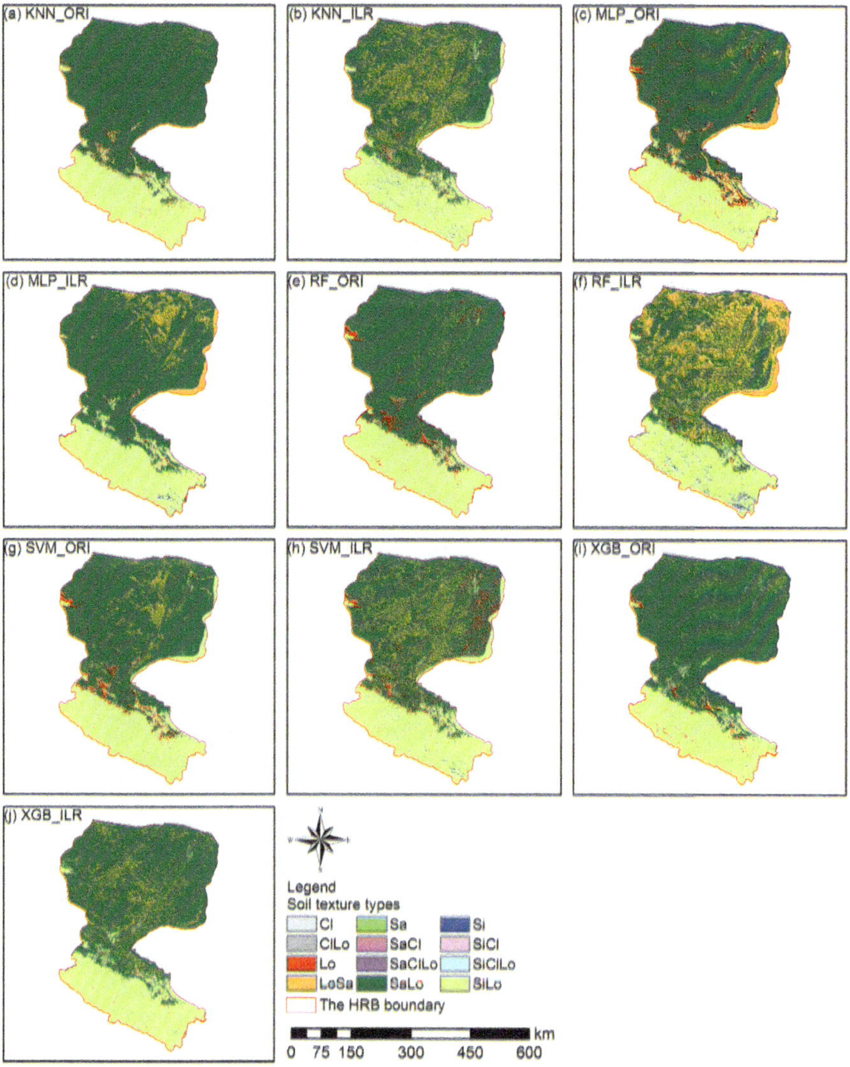


**Fig. 28** The prediction maps of soil texture classification by indirect methods using KNN, MLP, RF, SVM, and XGB with either ILR-transformed data (ILR) or original data (ORI)

(iv) **Comparison of total computing time for each model in soil texture classification and soil PSF interpolation**

We compared the run times of different machine-learning models in soil texture classification and soil PSF interpolation (Fig. 29). Because the run times of the ORI

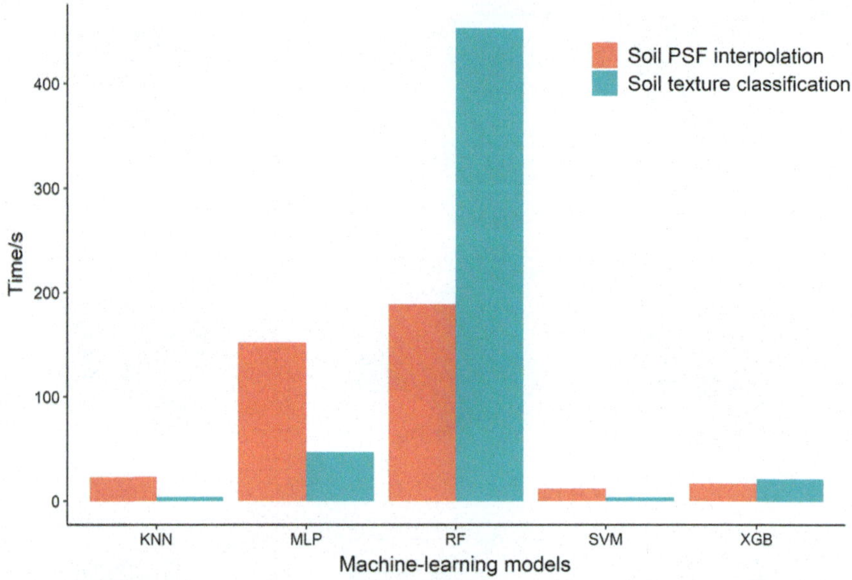


**Fig. 29** Average time spent running the KNN, MLP, RF, SVM, and XGB models 30 times for soil texture classification and soil PSF interpolation

and log-ratio methods were similar, we only selected the ILR for soil PSF interpolation in the comparison. For different models, RFs required the longest time for both classification (453.73s) and interpolation (188.87s). This may cause disadvantage compared with the other models when processing large datasets. KNN (classification, 4.2s; interpolation, 23.6s) and SVM (classification, 4.15s; interpolation, 12.4s) had shorter run times for both classification and interpolation. XGB (classification, 21.6s; interpolation, 17.13s) was much more stable and required less time; the data processes were also simple compared with MLP (classification: 47.28s, interpolation: 152.31s). Moreover, XGB delivered better performance than KNN and SVM in prediction maps, demonstrating that it is an effective way of dealing with large data sets.

## *10.4.4 Discussion*

### 10.4.4.1 The Systematic Comparison of the Five Machine-Learning Models

The range of applicability of the study is limited to independent modelling, i.e. the component-wise approaches. However, joint fractions modelling could lead to different results. We found that tree-based machine-learning models—RF and

XGB—delivered better performance than KNN, MLP, and SVM, which was also concluded by Heung et al. (2016). With respect to the total computing time, RF revealed the longest run time with respect to both the classification and interpolation mode. In addition, regarding trade-offs between the total computing time of the model and the accuracy, XGB was superior to the other four models, reducing the computing time significantly while maintaining acceptable accuracy. In fact, parallel calculations can be automatically executed during the training phase of the XGB model: this is a great advantage when working with large data sets, as the XGB can be more than 10 times faster than the existing gradient boosting models (Chen and Guestrin 2016). Therefore, XGB is recommended due to its speed (although this is at the expense of suboptimal accuracy) when researchers are dealing with large data sets in study areas. Moreover, some joint fractions approaches—compositional kriging (Wang and Shi 2017), high accuracy surface modelling (HASM; Yue et al. 2016, 2020) and the Dirichlet regression (Hijazi and Jernigan 2009)—can consider the multivariate treatment for soil PSFs using a joint model, but machine-learning models are more convenient for combining environmental covariables. For the machine-learning models in our study, KNN, MLP, RF, and SVM also can be applied to multivariate vectors combined with log-ratio methods. For example, the multivariate random forest (MRF) method, which is the extended version of RF, calculates predictions of all output features using a single model (Segal and Xiao 2011).

#### 10.4.4.2 The Systematic Comparison of the Models Using Log-Ratio-Transformed Data and Original Data

Log-ratio transformation methods can open the data and remove the "closure effect", which induces spurious correlation. The opened data can be interpolated into the mapping area, and the results can then be back-transformed using inverse equations. However, in the process of parameter optimization, the optimal parameters of different machine-learning models are obtained using log-ratio-transformed data, which cannot guarantee the most accurate back-transformed results. This is because the values of assessment indicators (e.g. MAEs and RMSEs) will remain stable with limited differences due to the small value range of log-ratio-transformed data. Therefore, when the prediction values of log-ratio methods are back-transformed to the real space, these indicator values will be enlarged.

Due to the contraction of the predicted values (Fig. 29), there were small numbers of predictions beyond the range of the original data values, including the negative predictions using ORI data. Although these few negative predictions can be eliminated by parameter adjustment in our study, there is still a drawback to using ORI data. Among the three log-ratio methods, ILR and CLR were superior to ALR, which can be explained by the fact that ILR and CLR are isometric transformations and they could preserve distances (Filzmoser and Hron 2009). Moreover, ALR has been criticized because the results were affected by the subjective choice of the denominator. In addition, ILR showed slightly better performance than CLR, because the geometric mean composed of all compositions of soil PSFs is the denominator in CLR, and one-to-one mapping of equations and soil PSFs could be implemented. Nevertheless, the sum of the dimensions of CLR is zero, and the problem of collinearity is still present. ILR transformed all of the information into D-1 orthogonal log contrasts (so-called balances) (Egozcue et al. 2003) and overcame the data collinearity and sub-compositional incoherence in CLR by using an appropriate choice of the basis (Egozcue and Pawlowsky-Glahn 2005). Moreover, in the ILR method, multiple sets of ILR-transformed data can be generated by permutations of components (different sequential binary partitions, SBPs) in compositional data, and different choices of ILR balances influenced the model accuracy.

The choice of a specific SBP for compositions is crucial for the intended interpretation of coordinates (Fiserova and Hron 2011). The choice of SBPs can be applied blindly (Fiserova and Hron 2011), can be based on a priori expert knowledge, or can be based on using a compositional biplot (Lloyd et al. 2012), and the best ILR balance also can be chosen using variograms and cross-variograms (Molayemat et al. 2018). The ILR balance chosen in our study was SBP1, because the ILR-transformed data using SBP1 were more symmetric than other two SBPs. However, there will be different results and prediction maps when different SBPs are used, which requires further research. Furthermore, each component of log-ratio or original soil PSF data was independently modelled using component-wise approaches (machine-learning methods), which may be suboptimal compared with the joint fractions approach under the circumstances (when dealing with the multivariate treatment). Therefore, ILR is recommended as a combination method with machine-learning models for component-wise modelling unless multivariate extensions of the methods (e.g., functional compositions) are considered.

#### 10.4.4.3 The Systematic Comparison of the Direct and Indirect Soil Texture Classification

Compared with the real soil texture distribution and environment of the HRB, SiLo overlaid the upper reaches of the HRB, and SaLo and Lo were present in the south of the upper reaches of the HRB (showing a strip distribution). Moreover, an uncovered area was detected in the northwest of the lower reaches of the HRB, where it cannot be predicted accurately due to a lack of input information in the model training process. The main soil texture types in the lower reaches of the HRB were SiLo, LoSa, and small areas of SaLo and Lo, which were distributed in the uncovered area. The main soil texture types predicted from direct classification using machine-learning models were SaLo and SiLo; RF and XGB delivered much more LoSa than other direct classification models. However, all of these models predicted that the main soil type in the lower reaches of the HRB was SaLo, which did not fit with the real environment (LoSa). In fact, LoSa and SaLo were obviously the most confusing. However, they are fairly similar to each other. In addition, due to the limitation of the training subsets, direct classification can only predict types that are contained in training subsets. In contrast, indirect classification broke such limitations, and new prediction types arose due to the transformation from soil PSFs to soil texture types. Moreover, more suitable matching performance with respect to the real environment should be considered such as the log-ratio methods of the MLP and RF models, KNN_ALR, KNN_ILR, and XGB_CLR.

### *10.4.5 Summary*

We systematically compared five machine-learning models using original data and three log-ratio-transformed data in the HRB for direct and indirect soil texture classification and soil PSF interpolation.

(i) As flexible and stable models, tree learners—RF models—delivered powerful performance in both classification and interpolation and were superior to the other machine-learning models mentioned above.
(ii) As a new and sub-optimal machine-learning method in soil science, XGB appeared to be more computationally efficient in processing large data sets. RF and XGB were recommended to evaluate the classification capacity of imbalanced data.
(iii) In addition, the log-ratio methods, especially ILR, had the advantage of modifying STRESS in soil PSF interpolation.
(iv) Moreover, the indirect methods for soil texture classification outperformed the direct methods, especially when combined with log-ratio transformations. The indirect methods for soil texture classification generated preferable results with respect to both the accuracy indicators and the prediction maps.
(v) The keys to improving the interpolator accuracy are using more appropriate interpolation techniques with environmental covariates, transforming soil PSF data using more efficient transformation methods, utilizing compositional data analysis in the multivariate studies, and using systematic parameter adjustment algorithms for compositional data.

## 10.5 Further Reading

(1) Wang Z, Shi WJ. 2017. Mapping soil particle-size fractions: A comparison of compositional kriging and log-ratio kriging. Journal of Hydrology 546: 526−541.
(2) Wang Z, Shi WJ. 2018. Robust variogram estimation combined with isometric log-ratio transformation for improved accuracy of soil particle-size fraction mapping. Geoderma 324: 56−66.
(3) Zhang M, Shi WJ, Xu ZW. 2020. Systematic comparison of five machine-learning models in classification and interpolation of soil particle size fractions using different transformed data, Hydrology and Earth System Sciences 24: 2505–2526.

## Exercises

(1) Why is surface modelling of soil particle-size fractions (including sand, silt, clay) or soil texture useful? Please give us an example of its application in a process model, like in a crop model, an ecosystem-service evaluation model, a hydrological model, a land surface process model and so on. You can introduce any model which needs the results of surface modelling of soil particle-size fractions or soil texture.
(2) What challenges of surface modelling of soil properties we are facing?
(3) What are the prospects of surface modelling of soil properties do you think in the future?

## References

Aitchison, J. 1986. *The statistical analysis of compositional data*, 416. London: Chapman & Hall.
Aitchison, J. 1990. Measures of variability for geological data—Comment. *Mathematical Geology* 22: 223–226.
Aitchison, J., C. Barcelo-Vidal, J.A. Martin-Fernandez, and V. Pawlowsky-Glahn. 2000. Logratio analysis and compositional distance. *Mathematical Geology* 32: 271–275.
Akpa, S.I.C., I.O.A. Odeh, T.F.A. Bishop, and A.E. Hartemink. 2014. Digital mapping of soil particle-size fractions for Nigeria. *Soil Science Society of America Journal* 78: 1953–1966.
Bedall, F.K. and H. Zimmermann. 1979. Algorithm as 143: The mediancentre. *J. Roy. Stat. Soc.*
Breiman, L. 2001. Random forests. *Machine Learning* 45: 5–32.
Breiman, L. 1996. Bagging predictors. *Machine Learning* 24: 123–140.
Brus, D.J., B. Kempen, and G.B.M. Heuvelink. 2011. Sampling for validation of digital soil maps. *Eur. J. Soil Sci.* 62: 394–407. https://doi.org/10.1111/j.1365-2389.2011.01364.x.
Burges, C.J.C. 1998. A tutorial on support vector machines for pattern recognition. *Data Mining in Knowledge Discovery* 2: 121–167.
Behrens, T., and T. Scholten. 2006. Chapter 25: A comparison of data mining techniques in predictive soil mapping. In *Developments in soil science*, ed. by P. Lagacherie, A.B. McBratney, and M. Voltz, vol. 31, 353–617. Elsevier.
Burrough, P.A., P.F.M. van Gaans, and R. Hootsmans. 1997. Continuous classification in soil survey: Spatial correlation, confusion and boundaries. *Geoderma.* 77: 115–135. https://doi.org/10.1016/S0016-7061(97)00018-9.
Coenders, G., J.A. Martin-Fernandez, and B. Ferrer-Rosell. 2017. When relative and absolute information matter: Compositional predictor with a total in generalized linear models. *Statistical Modelling* 17: 494–512.
Conrad, O., B. Bechtel, M. Bock, H. Dietrich, E. Fischer, L. Gerlitz, J. Wehberg, V. Wichmann, and J. Böhner. 2015. System for Automated Geoscientific Analyses (SAGA) v. 2.1.4. Geosci. Model Dev. 8: 1991–2007, https://doi.org/10.5194/gmd-8-1991-2015.
Cortes, C. and V. Vapnik. 1995. Support-vector networks. *Mach. Learn.* 20: 273–297.
Cover, T.M., and P.E. Hart. 1967. Nearest neighbor pattern classification. *IEEE Transactions on Information Theory* 13: 21–27.
Chen, T., T. He, M. Benesty, V. Khotilovich, Y. Tang, H Cho, K. Chen, R. Mitchell, I. Cano, T. Zhou, M. Li, J. Xie, M. Lin, Y. Geng, and Y. Li. 2018. Xgboost: Extreme gradient boosting, R package version 0.71.2.

Chen, T., and C. Guestrin. 2016. Xgboost: A scalable tree boosting system. Proceedings of the 22nd ACM SIGKDD International Conference on Knowledge Discovery and Data Mining, San Francisco, California, USA.

Cambardella, C.A. 1994. Field-scale variability of soil properties in central Iowa soils. *Soil Science Society of America Journal* 58: 1501–1511.

Egozcue, J.J., and V. Pawlowsky-Glahn. 2005. Groups of parts and their balances in compositional data analysis. *Mathematical Geology* 37: 795–828.

Elith, J., J.R. Leathwick, and T. Hastie. 2008. A working guide to boosted regression trees. *Journal of Animal Ecology* 77: 802–813.

Facevicova, K., K. Hron, V. Todorov, and M. Templ. 2018. General approach to coordinate representation of compositional tables. *Scandinavian Journal of Statistics* 45: 879–899.

Fiserova, E., and K. Hron. 2011. On the interpretation of orthonormal coordinates for compositional data. *Mathematical Geosciences* 43: 455–468.

Egozcue, J.J., V. Pawlowsky-Glahn, G. Mateu-Figueras, and C. Barceló-Vidal. 2003. Isometric logratio transformations for compositional data analysis. *Mathematical Geology* 35: 279–300.

Filzmoser, P., and K. Hron. 2008. Outlier detection for compositional data using robust methods. *Mathematical Geosciences* 40: 233–248.

Filzmoser, P., K. Hron, and C. Reimann. 2009. Univariate statistical analysis of environmental (compositional) data: Problems and possibilities. *Science of the Total Environment* 407: 6100–6108.

Filzmoser, P., and K. Hron. 2009. Correlation analysis for compositional data. *Mathematical Geosciences* 41: 905–919.

Hengl, T., J. Mendes de Jesus, G.B.M. Heuvelink, M. Ruiperez Gonzalez, M. Kilibarda, A. Blagotic, W. Shangguan, M.N. Wright, X. Geng, B. Bauer-Marschallinger, M.A. Guevara, R. Vargas, R.A. MacMillan, N.H. Batjes, J.G.B. Leenaars, E. Ribeiro, I. Wheeler, S. Mantel, and B. Kempen. 2017. SoilGrids250m: Global gridded soil information based on machine learning. PLoS One 12: e0169748.

Heung, B., H.C. Ho, J. Zhang, A. Knudby, C.E. Bulmer, and M.G. Schmidt. 2016. An overview and comparison of machine-learning techniques for classification purposes in digital soil mapping. Geoderma 265: 62–77. https://doi.org/10.1016/j.geoderma.2015.11.014.

Huete, A., K. Didan, T. Miura, E.P. Rodriguez, X. Gao, and L.G. Fer-reira. 2002. Overview of the radiometric and biophysical perfor-mance of the MODIS vegetation indices. *Remote Sensing Environment* 83: 195–213.

Huete, A.R. 1988. A soil-adjusted vegetation index (SAVI). *Remote Sensing Environment* 25: 295–309.

Hijazi, R., and R. Jernigan. 2009. Modelling compositional data using Dirichlet regression models. *Journal of Applied Probability and Statistics* 4: 77–91.

Hron, K., and L. Kubácek. 2011. Statistical properties of the total variation estimator for compositional data. *Metrika* 74 (2): 221–230 74 (2): 221–230.

Landis, J. R. and G.G. Koch.1977. Measurement of observer agreement for categorical data. Biometrics 33: 159–174. https://doi.org/10.2307/2529310.

Lloyd, C.D., V. Pawlowsky-Glahn, and Egozcue J. Jose. 2012. Compositional data analysis in population studies. *Annals of the Association of American Geographers* 102: 1251–1266.

Lark, R.M. 2000. A comparison of some robust estimators of the variogram for use in soil survey. *European Journal of Soil Science* 51: 137–157.

Lark, R.M. 2003. Two robust estimators of the cross-variogram for multivariate geostatistical analysis of soil properties. *European Journal of Soil Science* 54: 187–201.

Lark, R.M., and T.F.A. Bishop. 2007. Cokriging particle size fractions of the soil. *European Journal of Soil Science* 58: 763–774.

Martin-Fernandez, J.A., R.A. Olea-Meneses, and V. Pawlowsky Glahn. 2001. Criteria to compare estimation methods of regionalized compositions. Mathmatical Geosciences 33: 889–909.

Meul, M., and M. Van Meirvenne. 2003. Kriging soil texture under different types of nonstationarity. Geoderma 112: 217–233.

Menafoglio, A., A. Guadagnini, and P. Secchi. 2014. A kriging approach based on Aitchison geometry for the characterization of particle-size curves in heterogeneous aquifers. *Stochastic Environmental Research and Risk Assessment* 28: 1835–1851.

Metternicht, G.I. and J.A. Zinck. 2003. Remote sensing of soil salinity: Potentials and constraints. *Remote Sens. Environ.* 85: 1–20.

Mishra, S., and A. Datta-Gupta. 2018. Exploratory data analysis, in: Applied Statistical Modeling and Data Analytics, chap. 2, edited by: Mishra, S. and Datta-Gupta, A., Elsevier. 15–29. https://doi.org/10.1016/B978-0-12-803279-4.00002-X.

Menafoglio, A., P. Secchi, and A. Guadagnini. 2016a. A class-kriging predictor for functional compositions with application to particle-size curves in heterogeneous aquifers. *Mathematical Geosciences* 48: 463–485.

Menafoglio, A., A. Guadagnini, and P. Secchi. 2016b. Stochastic simulation of soil particle-size curves in heterogeneous aquifer systems through a Bayes space approach. *Water Resources Research* 52: 5708–5726.

Molayemat, H., F.M. Torab, V. Pawlowsky-Glahn, A.H. Morshedy, and Egozcue J. Jose. 2018. The impact of the compositional nature of data on coal reserve evaluation, a case study in Parvadeh IV coal deposit, Central Iran. *International Journal of Coal Geology* 188: 94–111.

Odeh, I.O.A., A.B. McBratney, and B.K. Slater. 1997. Predicting soil properties from ancillary information: non-spatial models compared geostatistical and combined methods. In: Geostatistics Wollongong, ed. E.Y. Baafi and N.A. Schofield, vol. 96, 1008–1019. Dordrecht: Kluwer Academic Publishers.

Odeh, I.O.A., A.J. Todd, and J. Triantafilis. 2003. Spatial prediction of soil particle-size fractions as compositional data. *Soil Sci.* 168 (7): 501–515.

Pawlowsky-Glahn, V., J.J. Egozcue, and R. Tolosana Delgado. 2007. Lecture notes on compositional data analysis. http://hal.handle.net/20256/297.

Saavedra, A., C. Ordonez, J. Taboada, and J. Armesto. 2010. Compositional kriging applied to the reserve estimation of a granite deposit. *Dyna-Colombia* 77: 53–60.

Segal, M., and Y.Y. Xiao. 2011. Multivariate random forests. *Wiley Interdisciplinary Reviews-Data Mining and Knowledge Discovery* 1: 80–87.

Sun, X.L., Y.J. Wu, H.L. Wang, Y.G. Zhao, and G.L. Zhang. 2014. Mapping soil particle size fractions using compositional kriging, cokriging and additive log-ratio cokriging in two case studies. *Mathematical Geosciences* 46: 429–443.

Song, X., D.J. Brus, F. Liu, D. Li, Y. Zhao, J. Yang, and G. Zhang. 2016. Mapping soil organic carbon con-tent by geographically weighted regression: A case study in the Heihe River Basin, China. *Geoderma* 261: 11–22.

Shi, W., J. Liu, Z. Du, Y. Song, C. Chen, and T. Yue. 2009. Surface modelling of soil pH. *Geoderma* 150: 113–119.

Shi, W., J. Liu, Z. Du, A. Stein, and T. Yue. 2011. Surface modelling of soil properties based on land use information. *Geoderma* 162: 347–357.

Shi, W.J., J.Y. Liu, Z.P. Du, and T.X. Yue. 2012. Development of a surface modeling method for mapping soil properties. *J. Geogr. Sci.* 22 (4): 752–760.

Small, C.G. 1990. A survey of multidimensional medians. Int. Stat. Rev. 58: 263–277. https://doi.org/10.2307/1403809

Streiner, D.L. 1996. Maintaining standards: Differences between the standard deviation and standard error, and when to use each. *Can. J. Psychiat.* 41: 498–502.

Taboada, J., A. Saavedra, C. Iglesias, and E. Giraldez. 2013. Estimating quartz reserves using compositional kriging. *Abstr. Appl. Anal.* 2013(1): 716593.

Tolosana-Delgado, R. 2006. Geostatistics for constrained variables: Positive data, compositions and probabilities. Applications to environmental hazard monitoring. Universitat de Girona.

Tolosana-Delgado, R., V. Pawlowsky-Glahn, and J.J. Egozcue. 2008. Indicator Kriging without order relation violations. *Mathematical Geosciences* 40: 327–347.

Vapnik V. 1998. The support vector method of function estimation. In: Nonlinear modeling: Advanced black-box techniques, ed. by J.A.K. Suykens and J. Vandewalle, 55–85.

Wang, Z., and W. Shi. 2017. Mapping soil particle-size fractions: A comparison of compositional kriging and log-ratio kriging. *Journal of Hydrology* 546: 526–541.

Wang, Z., and W. Shi. 2018. Robust variogram estimation combined with isometric log-ratio transformation for improved accuracy of soil particle-size fraction mapping. *Geoderma* 324: 56–66.

Wu, B., N. Yan, J. Xiong, W.G.M. Bastiaanssen, W. Zhu, and A. Stein. 2012. Validation of ET-Watch using field measurements at diverse landscapes: A case study in Hai Basin of China. J.

Yi, C., D. Li, G. Zhang, Y. Zhao, J. Yang, F. Liu, and X. Song. 2015. Criteria for partition of soil thickness and case studies. *Acta Pedologica Sinica* 52: 220–227.

Yang, R., G. Zhang, F. Liu, Y. Lu, F. Yang, F. Yang, M. Yang, Y. Zhao, and D. Li. 2016. Comparison of boosted regression tree and random forest models for mapping topsoil organic carbon concentration in an alpine ecosystem. *Ecological Indicators* 60: 870–878.

Yue, T.X., N. Zhao, Y. Liu, Y. Liu, Y.F. Wang, B. Zhang, Z.P. Du, Z.M. Fan, W.J. Shi, C.F. Chen, and M.W. Zhao. 2020. A fundamental theorem for eco-environmental surface modelling and its applications. *Science China-Earth Sciences* 63: 1092–1112.

Yue, T.X., Y. Liu, M.W. Zhao, Z.P. Du, and N. Zhao. 2016. A fundamental theorem of Earth's surface modelling. *Environmental Earth Sciences* 75: 1–12.

Zhang, S.W. 2013. Spatial interpolation of soil texture using compositional kriging and regression kriging with consideration of the characteristics of compositional data and environment variables. *Journal of Integrative Agriculture* 12: 1673–1683.

Zhang, X., H. Liu, X. Zhang, S. Yu, X. Dou, Y. Xie, and N. Wang. 2018. Allocate soil individuals to soil classes with topsoil spectral characteristics and decision trees. *Geoderma* 320: 12–22.

The manufacturer's authorised representative in the EU is Springer Nature Customer Service Centre GmbH, Europaplatz 3, 69115 Heidelberg, Germany. If you have any concerns regarding our products, please contact ProductSafety@springernature.com

Printed and bound by CPI Group (UK) Ltd, Croydon, CR0 4YY
28/07/2026
02179267-0002